Nutraceutical Potential of Millet-Based Food Products

Nutraceutical Potential of Millet-Based Food Products

Chemistry and Applications

Edited by

Parmjit S. Panesar
Department of Food Engineering and Technology, Sant Longowal Institute of Engineering and Technology, Longowal, Punjab, India

Charanjit S. Riar
Department of Food Engineering and Technology, Sant Longowal Institute of Engineering and Technology, Longowal, Punjab, India

Sushil Dhital
Bioresource Processing Research Institute of Australia, Department of Chemical and Biological Engineering, Monash University, Melbourne, Australia

WILEY

This edition first published 2026

Registered Offices
John Wiley & Sons, Inc., 111 River Street, Hoboken, NJ 07030, USA
John Wiley & Sons Ltd, New Era House, 8 Oldlands Way, Bognor Regis, West Sussex, PO22 9NQ, UK

For details of our global editorial offices, customer services, and more information about Wiley products visit us at www.wiley.com.

The manufacturer's authorized representative according to the EU General Product Safety Regulation is Wiley-VCH GmbH, Boschstr. 12, 69469 Weinheim, Germany, e-mail: Product_Safety@wiley.com.

Wiley also publishes its books in a variety of electronic formats and by print-on-demand. Some content that appears in standard print versions of this book may not be available in other formats.

Library of Congress Cataloging-in-Publication Data Applied for

Hardback ISBN: 9781394294701

Cover Design: Wiley
Cover Images: © Indian Food Images/Shutterstock, zhengzaishuru/Shutterstock

Set in 9.5/12.5pt STIXTwoText by Straive, Pondicherry, India

Contents

List of Contributors

Shiva Bakshi
Department of Dairy Science and Food Technology
Institute of Agricultural Sciences
Banaras Hindu University
Varanasi, India

Farhan M. Bhat
Department of Food Science
Nutrition and Technology
CSKHPKV-Palampur
Palampur, India

Durga S. Bunkar
Department of Dairy Science and Food Technology
Institute of Agricultural Sciences
Banaras Hindu University
Varanasi, India

Lucía Ortega Cabello
Departamento de Sistemas Biológicos
Universidad Autónoma Metropolitana-Unidad Xochimilco
Mexico City, Mexico

Narender K. Chandla
Department of Dairy Engineering
College of Dairy and Food Science Technology
Guru Angad Dev Veterinary and Animal Sciences University
Ludhiana, India

Chuanjie Chen
School of Chemical Sciences
The University of Auckland
Auckland, New Zealand

Muskan Dawra
Department of Food Technology and Nutrition
School of Agriculture
Lovely Professional University
Punjab, India

Gajanan P. Deshmukh
Department of Dairy Engineering
College of Dairy and Food Science Technology
Guru Angad Dev Veterinary and Animal Sciences University
Ludhiana, India

Souvik Giri
Department of Food Engineering and Technology
Sant Longowal Institute of Engineering and Technology
Longowal, Punjab, India

Neha Goyal
Department of Food Engineering and Technology
Sant Longowal Institute of Engineering and Technology
Longowal, Punjab, India

Suneel K. Goyal
Department of Agricultural Engineering
Institute of Agricultural Sciences
Banaras Hindu University
Varanasi, India

Gurjeet Kaur
Department of Processing and Food Engineering
Punjab Agricultural University
Ludhiana, India

Jaspreet Kaur
Department of Food and Human Nutritional Sciences
University of Manitoba
Winnipeg, Canada

Tanveen Kaur
Department of Agriculture and Environmental Science
National Institute of Food Technology Entrepreneurship and Management (NIFTEM)
Kundli, Sonipat
Haryana, India

Tejinder Kaur
Department of Food Engineering and Technology
Sant Longowal Institute of Engineering and Technology
Longowal, Punjab, India

Sunil K. Khatkar
Department of Dairy Technology
College of Dairy and Food Science Technology
Guru Angad Dev Veterinary and Animal Sciences University
Ludhiana, India

Zakir Showkat Khan
Department of Food Technology
Islamic University of Science and Technology
Awantipora, India

Satish Kumar
Department of Food Science & Technology
Dr. Y.S. Parmar University of Horticulture and Forestry
Nauni, Solan
Himachal Pradesh, India

Vikas Kumar
Indian Council of Agricultural Research
Indian Institute of Millets Research (IIMR)
Rajendranagar
Hyderabad, Telangana State, India

Vishal Kumar
Department of Dairy Science and Food Technology
Institute of Agricultural Sciences
Banaras Hindu University
Varanasi, India

José de Jesús Lira-Ricárdez
Departamento de Sistemas Biológicos
Universidad Autónoma Metropolitana-Unidad Xochimilco
Mexico City, Mexico

Parmjit S. Panesar
Department of Food Engineering and Technology
Sant Longowal Institute of Engineering and Technology
Longowal, Punjab, India

Vinod K. Paswan
Department of Dairy Science and Food Technology
Institute of Agricultural Sciences
Banaras Hindu University
Varanasi, India

Abhishek Rana
Department of Food Science & Technology
Dr. Y.S. Parmar University of Horticulture and Forestry
Nauni, Solan
Himachal Pradesh, India

Prasad Rasane
Department of Food Technology and Nutrition
School of Agriculture
Lovely Professional University
Punjab, India

Charanjit S. Riar
Department of Food Engineering and Technology
Sant Longowal Institute of Engineering and Technology
Longowal, Punjab, India

Luman Sang
College of Food Science and Nutritional Engineering
China Agricultural University
Beijing, China

Kawaljit S. Sandhu
Department of Food Science and Technology
MRSPTU
Bathinda, India

Anu Sharma
Department of Food Science and Technology
Khalsa College
Amritsar, India

Sachin Sharma
Department of Food Science & Technology
Dr. Y.S. Parmar University of Horticulture and Forestry
Nauni, Solan
Himachal Pradesh, India

Qun Shen
College of Food Science and Nutritional Engineering
China Agricultural University
Beijing, China

Amandeep Singh
Department of Food Engineering and Technology
Sant Longowal Institute of Engineering and Technology
Longowal, Punjab, India

Kanchan Suri
PG Department of Agriculture
Khalsa College
Amritsar, India

Divyanshi Thakur
Department of Food Engineering and Technology
Sant Longowal Institute of Engineering and Technology
Longowal, Punjab, India

Priyanka Thakur
Department of Agriculture and Environmental Science
National Institute of Food Technology Entrepreneurship and Management (NIFTEM)
Kundli, Sonipat
Haryana, India

Ritesh Verma
Department of Agriculture and Environmental Science
National Institute of Food Technology Entrepreneurship and Management (NIFTEM)
Kundli, Sonipat
Haryana, India

Nitin Wakchaure
Department of Dairy Economics and Business Management
College of Dairy and Food Science Technology
Guru Angad Dev Veterinary and Animal Sciences University
Ludhiana, India

Liangxing Zhao
College of Food Science and Nutritional Engineering
China Agricultural University
Beijing, China

Qingyu Zhao
College of Food Science and Nutritional Engineering
China Agricultural University
Beijing, China

Fan Zhu
School of Chemical Sciences
The University of Auckland
Auckland, New Zealand

Preface

Millets are a group of small, grained cereal food crops which are highly tolerant to drought and other extreme weather conditions and are grown with low chemical inputs such as fertilizers and pesticides. Millets rank sixth in the world cereal grain production. In Africa and Asia, these underutilized grains play a major role in the food security of millions of people. The Ministry of Agriculture and Farmers Welfare has recognized the importance of millets and declared them as "Nutri-cereals." Millets comprise sorghum, pearl millet, finger millet, and minor millets, i.e., foxtail millet, proso millet, kodo millet, barnyard millet, little millet, brown top millet, and two pseudo millets, i.e., buckwheat and amaranth. In view of this, Section I – "Fundamental Aspects of Millets" – comprises three chapters, which deal with introduction, classification, production, distribution of millets, along with their nutritional composition, health benefits, antinutritional factors, and their elimination methods.

Millets are considered to have immense significance because of their nutritional and nutraceutical potential. In addition to being a rich source of nutrients, millet grains have an abundance of phytochemicals, particularly phenolic compounds. Millets are rich sources of minerals, dietary fibers, and phenolic compounds, and offer health benefits such as antimicrobial, antidiabetic, anticancer, antiartherosclerogenic effects, antioxidant, and antiaging properties. Thus, Section II is focused on "Millet Bioactive Compounds and Characterization." In this section, Chapters 4 and 5 cover classification, chemical profiling, and extraction of phenolics, quality assessment, and health benefits of millets polyphenols. It has been established that dietary fibers have water-absorbing and bulking properties. These increase the transit time of food in the gut which helps in reducing the risk of inflammatory bowel disease and acts as a detoxifying agent in the body. In this context, Chapter 6 of the proposed book focuses on the extraction, modification, and quality assessment of millets dietary fibers. Intake of millet protein, especially uncooked extracted (UCE) and uncooked extracted enriched (UCEE) finger millet protein, helps to reduce body weight and increase the liver's relative weight. Millet starch is recognized to exhibit a low or medium glycemic index (GI) and is an excellent food option for people who are worried about the blood glucose response of food products. Accordingly, Chapters 7 and 8 are dedicated to the extraction, quality assessment, and health benefits of millet proteins and starch. Health benefits of phenolic compounds in foods vary with the level of absorption and bioaccessibility. Potential absorption of a compound can be determined as the quantity that passes through the cell membranes in the intestine and is available for action within the cells. Bioaccessibility is referred to as the amount of compound that is released from the solid food matrix into the gut. In this context, Chapter 9 covers the bioavailability studies of millet bioactives and the methods to enhance their bioaccessibility.

Processing and preparing millets for consumption can affect nutrients, reduce the nutrient content of grain, concentrate desirable grain components, and remove poorly digested compounds and contaminants, which increases digestibility of nutrients and phytochemicals. Section III of the book "Millet Processing and Applications" comprises various chapters including Chapter 10, which focuses on a detailed discussion of emerging technologies in millet processing. Millets are widely utilized as snacks, infant food, processed food, etc., due to their high nutrient content, and so are considered a complete food. In developing countries, sorghum and other millets contribute greatly to nutritional security. The millets are also known as miracle grains/nutri-cereals. In this context, Chapter 11 thoroughly covers the preparation of millet-based products and their commercialization. Fermented food contains probiotic organisms which help the gut for proper digestion of food and provides many advantages to the body. Fermented millet-based food products play an important role in good digestive health. Chapter 12 of the proposed book covers the nutritional and health benefits of millet-based fermented food products. Millets generate an enormous amount of by-products during harvesting/processing, and these contain a good amount of phytochemicals having potential applications. So, the final chapter covers the bioactive composition of millet by-products and their value addition.

In conclusion, this volume will provide public acquaintance of millets and their products, health advantages, and their adaptability for production in the present challenging circumstances brought by climate change and public awareness activities. This book will be a useful resource to academicians, industry personnel, and students working in the areas of food grain processing/technology/engineering and postharvest technology.

Parmjit S. Panesar
SLIET Longowal, Punjab, India
Charanjit S. Riar
SLIET Longowal, Punjab, India
Sushil Dhital
Monash University, Melbourne, Australia

Section I

Fundamental Aspects of Millets

1

Classification and Production of Millets

Charanjit S. Riar and Parmjit S. Panesar

Department of Food Engineering and Technology, Sant Longowal Institute of Engineering and Technology, Longowal, Punjab, India

1.1 Introduction

Nutrients from cereals are necessary for the sustainability of animal life, including human beings. Wheat, rice, maize, rye, millets, etc., are the major cereal crops produced in the world and belong to the family Poaceae. The word "millet" has its origin from the French word "mile" signifying a thousand grains, which means thousands of grains in a handful (Taylor & Emmambux, 2008). The original recorded descriptions of millet production date back to roughly 5,500 BC in the Chinese province (Crawford, 2014). Millets can be characterized as a significantly variable, small seed-bearing grass, extensively produced as a cereal crop or grain for feed and fodder worldwide. Millets rank the sixth most vital grain and nourish 33% of the world's population (Saleh et al., 2013). Millets belong to various genera such as *Pennisetum*, *Eleusine*, *Setaria*, *Panicum*, *Paspalum*, *Echinochloa*, and *Digitaria*. Millets do not represent a taxonomic class but are practically classified as an operational or agronomic class. Even though millets are aboriginal to several world nations, it is assumed that they have an evolutionary descent. Millets are cultivated in the *kharif* season. They are considered prominent crops in the semi-dry regions of Asia and Africa, principally India and Nigeria, including 97% of the millet cultivation among the developing nations (FAO, 2024). Millets are the oldest cereal grains that have emerged for use in feeding, but with the passage of times, this grain has lost its light. Its usage is restricted to the rural and underdeveloped nations of the world. Millet crop is favored due to its productivity and short producing time under temperate and high-temperature conditions. Pearl millet is the most widely grown millet and is a valuable crop in India and many areas of Africa. Finger, proso, and foxtail millets are other notable millet crops. In developed countries, millets have hardly any importance. In the United States, proso millet is barely grown, and it is primarily used as a birdseed (USDA, 2024). Previously considered as a poor man's crop, millet remains fulfilling prime nutritional requirements for the people of developing nations, including

Nutraceutical Potential of Millet-Based Food Products: Chemistry and Applications, First Edition.
Edited by Parmjit S. Panesar, Charanjit S. Riar, and Sushil Dhital.

Published 2026 by John Wiley & Sons Ltd.

India, Africa, and the Sahara region (UNICEF, 2023). They have been consumed in Asia and Africa since ancient times and contribute critically to the agriculture of their countries because they are mainly rain-fed crops and do not demand large amounts of water and pesticides. They are essential harvests in semi-dry regions where different crops do not endure. Millets are easy to cultivate, are naturally bio-diverse, and are capable of being grown together with diverse crops (Ahmed et al., 2013). Further conditions such as in case of peal millet that determine their preference in the regions where they are progressively cultivated include a brief crop accumulation period of 45–65 days (Bukhari et al., 2011).

Altering climatic situations is declining agrarian production, which is forcing the farmers to decide on altered crops for the persistence of mankind. Today's scientific agriculture is concentrating on lengthy food protection and development of crops which require less consideration during production with minimum production time. Millets are resistant to the environment and have prospective benefits with enhanced nutrition and health benefits. These draught-tolerant crops have a brief growing period, acting as a primary food source for the substitution of major cereals. Millets ought to reduce input while growing, as they have marginal water necessity, which lessens their production costs (Sharma et al., 2023); they are also recognized as C_4 crops (Singh et al., 2021). Millets are capable of growing alongside other crops since they are competent for numerous climates while being challenged to certain pests and diseases (Saleh et al., 2013). As recounted earlier, rice needs about 1,200 mm of water, whereas millets' requirement is only 350–450 mm of water. To pave the way and encourage nations and farmers, the 75th session of UN General Assembly declared the year 2023 as "The International Year of Millets-23" (IYM, 2023) in gratitude of millets as per the proposal suggested by the Indian Government. The IYM 2023 ropes the aim of the United Nations concerning sustainable development, predominantly the goals of zero hunger, good health and well-being, decent work, economic growth, responsible production and consumption, climate action, and life on land (FAO, 2023).

1.2 Changing Eating Habits Globally

Cereals are important food crops consumed globally because of their potential to provide excellent quality of nutrients. The cumulative assessment of wholesome living has encouraged the commodities in the search for more natural and favorable food prospects. Consumers have become more aware of foods with superior capacities of phytochemicals and dominance of proteins obtainable from inexpensive plant sources such as grain (millets), precisely because of the nutritional and health advantages (Abhilasha et al., 2022). Wheat, rice, and maize bring nutritional sanctuary, whereas millets give manifold securities for food, fiber, fodder, ecology, health, and livelihood while keeping the fertility of the soil as well, and offer sustainability owing to the increasing population and water catastrophes around the globe. Millets have become apparent as nutria-cereals, emerging as candidate crops, and are becoming the alternative of the modern era among the urban population as they possess distinctive properties, such as being gluten-free and hypoglycemic in nature. Having a nonglutinous nature, millets are considered harmless and suitable for people with gluten allergy or celiac disease. They are nonacid forming, easy to digest, and are nonallergenic (Saleh et al., 2013). These small-seeded grains are rich in nutrients and are commensurate with other cereal grains. When compared to major cereals like wheat, rice, and maize, they have a substantial nutritional level which makes them a substitute for other starchy cereals, giving huge opportunities for their utilization. All the millets have different nutritional profiles and therefore have rightly been termed as nutri-cereals (Bijalwan et al., 2016; Jideani & Diedericks, 2014). Millets have succeeded in making a comeback strongly in the daily platter after decades of ignorance. Practically altogether, the millets such as sorghum, pearl, finger, foxtail, little, barnyard, kodo, and common millet are eaten by humans in developing countries, but their application remains principally limited to animal forage in developed countries (Harish et al., 2024).

1.3 Millets' Unique Characteristics

Despite the unique nutritional quality, millets are experiencing difficulties in milling as they have less significant dehusked grain retrieval and produce coarse flour. High fiber content and the presence of some of the antinutritional components, such as phytates and tannins in millets, impact bioavailability of minerals. Some studies on humans have indicated that the absorption of iron tends to be minimal from millets than from other cereals like paddy or wheat (Rao et al., 1983). Before converting the millet grains into different food products, they are processed by different techniques like milling, germination, fermentation, roasting, flaking, and grinding to enhance the palatability of the grains (Saleh et al., 2013). Numerous processing practices are used to overcome the problems encountered in milling, and hydrothermal treatment is one of them. Hydrothermal treatment, also known as parboiling, was originated in India (Prashanth et al., 2024; Srilakshmi, 2003). It is widely used for the processing of paddy to improve the dehusking/decortication yield along with a reduction in antinutritional factors. This technique is also found to improve grain/flour storage stability of millet grain and diversify its potential use. Parboiling followed by decortication is known to decrease the antinutritional compounds reported in millets and increases the sensorial characteristics of millets and their developed products. For the uninterrupted supply of millet grain, they need to be processed efficiently to increase their nutritional and sensorial characteristics, maintain quality, and maximize the milling yield (Kaur et al., 2023, 2024).

As per the report of the Ministry of Agriculture and Farmers Welfare (MAFW), from 2016 to 2017, the part under the production of millet diminished to 60% and even lower with a cultivated area of 14.72 MHac, owing to variation in the utilization design, expansion of watered area for wheat and rice production, millet unavailability, low grain yield, changing dietary habits, and fewer demands (APEDA, 2023, 2024). This caused a fall in the uptake of essential nutrients such as vitamin A, proteins, Fe, and iodine (I) in the feeding of women and youngsters, resulting in undernourishment. The Global Hunger Index (GHI) score of 27.3 for India in the year 2024 ranked it 105th among 127 world nations based on relevant data involved in the calculation of the GHI score (Global Hunger Index, 2024). With this score, India's degree of hunger seems to be stern. India's GHI grade depends on the estimates of four component indicators (undernourished population, stunted children under five, wasted children under five, and deaths of children before their fifth birthday). India stands second in children malnutrition. This circumstance persists, while the public distribution system (PDS) and targeted PDS are operational for almost 50 years. The cause of this is because attention has been given to increasing share for crops like wheat and rice, while millet cultivation has been overlooked for a long time (FSSAI, 2023; India Science, Technology and Innovation Portal, 2024; Ministry of Agriculture and Farmer's Welfare, 2023). This indicates that for food crops, millets dominate a relatively lower position in the Indian crop-growing system, although their influence is substantial from the perspective of food security at the provincial and domestic intensity. In acknowledgment of this evidence, important highlights are given below to feature the significance of millets (APEDA, 2023; India Data Insights, 2023).

1) All true millets have nonacidic characteristics and are gluten-free, markedly nutritious, and straightforwardly assimilated. Having a lower glycemic index (GI) and not containing gluten, their benefits include protracted emancipation of monosaccharides especially glucose, across an extended time, thus lowering the impact of diabetes mellitus. People bearing celiac disease can certainly include millet-based food in their diets (NITI AAYOG, 2023a).
2) Millets are excellent sources of minerals such as Fe, K, Ca, Zn, P, and Mg. A considerable level of soluble (SDF) and insoluble dietary fibers (IDF) and vitamins such as folate, pyridoxine, β-carotenoid, and niacin (B_3) are also present in millet grains. The accessibility of large quantities of lecithin is beneficial for improving the health of the nervous system (India Data Insights, 2023; Vikaspedia, 2024).
3) Even though millets are enriched with phytochemicals like tannins, phytosterols, polyphenols, and antioxidants, some antinutritional substances are present in them, which are effectively reduced by various processing techniques like hydrothermal treatment, soaking, and milling (India Data Insights, 2023; Vikaspedia, 2024).
4) Millets have an adaptive capability for modification since they can be cultivated from the coastal provinces of India to fairly elevated heights of northeastern areas and mountainous regions. Millets can tolerate adaptations in humidity, temperature, and a variety of soils extending from substantial to filthy barren domains (India Data Insights, 2023; Vikaspedia, 2024).

Consequently, to safeguard the food and nutritional security of our country and developing countries in general, it is imperative to intensify the farming of millet and coarse grain crops and concurrently relapse the production control, distribution, and consumption regulation back to the society (India Data Insights, 2023; Vikaspedia, 2024).

1.4 Classification of Millets

Millet is used as a conventional term for all classifications, comprising major millets such as pearl (*Pennisetum glaucum*), sorghum (*Sorghum bicolor*), and finger (*Eleusine coracana*) and minor millets such as foxtail (*Setaria italica*), barnyard (*Echinochloa frumentacea*), little millet (*Panicum sumatrense*), browntop (*Urochloa ramosa*), kodo (*Paspalum scrobiculatum*), and proso (*Panicum miliaceum*) (FSSAI, 2024; India Science, Technology and Innovation Portal, 2024; Vikaspedia, 2024) (Table 1.1). Pseudo-millets (buckwheat and amaranth) are also included in the classification of millets by FSSAI (2024) as they are nutritionally identical to the true grains of millets; however, they do not belong to the Poaceae family. Pseudo-millets

Table 1.1 Scientific and common terminologies used for cultivated millets

Scientific names	Common names	Taxonomy (tribe)	Chief regions of cultivation
Cenchrus americanus	Pearl, bajra, munga, cattail, sanyo, seno, bulrush, and candlestick	Paniceae	India, Africa (Western, Central, and Eastern), Southern African region
Eleusine coracana	Finger, ragi, African, bird's foot, and rapoko	Eragrostideae	India, Eastern and Southern Africa, Uganda, Tanzania, Sri Lanka, Malaysia
Setaria italica	Foxtail, Italian, German, and Hungarian	Paniceae	China, India, Eastern Europe
Panicum miliaceum	Proso, common, hog, and broom	Paniceae	China, Russia, Kazakhstan, Ukraine, India, Japan
Panicum sumatrense	Little, blue panic, and heenmeneri	Paniceae	India
Paspalum scrobiculatum	Kodo, varagu, bastard, and ditch	Paniceae	India
Echinochloa crus-galli	Barnyard, Japanese, and sanwa	Paniceae	India, Japan, China, Malaysia
Eragrostis tef	Teff, Abyssinian	Paniceae	Ethiopia, Eritrea, Australia

Note: ICRISAT (2012), Serna-Saldivar and Rooney (1995), and Taylor and Emmambux (2008).

are consumed akin to the tiny grains of millets. Pseudo-millets are classified as millets as they require less water during cultivation and are gluten-free as millets (Ministry of Agriculture and Farmer's Welfare, 2023). Table 1.2 shows the generalized terminologies for millets utilized in various countries and their taxonomic classifications.

Millets are classified on the basis of size, shape, color, botanical family, caryopsis (seed coat), and significance. Each type of classification is summarized in Sections 1.4.1–1.4.6 to elaborate on the classification basis and characteristics (FSSAI, 2024; India Science, Technology and Innovation Portal, 2024; Vikaspedia, 2024).

1.4.1 Grain Size and Shape Classification

Based on the size of the grain, millets are categorized into two types—large or major millets and small or minor millets. Major (large) millets: sorghum (*jowar*), pearl millet (*bajra*), finger millet (*ragi*), foxtail millet (*kagni*), and proso millet (*cheena*) (FSSAI, 2024). The small (minor) millets: kodo millet (*kodra*), barnyard millet (*sama*), browntop millet (*hari kagni*), and little millet (*kutki*) (FSSAI, 2024; Vikaspedia, 2024). Millets have diminutive size (Ministry of Agriculture and Farmer's Welfare, 2023). These grains are round, oval, and elliptical in shape (Brink et al., 2006), existing in a variety of colors that flourish well in any soil environment of tropical or subtropical regions of the world (Africa and Asia) (Saleh et al., 2013). Millets, the shortest grains, are from South China (1.2 mm) and the Russian Far East (1.1 mm). The shape and size classification for millets is given in Table 1.3.

Table 1.2 Millet names in Indian languages

English	Pearl millet	Finger millet	Foxtail millet	Kodo millet	Little millet	Barnyard millet	Sorghum	Buckwheat millet	Amaranth millet	Broomcorn millet
Oriya	Bajra	Mandia	Kanghu/Kangam/Kora	Kodua	Suan Gurji	Khira	Juara	—	Kosala sag manji	China Bacari bangmu
Hindi	Bajra	Nachani/Mundua/Mandika/Marwah	Kangni/Kakum/Rala	Koden/Kodra	Kutki/Shavan	Jhangora/Sanwa	Jowar	Kuttu	Ramadana, Rajgira	Cena, Barri
Tamil	Kambu	Kezhvaragu/Kelvaragu/Keppai/Ragi	Thinai	Varagu	Saamai	Kuthiravali (Kuthiraivolly)	Cholam	—	Keerai vidai	Pani varagu
Telugu	Sajjalu	Ragula/Ragi Chodi	Korra	Arikelu/Arika	Sama/Samalu	Udalu, Kodisama	Jonna	Bukvit	Thotakoora ginjalu	Variga
Kannada	Sajje	Ragi	Navane	Harka	Saame/Save	Oodalu	Jola	Huruli	Danthu beeja	Baragu
Malayalam	Kambam	Panji Pullu	Thina	Koovaragu	Chama	Kavadapullu	Cholam	—	—	—
Marathi	Bajri	Nagli/Nachni	Kang/Rala	Kodra	Sava/Halvi/Vari	—	Jowari/Jondhala	Bataravhita	Chaulai beej	Vari
Punjabi	Bajra	Mandhuka/Mandhal	Kangni	Kodra	Swank	Swanks	Jowar	Baikavata	The Cavali biya	The Cheena
Gujarati	Bajri	Bavto/Nagli	Kang'	Kodra'	Gajro/Kuri	—	Jowari	Biyam satheno dano	Rajgaro	Cheno
Bengali	Bajra	Marwa	Kaon	Kodo	Sama	Shyama	Jowar	Baikavata	Amaratba bijja	Cheena

Note: Adapted from Serna-Saldivar and Rooney (1995), Taylor and Emmambux (2008), ICRISAT (2012), FSSAI (2024), and India Science, Technology and Innovation Portal (2024).

Table 1.3 Classification of millet based on shape, size, and origin

Millet grain	Area of origin	Shape and type of grain	Size	References
Sorghum	Sudan	Spherical caryopsis	4 mm long 2 mm wide	Serna-Saldivar and Espinosa-Ramírez (2019)
Pearl	Africa	Ovoid caryopsis	3–4 mm long	Yousaf et al. (2021)
Finger millet	Sudan	Spherical utricle	1.2–1.8 mm diameter	Shobana and Malleshi (2007)
Foxtail millet	China	Oval caryopsis	2 mm	Kaur et al. (2023)
Kodo	India	Elliptical to oval	3–4 mm	Taylor and Emmambux (2008)
Barnyard	Asia	Elliptical to oval caryopsis	1.3 mm	Chandi and Annor (2015)
Little Millet	Indian Peninsula	Elliptical to oval	1.8–1.9 mm long	Chapke et al. (2018)
Proso	China	Oval utricle	3 mm long 2 mm wide	Yousaf et al. (2021)
Teff	Ethiopia	Oval caryopsis	0.9–1.7 mm (length) 0.7–1.0 mm (diameter)	Gebremarian et al. (2014)
Fonio	West Africa	Elliptical/oval caryopsis	1.5 and 1.8 mm long and approximately 0.9 mm wide	Ballogou et al. (2015)
Buckwheat	China	Triangular/conical/ arrow/elliptical/oval	5.92–4.02 mm grain size	Deka et al. (2023)
Amaranth	America	Conical/cylindrical/ ellipsoidal	1.35–1.22 mm	Deka et al. (2023)

Note: FSSAI (2024), Vikaspedia (2024), and India Science, Technology and Innovation Portal (2024).

Minor millets comprise a collection of sedgy plants having short slender culms and miniature grains with notable capability to endure severe drought conditions, whereas major millets like pearl and sorghum are tall-growing and moderately drought-tolerant with the capability to endure under severe drought circumstances. Their subterranean root structure, with proficient soil moisture abstraction ability, makes them competent to survive under minimal to severe moisture conditions, where other cereal crops are unable to survive or develop grain. These are identified as a significant replacement for major cereal crops to handle the world food shortage and to encounter the constraints of growing population of both developing and developed countries (Veena et al., 2010).

1.4.1.1 Sorghum

The seeds vary widely in different types of colors, shape, and size. The seed size varies from 4 to 6 mm. Endosperm is normal, hetero-waxy, or waxy (India Science, Technology and Innovation Portal, 2024).

1.4.1.2 Pearl Millet

The pearl millet is termed "pearl" as the shape of the grains is pearl-like. However, the shapes vary from hexagonal and ovoid to globose-like (FSSAI, 2024). The grain has a thin testa and an aleurone layer which is occasionally pigmented. Its length varies from 3 to 5 mm. The seed type is caryopsis. A thousand-seed weight can be from 2.5 to 14 g, with a mean of 8 g (FSSAI, 2024; Vikaspedia, 2024).

1.4.1.3 Finger Millet

The finger millet seed is a utricle, and the grain is globose in shape. The testa of the grain is pigmented with five thick cell layers. The finger millet has the largest endosperm section comprising the kernel (Vikaspedia, 2024). The seed has a single aleurone layer that lies just beneath the seed coat and testa. Finger millet grains have a diameter of 1–2 mm, with the color ranging from brown to dark brown. The thousand-kernel weight is about 2.3 g (FSSAI, 2024; Vikaspedia, 2024).

1.4.1.4 Foxtail Millet

The seed type of foxtail millet (Italian millet, German millet, Hungarian millet) is caryopsis. The length of foxtail millet grains varies from 2 to 3 mm. The thousand-kernel weight is about 6.5 g (FSSAI, 2024; Vikaspedia, 2024).

1.4.1.5 Barnyard Millet

The length of the barnyard millet grain is about 3 mm, and the grain is concave to oval.

1.4.1.6 Proso Millet

Proso millet grains are spherical to oval, about 3 mm long and 2 mm in diameter. The seed type is utricle. The thousand-kernel weight is about 7.1 g (India Science, Technology and Innovation Portal, 2024).

1.4.1.7 Kodo Millet

Kodo millet grains are elliptical to oval. The grain is about 3–4 mm in length. The thousand-kernel weight is about 8.3 g (India Science, Technology and Innovation Portal, 2024).

1.4.1.8 Browntop Millet

Browntop millet grains are ellipsoid in shape and tan or white in color. The length of the browntop millet grains ranges from 4 to 5 mm, and the thousand-kernel weight is about 9.6 g (India Science, Technology and Innovation Portal, 2024).

1.4.1.9 Little Millet

As the name implies, it is the smallest among all millets. The length of the grain varies from 2 to 3 mm, and the shape varies from oblique to oval. The thousand-kernel weight is about 6.5 g (India Science, Technology and Innovation Portal, 2024; Vikaspedia, 2024).

1.4.2 Classification Based on Significance (Importance)

Based on their significance, millets are broadly classified as "major millets" and "minor millets" and are often used to discriminate between two groups of millets based on their importance in terms of cultivation, consumption, economic significance, production, and exploitation (McDonough et al., 2000). As per Food Safety and Standards Authority of India (FSSAI) Guidance note No. 12 (FSSAI, 2019), millets are subdivided into major, minor, and pseudo-millets (Figure 1.1). Major millets comprise pearl (bajra) millet, sorghum (jowar), and finger (ragi) millet, whereas minor millets consist of foxtail (kakum) millet, kodo (kodon) millet, barnyard (sanwa) millet, little (kutki/shavan) millet, and proso (chenna/barri) millet (FSSAI, 2024; Vikaspedia, 2024). Pseudo-millets are so called because they are not part of the Poaceae family, to which "true" grains belong; however, they are nutritionally similar and are used in similar ways to "true" grains. Examples of pseudo-millets include buckwheat and amaranth. Plants that have analogous flowers, reproductive structures, and other characteristics, and are evolutionarily narrated, are categorized into plant families (FSSAI, 2024; India Science, Technology and Innovation Portal, 2024; Vikaspedia, 2024).

1.4.3 Taxonomical Classification of Millets

"Millet" does not only represent a botanic classification for certain plants but reasonably an umbrella phrase for numerous small-seeded grasses sourced for human consumption (Harinarayanan et al., 2023). Species in the same plant family tend to have similar growth characteristics, nutrient needs, and often the same pests (pathogens, herbivores). All the millets are placed under Order: Poales and Family: Poaceae (*Gramineae*/true grasses). They belong to the subfamilies Chloridoideae and Panicoideae (Vetriventhan et al., 2020; Yang et al., 2017) (Figure 1.2).

Eragrostideae tribe (Chloridoideae subfamily): *Eleusine coracana*: finger millet, mawere (ragi, nachani, or mandwa in India); *Eragrostis tef*: teff (Vetriventhan et al., 2020; Yang et al., 2017).

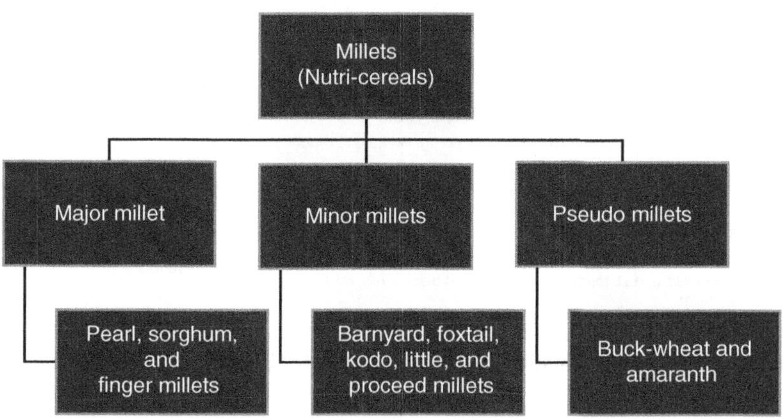

Figure 1.1 Millet classification according to the FSSAI advisory (Guidance note No. 12/2019)

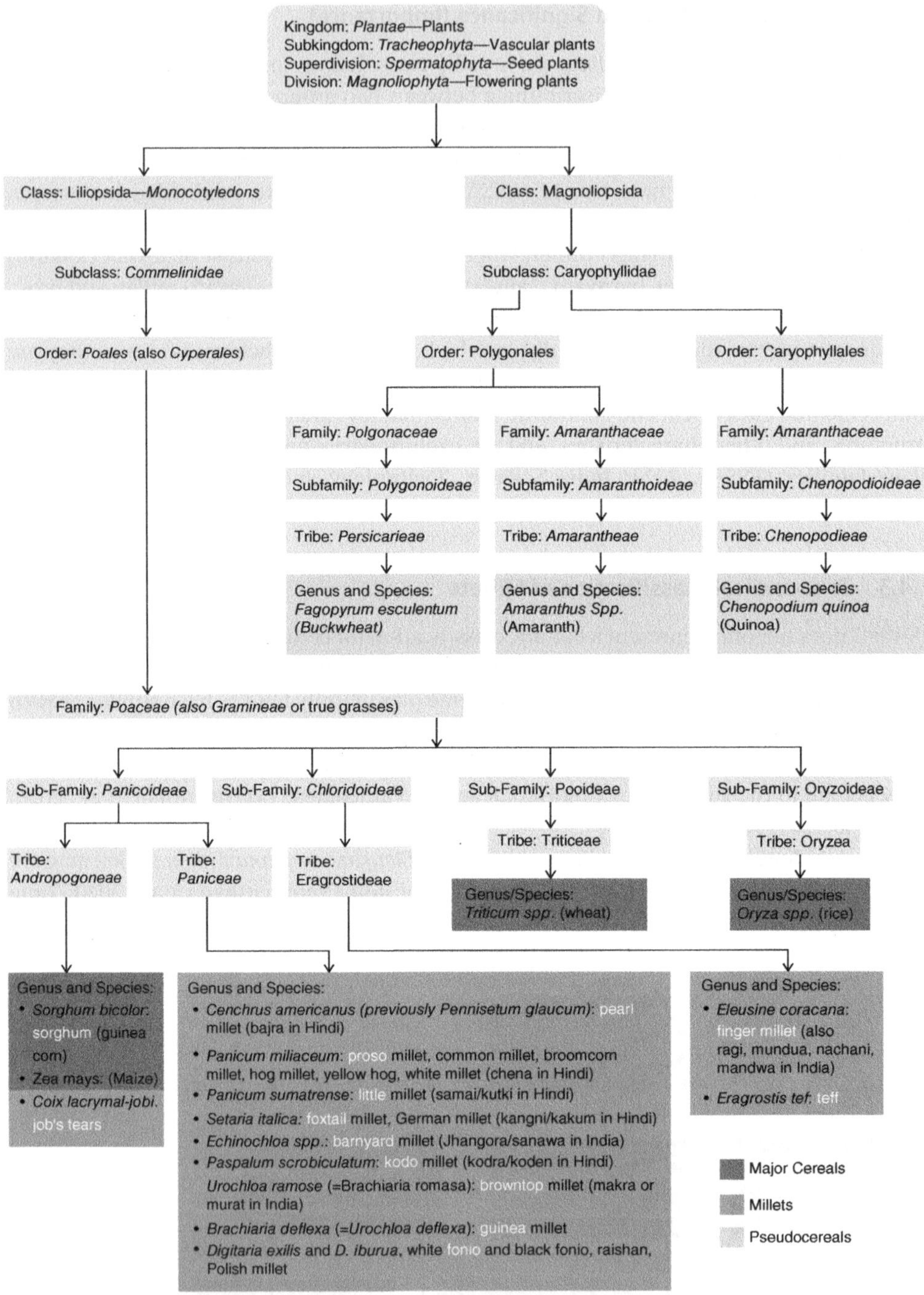

Figure 1.2 Taxonomical classification of millets. *Source:* Yang et al. (2017) and Vetriventhan et al. (2020)

Paniceae tribe (*Panicoideae* subfamily): *Panicum miliaceum*: proso millet, common millet, broom corn millet, hog millet, yellow hog, white millet (Vetriventhan et al., 2020; Yang et al., 2017).

- *Pennisetum glaucum*: pearl millet (kambu or bajra in India)
- *Setaria italica*: foxtail millet, German millet (thinai, kang, or rala in India)
- *Digitaria* spp.: white fonio, black fonio, raishan, Polish millet
- *Echinochloa* spp.: Japanese barnyard millet, Indian barnyard millet, sawa millet, burgu millet (kuthirai vaali, bhagar, or varai in India)
- *Panicum sumatrense*: little millet (samai in India)
- *Paspalum scrobiculatum*: kodo millet (varagu in India)
- *Urochloa* spp. (also known as *Brachiaria*): browntop millet (*U. ramosa*, dixie signalgrass), Guinea millet (Vetriventhan et al., 2020; Yang et al., 2017).

The millets are strictly interrelated to sorghum and maize, which are associated with PACMAD ancestor of grasses, and further indistinctly related to the cereals of the BOP (sometimes BEP clade is one of two major lineages (or clades) of undefined taxonomic rank in the grasses (Poaceae), containing more than 5,400 species, about half of all grasses. Well-known members of this clade include rice, some major cereals such as wheat, barley, oat, and rye, many lawn and pasture grasses, and bamboos. Its sister group is the PACMAD clade; in contrast to many species of that group that have evolved C4 photosynthesis, the BOP grasses use the C3 photosynthetic pathway) ancestor in association with wheat and barley (Wikipedia, 2024) (Figure 1.3).

1.4.4 Classification Based on Caryopsis (Covered or Naked)

As per Food and Agricultural Organization (FAO, 1995), the principal anatomical components of millets are the pericarp, endosperm, and germ (Figure 1.4). These grains acquire a distinctive morphological trait which triggers an obstacle in the decortication of the grains. They are either utricle-type or caryopsis-type grains. In utricle-type grains, the outer covering is loosely attached at one point, covering the pericarp like a sack. In these types of grains, the pericarp detaches easily. In the caryopsis-type grains (sorghum and pearl millet), the pericarp is completely fused with the endosperm and needs processing to detach the pericarp. The foxtail and finger millets comprise true caryopsis, while proso, fonio, teff, barnyard, and kodo millets are utricles (Taylor & Duodu, 2010).

Naked Grains: The types of millets that do not have a tough and indigestible exterior husk are included in the naked grains category (sorghum and pearl millet). These grains do not require processing once harvested and can be consumed after proper cleaning (FSSAI, 2024; India Science, Technology and Innovation Portal, 2024; Vikaspedia, Millets-The Nutri-Cereals, 2024).

Husked Grains: These indigestible seed-coated grains (foxtail, barnyard, and kodo) need to undergo mechanical processing, which was once done manually before being consumed. They are to a lesser extent prevalent than naked grains since they require considerable labor before consumption (FSSAI, 2024; India Science, Technology and Innovation

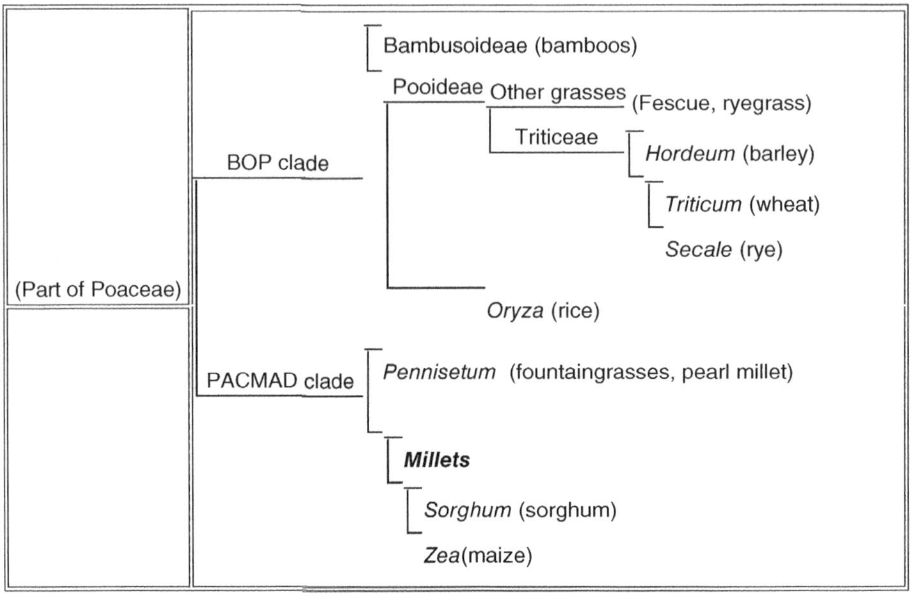

Figure 1.3 Botanical classification of the millets. *Source:* Adapted from Wikipedia (2024)

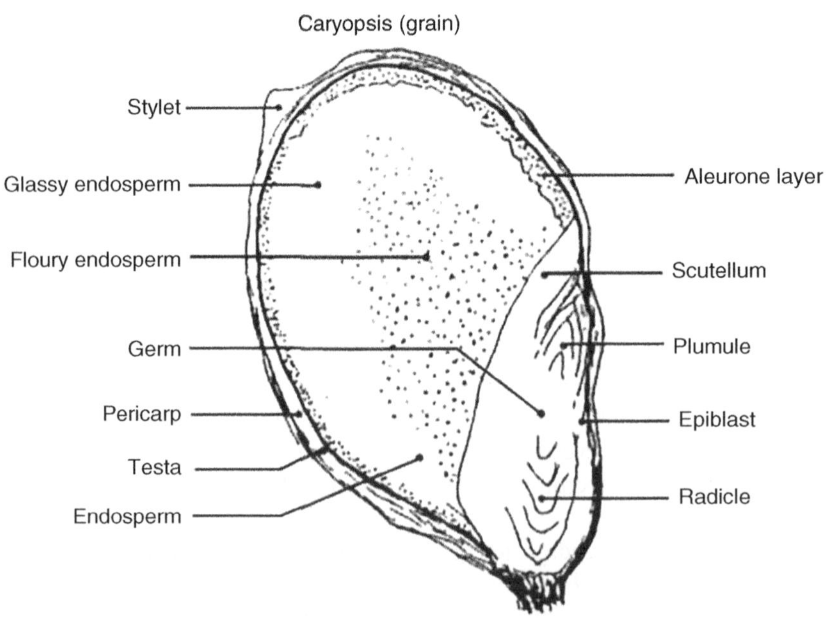

Figure 1.4 Millet grain morphology. *Source:* Adapted from Devisetti et al. (2014)

Portal, 2024; Vikaspedia, 2024). The pericarp of the grain millets is mostly rough, which is unfit for human consumption, and its elimination may decrease the concentrations of antinutritional factors (phytates and tannins), which may reduce the bioaccessibility of the nutrients (Devisetti et al., 2014). The germ portion of the grain is rich in fat, which may lead to rancidity and subsequent deterioration of grain quality with the loss of sensory properties. To enhance the utility of grains, certain milling technologies such as soaking, parboiling, and fermentation are needed for the removal of pericarp and germ from the millets, which hardens the endosperm and augments the milled grain yield, with a substantial decrease in broken grains. Softer endosperm and intact seed coat of millets are difficult to separate using common milling methods (Dharmaraj & Malleshi, 2011; Kaur et al., 2024; Malleshi, 2006).

1.4.5 Color Classification

Millets appear in different varieties and colors. The colored varieties are a valuable source of polyphenols, which are accountable for their antioxidant activity (FSSAI, 2023). Major millets like pearl and sorghum can be straightforwardly distinguished by their color, shape, and size, and similar distinction occur in finger millet (Vikaspedia, 2024). The seed coat and husk of proso, foxtail, and little millets are shiny in appearance, whereas barnyard and kodo millets bear multicolored seed coats. Generally, the seed coat of kodo millet is brownish, foxtail is yellowish, and rest of the millets are grayish, except the major millets (FSSAI, 2023; Ministry of Agriculture and Farmer's Welfare, 2023) (Table 1.4).

Sorghum: The color, size, and shape of sorghum grain vary corresponding to the variety. The color of the pericarp is white, yellow, or red, whereas the endosperm color is white, hetero-yellow, or yellow. There may be presence or absence of a pigmented testa (with or without tannins) (Vikaspedia, 2024).

Pearl Millet: The millet seed has a single thin testa and an aleurone layer, which are occasionally pigmented. The colors of the grain are almost whitish, brown, gray, pale yellow, slate blue, or purple (Vikaspedia, 2024).

Finger Millet: The color of finger millet grains varies from dark brown to brown. The seed testa is colored with a five-cell-thick layer.

Little Millet: The color ranges from gray to straw white.

Foxtail Millet: The color of the foxtail millet grains varies starting from orange red to pale yellow, brown, or black (Kalsi & Bhasin, 2023). The color of the glumes of seed of foxtail millet can be red, white, black, white, and yellow (India Science, Technology and Innovation Portal, 2024).

Barnyard Millet: Barnyard millet seeds consist of different colors, including straw white, gray, light gray, and dull white (Vikaspedia, 2024).

Proso Millet: The *proso* millet grain color varies from whitish cream to golden yellow, straw white, or brown. The seeds have noticeable longitudinal lines, which easily differentiate the seed from other millet grains (Vikaspedia, 2024).

Kodo Millet: The color varies from blackish brown to dark brown.

Browntop Millet: The grains are tan or white in color.

Table 1.4 Color classification of millet seeds

Common names	Scientific names	Color characteristics
Jowar millet	*Sorghum vulgare*	Deep red and brown
Bajra millet	*Pennisetum typhoides*	White or yellow
Finger millet	*Eleusine coracana*	Red or purple
Foxtail millet	*Setaria italica*	Whitish, yellow, red, brown, and blackish variants
Proso millet	*Panicum miliaceum*	Whitish, creamish, yellow, orange, red, brown, and black
Kodo millet	*Paspalum scrobiculatum*	Pink
Barnyard millet	*Echinochloa esculenta*	White and cream
Little millet	*Panicum miliare*	Cream along with off-white
Browntop	*Urochloa ramosa*	Green along with brownish at the top

Note: Vikaspedia (2024) and India Science, Technology and Innovation Portal (2024).

1.4.6 Nutri-Classification

The nutri-classification of millets is discussed in Sections 1.4.6.1–1.4.6.11 and depends on the type of millets.

1.4.6.1 Sorghum

The grain sorghum is nutritious, containing promising levels of protein and dietary fiber. Sorghum significantly contributes minerals as well as vitamins, including potassium, phosphorus, calcium, iron, and zinc. It is cultivated and consumed across many states in India, and rotis made with jowar are much easier to digest (Vikaspedia, 2024).

1.4.6.2 Foxtail Millet

Foxtail millet seeds have numerous nutritional and medicinal benefits and are recommended in Ayurvedic and Unani products by experts (Adekunle, 2012). Being nonglutinous and nonacid generating, millet-based foods are therefore considered easily digestible (Prashant et al., 2005). It also possesses higher levels of proteins and minerals and acts as a potentially functional food ingredient and a supplementary protein source for most cereals due to their high lysine content (Mohamed et al., 2009). It is a gluten-free cereal and has a lower glycemic index, which makes it a valuable food for people with diabetes and celiac disorder. Foxtail millet provides nutritional security, including carbohydrates, protein, fat, fiber, minerals, and phytochemicals. Major phytochemicals present in foxtail millet include phenols, ferulic acid, chlorogenic acid, and p-coumaric acid, flavonoids, carotenoids, tocopherol, and tocotrienol (Suma & Urooj, 2012). Regular intake of foxtail millet can reduce the risk of Type-II diabetes mellitus, hypertension, and abnormal cholesterol levels (Verma et al., 2020).

1.4.6.3 Finger Millet

Finger millets contain high levels of crude protein, mineral matter, dietary fiber, and phenolics. It is exceptionally rich in calcium and serves valuable function in budding infants, expecting women, and individuals with obesity, diabetes, and malnutrition. For proper kidney and brain function, finger millet provides the required amount of potassium (Chandra et al., 2016). Food prepared using decorticated foxtail millet kernels is easily digestible, is versatile, and contains all forms of nutrients. It can be boiled like rice, ground for porridge or flour making, or utilized in cake formulations (De Wet, 2006). Germinated millet grains are recommended for children and elderly individuals. Finger millet seeds are utilized for liquor (*arake* or *areki*, as in Ethiopia) and beer preparation worldwide, with by-products utilized as cattle feed (FAOSTAT, 2013).

1.4.6.4 Pearl Millet

Bajra provides sufficient nutrients, including carbohydrates, proteins, mineral matter, vitamins, and dietary fiber, and required calories (Ojo et al., 2022).

1.4.6.5 Proso (Broomcorn) Millet

Proso millet grains contain a good quantity of protein, minerals, vitamins, dietary fiber, and bioactive compounds. The flour is gluten-free and considered ideal for gluten-sensitive individuals. Due to the presence of high lecithin in proso millet, it helps in the proper functioning of the nervous system. It also contains vitamins (niacin, B-complex, folic acid), minerals (P, Ca, Zn, Fe), and essential amino acids (methionine and cysteine) in required amounts (Habiyaremye et al., 2017).

1.4.6.6 Kodo Millet

Kodo millet (*Paspalum scrobiculatum*) is primarily cultivated in the Indian subcontinent and in other countries. Kodo millet exhibits high in vitro antioxidant capacity (Prashant et al., 2005). In addition, an antidiabetic effect of whole kodo millet has been reported in alloxan-induced diabetic rats. Kodo millet contains total dietary fibers of about 37–38%, the highest among other cereals, and is rich in high poly-unsaturated fatty acids (PUFA) (Malleshi & Hadimani, 1993). It has a good amount of nutrients like protein (11%) and has low fat (4.2%) content. Kodo millet food is easily digested and is high in lecithin, which is essential for proper growth and functioning of the nervous system. Kodo millets are rich in vitamin B-complex, notably niacin, B_6, and folate, minerals including Ca, Fe, K, Mg, and Zn. Kodo millets are nonglutinous and suitably consumed by people with gluten intolerance. Regular intake of kodo millet is considered valuable for women suffering from post-menopausal issues and individuals with cardiovascular problems, like hypertension and higher cholesterol (Mamta et al., 2000).

1.4.6.7 Barnyard Millet

Barnyard millet (*Echinochloa frumentacea*), one of the minor millets, contains slowly digestible carbohydrates, which is a gift of nature to the progressive mankind (Ugare et al., 2014). Similar to rice, it is consumed as food and can be cooked, and it is also suitable for patients with allergies, including atopic dermatitis (Watanabe, 1999). As a valuable food, its demand is now increasing both in Korea and Japan. Barnyard millet provides sufficient total dietary fiber, with a balanced proportion of soluble and insoluble fractions, and is a good source of highly digestible protein (Malleshi & Hadimani, 1993). As reported by Daniel (2005), omega-3 and omega-6 fatty acids and antioxidants are present in large amounts in the outer layers of barnyard millet. The primary fatty acids include omega-3 fatty acid (linoleic acid) along with palmitic acid and omega-7 (oleic acid). This millet has a high degree of resistance to amylase, which enables the formation of resistant starches. Having this ability, the millet diet is considered beneficial for patients suffering from heart disorders and diabetes mellitus. Barnyard millet is reported to have the ability to reduce blood glucose levels and regulate lipid (cholesterol) levels (Renganathan et al., 2020).

1.4.6.8 Little Millet

Little (*Panicum miliare*) millet grain is rich in vitamin B-complex and consists desirable amounts of minerals such as zinc, potassium, calcium, and iron. Little millet is commonly consumed in southern India, especially in traditional dishes. It is a beneficial complementary food to rice and prevents increasing of bodyweight (Srilekha et al., 2019). The little millet has 36–37% dietary fiber, which is highest among other cereals, and contains higher PUFA contents (Malleshi & Hadimani, 1993).

1.4.6.9 Browntop Millet

The browntop millet *Urochloa ramosa* (formerly *Brachiaria ramosa*) is native to India and Africa. It has good levels of dietary fiber (12%), which helps in sustaining a healthy digestive system. It has a low glycemic index that helps check diabetes. The serotonin compound found in browntop millet grain benefits in calming the neuronal system. It is very helpful in bronchitis and asthma. Browntop millet also boosts kidney and liver function. It is a gluten-free, nutritious, and resourceful millet grass that is high in protein, iron, calcium, magnesium, and other minerals. Browntop millet is utilized to prepare khichdi, dosa, upma, and roti. Some people exclusively involved in celiac disorder or having gluten sensitivity may confront an allergic reaction to browntop millet; symptoms include itchiness, swelling, rashes, nausea, or vomiting. This ingredient, along with some other millet, has managed to secure a place in the pool of SMART food categorized by ICRISAT (International crops research Institute for the semi-arid tropics). The fiber present in browntop millet surpasses the fiber composition of all other millets available. They also contain flavonoids like tannins, flavonoids, and quinones (India Science, Technology and Innovation Portal, 2024; Vikaspedia, 2024).

1.4.6.10 Amaranth Millet

Amaranth is gaining its recognition as a breakfast cereal and snack among health-sentient Indians (Vikaspedia, 2024, FSSAI, 2020). It is also viewed as a future encouraging crop to feed populations. Its numerous nutritional advantages include an excellent amino acid profile with considerable concentration of essential fatty acids and rich mineral and vitamin content. Its excellent fiber content is demonstrated to aid in maintaining improved gastrointestinal condition and reducing the cholesterol levels in the individuals. It encompasses a high amount of starch indicating high glycemic response in both flaked and roasted amaranth (India Data Insights, 2023; Vikaspedia, 2024).

This high-energy cereal works as an essential nutrient basis for all age groups. Amaranth has an excellent protein content of 13%, making it among the richest sources of proteins for vegans. It contains a balanced level of amino acids, with significant level of lysine (essential amino acid) that is otherwise absent in major cereal grains like rice and wheat. Lysine facilitates the absorption of calcium, building muscles and producing energy (FSSAI, 2020; Vikaspedia, 2024). Its high protein content helps in healthy hair. Representing the superior protein quality, amaranth is considered a complete source of protein when consumed with another cereal grain (India Data Insights, 2023). Amaranth seeds are high in unsaturated fatty acids, about 77%, mainly linoleic acid, demonstrating its cholesterol-reducing properties and antihypertensive effects (FSSAI, 2020; Vikaspedia, 2024). Its superior bioactive compounds are advantageous in decreasing inflammation in disease conditions and support healthy growth and promotion in children. Being full of antioxidants, it helps in combating free radicals, inflammation, and is antihypertensive. Hence, it is a significant antiaging grain (Vikaspedia, 2024). Amaranth seeds are regarded as a storehouse of essential vitamins and minerals comprising vitamin B_6, vitamin C, folate, magnesium, phosphorus, potassium, and carotene. Amaranth has been used as an alternative to generally utilize grains in the development of food products targeting to provide health benefits to populations of all ages (India Science, Technology and Innovation Portal, 2024; Vikaspedia, 2024).

1.4.6.11 Buckwheat Millet

Buckwheat millet recognition has increased prominently in recent times due to its gluten-free quality and high fiber content. When in necessity of the goodness of antioxidants and fiber, buckwheat millet from the millet family is a prime ingredient (India Data Insights, 2023; Yummy Valley, 2024). The ever-growing and endless need for food has persuaded the consumption of climate-resistant, economical, and nutritionally adequate minor millets uniformly rich in minerals and nutrient bioavailability, which are referred to as pseudo cereals (India Data Insights, 2023; Vikaspedia, 2024).

1.5 Millets Distribution and Its Production in India

1.5.1 Distribution of Millets in India

India was the highest producer of millets, followed by Nigeria, during the year 2000–2009. This tendency has maintained till date to year ending 2024. India produces eight different types of millet seeds, including sorghum, foxtail, kodo, proso, pearl, finger, barnyard, and little millets, normally cultivated under rain-supported environments. Further, in each of the millet breeding areas, at least four to five seed types are cultivated as either primary or allied crops in grouping with the pulses, oilseeds, spices, and condiments (APEDA, 2023; FSSAI, 2024; Yummy Valley, 2024). For instance, while pearl millet and sorghum are main and allied crops, respectively in the desert regions of Rajasthan, but in the eastern parts of Rajasthan and Gujarat, it is the contrary. Similarly, sorghum is sown as chief crop in Telangana, Andhra Pradesh, Maharashtra, and central India, while it is counted as a fodder crop in some of the southern regions (APEDA, 2023; India Data Insights, 2023). Likewise, finger millet is a prime crop in Tamil Nadu and Gujarat, while the same is a minor crop in

Telangana. Hence, the spatial distribution of millets either as a primary or allied crop fundamentally depends on the growing habitat and the amount of rainfall in the region. While sorghum dominates in areas getting annual rainfall above 400 mm, pearl millet counterparts it in areas with annual rainfall of 350 mm (APEDA, 2023; India Data Insights, 2023). Further, small millets including barnyard, finger, foxtail, proso, and little millets are cultivated in most of the southern and central regions where yearly rainfall is below 350 mm, possibly where other cereal crops are unable to grow under such moisture stress (India Data Insights, 2023). However, regardless of a rich inter/intra-species multiplicity and broader climatic malleability, cultivation of distinct millet species/varieties is gradually restricting in the recent past. In a way, a deprivation of institutional backing for millet crops compared to the formal encouragement of rice and wheat remains to shrink the millet-growing region. Despite this, numerous clusters in the dry/rain-fed regions that have recognized the food qualities of millets over generations remain incorporating a range of millets in their age-old cropping patterns, which distinguish millet to be an indispensable ingredient in Indian cuisines (FAO, 2023; World Population Review, 2025). A brief description of various millets grown in India and worldwide is given as follows (APEDA, 2023; India Data Insights, 2023).

1.5.1.1 Sorghum

Sorghum (*Sorghum bicolor* [L.] Moench) (Figure 1.5) is a warm seasonal crop, having intolerance to low-temperature climate but unaffected by stern pests and bugs. It is recognized by an assortment of names (such as great millet and guinea corn in west Africa, Asia, and areas of Middle East) (Taylor & Duodu, 2010; Taylor & Emmambux, 2008). Most of the sorghum produced in North and Central America, South America, and Oceania is utilized for animal feed (FAO, 1995). The sorghum grain has a naked caryopsis, comprising pericarp, endosperm, and germ. Even though there is a substantial range of somatic diversity, sorghum is classified into four classes: (i) grass sorghum, (ii) grain sorghum, (iii) forage sorghum glum, and (iv) Sudan sorghums/broomcorn (Macrae et al., 1993).

1.5.1.2 Foxtail Millet

Foxtail millet (*Setaria italica* [L.] P. Beauv) (Figure 1.5), among the world ancient crop grown by the humans, is generally cultivated in the dry and semi-dry regions of Asia and Africa (Ren et al., 2016). This discrete emergence of the panicle is like fox's tail, earning the popular name "foxtail" (Panchal et al., 2023). Foxtail millet, originated from China, is a drought lenient crop (Yang et al., 2017). It is known by many vernacular names around the world, namely, kangni, tenai kakum, navane, German millet (Sharma et al., 2015), Italian millet, Siberian millet, and foxtail bristlegrass (Sheahan, 2014). It is a prime diet in several regions across the globe. The farming of foxtail millet in developing countries provides significant support, as it can bloom under hostile heat and constrained rainfall conditions (Pawar & Machewad, 2006). Its plant continues to grow up to a height of 2–5 ft. Foxtail millet is captivating the researcher's attention due to its unique agronomical traits, environmental assistances in terms of safety, and economic advantages to the farmers by cultivating foxtail millet and greater benefit for human health by the straight consumption or as a part of food products (Sachdev et al., 2021). Since ages,

Figure 1.5 Pictures of commonly used millet. *Source:* FSSAI (2020) / Public Domain.

the cultivation of foxtail millet is restricted only for the fodder and bird feed, and it was not chosen for human consumption owing to its lesser palatability and deprivation of processing techniques (World Population Review, 2025). It has been identified as a C_4 model crop plant for genetic studies (Singh et al., 2021). It has a short life cycle of 10–14 weeks and has a bristle-like emergence of panicles of about 5–30 cm in length with reddish or purple color (India Data Insights, 2023).

1.5.1.3 Finger Millet

Finger millet (*Eleusine coracana* [L.] Gaertn) is a cereal grass cultivated regularly for its grain use as food and fodder (Figure 1.5). The finger millet is a vigorous, cespitose, tillering yearly grass, grows up to 170 cm in height (De Wet, 2006; FAO, 2013; Quattrocchi, 2006). The inflorescence is a panicle with 4–19 finger-like spikes that look like a fist when mature, thus the name finger millet (de Wet, 2006; Quattrocchi, 2006). The spikes bear up to 70 alternate spikelets, bearing four to seven small seeds (Dida & Devos, 2006). The seed pericarp is separate from the kernel and can be simply separated from the seed coat (FAO, 2013). Finger millet is a prime food in a handful of African and south Asian countries (World Population Review, 2025). It is additionally deemed as a beneficial famine crop as it can be simply dumped for lean years (FAO, 2013). Finger millet grain is not broadly used for livestock: it is predominantly a food grain, although it has less significant quality for livestock than corn, jawar, and bajra. In India, it is occasionally employed for infant calves feeding, growing animals, and ill and recovering animals (Sampath, 1986).

1.5.1.4 Pearl Millet

Pearl millet (*Pennisetum glaucum* [L.] R. Br.) (Figure 1.5) originates in central tropical Africa and is extensively dispersed in the dry tropics and India. It was taken to western states around the 1850s and started growing as a minor forage in the southeast and Gulf Coast states. The crop had been cultivated as a food grain around 4,000–5,000 years ago alongside the southern borders of the central highlands of the Sahara (World Population Review, 2025). It was afterward normally expanded beyond the semiarid tropics of Africa and Asia. Pearl millet has traditionally been a dominant grain, forage, and stover crop predominantly in the barren and subtropical provinces of numerous developing countries. As pearl millet cultivation spreads into nontraditional zones in temperate and developed countries, production limitations from diseases are becoming more significant (Bukhari et al., 2011; Chapke et al., 2018). Dissemination of precise statistics on diseases of the crop has not stayed pace with the increased attention to pearl millet as a balanced crop in nontraditional areas (FAO, 2023). Pearl millet is greatly diversified to growing areas differentiated by drought, low soil fertility, and high temperatures (FAS-USDA, 2024). It flourishes well in soils with abnormal salinity or low pH. Because of its forbearance to complicated growing conditions, it can be cultivated in areas where other cereal crops, such as corn, wheat, and rice, could not withstand (APEDA, 2023; India Data Insights, 2023).

1.5.1.5 Proso (Broom Corn) Millet

Proso millet (*Panicum miliaceum* [L.]) is an annual grass, rising from seed every year (Figure 1.5). Its descent goes back in history as far as 2,000 BC, when it is testified to have been cultivated in the central regions of Europe (FAO, 2010; Wikipedia, 2024). This plant is remarkably well competent to dry weather of central Russia, Middle East, northern India, Africa, Manchuria, and the great plains of North America. Proso millet was first reportedly introduced to Canada in the 17th century and was used in a controlled way as a feed crop in the early 1900s (Wikipedia, 2024). Proso millet is a relatively low-demanding crop, and its disease capabilities are unknown; hence, proso millet is generally used in organic farming methods in the European countries. In America, it is frequently used as an intercrop (FAO, 2023), which can help prevent a failover in summer and accomplish permanent crop rotation. Its superficial root system and its resistance to atrazine residue make proso millet a good intercrop between two water- and pesticide-persistent crops (Devisetti et al., 2014; Habiyaremye et al., 2017).

1.5.1.6 Kodo Millet

Kodo millet (*Paspalum scrobiculatum* [L.]) is extensively dispersed in wet locales within the tropics and subtropics of the world (FSSAI, 2020; India Data Insights, 2023) (Figure 1.5). It is aboriginal to India and is grown presently in Uttar Pradesh, in north, Tamil Nadu, and Kerala in the south (India Data Insights, 2023). This crop is also documented as *varagu*, *haraka*, and *arakalu* (APEDA, 2020).

1.5.1.7 Barnyard Millet

Barnyard millet (*Echinochloa crus-galli* [L.] P. Beauvois) is a multi-persistence crop for both food and feed (Figure 1.5), and is also known as *Japanese* barnyard millet, *ooda*, *oodalu*, *sawan*, *sanwa*, and *sanwank* (FSSAI, 2020; India Data Insights, 2023). The leaves are 1.5 cm in width and often have margin slightly wavy. The inflorescences spread out vertically across the preceding leaves, which are green or purple tinged and encompass 5–50 cm long (Renganathan et al., 2020).

1.5.1.8 Little Millet

Little millet (*Panicum sumatrense*) has been cultivated in India (Figure 1.5). It is grown all through India to a limited stretch up to an altitude of 2,100 m; however, it is of minimum importance elsewhere. The plant seeds of little millet are miniature when compared with other common millets. It is analogous in characteristics to the proso millet but is smaller (FSSAI, 2024; India Data Insights, 2023). It is a yearly herbaceous plant, which continues straight or with fold-over blades to a height of 30 cm to 1 m. The leaves are linear, sporadically with hairy lamina and membranous hairy ligules (Wikipedia, 2024). The panicles are from 4 to 15 cm in length with 2–3.5 mm long awn. The grain is round and smooth, 1.8–1.9 mm long. It can withstand both scarcity and water logging. It can be grown up to 2,000 m above sea level (India Data Insights, 2023). Little millet is an additional dependable catch crop in the assessment of its earliness and moreover endurance to hostile agro-climatic environments. The stover is a suitable forage for the cattle (APEDA, 2020; FSSAI, 2024).

1.5.1.9 Browntop Millet

Browntop millet (Figure 1.5) is indigenous to India, specifically in the dry areas of Karnataka and Andhra Pradesh. Browntop millet can flourish in a diversity of climates, involving tropical, subtropical, and adverse environments. Browntop millet is additionally recognized as palapul (Tamil), Korale (Kannada), and Andhra Korra (Telugu). Browntop millet (BTM) is an annual spikelet grass majorly planted in Africa, Australia, Western Asia, China, and Arab countries. It is grown in southern parts of India involving states such as Karnataka, Tamil Nadu, and Andhra Pradesh. It has a proficient fighting against antifungal action in the growth of seedlings (APEDA, 2023; FSSAI, 2024; India Data Insights, 2023; Yummy Valley, 2024).

1.5.1.10 Buckwheat Millet

The term pseudo cereals is the name given to amaranth (*Amaranthus* sp.) and buckwheat millet (*Fagopyrum esculentum*) (FSSAI, 2020) (Figure 1.6) as their seeds are frequently confused with conventional cereals. These grains are distinct from millets in their morphology but similar in analogous characteristics such as high nutritional profile and capability to mount food security (India Data Insights, 2023). Buckwheat millet is generally produced in high-altitude terrains like the Himalayan region. Due to its excellent weather resistance, the millet is competent for cultivation in fringe lands. Furthermore, it yet yields harvests in a quicker time, strengthening the livelihood of the hilly population. It comprises an extremely popular pseudo millet in India and is locally raised as “Kuttu.” Native to central Asia, it is also produced in China and other eastern countries (FSSAI, 2024; Yummy Valley, 2024).

1.5.1.11 Amaranth Millet

Amaranth is among the two pseudo cereals generally utilized for their extraordinary nutritional profile and prominent production sustainability in reserve for modest countries. It is morphologically distinct from other cereals and millets but analogous in its high protein and mineral composition of rice (FAO, 2010; India Data Insights, 2023). Pseudo cereals are exceptionally important for their superior climate endurance and sustainability in enhanced salinity and even scarcity conditions. Amaranth is scientifically indicated as *Amaranthus* L. and locally mentioned in India as ragira, seel, ramdana, or cholaibeej. This millet is nimble and puffy and can be both dark and light in color (NITI AAYOG, 2023a). Amaranth’s crop is known to be adjustable to several ecological areas and has encouraging agronomic properties, being resistant to various climatic stressors. These adaptable food grains are a notable segment of the food chain in the developing nations to assert food and

Figure 1.6 Pictures of commonly used millet. *Source:* FSSAI (2020) / Public Domain.

nutrition security and function as a nutrient vehicle for starving women and children. Amaranth is a superior energy pseudo cereal, analogous to major cereal grains identical to wheat and rice (FSSAI, 2023; India Data Insights, 2023).

1.5.2 Production of Millets

1.5.2.1 Advantage of the Production of Millet Crop

Millets have normally been termed as the coarse grain cereals; however, because of their nutritional assistance, they are promptly being denoted as "nutria-millets or nutria-cereals." Below are certain benefits of production of millets in India (FSSAI, 2024; India Data Insights, 2023; NITI AAYOG, 2023b).

1) Millets are designated as the "miracle grains" or "crops of the future" as they can grow underneath bleak conditions and are drought-resistant crops that demand limited exterior involvements (Bukhari et al., 2011; Malleshi & Hadimani, 1993).
2) Millets are dual-purpose crops. They are planted both as food and fodder, thus supplying food/livelihood protection to millions of households and promoting the monetary proficiency of farming (FAO, 2024; NITI AAYOG, 2023b; World Population Review, 2025).
3) Millets add to abating environmental adjustment as they help to decrease the atmospheric CO_2. On the converse, wheat being a thermally sensible crop and paddy is a major provider to climate switch through methane emission (FAO, 2024; World Population Review, 2025).
4) Production of millets does not require the use of chemical fertilizers. The millet crops do not invite pests and are not influenced by storage (India Data Insights, 2023).
5) Millets are noteworthy in their nutritive worth of vitamins, minerals, dietary fiber, or other nutrients. They are nearly three to five times nutritionally better than wheat and rice. Sorghum (Jowar) is a valuable source of polyphenols, antioxidants, and cholesterol-lowering waxes (Malleshi & Hadimani, 1993; Saleh et al., 2013).
6) Millets help in reducing obesity, decrease the risk of hypertension, CVDs, T2DM, cancers, as well as benefit in inhibiting constipation due to their high dietary fiber contents with low glycemic index (Malleshi & Hadimani, 1993; Saleh et al., 2013).

It is anticipated that there is around 51% expansion in the worldwide diabetic population by the year 2045. As mentioned earlier, ingesting millets decreases postprandial glucose levels in the millets (International Diabetes Federation, 2025). It is also marked that all millets do not have the same glycemic index; fonio, foxtail, barnyard, and teff have lower GI (<55), which means they are (35–79%) more applicable in lowering dietary GI comparatively (Anitha et al., 2021). Sustainable Development Goal-2 aims to achieve "zero hunger." It consists of one of the 17 viable development goals launched by the United Nations in 2015 (FAO, 2023). The official wording is, "End hunger, achieve food security and improved nutrition and promote sustainable agriculture." A thoughtful transformation in the global food and agricultural approach is required to sustain today's 800 million people (World Population Review, 2025). It can be feasible by concentrating on millet production. Nearly 40% of the global land apparently is dryland. Millets are the best-fit crop for dryland agriculture (FAO, 2023, 2024). Production of millets around the globe has been achieving new elevations. Millets are the sixth highest yielding crops in the world. Millet production is substantial on a global scale, with numerous countries contributing to its cultivation. The top millet (millet, sorghum, and buckwheat)-producing countries are India, the United States, Nigeria, China, Ethiopia, Niger, Mexico, Mali, Sudan, and Brazil. The production amounts vary each year depending on the circumstances such as weather conditions, government policies, and market challenges (FAO, 2024; World Population Review, 2025).

1.5.2.2 Global Millet Production

Corresponding to FAO statistics (FAO, 2010), the worldwide production of millets was 26 MMT involving a land area of about 33.6 million hectares. In the year 2002, almost a decade earlier, the production growth of the millets worldwide was down to 23.3 MMT from a cultivation area of 33.3 million hectares. Africa was the topmost contributor of millet in the year 2009 (20.6 MMT), followed by Asia (12.4 MMT), whereas India contributes about 10.5 MMT (FAO, 2024; FAS-USDA, 2024). Compared to major crops such as rice, wheat, barley, and corn, sorghum places fifth in significance, based on production and planted area, which accounts for 5% of the total crop production worldwide (Obilana & Manyasa, 2002). According to the APEDA-25 (Agricultural and Processed Food Products Export Development Authority), overall, India's contribution of millet is 17.68% during the year 2020 globally. Other countries include Brazil (25%), Sudan (3%), Mali (4%), Mexico (4%), Niger (5%), and Senegal (6%) (Figure 1.7). The global production of millets is 32.12 MMT (FAS-USDA, 2024). India ranks first in the production of millets and has a 41% stake in the global production of millets, i.e., 12.2 MMT in 2023–2024. Table 1.5 shows the data of topmost millet-growing countries of the world (FAO, 2023; FAS-USDA, 2024; India Data Insights, 2023).

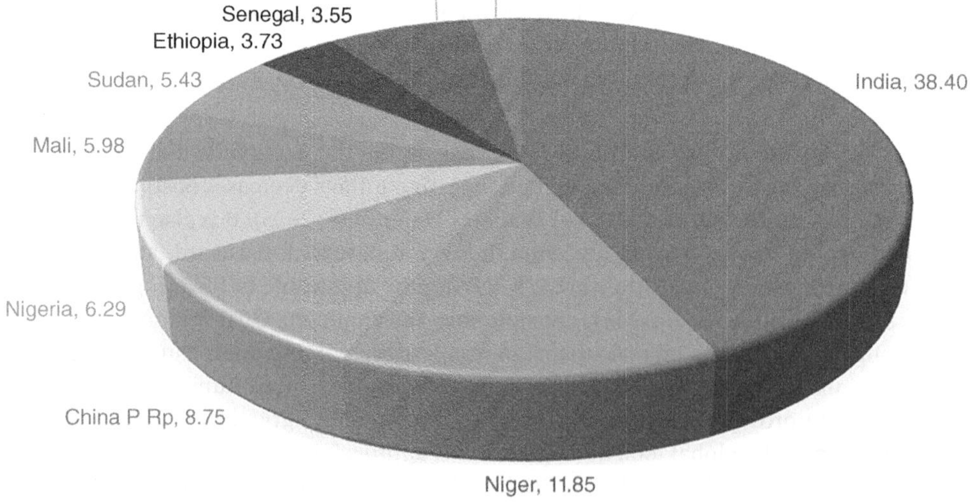

Figure 1.7 Worldwide growth of various millets. *Source:* Ministry of Commerce & Industry, GOI (2024)

Table 1.5 The leading millet producing countries of the world

Millet production around the world, in 1,000 MT				
Countries	**2022–2023**	**Percentage**	**2023–2024**	**Difference (%)**
India	13,506	41	12,220	−9.7
Niger	3,657	11	3,500	−7.0
China	2,700	9	2,750	0.0
Mali	1,833	7	2,170	−1.8
Nigeria	2,030	5	1,580	−1.5
Senegal	1,097	5	1,460	−8.8
Ethiopia	942	4	1,220	16.8
Burkina Faso	908	3	1,080	10.1
Sudan	1,675	2	694	−4.5
Chad	694	2	653	0.9
Others	3,062	11	3,475	7.8
Total	32,104	100	29,870	−4.1

Note: FAS-USDA (2024).

Corresponding to the latest data from the USDA, the worldwide production of millet remains at 30.802 MMT in the year 2023–2024, which is a lesser production of 4.1% compared to last year. The leading millet production nations such as India (12.2 MMT; −10%) and Niger (3.4 MMT; −7%) are supposed to partake a decline in production, while in Nigeria (2 MMT; −1.5%), Mali (1.8 MMT; −1.8%), and Sudan (1.6 MMT; −4.5%), the millet production seems plummeting. In the interlude, the millet production of China is projected to remain around 2.7 MMT, as of previous year, while in the Ethiopia (1.1 MMT; +17% to) and Burkina Faso (1 MMT; +10%), there is an increase in crop production (FAO, 2023; India Data Insights, 2023).

The global millets marketplace (Figure 1.8) has expanded strongly in current years. It grows from $10.99 billion in 2023 to $11.67 billion in 2024 at a compound annual growth rate (CAGR) of 6.2% (APEDA, 2025; FAO, 2024). The growth in the historic period can be credited to growing nutritional awareness, increasing health and wellness tendencies, and increase in viable agriculture and government initiatives (UNICEF, 2023). The global market of millets is estimated to have a sturdy growth in the next few years. It will nurture to $14.39 billion in 2028 at a CAGR of 5.4%. The growth in the anticipated period can be credited to global population growth, climate transformation pliability, urbanization and accessibility, research, and development (FAO, 2023; India Data Insights, 2023). Major tendencies in the projected period comprise consumer learning, research and innovation, culinary diversity, gluten-free, and allergen-friendly foods.

India with the millet production of 10.91 MMT stands as the chief millet-producing country in the world (Table 1.6). Millet is connected to Indian cuisine, with the country's distinct agro-climatic sectors assisting its vigorous cultivation (India Data Insights, 2023; NITI AAYOG, 2023b). Rajasthan surfaces as a substantial contributor, producing 27% of India's millet notwithstanding its desert conditions. Bajra and pearl millets are key

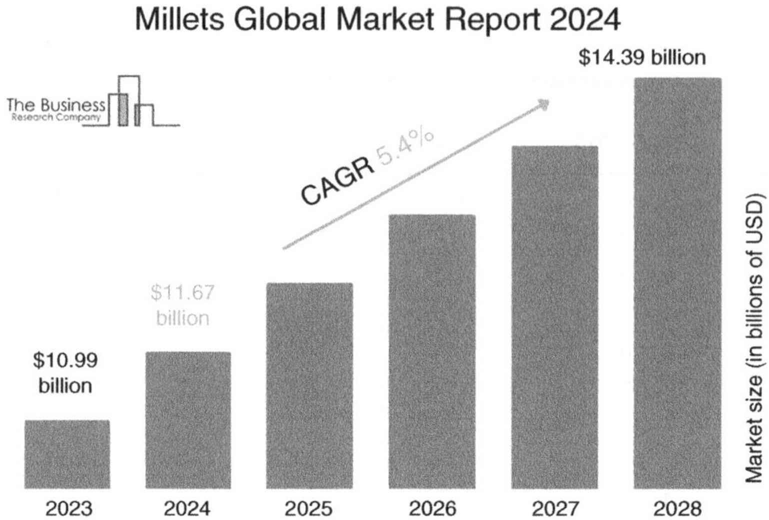

Figure 1.8 The global millets market statistics. *Source:* Ministry of Commerce & Industry, GOI (2024)

varieties cultivated by Rajasthan's resilient rural communities. This emphasizes Rajasthan's paramount role in standing millet farming and showcasing viable agricultural methods amid confronting circumstances, suggesting resolutions to water paucity and climate transformation. Nigeria ranks second globally in millet production, yielding 5 MMT annually. Millet comprises a basic carbohydrate source in the diet of northern Nigerians, indicating the country's agricultural promise and conventional crop focus. Nigeria's emphasizing of natural agriculture is apparent in millet's adaptability to numerous climates, safeguarding steadfast food sources. The nation's dynamic agriculture sector has forced it to the forefront of global millet production, notwithstanding changeability in productivity over the years. Niger ranks third globally in millet production, yielding 2.955 MMT annually. Millet reckons cultural and economic importance in Niger, amalgamated into traditional dishes and festivities. The country's allegiance to sustainable agriculture and millet's drought lenience contribute to its high yield. Niger's subterranean cultural connection to millet is apparent in its convention in daily life and rituals, sustaining agricultural ecologies and adopting global food security attempts regardless of the challenges of farming in dry conditions (FAO, 2023; India Data Insights, 2023).

1.5.2.3 Production Status of Indian Millet

There is a requirement for intensive and repeated efforts to foster millets and reinforce their capability for food and nutritional security, and viable agriculture in the long run. In the recent past, a revived focus and prominence have been allocated to the millets by the state and central governments. Successively, various programs and plans were initiated to push millet production (APEDA, 2024). Millets have pronounced projections for cultivation in India, given their compliance with various agro-climatic conditions and their manifold benefits (Sreekala et al., 2023). India stands in the top place in cultivation and food use

Table 1.6 Global millet production trend (2018–2022)

		Country-wise millets production (all millets) (2018–2022)						
		Production in (000) MT						
S. No.	Country	2018	2019	2020	2021	2022	Share (%)	Growth rate (%) in 2022 over 2021
1	India	16,440.00	13,711.24	17,260.00	18,020.00	15,999.76	18.1	−11.2
2	Niger	5,956.53	5,167.09	5,641.20	3,353.95	5,757.66	6.5	71.7
3	China P Rp	3,757.99	5,900.00	5,850.00	5,700.00	5,880.00	6.6	3.2
4	Nigeria	2,240.74	2,000.00	2,000.00	1,922.00	1,941.22	2.2	1.0
5	Mali	3,310.01	3,389.64	3,743.86	2,727.34	3,448.05	3.9	26.4
6	Sudan	2,647.00	1,133.00	484.96	1,500.00	1,675.00	1.9	11.7
7	Ethiopia	982.96	1,125.96	1,218.58	1,000.00	1,150.00	1.3	15.0
8	Senegal	717.02	1,077.21	1,522.18	1,392.33	1,460.19	1.7	4.9
9	Burkina Faso	3,118.91	2,841.97	2,797.00	2,335.00	2,921.61	3.3	25.1
10	Chad	1,744.18	1,690.14	1,656.82	1,517.15	1,660.77	1.9	9.5
11	Nepal	313.99	314.23	320.95	326.44	339.46	0.4	4.0
12	Tanzania Rep	1,136.01	1,117.84	1,075.00	1,412.00	1,335.00	1.5	−5.5
13	Russia	266.33	538.47	449.51	483.55	434.99	0.5	−10.0
14	Pakistan Ir	498.74	504.91	362.05	290.26	305.03	0.3	5.1
15	Guinea	275.32	285.61	276.53	280.59	284.71	0.3	1.5
16	USA	326.34	376.66	208.88	348.72	213.26	0.2	−38.8
17	Ghana	497.80	535.00	526.00	570.00	597.81	0.7	4.9
18	Myanmar	466.29	303.54	287.56	490.00	316.87	0.4	−35.3
19	Cameroon	104.23	105.27	97.59	98.21	100.00	0.1	1.8

(Continued)

Table 1.6 (Continued)

		Country-wise millets production (all millets) (2018–2022)						
		Production in (000) MT						
S. No.	Country	2018	2019	2020	2021	2022	Share (%)	Growth rate (%) in 2022 over 2021
20	Uganda	508.23	643.10	461.30	278.25	321.86	0.4	15.7
21	Ukraine	274.44	361.76	362.61	378.15	130.89	0.1	−65.4
22	Uzbekistan	13.29	13.67	13.11	121.90	94.34	0.1	−22.6
23	Korea Dp Rp	114.33	110.02	111.02	111.52	111.45	0.1	−0.1
24	Cote D Ivoire	130.22	135.00	141.66	137.00	143.46	0.2	4.7
25	Kenya	289.78	423.00	468.00	198.00	180.00	0.2	−9.1
26	Yemen Republic	57.84	50.39	36.00	30.00	50.00	0.1	66.7
27	Congo D. Rep.	45.15	46.68	50.39	48.14	57.08	0.1	18.6
28	South Sudan	5.00	5.00	6.00	47.85	48.02	0.1	0.4
29	Namibia	87.56	19.70	98.28	88.57	50.06	0.1	−43.5
30	Malawi	114.17	171.48	190.20	164.39	159.67	0.2	−2.9
31	Sierra Leone	75.46	88.00	84.00	83.38	103.00	0.1	23.5
32	Zimbabwe	119.25	76.00	186.47	174.10	125.63	0.1	−27.8
33	Gambia	128.10	65.00	110.00	42.00	42.64	0.0	1.5
34	Kazakhstan	42.55	49.23	44.01	40.35	55.75	0.1	38.2
35	Australia	35.92	36.38	36.44	36.62	36.64	0.0	0.1
36	Angola	114.38	83.31	115.00	69.45	70.55	0.1	1.6
37	Guinea Bissau	33.40	42.00	45.00	48.10	55.54	0.1	15.5
38	Togo	26.08	26.81	30.00	28.00	27.91	0.0	−0.3
39	Eritrea	23.92	25.00	25.00	25.00	25.00	0.0	0.0
40	Zambia	45.41	31.52	65.02	53.07	39.06	0.0	−26.4
41	Benin	26.14	25.00	27.29	23.46	23.99	0.0	2.3

42	Mozambique	17.00	12.83	22.10	25.00	23.52	0.0	−5.9
43	Belarus	19.93	18.02	15.64	17.45	16.06	0.0	−8.0
44	Congo P Rep	14.23	14.50	13.57	14.09	13.60	0.0	−3.5
45	Bangladesh Pr	12.80	9.29	10.08	9.68	12.49	0.0	29.0
46	Iran	17.57	19.14	12.27	12.34	12.21	0.0	−1.1
47	Saudi Arab	8.32	11.40	11.70	12.16	11.75	0.0	−3.4
48	Burundi	38.44	18.68	36.00	36.00	36.00	0.0	0.0
49	C Afri Rep	39.82	40.00	40.00	40.00	27.91	0.0	−30.2
50	Argentina	2.25	4.14	5.44	9.51	9.48	0.0	−0.3
51	Rwanda	5.12	5.20	5.07	5.07	7.32	0.0	44.4
52	South Africa	5.03	5.16	6.33	6.33	6.33	0.0	0.0
53	Libya	6.68	6.00	6.00	6.00	6.00	0.0	0.0
54	Morocco	7.97	7.34	9.44	10.10	10.50	0.0	4.0
55	Botswana	18.99	36.86	95.15	49.34	40.00	0.0	−18.9
56	Turkey	4.40	4.76	5.71	4.35	4.19	0.0	−3.7
57	Mauritania	3.29	3.22	2.63	1.00	3.44	0.0	244.0
58	Azerbaijan	2.99	5.73	4.57	2.80	3.09	0.0	10.4
59	Iraq	23.69	3.84	7.14	5.03	4.92	0.0	−2.2
60	Afghanistan Tis	11.64	2.71	1.90	2.62	1.67	0.0	−36.3
61	Vietnam Soc Rep	3.37	3.40	2.00	2.11	1.46	0.0	−30.8
62	Korea Rp	4.31	3.74	3.75	3.84	3.85	0.0	0.3
63	Switzerland	0.93	1.47	0.80	0.91	0.87	0.0	−4.4
64	Bermuda	0.00	0.00	0.00	0.00	0.60	0.0	
65	Taiwan	2.73	2.65	2.52	2.65	4.50	0.0	69.8
66	Sri Lanka Dsr	8.23	0.21	0.45	0.67	0.40	0.0	−40.3
67	Mexico	0.22	0.28	0.28	0.28	0.27	0.0	−3.6
68	Japan	0.20	0.21	0.25	0.25	0.25	0.0	0.0

(Continued)

Table 1.6 (Continued)

		Country-wise millets production (all millets) (2018–2022)						
		Production in (000) MT						
S. No.	**Country**	**2018**	**2019**	**2020**	**2021**	**2022**	**Share (%)**	**Growth rate (%) in 2022 over 2021**
69	Tajikistan	0.20	0.13	0.22	0.24	0.25	0.0	4.2
70	Serbia	8.08	8.06	8.19	8.17	8.30	0.0	1.6
71	Kyrghyzstan	0.14	0.19	0.35	0.10	0.09	0.0	−10.0
72	Syria	5.21	5.93	1.32	0.80	10.38	0.0	1,197.5
73	Macedonia	0.07	0.07	0.07	0.07	0.07	0.0	0.0
74	Maldives	0.05	0.06	0.05	0.06	0.06	0.0	0.0
75	Moldova	0.01	0.17	0.02	0.06	0.01	0.0	−83.3
76	Slovak Rep	1.13	0.00	0.00	0.00	0.00	0.0	
77	Slovenia	0.59	0.51	0.24	0.53	0.08	0.0	−84.9
78	Spain	1.76	0.00	0.00	0.00	0.00	0.0	
79	Poland	21.94	0.00	0.00	0.00	0.00	0.0	
80	Jordan	13.44	0.30	0.98	2.52	0.79	0.0	−68.7
81	Hungary	3.40	0.00	0.00	0.00	0.00	0.0	
82	Greece	0.13	0.00	0.00	0.00	0.00	0.0	
83	Croatia	0.36	0.00	0.00	0.00	0.00	0.0	
84	Czech Republic	1.02	0.00	0.00	0.00	0.00	0.0	
85	France	40.91	0.00	0.00	0.00	0.00	0.0	
86	Bhutan	1.75	1.24	1.61	1.12	0.00	0.0	−100.0
87	Bulgaria	4.90	0.00	0.00	0.00	0.00	0.0	
88	Austria	10.86	0.00	0.00	0.00	0.00	0.0	
	Total	90,361.25	86,264.98	89,169.34	91,454.39	88,441.46	100.0	−3.3

Note: Adapted from FAO (2023).

of millets around the world. Millets have remained conventionally grown and consumed in several regions of the country for centuries (India Data Insights, 2023; NITI AAYOG, 2023a). Millet was substantially produced some 50 years back; however, due to adaptation of western development patterns, India has disregarded its conventional perception. Millets have been considered archaic and rough grains, propagated as the food of rural people or ancestors; likewise the green revolution negatively had impact on the millet production (India Data Insights, 2023). Before the green revolution, millets contributed 40% of all the grain produced. Meanwhile, Indian production of millets is about 170 lakh tons (about 20% of the global output). The average yield at the global level is 1,229 kg per hectare, whereas the average yield of millets in India is 1,239 kg per hectare (APEDA, 2025).

A variety of millet seeds such as bajra (pearl millets), jowar (sorghum), ragi (finger millet), canary seeds, buckwheat, and other cereal have been produced in India. Pearl (bajra) and sorghum (jowar) contribution to total production of millet in India is highest. The three crops of millets namely bajra, sorghum, and buckwheat collectively supply more than 18% to world production in 2022. India's bajra production accounts for 38.4%, followed by sorghum 7.21% and buckwheat. The foremost millets producing states in India are Rajasthan, Karnataka, Maharashtra, Uttar Pradesh (UP), state of Haryana, Gujarat, Madhya Pradesh, Tamil Nadu, Andhra Pradesh (AP), and Uttarakhand (Figure 1.9). At present, these 10 territories in total account for around 98% millet production in India during the period 2023–2024 (second advance estimates). Six states namely Rajasthan, Uttar Pradesh, Karnataka, Maharashtra, Madhya Pradesh, and

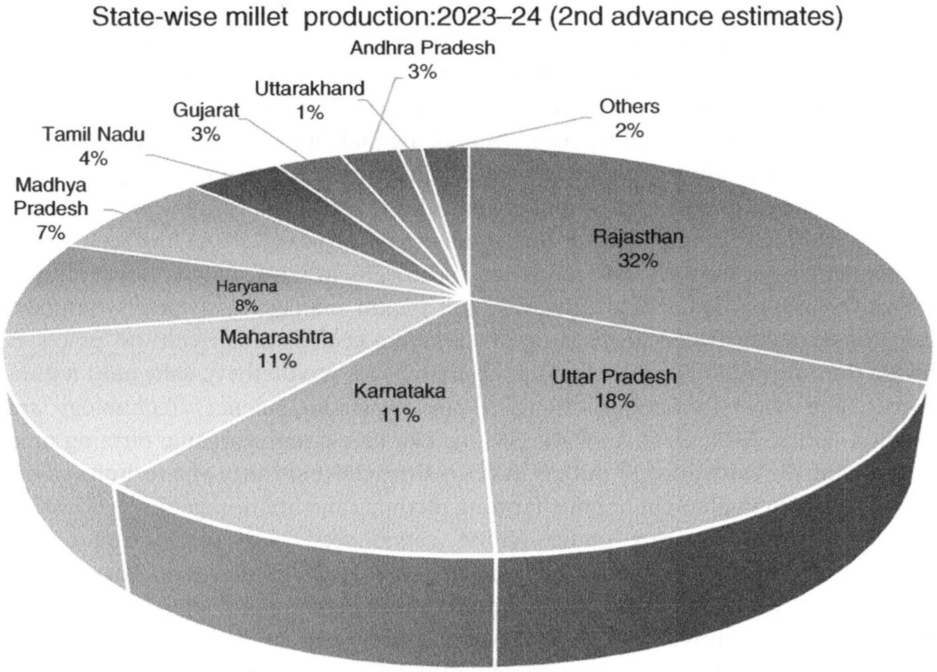

Figure 1.9 State-wise cultivation of millets in India. *Source:* Ministry of Commerce & Industry, GOI (2024)

Haryana contribute about 79.6% share of millet produced totally. Rajasthan contributes 31.3% of the total millet production in India. According to the available data, India's millet production receives a positive trend in latest years (Table 1.7) (APEDA, 2024; India Data Insights, 2023). India produces around 12 MMT of millets annually, according to Ministry of Agriculture and Farmers Welfare data. Pearl millet (bajra) is the chief (61%) millet variety cultivated in India, followed by sorghum (jowar) (27%), finger millet (ragi) 9%, and small millet 2% (Figure 1.10) (APEDA, 2024; India Data Insights, 2023; NITI AAYOG, 2023a).

1.6 Indian Government Schemes to Promote Millets

The government of India has been carrying out several projects to assist millet cultivation and consumption via various programs (India Data Insights, 2023). The objective of these efforts is to enhance millet production, boost the income of farmers, and focus on nutritional challenges. Indian Millets Research Centre (Hyderabad) is being progressed as a center of excellence so that it can outshine at the international level (NITI AAYOG, 2023a).

The National Food Security Mission (NFSM) was launched in October 2007 for warranting universal development of agriculture and associated sectors. The NFSM promotes the production of millet crop as one of its components. It aims to strengthen the millet cultivation along with other crops to improve food security and expand the income of farmers (NITI AAYOG, 2023a). The Rashtriya Krishi Vikas Yojana (RKVY) launched in the year 2007 thrusted as an umbrella program for safeguarding holistic upliftment of agriculture and associated segments. Restructured as RKVY cafeteria system from 2022 to 2023, reunion some plans such as Paramparagat Krishi Vikas Yojana (PKVY), Crop Diversification Programme, etc. RKVY sponsors the cultivation of millets and other crops by offering monetary assistance to states (India Science, Technology and Innovation Portal, 2024; NITI AAYOG, 2023b). It promotes the application of appropriate farming practices, technologies, and infrastructure extension for millet cultivation (India Data Insights, 2023).

The National Mission intended for Sustainable Agriculture (NMSA) has been articulated for augmenting agricultural production specifically in rainfed areas, concentrating on cohesive farming, water-use efficiency, soil health management, and synergizing reserve conservation. NMSA focuses on encouraging ecological agricultural practices, involving the cultivation of millets. It intends to improve productivity, safeguard natural resources, and boost farmers in changing climate (India Science, Technology and Innovation Portal, 2024; NITI AAYOG, 2023b). The PKVY urges organic farming practices, involving the cultivation of millets. It offers financial assistance and technical assistance to farmers to implement organic farming methods and attain organic certification. Sub-mission on nutri-cereals under NFSM commenced with an expenditure of Rs. 300 crores for 2018–2019 (India Science, Technology and Innovation Portal, 2024; NITI AAYOG, 2023b). It aimed to obtain a strategy for attending problems involving production, demand, and research with a market-oriented approach. Improving production of nutri-cereals through area extension and productivity augmentation in a viable manner in the associated districts of the country (India Science, Technology and Innovation Portal, 2024; Yummy Valley, 2024), boosting seed stock system of nutri-cereals. Developing

Table 1.7 State-wise millets production progress in India (2019–2024)

State-wise production of millets in India															
	Area (Lac ha)					Production (Lac tons)					Yield (kg/Ha)				
States	2019–20	2020–21	2021–22	2022–23	2023–24	2019–20	2020–21	2021–22	2022–23	2023–24	2019–20	2020–21	2021–22	2022–23	2023–24
Rajasthan	49.40	49.15	43.63	52.12	47.75	51.47	51.56	42.80	56.74	48.09	1,042	1,049	981	1,088	1,007
UP	10.93	10.93	10.83	11.55	13.47	21.72	22.98	22.26	23.69	26.98	1,987	2,103	2,055	2,051	2,003
Karnataka	18.39	17.83	16.38	15.27	14.25	25.56	25.69	20.54	20.33	17.49	1,390	1,441	1,254	1,331	1,227
Maharashtra	30.87	28.85	24.57	20.35	20.82	24.29	25.14	23.05	18.99	17.15	787	871	938	933	824
Madhya Pradesh	4.88	5.17	5.55	6.05	6.40	8.96	10.24	11.81	12.54	12.68	1,836	1,981	2,129	2,072	1,982
Haryana	5.23	6.00	5.06	5.53	5.90	10.35	13.67	11.32	12.14	11.94	1,980	2,276	2,235	2,195	2,024
Tamil Nadu	6.28	5.80	5.55	4.90	4.69	10.17	9.05	7.65	6.30	6.23	1,620	1,560	1,380	1,287	1,327
Gujarat	4.96	5.20	5.10	5.52	2.64	9.90	10.92	11.79	13.64	4.51	1,997	2,099	2,310	2,473	1,707
AP	2.36	2.06	1.52	1.33	1.46	5.14	5.41	3.59	3.76	3.90	2,179	2,624	2,363	2,826	2,672
Uttarakhand	1.37	1.38	1.33	1.15	1.09	1.91	2.01	2.00	1.76	1.61	1,395	1,455	1,507	1,528	1,481
Telangana	1.00	1.02	0.74	0.73	0.62	1.39	1.66	1.23	1.32	1.15	1,392	1,631	1,659	1,814	1,851
Odisha	0.76	0.83	1.01	0.93	1.04	0.48	0.55	0.68	0.76	0.72	629	662	674	820	694
Others					0.67					0.63					939
Jharkhand	0.17	0.21	0.21	0.17	0.32	0.14	0.18	0.18	0.14	0.30	859	853	872	781	922
Chhattisgarh	0.72	0.92	0.57	0.48	0.53	0.25	0.26	0.28	0.24	0.20	341	285	495	495	387
Bihar	0.09	0.10	0.08	0.07	0.08	0.08	0.10	0.09	0.07	0.08	930	993	1,004	1,003	1,001
West Bengal	0.09	0.07	0.07	0.06	0.05	0.10	0.07	0.08	0.06	0.06	1,068	991	1,126	1,114	1,103
Assam	0.05	0.05	0.05	0.06	0.05	0.03	0.03	0.03	0.05	0.04	619	656	664	799	705
Himachal Pradesh	0.07	0.04	0.03	0.03	0.03	0.07	0.03	0.02	0.02	0.02	961	886	737	712	718

(Continued)

Table 1.7 (Continued)

State-wise production of millets in India															
	Area (Lac ha)					Production (Lac tons)					Yield (kg/Ha)				
States	**2019–20**	**2020–21**	**2021–22**	**2022–23**	**2023–24**	**2019–20**	**2020–21**	**2021–22**	**2022–23**	**2023–24**	**2019–20**	**2020–21**	**2021–22**	**2022–23**	**2023–24**
Kerala	0.01	0.01	0.00	0.00	0.02	0.01	0.01	0.01	0.00	0.01	971	1,118	1,518	1,588	367
Punjab	0.01	0.00	0.01	0.01	0.01	0.00	0.00	0.01	0.00	0.00	635	640	635	690	654
And & Nic			0.00					0.00					0		
Arun Pradesh	0.27	0.27	0.27	0.27		0.27	0.28	0.28	0.28		1,025	1,030	1,033	1,036	
Chandigarh			0.00					0.00					0		
Dad & Nag Hav	0.01	0.01	0.01	0.01		0.01	0.01	0.01	0.02		1,653	1,652	1,667	1,605	
D&D	0.00	0.00	0.00			0.00	0.01	0.01			1,150	1,509	1,530		
Delhi	0.05	0.04	0.04	0.04		0.06	0.08	0.06	0.08		1,349	1,907	1,527	1,726	
Goa			0.00					0.00					0		
J&K	0.18	0.21	0.11	0.18		0.06	0.09	0.06	0.11		329	405	558	586	
Ladakh			0.00					0.00					0		
Lakshadweep			0.00					0.00					0		
Manipur			0.00					0.00					0		
Meghalaya	0.03	0.03	0.03	0.03		0.03	0.03	0.03	0.03		941	941	960	921	
Mizoram								0.00							
Nagaland	0.10	0.10	0.09	0.13		0.11	0.11	0.08	0.14		1,112	1,112	948	1,118	
Puducherry	0.00	0.00	0.00	0.00		0.00	0.00	0.00	0.00		2,358	2,412	2,221	2,457	
Sikkim	0.02	0.02	0.02	0.00		0.02	0.02	0.02	0.00		1,037	1,038	1,041	1,043	
Tripura	0.01	0.02	0.02	0.01		0.01	0.01	0.01	0.01		822	806	812	984	
All India	138.29	136.33	122.89	126.98	121.88	172.61	180.21	160.00	173.21	153.79	1,248	1,322	1,302	1,364	1,262

Note: APEDA (2024).

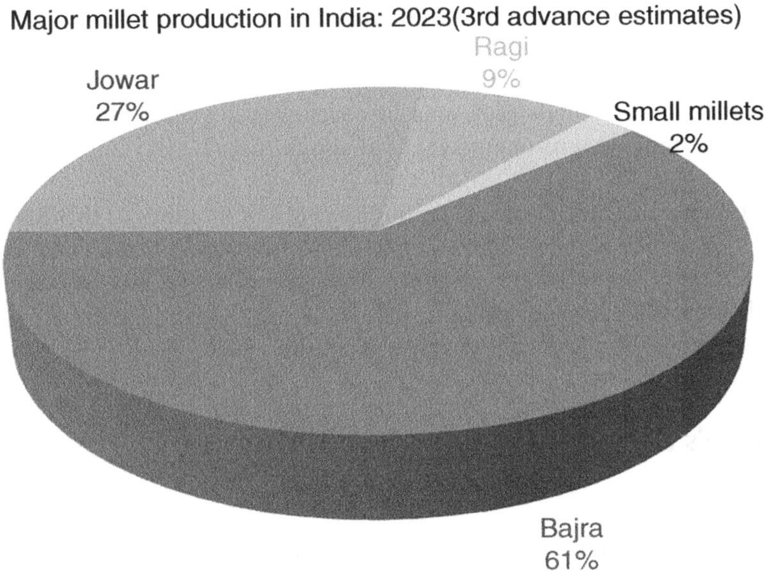

Figure 1.10 Millet production in India. *Source:* Ministry of Commerce & Industry, GOI (2024)

postharvest value accumulation at farm gate for competent price realization to farmers with capable market linkages. The Indian Council of Agricultural Research (ICAR) has introduced a Millets Development Program (MDP) to support research, development, and encouragement of millets. The program focuses on improving millet varieties, agronomic practices, and postharvest technologies (India Data Insights, 2023; NITI AAYOG, 2023a).

1.7 India Government's Steps to Intensify Millet Production

Regardless of several abilities, millet consumption as an article of food consumption remains restrained to the ethnic users, i.e., ancestral inhabitants. This was primarily because there was nonavailability of buyer approachable ready-to-eat millet-based foodstuffs. Recently, millets have achieved attention, and attempts are in progress to acquire their convenient and value-added processed products. Since several households in dryland and hilly districts depend on millets to meet their food needs (India Data Insights, 2023; India Science, Technology and Innovation Portal, 2024), it has currently remained projected to widen the food storage by incorporating millets, correspondingly pearl, sorghum, finger, kodo millets, etc., in the PDS. Government has realized the role of millets in the food chain. Under the NFSM of the primary targets for increasing food grain production by an additional 25 million tons, the share allocated for millets is 2 MT, i.e., 8% of the increased food grain production (APEDA, 2024). The Indian officials reorganized their courtesy concerning millet agrobusiness schemes and proclaimed policies to initiate a supporting environment for the farmers. Regarding millet production, some of the existing schemes by the Government of India include the following (India Data Insights, 2023; NITI AAYOG, 2023a):

1) The integrated cereals development programs in coarse cereals (ICDP-CC)-based cropping system areas under macro-management of agriculture (MMA) (India Data Insights, 2023).
2) Inventiveness for nutritional safety by adopting Intensive Millet Promotion (INSIMP) is a segment of RKVY, which is the solitary complete inventiveness to sustenance of millet cultivation (India Data Insights, 2023; India Science, Technology and Innovation Portal, 2024; NITI AAYOG, 2023a).
3) Rainfed Area Development Programme (RADP) is a module of RKVY.

1.8 The Company's Business Associated with Millet Processing

Companies within the millet market are aggressively presenting planned investments and discovering prospects within the food industry to progress and foster products in this manner sustaining their market perception. These capital spending play an important role in advancing the millet market by advocating its growth, undertaking environment and food collateral tasks, and promoting the health gains linked with millets.

In February 2022, Wholsum Foods Pvt. Ltd., an India-based food startup, obtained $7 million in sponsoring to increase its range of millet-based offerings. This venture is hovering to significantly provide to Wholsum Foods' objective of improving approachability and attraction of millets to consumers. In March 2023, Ivar Agro, another India-based food startup, assigned an investment of 27.5 million (equivalent to 200 crores) to set up a project for millet-based food products in Uttarakhand. This project, predictable for completion by 2024, is set to have a monthly production capacity of 1,000 metric tons of millet-based products, including biscuits, cookies, and flour. Similarly, in February 2021, Tata Consumer Products Limited, an FMCG company created in India, bought the millet-based brand Soulfull from Kottaram Agro Foods Pvt. Ltd. for a business deal value of $18.8 million. This possession permits Tata Consumer Products Limited to discover new consumption circumstances and expand its product scope into the promptly growing on-the-go market segments.

The Kottaram Agro Foods Pvt. Ltd. is an Indian food manufacturing company renowned for making millet-based products. The foremost players in the millets market are Cargill Inc., Nestle S.A., The Archer Daniels Midland Company (ADMC), Bayer Crop Science AG, E.I. DuPont De Nemours and Company, Glanbia Nutritionals Inc., Warburtons Ltd., Bob's Red Mill UK Ltd., Bay State Milling Company, Grain Millers Inc., Woodland Foods, Eden Foods Inc., Sresta Natural Bioproducts Pvt. Ltd., Ernst Conservation Seeds, Sydler India Pvt. Ltd., Brett-Young Seeds Limited, Minn-Dak Growers Ltd., Wise Seed Company Inc., Avena Foods Limited, Roundstone Native Seed Company, To Your Health Sprouted Flour Co., Dharani FaMCooP Ltd., Nature's Logic, Arrowhead Mills Inc., Navan Foods LLC, and Mayoora Foods LLC.

Asia-Pacific is the prime section now in the millets business. Africa is projected as the fastest-budding zone in the coming year. The provinces covered in millets information bulletin remain in Asia-Pacific, western and eastern Europe, North and South America,

Middle East region, and the African continent. The nations included in the millets market information are Australasia, Brazil, China, France, Germany, India, Indonesia, Japan, Russia, South Korea, the United Kingdom, America, Canada, Italy, and Spain. The chief varieties of millets include pearl millet, foxtail millet, sorghum, finger millet, etc. These grains are accessible to consumers through trade associations, supermarkets, grocery stores, online platforms, and other outlets, providing to various end users such as those producing ready-to-eat meals, direct consumption goods and bakery products. The millets market also includes sales of buckwheat millets, amaranth millets, little millets, barnyard millets, broomcorn millets, millet snacks, and kodo millets.

1.9 Conclusion

The year 2023 was declared "The international year of millets (IYM-2023)" by the general assembly of United Nations, Food and Agriculture Organization (FAO), Rome, upon the request from India. Similar programs by national and international multidisciplinary public sector organizations are necessitated in their campaign by other key millet-growing countries for boosting millet utilization. Nonetheless, millet does not comprise a chief crop for food use in the developed nations, but it has an important place, providing a major contribution in the nutrition of a number of developing countries. The diverse categories of millets are mentioned in the *Ayurveda* by the names such as *Shyamaka* (barnyard millet), *Kodrav/Kordush/Uddalaka* (kodo millet), *Kangni* (foxtail millet), *Yavanala/Raktika/Sugandhika* (great millet), *Madhulee* (finger millet), *Naalikaa* (pearl millet), *Cheena* (proso millet), etc. Millet cultivation has continued to be steady from the last few years. Millet production, which was 28 MMT in 2020, has increased to 30.802 MMT in 2023/2024. Largely, the millet is cultivated in the areas of African subcontinent, along with Asia. India remains the major producer of the millet, while Nigeria and China hold the second and third places, respectively. Other major millet-producing countries include Burkina Faso, Mali, and Senegal. In India, millet growth upsurges in the latest years. Indian agriculturists are progressively seeding millet, considering it as a drought-resilient crop. Moreover, millets can grow with less fertilizer and irrigation constraints on poor soils and under hostile and arid situation compared to other major cereals. These crops also have splendid resistance to pests and diseases. Millet has the capability to create livelihoods and improve the farmers' earnings and can ease the opposing climate change and food security. The Indian authority is also pushing for enhanced millet cultivation as part of the NFSM. The millet growth in India is anticipated to remain to flourish in the future because of these considerations. Millet cultivation helps to enhance the farm earnings through assisting crop divergence by introducing millets in the belts where farmers' accusations are evident. Committed sequencers and capacity-building schemes with appropriate training impulse can be an appropriate plan to persuade farmers toward crop divergence via millets cultivation. Millet is categorized as a dry spill resistance and lenient crop that can be produced in dry and arid climatic zones where other crops could not survive and is also a nutritious grain having dietary fiber and valuable minerals in required quantity. Taking this into consideration, millets will remain a significant food crop in the coming years.

References

Abhilasha, A., Kaur, L., Monro, J., Hardacre, A., & Singh, J. (2022). Effects of hydrothermal treatment and low-temperature storage of whole wheat grains on in vitro starch hydrolysis and flour properties. *Food Chemistry, 395*, 133516.

Adekunle, A. A. (2012). *Agricultural Innovation in Sub-Saharan Africa: Experiences from multiple stake holder approaches*. Forum for Agricultural Research in Africa, Ghana. ISBN: 978-9988-8373-2-4

Agricultural & Processed Food Products Export Development Authority (APEDA). (2025). Website of Agricultural and processed food products export development authority, Ministry of Commerce and Industry, Government of India. Retrieved July 12, 2023, from https://apeda.gov.in/apedawebsite.

Ahmed, S. M., Saleh, Q. Z., Jing, C., & Qun, S. (2013). Millet grains: Nutritional quality. *Processing-and Potential Health Benefits, Comprehensive, 12*(3), 281–295.

Anitha, S., Kane-Potaka, J., Tsusaka, T. W., Botha, R., Rajendran, A., Givens, D. I., Parasannanavar, D. J., Subramaniam, K., Prasad, K. D. V., Vetriventhan, M., & Bhandari, R. K. (2021). A systematic review and meta-analysis of the potential of millets for managing and reducing the risk of developing diabetes mellitus. *Frontiers in Nutrition, 8*, 687428.

APEDA. (2020). Country wise production of millet 2020 source. Retrieved from https://apeda.gov.in/

APEDA. (2023). Indian Millets. Retrieved from https://apeda.gov.in/milletportal/Production.html

APEDA. (2024). Agricultural and processed food products export development authority. Retrieved from https://apeda.gov.in/apedawebsite/SubHead_Products/Indian_Millets.htm

Ballogou, V. Y., Sagbo, F. S., Soumanou, M. M., Manful, J. T., Toukourou, F., & Hounhouigan, J. D. (2015). Effect of processing method on physico-chemical and functional properties of two fonio (*Digitaria exilis*) landraces. *Journal of Food Science and Technology, 52*, 1570–1577.

Bijalwan, V., Ali, U., Kesarwani, A. K., Yadav, K., & Mazumder, K. (2016). Hydroxycinnamic acid bound arabinoxylans from millet brans-structural features and antioxidant activity. *International Journal of Biological Macromolecules, 88*, 296–305.

Brink, M., Belay, G., & De Wet, J. M. J. (2006). Cereals and pulses. In M. Brink & G. Belay (Eds.), *Plant resources of tropical Africa 1: Cereals and pulses* (pp. 297). PROTA Foundation, Wageningen, Netherlands/Backhuys Publishers. Leiden, Netherlands/CTA, Wageningen, Netherlands

Bukhari, M. A., Ayub, M. A., Ahmed, R., & Wagas, R. (2011). Impact of different harvesting intervals on growth, forage yield and quality of three pearl millet *Pennisetum americanum* L cultivars. *International Journal for Agro Veterinary and Medical Sciences, 5*(3), 307.

Chandi, G. K., & Annor, G. A. (2016). Millet minor: Overview. In *Encyclopedia of food grains*, The world of food grains (Second ed., Vol. 1, pp. 199–208). https://doi.org/10.1016/B978-0-12-394437-5.00012-7

Chandra, D., Chandra, S., & Sharma, A. K. (2016). Review of Finger millet (*Eleusine coracana* (L.) Gaertn): A power house of health benefiting nutrients. *Food Science and Human Wellness, 5*, 149–155. https://doi.org/10.1016/j.fshw.2016.05.004

Chapke, R. R., Prabhakar, S. G., Das, I. K., & Tonapi, V. A. (2018). Improved millets production technologies and their impact Technology Bulletin. *Indian Institute of Millets Research, Hyderabad, India*, 84.

Crawford, G. W. (2014). Millet in China. In H. Selin (Ed.), *Encyclopaedia of the history of science, technology, and medicine in non-western cultures* (Vol. 4706, pp. XXXVII). Springer Science+Business Media Dordrecht. https://doi.org/10.1007/978-94-007-7747-7

Daniel, B. (2005). Matering fiber in gastrointestinal foods. *Proceedings of Healthy Foods European*.

de Wet, J. M. J. (2006). *Eleusine coracana* (L.) Gaertn. Record from protabase. In M. Brink & G. Belay (Eds.), *PROTA (Plant Resources of Tropical Africa/Ressources végétales de l'Afriquetropicale)*. Feedipedia: Animal feed resources information system, INRAE CIRAD AFZ and FAO © 2012-2025

Deka, G., Dutta, H., & Sharma, P. (2023). A review: Hydrothermal processing of millets and effect on techno-functional and nutritional properties. *Journal of Food Process Engineering*, *46*(9), e14408.

Devisetti, R., Yadahally, S. N., & Bhattacharya, S. (2014). Nutrients and antinutrients in foxtail and proso millet milled fractions: Evaluation of their flour functionality. *LWT-Food Science and Technology*, *59*(2), 889–895.

Dharmaraj, U., & Malleshi, N. G. (2011). Changes in carbohydrates, proteins and lipids of finger millet after hydrothermal processing. *LWT-Food Science and Technology*, *44*(7), 1636–1642.

Dida, M. M., & Devos, K. M. (2006). Finger millet. In K. Chittarajan (Ed.), *Genome mapping and molecular breeding in plants* (Vol. 1, pp. 333–344). Springer-Verlag.

FAO. (1995). *Sorghum and millets in human nutrition*. FAO.

FAO. (2010). Economic and social department: The statistical division. Statistics Division.

FAO. (2013). Economic and social department: The statistical division. Statistics Division.

FAO. (2023). Agrifood Economics, *World Food and Agriculture – Statistical Yearbook 2023*. 10.4060/cc8166en

FAO. (2024). International Year of Millets 2023 Final Report, Food and Agriculture Organization of the United Nations Rome, 2024. https://openknowledge.fao.org/server/api/core/bitstreams/d1c289c5-b355-48a1-970a-2f3147c269e7/content

FAO FAOSTAT. (2013). Food and agriculture organization of the United Nations. Rome. (pp. 403–403). Retrieved from http://faostat.fao.org.

FAS-USDA. (2024). Production trends, millet Food & Agricultural Organisation (FAO) (Updated as on December 01, 2023).

FSSAI. (2019). Guidance note no. 12/2019. Millets – The Nutricereals. Retrieved from https://www.fssai.gov.in/upload/uploadfiles/files/Guidance_Notes_Version_2_Millets_29_01_2020.pdf

FSSAI. (2020). FSSAI Guidance Note on Millets – The Nutri –cereals. https://foodsafetyhelpline.com/fssai-guidance-note-on-millets-the-nutri-cereals/

FSSAI. (2023). FSSAI guidance notes on millets. Retrieved from fssai.gov.in/upload/uploadfiles/files/Guidance_Notes_Version_2_Millets_29_01_2020.pdf

FSSAI. (2024). Millets-2023. Retrieved from https://www.globalhungerindex.org/india.html

Global Hunger Index. (2024). How gender justice can advance climate resilience and zero hunger. Retrieved from https://www.globalhungerindex.org/india.html.

Gebremarian, M. M., Zarnkow, M., & Becker, T. (2014). Teff (Eragrostis tef) as a raw material for malting, brewing and manufacturing of gluten-free foods and beverages: A review. *Journal of Food Science and Technology*, *51*, 2881–2895. https://doi.org/10.1007/s13197-012-0745-5

Habiyaremye, C., Matanguihan, J. B., D'Alpoim Guedes, J., Ganjyal, G. M., Whiteman, M. R., Kidwell, K. K., & Murphy, K. M. (2017). Proso millet (*Panicum miliaceum* L.) and its potential for cultivation in the Pacific Northwest, U.S.: A review. *Frontiers in Plant Science, 7*, 1961. https://doi.org/10.3389/fpls.2016.01961

Harinarayanan, C. M., Haritha, V., Geetha, S. P., & Balachandran, I. (2023). A review on botanical, phytochemical, nutritional and pharmacological properties of major millets in India. *Aryavaidyan, 36*(3), 41–44.

Harish, M. S., Bhuker, A., & Chauhan, B. S. (2024). Millet production, challenges, and opportunities in the Asia-pacific region: A comprehensive review. *Frontiers in Sustainable Food Systems, 8*, 1386469.

India Data Insights. (2023). Agriculture: Millet cultivation in India: History and trends. Retrieved from https://idronline.org/article/agriculture/millet-cultivation-history-and-trends/

India Science, Technology and Innovation Portal. (2024). Latest from the India Science, Technology & Innovation. Retrieved from https://www.indiascienceandtechnology.gov.in/listingpage/millets-future-food

International Diabetes Federation. (2025). *Global Diabetes Data Report 2010–2045*. Retrieved from diabetesatlas.org

ICRISAT. (2012). *Annual Report 2012, International Crops Research Institute for the Semi-Arid Tropics* (pp. 40). ISSN 1017-9933

Jideani, V. A., & Diedericks, C. F. (2014). Nutritional, therapeutic, and prophylactic properties of *Vigna subterranea* and *Moringa oleifera*. In O. Oguntibeju (Ed.), *Antioxidant-antidiabetic agents and human health* (Vol. 9, pp. 187–201). INTECH Open.

Kalsi, R., & Bhasin, J. K. (2023). Nutritional exploration of foxtail millet (*Setaria italica*) in addressing food security and its utilization trends in food system. *eFood, 4*(5), e111.

Kaur, T., Panesar, P. S., & Riar, C. S. (2023). Combined effect of hydrothermal treatments and drying temperature on the physicochemical, milling, and nutritional characteristics of foxtail millet. *Journal of Cereal Science, 114*, 103788.

Kaur, T., Panesar, P. S., & Riar, C. S. (2024). Optimization of hydrothermal (soak-boil and soak-steam) treatments of foxtail (*Setaria italica* L.) millet and their effect on milling, antioxidant, pasting, and drying characteristics. *Journal of Food Measurement and Characterization, 18*(1), 500–515.

Macrae, R., Robinson, R. K., & Sadler, M. J. (1993). *Encyclopedia of food science, food technology and nutrition*. Academic Press.

Malleshi, N. G. (2006). Decorticated finger millet (*Eleusine coracana*) and a process for the preparing the decorticated finger millet. US Patent No. 7029720.

Malleshi, N. G., & Hadimani, N. A. (1993). Nutritional and technological characteristics of small millets and preparation of value-added products from them. In K. W. Riley, S. C. Gupta, A. Seetharam, & J. N. Mushonga (Eds.), *Advances in small millets* (pp. 271–287). Oxford and IBH Publishing Co Pvt. Ltd.

Mamta, M., Shukla, Y. N., & Sushil, K. (2000). Chemistry and biological activity of Paspalum scrobiculatum-a review. *Journal of Medicinal and Aromatic Plant Sciences, 22*(2/3), 288–292.

McDonough, C. M., Rooney, L. W., & Serna-Saldivar, S. O. (2000). The millets. In K. Kulp (Ed.), *Handbook of cereal science and technology, revised and expanded* (pp. 177–201). CRC Press.

Ministry of Agriculture and Farmer's Welfare. (2023). Department of Agriculture and Farmers Welfare, Govt. of India, Retrieved from https://static.pib.gov.in/WriteReadData/specificdocs/documents/2023/mar/doc2023318173501.pdf

Ministry of Commerce & Industry, GOI. (2024). Department of Commerce, Annual Report 2024–25, https://www.commerce.gov.in/publications-reports/annual-report-2024-25/

Mohamed, T. K., Zhu, K., Issoufou, A., Fatmata, T., & Zhou, H. (2009). Functionality, in vitro digestibility and physicochemical properties of two varieties of defatted foxtail millet protein concentrates. *International Journal of Molecular Sciences*, *10*(12), 5224–5238.

NITI AAYOG. (2023a). ANNUAL REPORT 2022–23, Retrieved from https://niti.gov.in/sites/default/files/2023-06/Report-on-Promoting-Best-practices-on-Millet-26_4_23.pdf

NITI AAYOG. (2023b). Promoting millets in diets: Best practices across states/UTs of India.

Obilana, A. B., & Manyasa, E. (2002). Millets. In P. S. Belton & J. R. N. Taylor (Eds.), *Pseudocereals and less common cereals: Grain properties and utilization potential* (pp. 177–217). Springer-Verlag.

Ojo, O. A., Ojo, A. B., Barnabas, M., Iyobhebhe, M., Elebiyo, T. C., Evbuomwan, I. O., Michael, T., Ajiboye, B. O., Oyinloye, B. E., & Oloyede, O. I. (2022). Phytochemical properties and pharmacological activities of the genus Pennisetum: A review. *Scientific African*, *16*, e01132. https://doi.org/10.1016/j.sciaf.2022.e01132

Panchal, A., Singh, R. K., & Prasad, M. (2023). Recent advancements and future perspectives of foxtail millet genomics. *Plant Growth Regulation*, *99*(1), 11–23.

Pawar, V. D., & Machewad, G. M. (2006). Processing of foxtail millet for improved nutrient availability. *Journal of Food Processing and Preservation*, *30*(3), 269–279.

Prashant, S. H., Namakkal, S. R., & Chandra, T. S. (2005). Effects of the antioxidant properties of millet species on oxidative stress and glycemic status in alloxan-induced rats. *Nutrition Research*, *25*, 1109–1120.

Prashanth, P., Joshi, T. J., Singh, S. M., & Rao, P. S. (2024). Hydrothermal treatment of pearl millet grains: Effects on nutritional composition, antinutrients and flour properties. *Grain & Oil Science and Technology*, *7*, 87–95.

Quattrocchi, U. (2006). *CRC world dictionary of grasses: Common names, scientific names, eponyms, synonyms, and etymology*. CRC Press, Taylor and Francis Group.

Rao, B. S. N., Vijayasarathy, C., & Prabhavathi, T. (1983). Iron absorption from habitual diets of Indians studied by the extrinsic tag technique. *Indian Journal of Medical Research*, *77*, 648–657.

Ren, X., Chen, J., Molla, M. M., Wang, C., Diao, X., & Shen, Q. (2016). In vitro starch digestibility and in vivo glycemic response of foxtail millet and its products. *Food & Function*, *7*(1), 372–379.

Renganathan, V. G., Vanniarajan, C., Karthikeyan, A., & Ramalingam, J. (2020). Barnyard millet for food and nutritional security: Current status and future research direction. *Frontiers in Genetics*, *11*, 500.

Sachdev, N., Goomer, S., & Singh, L. R. (2021). Foxtail millet: A potential crop to meet future demand scenario for alternative sustainable protein. *Journal of the Science of Food and Agriculture*, *101*(3), 831–842.

Saleh, A. S., Zhang, Q., Chen, J., & Shen, Q. (2013). Millet grains: Nutritional quality, processing, and potential health benefits. *Comprehensive Reviews in Food Science and Food Safety*, *12*(3), 281–295.

Sampath, S. R. (1986). Scope for using small millets as forage in India. In A. Seetharam, K. W. Riley, & G. Harinarayana (Eds.), *Small millets in global agriculture. Proceedings of the 1st International Small Millets Workshop*. Bangalore, India, October 29–November 2, 1986

Serna-Saldivar, S. O., & Espinosa-Ramírez, J. (2019). Grain structure and grain chemical composition. In J. R.n. Taylor & K. G. Duodu (Eds.), *Sorghum and millets: Chemistry, technology and nutritional attributes* (Second ed., pp. 85–129). Woodhead Publishing and AACC International Press. ISBN: ISBN: 978-0-12-811527-5

Serna-Saldivar, S. O., & Rooney, L. W. (1995). Structure and chemistry of sorghum and millets. In D. A. V. Dendy (Ed.), *Sorghum and Millets: Chemistry and Technology* (pp. 69–124). American Association of Cereal Chemists.

Sharma, N., Sahu, J. K., Bansal, V., Esua, O. J., Rana, S., Bhardwaj, A., Bangu, S. P., & Adedeji, A. A. (2023). Trends in millet and pseudomillet proteins-Characterization, processing and food applications. *Food Research International, 164*, 112310.

Sharma, S., Saxena, D. C., & Riar, C. S. (2015). Antioxidant activity, total phenolics, flavonoids and antinutritional characteristics of germinated foxtail millet (*Setaria italica*). *Cogent Food & Agriculture, 1*(1), 1081728.

Sheahan, C. M. (2014). Plant guide for Foxtail millet (Setaria italica). USDA-Natural Resources Conservation Service, Cape May Plant Materials Center.

Shobana, S., & Malleshi, N. G. (2007). Preparation and functional properties of decorticated finger millet (*Eleusine coracana*). *Journal of Food Engineering, 79*(2), 529–538.

Singh, R. K., Muthamilarasan, M., & Prasad, M. (2021). Biotechnological approaches to dissect climate-resilient traits in millets and their application in crop improvement. *Journal of Biotechnology, 327*, 64–73.

Sreekala, A. D. S., Anbukkani, P., Singh, A., Dayakar Rao, B., & Jha, G. K. (2023). Millet production and consumption in India: Where do we stand and where do we go? *National Academy Science Letters, 46*, 65–70. https://doi.org/10.1007/s40009-022-01164-0

Srilakshmi, B. (2003). *Food science* (3rd ed., pp. 403). New Age International.

Srilekha, K., Kamalaja, T., Uma Maheswari, K., & Neela Rani, R. (2019). Nutritional composition of little millet flour. *International Research Journal of Pure and Applied Chemistry, 20*(4), 1–4.

Suma, P. F., & Urooj, A. (2012). Antioxidant activity of extracts from foxtail millet (*Setaria italica*). *Journal of Food Science and Technology, 49*(4), 500–504.

Taylor, J. R. N., & Duodu, K. G. (2010). Sorghum and millets: Characteristics and quality requirements. In C. W. Wrigley & I.l. Batey (Eds.), *Cereal grains assessing and managing quality* (pp. 237–263). Woodhead Publishing Series in Food Science, Technology and Nutrition. ISBN: ISBN: 978-1-84569-563-7

Taylor, J. R. N., & Emmambux, M. N. (2008). Products containing other specialty grains: Sorghum, the millets and pseudocereals. In B. R. Hamaker (Ed.), *Technology of functional cereal products* (pp. 281–335). Woodhead Publishing Series in Food Science, Technology and Nutrition. https://doi.org/10.1533/9781845693886.2.281

Ugare, R., Chimmad, B., Naik, R., Bharati, P., & Itagi, S. (2014). Glycemic index and significance of barnyard millet (*Echinochloa frumentacae*) in type II diabetics. *Journal of Food Science and Technology, 51*, 392–395.

UNICEF. (2023). Everything about Millets (POSHAN) power. Top Producing Countries. Retrieved from https://knowledge.unicef.org/resource/everything-about-millets-poshan-power

USDA. (2024). Production-Millet, 2024/2025 Millet Production, Top Producing Countries. Retrieved from https://fas.usda.gov/data/production/commodity/0459100

Veena, B., Chimmad, B. V., Naik, R. K., & Malagi, U. (2010). Development of barnyard millet based traditional foods. *Karnataka Journal of Agricultural Sciences*, *17*(3), 522–527.

Verma, K. C., Joshi, N., Rana, A. S., & Bhatt, D. (2020). Quality parameters and medicinal uses of foxtail millet (*Setaria italica* L.): A review. *Journal of Pharmacognosy and Phytochemistry.*, *9*(4), 1036–1038.

Vetriventhan, M., Azevedo, V., Upadhyaya, H. D., & Tonapi, V. (2020). Genetic and genomic resources, and breeding for accelerating improvement of small millets: Current status and future interventions. *Nucleus*, *63*(3), 1–22. https://doi.org/10.1007/s13237-020-00322-3

Vikaspedia. (2024). Millets-The Nutri-Cereals Government of India. Retrieved from https://agriculture.vikaspedia.in/viewcontent/health/nutrition/nutritive-value-of-foods/-of-cereals-and-millets/millets-the-nutri-cereals?lgn=en

Watanabe, M. (1999). Antioxidative phenolic compounds from Japanese barnyard millet (*Echinochloa utilis*) grains. *Journal of Agricultural and Food Chemistry*, *47*, 4500–4505.

Wikipedia. (2024). Millet. Retrieved from https://en.wikipedia.org/wiki/Millet

World Population Review. (2025). Millet production by Country 2025. Retrieved from https://worldpopulationreview.com/country-rankings/millet-production-by-country

Yang, Y. B., Jia, G. Q., Deng, L. G., Ling, Q. I. N., Chen, E. Y., Cong, X. J., Zou, R. F., Wang, H. L., Zhang, H. W., Liu, B., Guan, Y. A., Diao, X. M., & Yin, Y. P. (2017). Genetic variation of yellow pigment and its components in foxtail millet (*Setaria italica* (L.) P. Beauv.) from different eco-regions in China. *Journal of Integrative Agriculture*, *16*(11), 2459–2469.

Yousaf, L., Hou, D., Liaqat, H., & Shen, Q. (2021). Millet: A review of its nutritional and functional changes during processing. *Food Research International*, *142*, 110197.

Yummy Valley. (2024). Different types of millets and their benefits. Retrieved from https://yummy-valley.com/blog/types-ofmillet/?srsltid=AfmBOoqqvd1onxe5Ngop39mauQZUa0MogWOAyTjokGnMvzYlkg9ZeTg7

2

Millets as Nutri-cereals: Nutritional Composition and Health Benefits

Muskan Dawra[1], Jaspreet Kaur[2], Vikas Kumar[3], and Prasad Rasane[1]

[1] Department of Food Technology and Nutrition, School of Agriculture, Lovely Professional University, Punjab, India
[2] Department of Food and Human Nutritional Sciences, University of Manitoba, Winnipeg, Canada
[3] Indian Council of Agricultural Research, Indian Institute of Millets Research (IIMR), Rajendranagar, Hyderabad, Telangana State, India

2.1 Introduction

Millets are one of the oldest cultivated grains and are packed with nutrients. These small-seeded grains belong to the grass family and come in different varieties, each with unique health benefits. Some of the most well-known types include finger millet, proso millet, pearl millet, foxtail millet, barnyard millet, fonio, tef, and kodo millet. The word "millet" originates from "mil," meaning "thousand," representing the remarkable ability of a single seed to produce an abundance of seeds. This highlights the efficiency and productivity of millets, making them a valuable and resilient crop option (Tripathi et al., 2021). These are known for their nutritional value, resilience, and versatile uses in various cuisines and food products (Almutairi, 2022). Millets are known to flourish in the diverse landscapes of Asian and African countries, as well as in selected regions of Europe. These ancient grains are cultivated across the globe's expansive tropical and subtropical areas. Renowned as the earliest domesticated cereals, millets are well known for their remarkable ability to thrive in less fertile soil, their impressive ability to withstand drought and pests, and their relatively short growing season, typically spanning 45–60 days (Nithiyanantham et al., 2019). Moreover, they are highly favored over other grains due to their unique characteristics. Their short life cycle allows for quicker harvests, reducing the time and resources required for cultivation. Furthermore, millets have low input requirements, making them more sustainable and cost-effective (Tripathi et al., 2021). Millets exhibit remarkable abiotic stress tolerance, attributed to their genetic adaptability. They can withstand extreme climatic conditions, including drought and high temperatures, while also thriving in low-fertility soil, making them an essential crop for sustainable agriculture and food security in climate-affected regions (ICRISAT, 2020; Saleh et al., 2013).

Nutraceutical Potential of Millet-Based Food Products: Chemistry and Applications, First Edition.
Edited by Parmjit S. Panesar, Charanjit S. Riar, and Sushil Dhital.

Published 2026 by John Wiley & Sons Ltd.

A nutritious and gluten-free option, millet grains play an important role in maintaining overall health. Being nonacidic and easily digestible, they are an excellent choice for people with celiac disease and diabetes. Additionally, millet grains are packed with dietary fiber, complex carbohydrates, iron, and calcium, making them suitable for a balanced diet. Also, they are packed with high levels of vital minerals such as magnesium and phosphorus, enhancing overall wellness and bodily function (Abah et al., 2020a). They are a gluten-free grain that helps regulate blood sugar, promote digestive wellness, and deliver essential benefits, making them an excellent source (Dhaka et al., 2021).

2.2 Nutritional Composition of Different Types of Millets

As an ancient grain, millet offers enhanced nutrition and resilience to environmental conditions compared to other grains like rice and wheat. Millets are a powerhouse of protein, fiber, vitamins, and minerals, making them a wholesome dietary option. Their exceptional antioxidant and bioactive compound levels support chronic disease prevention and overall vitality (Nithiyanantham et al., 2019). Table 2.1 highlights the nutrient content of various millet varieties along with their unique health benefits.

2.3 Importance of Millets

2.3.1 Nutritional Powerhouse

There has been an increased trend in the area of food and agriculture toward replacing conventional staple grains like wheat, maize, and rice with resilient alternatives like millets. The transformation is driven by urgent global issues, including the adverse effects of climate change, declining freshwater availability, rapid population expansion, and the increasing burden of cardiovascular diseases. To ensure food security and sustainability, alternative crops that require fewer resources while maintaining high nutritional value are becoming essential (Abah et al., 2020b).

As a nutrient-dense grain, millets contribute significantly to dietary health by supplying essential phytochemicals, vitamins, minerals, and nonstarch polysaccharides (dietary fiber). These beneficial bioactive constituents play a key part in improving metabolic health, aiding in the management of type 2 diabetes, and promoting overall health (Nithiyanantham et al., 2019). Millets are a nutritionally dense grain, offering significant amounts of B-complex vitamins, including riboflavin, thiamin, and niacin. These B vitamins play crucial roles in energy metabolism, supporting neuron function, and promoting overall health. Additionally, millets boast significant levels of essential minerals, including magnesium, phosphorus, and manganese. They play a crucial part in maintaining health due to their mineral-rich composition, which aids in bone strengthening, muscle function regulation, and oxidative stress reduction. Additionally, the presence of antioxidants like carotenoids and phenolic compounds help to safeguard the body against free radical damage and its harmful effects (Kumar et al., 2024). Millets, with a protein content of 6–12%, are unique among grains for providing all essential amino acids, including sulfur-rich cysteine and methionine. Millets are also a source of 1–5% fat, mainly consisting of essential fatty acids such as linoleic,

Table 2.1 Nutritional composition and health benefits

Nutrient	Amount	Source	Health benefits	References
Protein	6–14 g/100 g	Seeds, endosperm	Supports muscle growth, repair tissues, and boosts immune function	Anitha et al. (2021a)
Carbohydrates	55–82 g/100 g	Seeds, bran	Provides sustained energy, helps in maintaining blood sugar levels	Almutairi (2022)
Dietary fiber	1–14 g/100 g	Bran	Improves digestion, lowers cholesterol, and aids in blood sugar control	Kumar et al. (2024)
Calcium	10–410 mg/100 g	Millet grain	Strengthens bones and teeth, supports heart and nerve function	Nithiyanantham et al. (2019)
Iron	0.7–20 mg/100 g	Seed coat, endosperm	Prevents anemia, enhances oxygen transport in the blood	Singh et al. (2020)
Magnesium	100–166 mg/100 g	Seed germ	Aids in muscle and nerve regulation while promoting balanced blood pressure and blood sugar levels	
Phosphorus	215–379 mg/100 g	Grain	Supports bone health and helps in energy production at the cellular level	Kumar et al. (2024)
Antioxidant	2.5–10.3%	Bran, seed coat	Reduces oxidative stress, lowers inflammation, and protects against chronic diseases	Almutairi (2022)
Vitamin B complex	0.1–10 mg/100 g	Whole grain, germ	Improves energy production, brain function, and maintains healthy red blood cells	Geetha et al. (2020)
Phytochemicals (polyphenols)	Varies per compound	Seed coat	Provides anti-inflammatory, antidiabetic, and anticancer properties	Nithiyanantham et al. (2019)
Resistant starch	Varies per compound	Bran, seed	Promotes gut health, supports weight management, improves insulin sensitivity, and helps in blood sugar control	Kumar et al. (2024)
Lipids	1–4 g/100 g	Germ seed	Supports brain health, helps in cell membrane formation, and provides energy	Singh et al. (2020)

linolenic, and arachidonic acids, all of which support heart health. Additionally, their intake has been associated with the regulation of blood sugar levels and reduced inflammation, making them useful for diabetes and cardiovascular health (Dhaka et al., 2021).

2.3.2 Gluten-Free Alternatives

As a hardy, drought-tolerant grain, millet flourishes in dry conditions and offers a highly nutritious, gluten-free option for managing celiac disease. Gluten is composed of two main protein types: prolamins and glutenin. Different cereals contain distinct types of prolamins, which are named accordingly. For example, barley contains hordeins, wheat has gliadins, oats include avenins, and rye contains secalins. These proteins have a lot of proline and glutamine. This makes them hard to break down in the gut and more likely to be changed by tissue transglutaminase (Dhaka et al., 2021). When people with celiac disease eat these proteins, it can cause inflammation, shrinking, and abnormal growth in the small intestine. However, celiac disease doesn't just affect the gut. It's a systemic disease that can also hurt the liver, skin, brain, joints, heart, and other organs. For those with gluten sensitivity, consuming gluten-containing grains can lead to symptoms like bloating, abdominal discomfort, and fatigue, much like celiac disease (Singh et al., 2020).

2.3.3 Low Glycemic Index

The glycemic index quantifies the postprandial glucose response of carbohydrate-containing foods. Low-glycemic index (GI) foods less than 55 exhibit slower digestion and absorption rates, facilitating a gradual increase in blood glucose levels. This controlled glycemic response aids in optimal diabetes management, emphasizing the importance of dietary modifications in regulating glucose metabolism and mitigating diabetes risk (Abah et al., 2020b; Akanbi et al., 2021). Selecting foods with a low glycemic index can help individuals with diabetes manage their blood sugar more effectively and minimize sudden glucose level fluctuations (Geetha et al., 2020). Scientific studies confirm that incorporating millets into the diet can lead to better glycemic control by reducing fasting and postprandial glucose levels, improving insulin sensitivity, and lowering HbA1c levels, making them beneficial for diabetes management (Anitha et al., 2021a). Finger millet is a notable source of phytic acid, which contributes to carbohydrate malabsorption, resulting in reduced postprandial glycemic response. Short-chain fatty acids and probiotics have demonstrated potential in providing benefits for individuals with diabetes and modulating gut microbiota. Millet grains have different effects on blood sugar and varying nutritional content. Foxtail millet has a low GI of 50–60, making it good for managing blood sugar, and it contains a high protein content of 12.3%. Kodo and proso millet have moderate GIs ranging from 50 to 65, providing balanced energy and nutrients with protein levels between 7.7% and 11.5%. Similarly, barnyard millet falls in this range, with a GI of 50–65 and a protein content of 11.2%. Whereas, finger millet, pearl millet, and sorghum have higher GIs (70–85), causing a quicker increase in blood sugar (Agrawal et al., 2023).

2.3.4 Environment Sustainability

Millets are highly resilient grains with remarkable drought tolerance, allowing them to thrive in areas with rainfall below 450 mm. Their remarkable ability to thrive in marginal lands with undulating terrain demonstrates their resilience to climate vulnerability. Notably, millets are categorized as C4 crops, which set them apart due to their significantly higher photosynthesis efficiency compared to rice and wheat. This enhanced photosynthesis efficiency allows millets to be highly productive, especially in hot and dry climatic conditions (Geetha et al., 2020). Furthermore, their cultivation requires nominal water resources, making them an excellent option for regions with limited access to water. Additionally, the low demand for external inputs further highlights millets as a sustainable and resource-efficient crop (Pandey & Bolia, 2023). Growing millets play a crucial role in supporting biodiversity by providing a rich habitat and food source for various organisms. The diverse range of millet varieties show remarkable adaptability to different climatic conditions, contributing significantly to agrobiodiversity. Furthermore, millet crops contribute to natural pest control by attracting beneficial insects and organisms that prey on pests, thereby reducing the reliance on chemical pesticides that can have detrimental effects on the environment and human health (Dwivedi et al., 2023; Mahendra et al., 2021).

2.4 Nutrients from Millets

Globally, the current focus is on increasing millet production and consumption through sustainable cultivation, nutritional awareness, broader acceptance, and enhanced dietary inclusion (Maharajan et al., 2024). An extensive variety of millets are grown worldwide, such as sorghum and pearl millet, along with seven lesser-known varieties, including finger millet, foxtail millet, proso millet, barnyard millet, fonio, teff, and Job's tear.

2.4.1 Sorghum

India is the sixth-largest producer of sorghum globally, with the cultivation of *Sorghum bicolor* becoming increasingly popular among consumers. This surge in popularity is due to the crop's high nutrient content, including essential vitamins and minerals, as well as its potential to prevent various diseases. Sorghum is recognized for its significant levels of antioxidants, dietary fiber, and other beneficial compounds (Li et al., 2020). Its outer layer, known as the pericarp, is rich in non-starch polysaccharides—complex carbohydrates that resist easy digestion. It also contains phenolic compounds like 3-deoxyanthocyanidins, recognized for their antioxidant properties, along with tannins and phenolic acids, which offer various health benefits. Furthermore, the pericarp contains carotenoids, which are organic pigments that act as antioxidants and are important for eye health. On the other hand, the germ of sorghum, which is the embryo of the grain, is composed of lipids (fats), fat-soluble vitamins such as vitamin E and vitamin K, B-complex vitamins, and various minerals such as phosphorus and magnesium. These nutrients are essential for overall health and well-being. The endosperm, the innermost part of the grain, is a rich source of carbohydrates, making it an excellent provider of energy. It also contains proteins, B-complex vitamins like niacin and pantothenic acids, and essential minerals (Fe, Zn) (Mawouma et al., 2022).

These are nutritionally dense grains with a varied composition of macronutrients, micronutrients, and bioactive compounds. The protein content in sorghum can range from 4.40% to 21.10%, while water-soluble protein, an important fraction of total protein, ranges between

0.30% and 0.90%. These grains are rich in carbohydrates, particularly starch, which constitutes between 55.60% and 75.20% of their dry weight. The amylose content, a component of starch, ranges from 21.20% to 30.20%, which contributes to its digestibility and glycemic response. In terms of sugars, they contain 0.70–4.20% soluble sugars and 0.05–0.53% reducing sugars. The fiber content is also notable, with crude fiber levels ranging from 1.00% to 3.40%, contributing to digestive health. Additionally, they are rich in essential minerals such as calcium, phosphorus, and iron. The calcium content ranges widely from 11.00 to 586.00 mg per 100 g, depending on the variety. Phosphorus levels range from 167.00 to 751.00 mg per 100 g, and iron content varies between 0.90 and 20.00 mg per 100 g, making them beneficial for preventing micronutrient deficiencies. These also contain essential vitamins, particularly those in the B-complex group. Thiamine content ranges from 0.24 to 0.54 mg per 100 g, niacin (vitamin B3) from 2.90 to 6.40 mg per 100 g, and riboflavin (vitamin B_2) from 0.10 to 0.20 mg per 100 g. Despite their rich nutrient profile, certain antinutritional factors present in millets can affect the bioavailability of nutrients. Sorghum has tannin concentration ranging from 0.1% to 7.22% and phytic acid content between 875.00 and 2211.90 mg per 100 g (Abah et al., 2020a).

2.4.2 Pearl Millet

Pearl millet (*Pennisetum glaucum*) has been grown for centuries in the Indian subcontinent and Sahelian West Africa. This resilient crop is well-suited to the challenging agricultural conditions of these regions, where other staple crops often struggle to thrive and produce viable grain yields. Pearl millet holds significant importance as a food crop on a global scale, ranking after maize, wheat, rice, sorghum, and barley. Its adaptability and resilience make it an essential and environmentally friendly field crop in traditional farming practices, contributing to human nutrition and the overall health of the planet (Srivastava et al., 2022). Well-suited to arid climates, it is primarily grown in areas with low agronomic potential, where rainfall is typically between 200 and 500 mm, and soils are marginal (Passot et al., 2016). It is an incredibly nutritious grain that offers a wide range of essential nutrients. Millets are an exceptional energy-dense grain, offering a diverse range of nutrients. They provide carbohydrates, proteins (8–19%), and fats (3–8%) with superior digestibility, ensuring efficient nutrient absorption. The presence of resistant starch, soluble and insoluble dietary fibers (1.2 g per 100 g), crude fibers, antioxidants, and essential minerals further enhances their nutritional significance (Geetha et al., 2020). Compared to other major cereals, pearl millet also provides significant amounts of iron and zinc. Moreover, it is rich in vitamins such as niacin, thiamine, riboflavin, along with minerals (2.3 mg/100 g) such as potassium, magnesium, phosphorus, zinc, iron, manganese, and copper (Li et al., 2020). Millets boast high protein digestibility, attributed to their low levels of cross-linked prolamins. Their fat composition is particularly noteworthy, with 74% polyunsaturated fatty acids and a significant quantity of omega-3 fatty acids, both of which support cardiovascular and neurological functions. The primary fatty acids found in millet—linoleic acid (45%), oleic acid (25%), and linolenic acid (4%)—play an essential role in reducing inflammation, improving lipid metabolism, and promoting overall well-being (Satyavathi et al., 2021).

2.4.3 Finger Millet

Finger millet, scientifically known as *Eleusine coracana*, and also commonly referred to as "ragi" or "tamba," is a cereal grain belonging to the millet group. It is named "finger millet" due to the characteristic growth form of its panicles, which resemble fingers (Bhatt et al., 2023). This grain is a member of the Poaceae family, which is also stated as the grass family. Originating in Ethiopia, finger millet falls within the sub-family Chloridoideae, which is a diverse group of grasses found in various habitats worldwide (Ramashia et al., 2019). Ragi is known for its impressive nutritional composition. Its high fiber and mineral content make it superior to wheat and rice. Among millets, finger millet is distinguished by its well-balanced protein profile, containing higher amounts of lysine, threonine, and valine (Srivastava et al., 2022). Unlike many starch foods, finger millet contains high levels of essential amino acids such as isoleucine, leucine, and methionine, contributing to a well-balanced diet (Gull et al., 2014). Additionally, millets provide an abundance of B vitamins like niacin and B6, along with folic acid, iron, potassium, calcium, magnesium, and zinc (Bhatt et al., 2023).

2.4.4 Barnyard Millet

Barnyard millet is an ancient cereal crop widely cultivated in warm and temperate regions globally. Predominantly grown in India, China, Japan, and Korea, it ranks as the fourth most cultivated minor millet. This millet contributes significantly to food security, particularly for low-income populations. Its resilience to diverse climatic conditions makes it a valuable candidate for sustainable agricultural practices, particularly in stress-prone environments (Renganathan et al., 2020). With 8.74% moisture, it can be kept for prolonged periods under dry conditions. The millet contains approximately 10.1% protein, offering a reliable source of plant-based protein. Its 3.9% fat content contributes to its caloric value, while the crude fiber content (6.7%) helps support digestive health by improving bowel movements and avoiding constipation. It is rich in total minerals at 2.1%; barnyard millet is also a robust source of carbohydrates at 68.8%, providing sustained energy. This makes it an ideal staple for individuals requiring long-lasting fuel, particularly in rural and agricultural communities (Nithyashree & Vijayalaxmi, 2022).

Barnyard millet is rich in dietary fiber, with a total fiber content of 12.5%. It comprises 8.4% insoluble fiber, which aids digestion, and 4.2% soluble fiber, which helps manage cholesterol levels and stabilize blood sugar. Additionally, it is rich in minerals, making it a valuable nutritional source. It provides 281 mg of phosphorus, an essential nutrient for maintaining strong bones and supporting various cellular functions (Srivastava et al., 2022). It also provides 5 mg of iron, aiding in preventing anemia and supporting oxygen transport in the blood. Additionally, with 83 mg of magnesium, vital for muscle and nerve function, and 19 mg of calcium, contributing to strong bones and teeth, barnyard millet offers a comprehensive array of essential minerals. Barnyard millet is a highly nutritious food source, especially for individuals seeking a high-fiber, gluten-free grain with an excellent profile of essential minerals and macronutrients (Kaur & Sharma, 2020).

2.4.5 Proso Millet

Proso millet (*Panicum miliaceum*), also known as hog millet, broomcorn millet, or golden millet, is a C4 grass crop mainly cultivated in regions of China, India, Russia, and North America. In India, it is majorly grown in Uttar Pradesh, Punjab, Haryana, Madhya Pradesh, Karnataka, and Maharashtra. Due to its drought tolerance, proso millet has been a key component of traditional dryland farming systems. Nutritionally, 100 g of proso millet provides 354 kcal, 70.4 g carbohydrates, 1.1 g fat, 12.5 g protein, and 5.2 g fiber, along with essential micronutrients such as niacin, calcium, and phosphorus (206 mg), highlighting its importance in human nutrition (Nithyashree and Vijayalaxmi, 2022). Moreover, proso millet exhibits significant antioxidant activity and is particularly rich in total carotenoids. The dietary protein found in proso millet aids in cholesterol metabolism and can help prevent vitamin B3 deficiency, known as pellagra. Including proso millet in the daily diet, alongside cereals, can contribute to improving nutritional security by expanding the food basket and enhancing overall dietary micronutrient intake (Kaur & Sharma, 2020).

2.4.6 Kodo Millet

India is widely recognized as the birthplace of kodo millet, where the domestication of araka or kodo millet is believed to have taken place. This highly nutritious grain is well adapted to tropical and subtropical climates, thriving even in nutrient-deficient soils. It is widely grown across arid and semi-arid regions in India and various African countries. Kodo millet (*Paspalum scrobiculatum*), a crop of high agricultural importance in India, is primarily grown in the Deccan plateau. Its cultivation is concentrated in states such as Gujarat, Karnataka, Chhattisgarh, eastern Madhya Pradesh, and Tamil Nadu (Deshpande et al., 2015; Bunkar et al., 2021). It is also known as cow grass, rice grass, native paspalum, ditch millet, and Indian crown grass. This millet has a strong nutritional profile that makes it a highly beneficial cereal. It contains 14.23 g of moisture, which ensures freshness and longevity when properly stored. Kodo millet contains 8.92 g of protein, making it an excellent plant-based protein source that supports growth, repairs tissue damage, and overall metabolic functions. The fat content is relatively low (2.55 g/100 g), contributing to a balanced diet with moderate caloric intake. It contains 1.72 g of ash, indicating its mineral-rich composition. The carbohydrate content is 66.19 g/100 g, providing a significant source of energy, making it suitable for individuals requiring sustained energy release throughout the day (Kaur & Sharma, 2020). The total energy value of kodo millet is 1388 KJ, highlighting its role as a high-energy food. Moreover, it contains 6.39 g of dietary fiber per serving, which contributes to digestive health by enhancing bowel regularity and reducing the risk of constipation. This nutrient profile makes kodo millet a versatile and nutritious grain, especially suitable for those looking for plant-based, high-fiber alternatives in their diet (Dey et al., 2022).

2.5 Health Benefits of Millets

2.5.1 Probiotic and Prebiotic Benefits

Probiotics are beneficial live bacteria and yeasts that support overall health, particularly digestive well-being. Commonly known as "good" or "friendly" bacteria, they play a key role in maintaining a healthy gut. When taken in sufficient amounts, probiotics offer several health benefits. Prebiotics, on the other hand, are a type of fiber that the human body cannot digest (Di Stefano et al., 2017). Consuming prebiotic-rich foods such as legumes, bananas, asparagus, and chicory root can significantly enhance gut flora composition, improving both digestive and metabolic processes (Sarita & Singh, 2016). The dietary fiber in millet benefits gut health by interacting with gut microbiota and enhancing fermentation, which leads to the production of short-chain fatty acids (SCFAs). SCFAs help lower the gut pH, which thwarts pathogen invasion and colonization (Figure 2.1). These specific dietary fibers are commonly referred to as "prebiotics" (Gupta et al., 2023).

When we consume millets, the dietary fibers contained within these grains take an interesting journey through our digestive system. Unlike most foods, these fibers resist digestion in the upper gastrointestinal tract and reach the colon intact. Here, they undergo a vital fermentation process facilitated by the diverse community of gut microbiota—trillions of microorganisms that call our intestines home. This fermentation leads to the flourishing of beneficial bacteria, specifically species such as *Bifidobacterium* and *Lactobacillus*, which eagerly feast on the millet fibers. In this process, these fibers are broken down into SCFAs, primarily acetate, propionate, and butyrate (Divisekera et al., 2019). Butyrate is worth highlighting as it serves as a crucial energy source for colonocytes, the specialized cells lining the colon. By nourishing these cells, butyrate helps fortify the gut barrier and maintains the integrity of the epithelial layer (Plaza-Diaz et al., 2019). This means it helps in preventing harmful substances from entering the bloodstream, thereby reducing inflammation and bolstering our immune response. Additionally, the production of SCFAs results in a lowered pH within the colon, creating an acidic environment that is inhospitable to pathogenic bacteria. This unfavorable setting makes it difficult for harmful microbes to establish

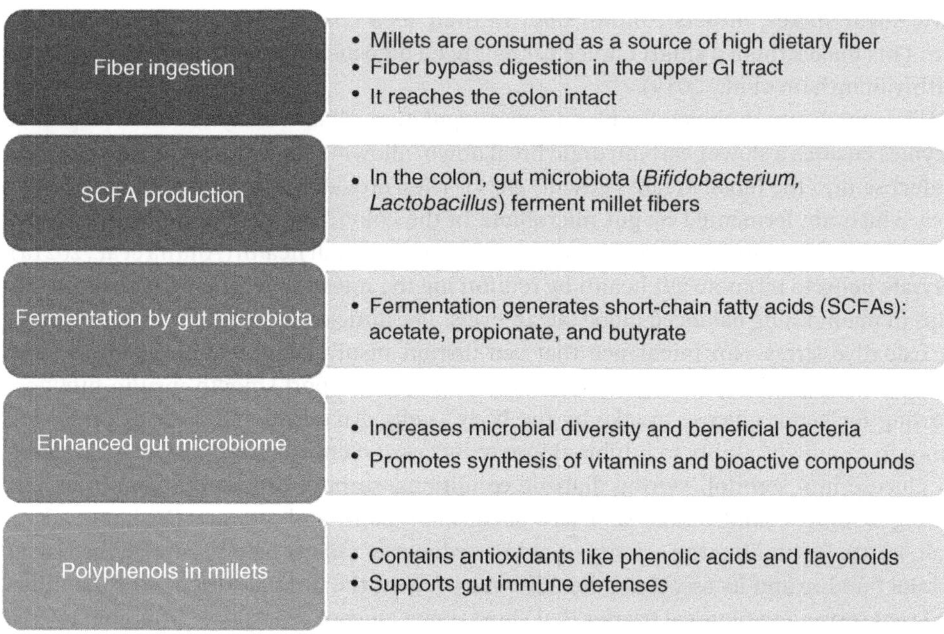

Figure 2.1 Role of millets as probiotics (Edwin, 2024)

themselves and thrive, while at the same time promoting the growth of beneficial microbes and helping to maintain a balanced gut microbiome. By nurturing the population of beneficial bacteria, millets indirectly facilitate the synthesis of various vitamins and bioactive compounds (Mazziotta et al., 2023; Nambiar et al., 2020). This enhances nutrient absorption, optimizing our overall gut health. Additionally, millets are abundant in distinctive polyphenolic compounds, which exhibit strong antioxidant and anti-inflammatory effects. These compounds provide an extra layer of support to the gut's immune defenses. Incorporating millets into our diet on a regular basis can significantly promote gut health. They not only enhance microbial diversity and reduce inflammation but also offer protection against a range of gastrointestinal disorders. The prebiotic properties of millets contribute to a broader spectrum of health benefits, making them an invaluable addition to any balanced diet, encouraging a thriving and resilient gut ecosystem (Plaza-Diaz et al., 2019).

2.5.2 Regulating Blood Sugar Levels

Millets help regulate blood sugar levels by inhibiting α-glucosidase and pancreatic amylase, enzymes that break down complex carbs, thereby reducing post-meal glucose spikes (Figure 2.2). They also contain aldose reductase, which prevents sorbitol accumulation linked to diabetic cataracts. Additionally, their antioxidants aid in dermal wound healing (Ambati & Sucharitha, 2019). These antioxidants help in combating oxidative stress and promoting tissue repair, making millets a valuable dietary addition for individuals looking to manage blood glucose levels and support overall health. The seed coat is composed of pectin, cellulose, and hemicelluloses, which are types of dietary fiber that resist breakdown by digestive enzymes. Unlike refined grains, which can cause sugar spikes, millets contain fiber in their seed coat that slows glucose absorption. This makes them a smart choice for balanced digestion and blood sugar regulation (Nithiyanantham et al., 2019).

Millets are a smart choice for blood sugar regulation. Their ability to inhibit digestive enzymes ensures a slower carbohydrate breakdown, allowing for a more controlled release of glucose into the bloodstream (Edwin, 2024). They provide resistant starch and dietary fiber, which are fermented by gut microbiota in the colon, and further helps to improve insulin sensitivity, support glucose regulation, and promote gut health (Anitha et al., 2021a). Butyrate helps to promote gut health by reinforcing the intestinal barrier, which indirectly helps in maintaining balanced blood sugar levels. The antioxidants present in millets combat oxidative stress—an imbalance that can disrupt insulin signaling and contribute to insulin resistance. By reducing oxidative stress, millets support smooth insulin function, allowing for better glucose uptake by the body's cells. An additional layer of protection comes from millets' ability to inhibit aldose reductase, an enzyme responsible for converting glucose into sorbitol. During diabetic conditions, sorbitol can accumulate in excess, leading to severe complications such as diabetic neuropathy and cataracts (Vedamanickam et al., 2020). By limiting the activity of aldose reductase, millets help minimize the risk of sorbitol buildup and its associated health issues. Lastly, the unique structure of the millet seed coat serves as a physical barrier that slows down glucose absorption. This characteristic ensures a more controlled release of glucose into the bloodstream. Together, these

Figure 2.2 Role of millets in reducing blood sugar level. *Source:* Anitha et al. (2021a)/MDPI/ CC BY 4.0

multifaceted mechanisms empower millets to maintain stable blood sugar levels and significantly reduce the risk of postprandial hyperglycemia (Edwin, 2024).

2.5.3 Improves Cardiovascular Health

Foxtail millet contains beneficial compounds such as catechin, quercetin, apigenin, and kaempferol. These compounds have been found to be beneficial in combating diabetes, reducing the risk of cardiovascular disease, and helping to maintain healthy lipid levels in the body. Additionally, the magnesium content in foxtail millet contributes to its reputation as a heart-healthy food. Furthermore, foxtail millet exhibits antimicrobial properties, which can help in fighting off harmful microorganisms in the body. It also possesses antitumorigenic properties, indicating potential benefits in inhibiting tumor growth. Moreover, the grain supports natural detoxification processes in the body, thereby contributing to overall health (Saini et al., 2021).

Millets are known to significantly support cardiovascular health through various mechanisms. Packed with powerful antioxidants like catechin, apigenin, and kaempferol, millets combat oxidative stress and inflammation—key factors in cardiovascular disease. These antioxidants protect blood vessels and the heart from damage, preventing plaque formation and maintaining arterial flexibility. Additionally, millets are rich in soluble fiber, which binds to cholesterol in the digestive tract, reducing absorption into the bloodstream (Anitha et al., 2021a). This leads to lower cholesterol levels, decreasing the risk of atherosclerosis and heart-related issues. Millets also provide a significant amount of magnesium, which helps regulate blood pressure by relaxing blood vessels, and support heart rhythm and muscle function. Furthermore, the antimicrobial properties of some compounds in millets protect against harmful microorganisms, contributing to a healthier body. Their ability to stabilize blood sugar levels further reduces cardiovascular risk, making them an excellent choice for promoting heart health (Edwin, 2024).

2.5.4 Improves Bone Strength

It's important to recognize that finger millet offers significant potential in addressing calcium deficiency. This is particularly true in regions where finger millet is cultivated, given its high calcium content and cost-effectiveness. Millets are increasingly being acknowledged as "smart food" due to their multifaceted benefits. They are not only nutritionally rich and beneficial for human health, but also environmentally sustainable, requiring less water and contributing to a lower carbon footprint. Moreover, they offer advantages for farmers by being resilient to climate variations. Millets contain abundant amounts of essential minerals like calcium, iron, zinc, selenium, and magnesium, surpassing the levels found in milled rice and refined wheat by several folds (Anitha et al., 2021b). This makes them an attractive dietary option for addressing nutritional deficiencies and promoting overall well-being. Finger millet consistently contains high levels of calcium (364 mg/100 g) and is also rich in zinc and magnesium (Kaur & Sharma, 2020).

2.5.5 Supports Immune System

Among the various health-promoting constituents of millets, phenolic compounds are particularly notable for their immune-boosting properties. These include ferulic acid, p-coumaric acid, quercetin, kaempferol, and tannins, all of which are known for their antioxidant and anti-inflammatory benefits. These compounds support immune health by preventing oxidative stress and inflammation, both of which can weaken the immune response. Including millets in the diet can, therefore, provide long-term health benefits by naturally strengthening immunity (Gupta et al., 2023). A diverse range of flavonoids, such as luteolin, quercetin, catechin, naringenin, daidzein, apigenin, and kaempferol, has been identified in millets. These flavonoids are known for their powerful antioxidant activity and their role in promoting health. Regular intake of millets, which are abundant in both flavonoids and tannins, has been linked to a decreased risk of developing chronic diseases, including cancer, cardiovascular ailments, diabetes, and neurological disorders. The combined presence of flavonoids and tannins in millets is believed to contribute to their potential protective benefits (Anitha et al., 2021b).

2.5.6 Anticancer Properties

Scientific studies have established that bioactive peptides obtained from various natural sources possess remarkable anticancer properties. A 4 KDa peptide from buckwheat has been identified for its dual benefits—it exhibits antifungal activity and inhibits the growth of certain cancer lines. Research indicates that they effectively suppress HepG2 (liver cancer), MCF-7 (breast cancer), and WRL 68 (liver embryonic) cells. These findings suggest its potential application in cancer therapy. Additionally, peptides obtained from quinoa, including LWREGM (F-1), DKDYPK (F-2), RELGEWGI (F-3), DVYSPEAG, and IFQEYI, have been recognized for their antioxidant properties and chemoprotective potential, making them promising candidates for anticancer treatments (Majid et al., 2020). The antioxidant potential of different millet grains and their respective white varieties has been extensively studied. The presence of antioxidants and phenolic compounds in millet grains is of particular interest due to their potential role as antiaging agents and their ability to protect cells against metabolic syndrome. This suggests that incorporating millet grains into the diet may offer potential health benefits related to their antioxidant and antiaging properties (Nithiyanantham et al., 2019).

2.6 Culinary Uses and Cultural Relevance

In recent years, there has been a growing interest in millets as consumers and researchers recognize their health and environmental benefits. The resurgence of the "back-to-basics" food movement and a shift toward whole grains have particularly fueled this trend among urban and health-conscious populations. Moreover, millets are being promoted as "superfoods," which has further increased global interest in them, positioning them as functional foods with health-promoting properties (Edwin, 2024). Governments and international organizations, including the United Nations, are advocating for the inclusion of millets in diets to diversify food systems and promote sustainable agriculture. Researchers and food industry experts are developing value-added millet-based products such as flours, ready-to-eat snacks, breakfast cereals, and gluten-free options (FAO, 2020; Gupta et al., 2023). Millet-based foods align well with dietary trends that favor gluten-free, low-glycemic, and nutrient-dense choices. By reintroducing millets into mainstream diets and expanding their applications in modern culinary practices, they have the potential to reclaim their status as staple grains that nourish individuals and support global sustainability efforts (Anitha et al., 2021a).

As functional foods continue to gain importance in dietary recommendations, millets have been recognized for their impressive nutritional composition and health benefits. Unlike refined grains, which lose a significant portion of their essential nutrients through processing, millets remain rich in dietary fiber, high-quality protein, and a balanced spectrum of essential amino acids. These grains also provide key minerals, which play fundamental roles in various physiological processes, such as bone development, muscle function, and metabolic regulation. Additionally, millets are packed with bioactive compounds like phenolic acids, which possess strong antioxidant properties. These compounds help reduce oxidative stress, thereby lowering the risk of inflammation-related diseases and enhancing overall cellular health (Banerjee et al., 2020; Latha Ravi and Rana, 2024). Different types of millet exhibit distinct nutrient profiles, making them valuable additions to the diet. For instance, finger millet is notably rich in calcium, which is crucial for maintaining strong bones and may help prevent osteoporosis. In contrast, pearl millet is celebrated for its high iron content, an important factor in preventing and managing anemia, particularly in populations at risk. Furthermore, the impressive fiber content in millets promotes the regulation of blood sugar levels, enhances digestive health by aiding in regular bowel movements, and fosters a sense of satiety that can assist in weight management. These attributes make millets particularly effective in the management of chronic conditions. Culinarily, millets are remarkably versatile and can be adapted to a wide variety of dishes and cooking methods (Anitha et al., 2021b). They are often utilized in traditional recipes, such as wholesome porridges, which are popular breakfast staples in many African and Indian households. In South Indian cuisine, millets play a role in the preparation of fermented foods like dosa and idli. The fermentation process is particularly beneficial as it facilitates the production of beneficial probiotics that support gut health, promoting a balanced gut microbiome (Mishra et al., 2020). Beyond porridges and fermented items, millets can be creatively incorporated into various types of dishes. They can be used to make flatbreads, such as rotis or chapatis, as well as baked goods like bread and muffins. Millets can also be transformed into nutritious snacks, porridge bars, and even beverages like millet-based smoothies or milk. This adaptability not only caters to different culinary traditions but also accommodates diverse dietary preferences and restrictions, making millets a fantastic option for those seeking a healthier lifestyle. With their growing popularity, millets are making a comeback in modern kitchens, showcasing their timeless value and versatility (Bunkar et al., 2021).

2.7 Conclusion

Millets stand as a perfect blend of nutrition and sustainability, earning them the title of nutri-cereals. Rich in fiber, essential minerals, and antioxidants, they help combat diabetes, cardiovascular conditions, and digestive disorders. Their gluten-free composition makes them a suitable option for individuals with gluten sensitive enteropathy. Beyond their health benefits, millets are an environmentally resilient crop, thriving in drought-prone areas with minimal resources. As global food security concerns grow, incorporating millets into mainstream diets can help build a more sustainable and nutritious future.

References

Abah, C. R., Ishiwu, C. N., Obiegbuna, J. E., & Oladejo, A. A. (2020a). Sorghum grains: Nutritional composition, functional properties and its food applications. *European Journal of Nutrition & Food Safety*, *12*(5), 101–111.

Abah, C. R., Ishiwu, C. N., Obiegbuna, J. E., & Oladejo, A. A. (2020b). Nutritional composition, functional properties and food applications of millet grains. *Asian Food Science Journal*, *14*(2), 9–19.

Agrawal, P., Singh, B. R., Gajbe, U., Kalambe, M. A., & Bankar, M. (2023). Managing diabetes mellitus with millets: A new solution. *Cureus*, *15*(9), e44908.

Akanbi, O. A., Kaniyamattam, K., & Othman, A. (2021). Glycemic index of cereals and millets: A comparative study. *Journal of Food Science and Nutrition*, *9*(3), 456–468.

Almutairi, Z. M. (2022). Genetic diversity and phylogeny of pearl millets [Pennisetum glaucum (L.) R. Br.] based on chloroplast trnL-F region. *Genetic Resources and Crop Evolution, 69*(8), 2849–2859.

Ambati, K., & Sucharitha, K. V. (2019). Millets-review on nutritional profiles and health benefits. *International Journal of Recent Scientific Research, 10*(7), 33943–33948.

Anitha, S., Botha, R., Kane-Potaka, J., Givens, D. I., Rajendran, A., Tsusaka, T. W., & Bhandari, R. K. (2021a). Can millet consumption help manage hyperlipidemia and obesity?: A systematic review and meta-analysis. *Frontiers in Nutrition, 8*, 700778.

Anitha, S., Givens, D. I., Botha, R., Kane-Potaka, J., Sulaiman, N. L. B., Tsusaka, T. W., Subramaniam, K., Rajendran, A., Parasannanavar, D. J., & Bhandari, R. K. (2021b). Calcium from finger millet—A systematic review and meta-analysis on calcium retention, bone resorption, and in vitro bioavailability. *Sustainability, 13*(16), 8677.

Banerjee, S., Kumar, D., & Jadhav, P. (2020). Modern processing techniques for millets: Enhancing nutritive value and appeal. *Advances in Food Research and Technology, 22*(1), 231–249.

Bhatt, D., Rasane, P., Singh, J., Kaur, S., Fairos, M., Kaur, J., Gunjal, M., Mahato, D. K., Mehta, C., Avinashe, H., & Sharma, N. (2023). Nutritional advantages of barnyard millet and opportunities for its processing as value-added foods. *Journal of Food Science and Technology, 60*(11), 2748–2760.

Bunkar, D. S., Goyal, S. K., Meena, K. K., & Kamalvanshi, V. (2021). Nutritional, functional role of kodo millet and its processing: A review. *International Journal of Current Microbiology and Applied Sciences, 10*(01), 1972–1985.

Deshpande, S. S., Mohapatra, D., Tripathi, M. K., & Sadvatha, R. H. (2015). Kodo millet-nutritional value and utilization in Indian foods. *Journal of Grain Processing and Storage, 2*(2), 16–23.

Dey, S., Saxena, A., Kumar, Y., Maity, T., & Tarafdar, A. (2022). Understanding the antinutritional factors and bioactive compounds of kodo millet (Paspalum scrobiculatum) and little millet (Panicum sumatrense). *Journal of Food Quality, 2022*(1), 1578448.

Dhaka, A., Singh, R. K., Muthamilarasan, M., & Prasad, M. (2021). Genetics and genomics interventions for promoting millets as functional foods. *Current Genomics, 22*(3), 154–163.

Di Stefano, E., White, J., Seney, S., Hekmat, S., McDowell, T., Sumarah, M., & Reid, G. (2017). A novel millet-based probiotic fermented food for the developing world. *Nutrients, 9*(5), 529.

Divisekera, D. M. W. D., Samarasekera, J. K. R. R., Hettiarachchi, C., Gooneratne, J., Choudhary, M. I., Gopalakrishnan, S., & Wahab, A. T. (2019). Lactic acid bacteria isolated from fermented flour of finger millet, its probiotic attributes and bioactive properties. *Annals of Microbiology, 69*, 79–92.

Dwivedi, N., Rathore, V., & Sharma, K. (2023). A review of millet crops for agricultural sustainability in India. *Asian Journal of Agricultural Extension, Economics & Sociology, 41*(10), 216–224.

Edwin, M. R. (2024, November). From ancient grains to modern solution: Exploring the significance of millets in reversing imbalance of blood sugar levels. In *Revitalizing health through humanities: Foregrounding unheard trends: Proceedings of the international conference on health and humanities: Exploring uncharted terrains in literary paradigms* (pp. 166). Taylor & Francis.

FAO. (2020). *Millets and cereals: Global trade and production*. Food and Agriculture Organization of the United Nations.

Geetha, K., Yankanchi, G. M., Hulamani, S., & Hiremath, N. (2020). Glycemic index of millet based food mix and its effect on pre diabetic subjects. *Journal of Food Science and Technology, 57*, 2732–2738.

Gull, A., Jan, R., Nayik, G. A., Prasad, K., & Kumar, P. (2014). Significance of finger millet in nutrition, health and value added products: A review. *Magnesium (mg), 130*(32), 120.

Gupta, M., Asfaha, D. M., & Ponnaiah, G. (2023). Millets: A nutritional powerhouse with anti-cancer potential. *Cureus, 15*(10), e47769.

ICRISAT. (2020). *Millets for food security: A policy review*. International Crops Research Institute for the Semi-Arid Tropics.

Kaur, H., & Sharma, S. (2020). An overview of Barnyard millet (Echinochloa frumentacea). *Journal of Pharmacognosy and Phytochemistry, 9*(4), 819–822.

Kumar, V., Yadav, M., Awala, S. K., Valombola, J. S., Saxena, M. S., Ahmad, F., & Saxena, S. C. (2024). Millets: A nutritional powerhouse for ensuring food security. *Planta, 260*(4), 101.

Latha Ravi, J., & Rana, S. S. (2024). Maximizing the nutritional benefits and prolonging the shelf life of millets through effective processing techniques: a review. *ACS Omega, 9*(37), 38327–38347.

Li, C., Jeong, D., Lee, J. H., & Chung, H. J. (2020). Influence of germination on physicochemical properties of flours from brown rice, oat, sorghum, and millet. *Food Science and Biotechnology, 29*, 1223–1231.

Maharajan, T., Krishna, T. P. A., Krishnakumar, N. M., Vetriventhan, M., Kudapa, H., & Ceasar, S. A. (2024). Role of genome sequences of major and minor millets in strengthening food and nutritional security for future generations. *Agriculture, 14*(5), 670.

Mahendra, T., Singh, A., & Gupta, P. (2021). Environmental sustainability of millet cultivation. *Journal of Environmental Sustainability, 12*(4), 159–167.

Majid, A., Priyadarshini, C. G., & P. (2020). Millet derived bioactive peptides: A review on their functional properties and health benefits. *Critical Reviews in Food Science and Nutrition, 60*(19), 3342–3351.

Mawouma, S., Condurache, N. N., Turturică, M., Constantin, O. E., Croitoru, C., & Rapeanu, G. (2022). Chemical composition and antioxidant profile of sorghum (Sorghum bicolor (L.) moench) and pearl millet (Pennisetum glaucum (L.) R. Br.) grains cultivated in the far-North region of Cameroon. *Foods, 11*(14), 2026.

Mazziotta, C., Tognon, M., Martini, F., Torreggiani, E., & Rotondo, J. C. (2023). Probiotics mechanism of action on immune cells and beneficial effects on human health. *Cells, 12*(1), 184.

Mishra, N., Chaturvedi, M., & Jha, S. (2020). Traditional and modern uses of millets in Indian cuisine. *Journal of Ethnic Foods, 7*(2), 45–56.

Nambiar, V. S., Rajiv, J., & Vasudevan, V. (2020). Bioactive components in millets and their health benefits. *Journal of Nutraceuticals and Functional Foods, 13*(3), 231–240.

Nithiyanantham, S., Kalaiselvi, P., Mahomoodally, M. F., Zengin, G., Abirami, A., & Srinivasan, G. (2019). Nutritional and functional roles of millets—A review. *Journal of Food Biochemistry, 43*(7), e12859.

Nithyashree, K., & Vijayalaxmi, K. G. (2022). Effect of popping on nutrients and anti-nutrients in proso millet. *Mysore Journal of Agricultural Sciences, 56*(4), 313–319.

Pandey, A., & Bolia, N. B. (2023). Millet value chain revolution for sustainability: A proposal for India. *Socio-Economic Planning Sciences, 87*, 101592.

Passot, S., Gnacko, F., Moukouanga, D., Lucas, M., Guyomarc'h, S., Ortega, B. M., Atkinson, J. A., Belko, M. N., Bennett, M. J., Gantet, P., Wells, D. M., Guédon, Y., Vigouroux, Y., Verdeil, J.-L., Muller, B., & Laplaze, L. (2016). Characterization of pearl millet root architecture and anatomy reveals three types of lateral roots. *Frontiers in Plant Science*, *7*, 829.

Plaza-Diaz, J., Ruiz-Ojeda, F. J., Gil-Campos, M., & Gil, A. (2019). Mechanisms of action of probiotics. *Advances in Nutrition*, *10*, S49–S66.

Ramashia, S. E., Anyasi, T. A., Gwata, E. T., Meddows-Taylor, S., & Jideani, A. I. O. (2019). Processing, nutritional composition and health benefits of finger millet in sub-saharan Africa. *Food Science and Technology*, *39*, 253–266.

Renganathan, V. G., Vanniarajan, C., Karthikeyan, A., & Ramalingam, J. (2020). Barnyard millet for food and nutritional security: Current status and future research direction. *Frontiers in Genetics*, *11*, 500.

Saini, S., Saxena, S., Samtiya, M., Puniya, M., & Dhewa, T. (2021). Potential of underutilized millets as Nutri-cereal: An overview. *Journal of Food Science and Technology*, *58*, 1–13.

Saleh, A. S., Zhang, Q., Chen, J., & Shen, Q. (2013). Millet grains: Nutritional quality, processing, and potential health benefits. *Comprehensive Reviews in Food Science and Food Safety*, *12*(3), 281–295.

Sarita, E. S., & Singh, E. (2016). Potential of millets: Nutrients composition and health benefits. *Journal of Scientific and Innovative Research*, *5*(2), 46–50.

Satyavathi, C. T., Ambawat, S., Khandelwal, V., & Srivastava, R. K. (2021). Pearl millet: A climate-resilient nutricereal for mitigating hidden hunger and provide nutritional security. *Frontiers in Plant Science*, *12*, 659938.

Singh, A., Kumar, M., & Shamim, M. (2020). Importance of minor millets (Nutri Cereals) for nutrition purpose in present scenario. *International Journal of Chemical Studies*, *8*(1), 3109–3113.

Srivastava, R. K., Yadav, O. P., Kaliamoorthy, S., Gupta, S. K., Serba, D. D., Choudhary, S., Govindaraj, M., Kholová, J., Murugesan, T., Satyavathi, C. T., Gumma, M. K., Singh, R. B., Bollam, S., Gupta, R., & Varshney, R. K. (2022). Breeding drought-tolerant pearl millet using conventional and genomic approaches: Achievements and prospects. *Frontiers in Plant Science*, *13*, 781524.

Tripathi, M. K., Mohapatra, D., Jadam, R. S., Pandey, S., Singh, V., Kumar, V., & Kumar, A. (2021). Nutritional composition of millets. *Millets and Millet Technology*, 101–119.

Vedamanickam, R., Anandan, P., Bupesh, G., & Vasanth, S. (2020). Study of millet and non-millet diet on diabetics and associated metabolic syndrome. *Biomedicine*, *40*(1), 55–58.

3

Antinutritional Factors in Millets and Their Elimination Methods

Amandeep Singh[1]*, Tejinder Kaur*[1]*, Souvik Giri*[1]*, Kawaljit S. Sandhu*[2]*, and Charanjit S. Riar*[1]

[1] *Department of Food Engineering and Technology, Sant Longowal Institute of Engineering and Technology, Longowal, Punjab, India*
[2] *Department of Food Science and Technology, MRSPTU, Bathinda, India*

3.1 Introduction

Millets are referred to as "yesterday's coarse grains and today's nutri-cereals." They are the crop of antiquity, which can create tiny grains from a variety of naturally occurring grass species—also known as "coarse" or "minor" cereals. Historically, millet cultivation has been practiced in the tribal, highlands, and drylands of India for over 5,000 years. They can be grown on less productive lands, in arid regions. The primary millets grown in India include jowar (sorghum), bajra (pearl millet), ragi (finger millet), sawan (barnyard millet), kodo (kodo millet), kangni (foxtail millet), and kutki or samai (little millet). It plays a significant role in ensuring sustenance, livestock feed, and dietary well-being on farms and within households (Dayakar Rao et al., 2016). Millets serve as an affordable nutrient source for balanced diets, thriving in diverse contrary conditions and dry regions with limited external inputs. Their strong resistance to drought and short growing seasons makes them an excellent choice for rotational cover cropping. They are also known as the "Supergrains," and it is scientifically proven that millet has better nutritional properties than cereals. Due to their superior adaptability to diverse temperature and moisture conditions, millets are becoming a preferred dietary option for the general populace, especially in developing countries. Research has documented and confirmed the nutraceutical properties of fermented millet products and is considering replacing other cereals with millets, which are equally enriched and offer a range of beneficial effects (Balli et al., 2020). Millets demonstrate substantial nutritional importance and can be utilized for the creation of a range of value-added products; nevertheless, the existence of antinutritional elements within millets often impairs the bioaccessibility and bioavailability of essential minerals. Antinutrients are substances

Nutraceutical Potential of Millet-Based Food Products: Chemistry and Applications, First Edition.
Edited by Parmjit S. Panesar, Charanjit S. Riar, and Sushil Dhital.

Published 2026 by John Wiley & Sons Ltd.

that are often produced in plant materials by numerous biological processes. These are compounds that inhibit and obstruct the accessibility of minerals and other important biomolecules. Most antinutrients can trigger harmful biological effects; however, some are utilized for providing nutrition and medicinally active ingredients (Popova and Mihaylova, 2019; Soetan and Oyewole, 2009). Some major antinutritional factors (ANFs) present in millets and cereals are phytate, oxalate, tannins, lectins, and goitrogens (Samtiya et al., 2020). These antinutrients can be eliminated by using various techniques such as hydrothermal treatment (parboiling), soaking, malting, germination, cooking, and fermentation (Sweta Kumari et al., 2015). Therefore, in this chapter, we will foreground the ANFs and their elimination methods to maximize millet utility and enhance the efficacy of millet and products developed from them.

3.2 Nutritional Composition of Millets

Millets are frequently likened to cereals because of their high content of carbohydrates, dietary proteins, and fibers. The grains primarily consist of carbohydrates and varying amounts of proteins, fiber, and lipids (Devi et al., 2014). Starch serves as the primary carbohydrate in millet grains and plays a key role in regulating the digestibility and glycemic impact of millet. Millet protein is rich in cysteine, lysine, and methionine, which are known essential amino acids and not commonly found in cereals. They are rich in protein, polyphenols, and micronutrients (Anitha et al., 2020). They typically comprise 7–15% protein, 2–6% fat, 66–75% carbohydrates, and 3–10% crude fiber. Millets contribute to antioxidant activity with polyphenols, anthocyanins, phytosterols, and pinacosanols present in them, having an important role in ageing and metabolic diseases. All millets possess high antioxidant activities (Kheya et al., 2023). Millets are abundant in essential micronutrients, including calcium (ranging from 10 to 348 mg per 100 g), iron (2.2–17.7 mg per 100 g), zinc (0.4–2.8 mg per 100 g), and phosphorus (189–293 mg per 100 g). They also provide various vitamins, such as thiamine (0.15–0.60 mg per 100 g), niacin (0.89–4.6 mg per 100 g), and riboflavin (0.9–0.28 mg per 100 g) (Table 3.1).

3.3 Antinutrients in Millets

Antinutrients are substances that can intervene with the absorption of nutrients, making them less available for the body to utilize. However, it is fascinating that these antinutritional elements can be diminished through simple household food processing methods such as fermentation, cooking, and germination (Manzoor et al., 2021). With their rich array of biologically active compounds, millets hold great potential as a future food source and a natural means to enhance human health, particularly in regions facing food shortages and economic challenges. Moreover, traditional food processing methods have demonstrated encouraging results in boosting the nutritional profile of cereals, including millet-based products, straightforwardly and cost-effectively. Below are highlighted several common antinutrients present in millets. The chemical structures of different antinutrients in millets are shown in Table 3.2.

Table 3.1 Nutritional composition of different millets in comparison to rice and wheat/100 g

Nutrients (%)	Finger millet	Foxtail millet	Pearl millet	Barnyard millet	Kodo millet	Little millet	Proso millet	Rice	Wheat
Carbohydrate	68	75.86	69.74	67.48	66.89	67.86	69.93	82.86	69.88
Protein	7.31	15.28	10.23	10.45	9.07	9.44	12.61	6.4	11.81
Fat	2.73	5.60	5.69	3.50	3.22	3.51	3.08	1.90	2.81
Fiber	3.03	8.12	2.63	10.10	9.31	7.61	2.51	1.63	1.77
Calcium (mg)	344	31	42	20	27	17	14	2	29
Iron (mg)	3.91	1.11	2.81	15.20	0.61	9.31	3.81	0.61	3.91
Zinc (mg)	2.81	1.11	1.71	0.40	1.21	1.21	1.71	0.91	2.71
Phosphorus (mg)	283	258	285	293	189	207	277	52	135
Thiamine (mg)	0.42	0.59	0.27	0.33	0.15	0.3	0.41	0.4	0.24
Niacin (mg)	1.10	3.21	0.89	4.20	2.11	3.21	4.61	1.61	5.51
Riboflavin (mg)	0.19	0.11	0.15	0.11	0.09	0.09	0.28	0.03	0.13

Note: Prashanth et al. (2024), Gull et al. (2016), Goudar et al. (2023), Kumar et al. (2018), and Kaur et al. (2023).

Table 3.2 Chemical structures of different antinutrients in millets

Antinutrients	Structures
1) Phytate	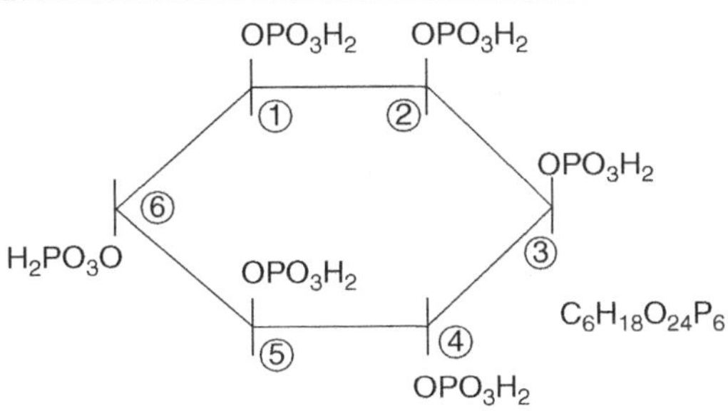
2) Tannins	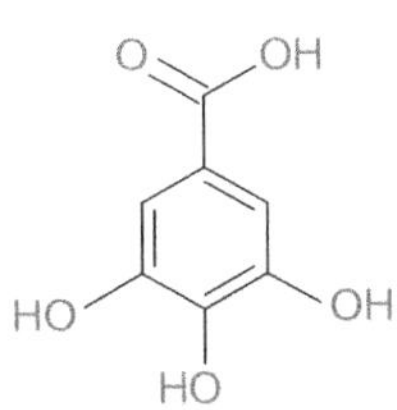

(*Continued*)

Table 3.2 (Continued)

Antinutrients	Structures
3) α-amylase inhibitors	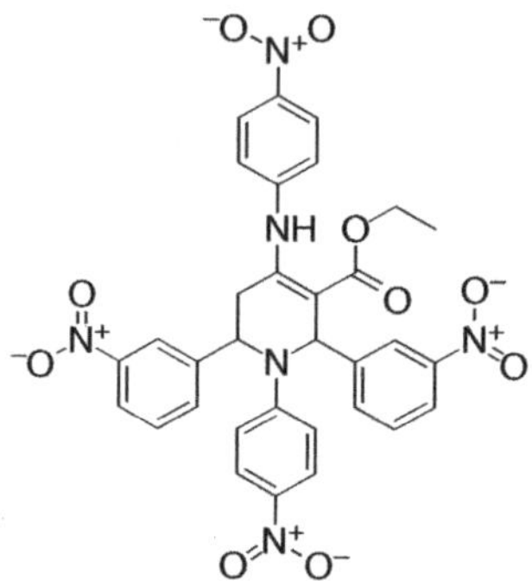
4) Protease inhibitor	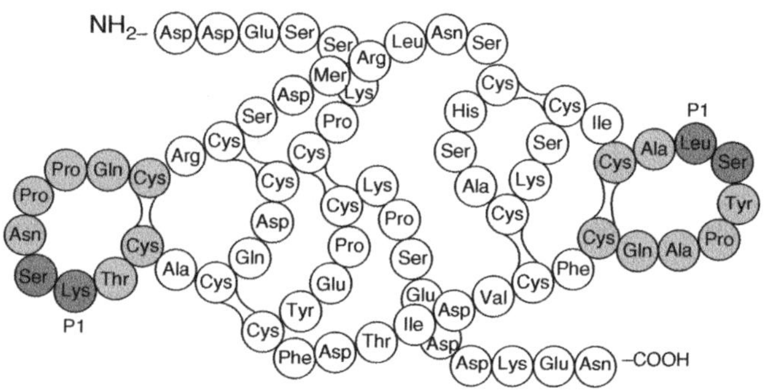
5) Trypsin Inhibitor	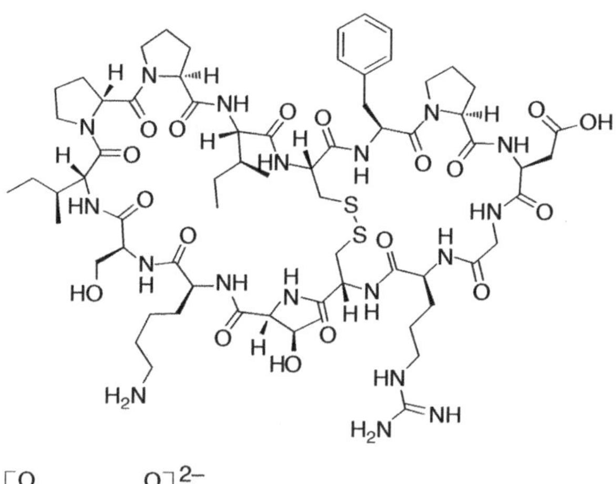
6) Oxalates	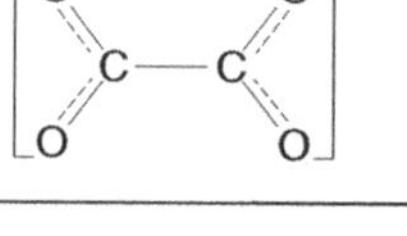

3.3.1 Phytates

Phytic acid, also referred to as phytate, is chemically defined as myoinositol 1,2,3, 4,5,6-hexakis dihydrogen phosphate. It serves as the principal storage form of phosphorus, constituting approximately 1–5% of the weight in various food sources, including cereals, legumes, nuts, and oilseeds. This compound functions as a secondary metabolite, acting as a nutrient reserve that accumulates in seeds during the ripening process (Nadeem et al., 2010; Gupta et al., 2015). Notably, humans are unable to absorb phytate or hydrolyze it, which contributes to its role in reducing the bioavailability of essential minerals. As a negatively charged ion, phytate interacts in pH-sensitive environments, adversely affecting the bioavailability of positively charged divalent and trivalent mineral ions such as Zn^{2+}, Fe^{2+}/Fe^{3+}, Ca^{2+}, and Mg^{2+} (Gemede & Ratta, 2014). The concentration of phytate varies significantly among grains, legumes, nuts, and seeds, influenced by factors such as cultivation conditions, processing techniques, harvesting practices, and analytical methods. It is widely recognized as a significant inhibitor of mineral absorption (Coulibaly et al., 2011). Several biofortified crop varieties have been developed to mitigate the adverse effects of phytates, including Dhanashakti, ICMH 1201, ICMH 1301, ICTP 8203, and ICMV 221 (Samtiya et al., 2020).

3.3.2 Tannins

Tannins represent a significant category of antioxidant polyphenols commonly found in various food and beverage products. Following cellulose, hemicellulose, and lignin, they are the fourth most prevalent component in plant matter. Characterized as polyphenolic biomolecules, tannins possess astringent and bitter flavors, with molecular weights ranging from 500 to 3,000 Da. These compounds can be classified as organic, nitrogenous, or inorganic, and they play a crucial role in plant defense mechanisms (Jaiswal et al., 2018). In humans, tannins exhibit multifunctional properties that have garnered considerable research interest (Singh & Kumar, 2020). However, they can adversely impact the nutritional quality of food. Tannins interact with proteins to form complexes, which can diminish food efficiency, hinder growth, and impair iron absorption. Excessive intake of tannins may lead to serious health issues, including gastrointestinal damage and the depletion of specific tissues, proteins, and essential amino acids. Therefore, it is advisable to limit their levels to ensure safety for human consumption (Adeyemo & Onilude, 2013; Reddy & Pierson, 1994).

Dietary assessments indicate that the average daily intake of tannins in India ranges from 1,500 to 2,500 mg. Consumption levels below 1.5–2.5 g per day are generally considered safe, while exceeding this threshold may result in health complications such as anemia, osteoporosis, and worsen cancer conditions (Sharma et al., 2021; Rao & Prabhavathi, 1982; Ricardo-da-Silva et al., 1991). Additionally, reduced serotonin levels, which can intensify migraine symptoms, are linked to the binding of starch by tannins, as starch serves as a precursor for serotonin (Mather, 1997). While tannins may inhibit cancer cell proliferation, certain carcinogenic compounds associated with them can induce cancer and lead to hepatic cell necrosis (Chung et al., 1998). The precipitation of proteins by tannins creates an astringent sensation, which can detract from the palatability of food products, posing challenges for the beverage industry (Sharma et al., 2021). Despite these drawbacks, tannins have diverse industrial applications due to their antioxidant and antibacterial properties, finding use in the food sector, preventive medicine, as well as in the leather and chemical industries. Hydrolysable tannins are specifically identified as esters formed from sugars and phenolic acids or their derivatives. The composition includes gallic acid, ellagic acid, hexahydroxy diphenic acid, and penta galloyl glucose, characterized by a central glucose molecule linked to multiple gallic acid units (Singh & Kumar, 2020). Gallotannins, which are the most basic form of hydrolysable tannins, comprise both a polyphenolic and a polyol component. The formation of ellagitannins occurs through the oxidative coupling of at least two galloyl units (Khanbabaee & Van Ree, 2001). Gallotannins are predominantly located in grapes, berries, pomegranates, and persimmons, while ellagitannins are found in tea, coffee, various fruits, nuts, and wine aged in fermented oak barrels. Condensed and complex tannins are typically present in fruits such as apples, cranberries, and pears, as well as in chocolate, coffee, cocoa, legumes (including red kidney beans, chickpeas, black-eyed peas, and lentils), green grapes, nuts, red wine, and tea (Martinez et al., 2017).

3.3.3 Enzyme Inhibitors

Different categories of enzyme inhibitors include protease inhibitors, alpha-amylase inhibitors, and trypsin inhibitors (Wisessing & Choowongkomon, 2012). These substances serve as a defense mechanism for plants against pests and microorganisms. To manage protein levels and protect themselves from herbivores, plants produce a range of protein inhibitors, primarily found in seeds and tubers, which are referred to as storage proteins. The concentration of these proteins typically varies between 6% and 10%. Trypsin inhibitors specifically target trypsin proteins, which interact with peptides following lysine and arginine residues. They play a crucial role in reducing the activity of trypsin and chymotrypsin, thereby hindering protein digestion (Popova & Mihaylova, 2019; Sowbaghya et al., 2019).

3.3.3.1 Protease Inhibitors

Protein enzymes are known as proteases that catalyze the hydrolytic cleavage of peptide bonds in the target proteins. These proteases play a crucial role in various proteolytic processes that help regulate and mitigate excessive or undesirable protein degradation. The activity of proteases can be influenced by specific inhibitors referred to as protease inhibitors. Typically, smaller proteins, these inhibitors are found in high concentrations within storage tissues and can constitute up to 10% of the total protein content (Clemente et al., 2019; Habib & Fazili, 2007). Protease inhibitors are vital for numerous physiological functions in plants, including the regulation of storage proteins, modulation of endogenous proteolytic activity, and serving as protective agents that enhance plant resistance to pests and pathogens. These inhibitors slow the digestion of proteins in the small intestine, which in turn diminishes the bioavailability of sulfur-containing amino acids, such as cysteine and methionine (Nikmaram et al., 2017). Furthermore, protease inhibitors can modulate protease activity, thereby affecting various biological processes in mammals, including apoptosis, blood coagulation, fibrinolysis, inflammation, and hormonal regulation. Consequently, there is a pressing need to either eliminate or significantly reduce the levels of protease inhibitors in food products. Nevertheless, protease inhibitors also possess biotechnological applications, showing promise as insecticides, anticancer agents, and antibacterial compounds (Alarcón et al., 2001; Mehrabadi et al., 2012; Paiva et al., 2013).

3.3.3.2 Amylase Inhibitors

Amylase inhibitors can act like alpha-amylases found in mammals and insects, attaching to alpha-amylase and rendering it inactive. These inhibitors are categorized into two types: proteinaceous and nonproteinaceous. In industrial applications, alpha-amylase inhibitors are utilized in bakery products and during malting to control the rate of starch breakdown and enhance flour characteristics. Additionally, α-amylase inhibitors may improve the stability of beer foam by interacting with iso-α-acid derived from hops. Studies have also indicated their significant role in diabetes management (Popova & Mihaylova, 2019; Wisessing & Choowongkomon, 2012).

3.3.3.3 Trypsin Inhibitors

This is a type of protein that diminishes the function of trypsin, an enzyme responsible for protein degradation in the human digestive system. These inhibitors can be preserved at 2–8 °C or frozen at −20 °C. The primary categories of trypsin inhibitors include Bowman-Birk inhibitors and Kunitz-type inhibitors. Furthermore, trypsin inhibitors hold promise for use in packaging materials, and research has indicated that the Bowman-Birk trypsin inhibitor extracted from foxtail millet bran demonstrates anti-colorectal cancer (CRC) properties (Zhang et al., 2024).

3.3.4 Oxalates

Oxalate, or oxalic acid, functions as an antinutrient by binding it with calcium and other mineral ions, inhibiting their nutritional absorption (Chakraborty et al., 2013; Weaver et al., 1987). It is a significant constituent in various plants, yet the mechanisms underlying its synthesis and accumulation in most crop species remain inadequately understood. Oxalic acid interacts with minerals such as calcium, magnesium, sodium, and potassium to form oxalate salts. Among these, salts from potassium and sodium are hydrophilic, while calcium oxalate salts are insoluble in nature (Yang et al., 2012). For calcium (Ca), the compounds it is bound to can exist in various types, with some being soluble; however, calcium oxalates are predominantly insoluble (Gemede & Ratta, 2014). The highest concentrations of bound calcium typically have crystalline structures, e.g., calcium oxalate (Ca-Ox), which is recognized as a primary antinutrient that diminishes calcium bioavailability in plants (D'Imperio et al., 2016). This phenomenon also impacts calcium absorption and utilization in animal systems. Conversely, oxalates serve as a defensive mechanism in plants. For instance, spinach, which has low calcium and high oxalate content, exhibits reduced calcium bioavailability, whereas kale, with its high calcium and low oxalate levels, demonstrates enhanced calcium bioavailability (Weaver et al., 1987). Research has identified three distinct biosynthetic pathways for oxalic acid across various organisms, with the ascorbate-oxalate and glyoxalate-oxalate conversion pathways being commonly observed in numerous plants (Amalraj & Pius, 2015; Nakata & McConn, 2000). These oxalate ions subsequently bind with minerals, proteins, and other nutrient ions, further hindering their absorption.

3.4 Techniques of Elimination of Millet Grain Antinutritional Factors

A variety of processing methodologies, such as decortication, parboiling, thermal treatment, fermentation, steeping, and malting, have been utilized over an extensive period to reduce the level of antinutrients found in plant-based foods and to enhance the nutritional quality of these foods by increasing the bioavailability of specific cations, including calcium (Ca), iron (Fe), and zinc (Zn), while also facilitating the absorption of protein. Table 3.3 represents the effect of various elimination methods on the ANFs of millets.

3.4.1 Decortication

Decortication is a crucial step during processing of millet, involving the removal of the husk, and the seed coat adhering to the grain's surface. Improper removal of the bran layer results in food products with a dark color, chewy texture, and a distinct musty odor, significantly lowering their appeal to nonmillet consumers. Due to uneven nutrient composition in millets, decortication has a significant impact on the phytochemical and nutritional characteristics. In a study by Rani et al. (2018), removal of 17.5% of outer layers from pearl millet grains significantly affected the nutritional composition and digestibility of other major nutrients such as carbohydrates and proteins. However, further removal of seed coat resulted in a significant reduction in protein content. Pushparaj and Urooj (2011) studied the effect of milling on pearl millet; the grains containing more bran showed higher in vitro protein digestibility. A majority of the polyphenolic constituents and ANFs that induce protein precipitation and hinder protein digestibility are predominantly located in the hull of millet grains; the process of decortication significantly mitigates their presence, thereby enhancing the digestibility of proteins. Babiker et al. (2018) evaluated the effect of decortication on pearl millet. They reported a decrease of 49% and 74% in tannins and phytates, respectively, after decortication using modern methods (using mechanical dehuller) in pearl millet.

3.4.2 Parboiling (Hydrothermal Treatments)

Parboiling is a conventional method employed to harden the grain endosperm, thus reducing milling losses. The parboiling improved the milling quality of different millet. Parboiling is widely recognized for minimizing nutrient loss during milling and enhancing milling quality. The process involves three main steps: soaking, boiling or steaming, and drying. Sene et al. (2018) examined the effects of decortication and hydrothermal treatment on two varieties of pearl millet, finding a reduction in phytate content. They observed that raw, un-parboiled, and un-decorticated grains had the highest phytate content, which decreased after decortication and showed a significant decline following hydrothermal treatment due to soaking, leading to phytate leaching. Sihag et al. (2015) reported that soaking pearl millet decreased its phytate content from 683.07 to 616.72 mg/100 g, with the reduction becoming more pronounced as soaking time

Table 3.3 Effect of various elimination methods on the antinutritional factors of millets

Treatments	Type of millets	Effect on antinutritional factors	References
Decortication	Pearl millet	Tannins and phytic acid reduced by 49% and 74%, respectively	Babiker et al. (2018)
Hydrothermal treatment	Pearl millet	Phytate content decreased from 683.07 to 616.72 mg/100 g	Sihag et al. (2015)
Parboiling	Finger millet	Reduction in phytic acid content by 60.03%	Kushwaha et al. (2019)
Hydrothermal treatment	Finger millet	Hydrothermal treatment reduced tannin content by 76.31%	Onyango et al. (2020)
Roasting	Barnyard millet	25% reduction in tannin content	Shinde et al. (2023)
Pressure cooking and microwave heating	Pearl millet	Phytic acid content was reduced by 38–55% and 31–36% after pressure cooking and microwave heating, respectively	Hithamani and Srinivasan (2014)
Soak-boil	Foxtail millet	4.51% reduction in phytic acid	Zhang et al. (2017)
Soaking	Finger millet	A reduction of 78% in phytic acid and 39% in tannin content	Singh et al. (2017)
Germination	Pearl, foxtail, and finger millet	After germination, phytic acid content was reduced by 41%, 60%, and 36%, whereas 48%, 52%, and 44% reduction in tannin content in foxtail millet, pearl, and finger millets, respectively	Mane et al. (2023)
Germination	Sorghum	Phytate content reduced by 32.7–79.2% depending on variety and treatment time	Usman (2018)
Fermentation	Foxtail millet	Reduction in phytic acid from 591.26 mg/100 g (RFMF) to 414.03 mg/100 g after 36 hours of fermentation	Jan et al. (2022a)
Fermentation	Finger millet	Oxalate content, phytate content, trypsin inhibitor, and tannin content reduced by 44.96%, 77%, 56.16%, and 52.41%, respectively, after treatment	Abioye et al. (2021)
Irradiation	Sorghum millet	90% reduction in phytic acid with a gamma radiation dose of 30 kGy	Rousta et al. (2014)
Cold plasma	Barnyard and pearl millet	• Tannin content was reduced by 24% and 18% in barnyard and pearl Millet, respectively • Phytic was reduced by 61% and 57% in barnyard and pearl Millet, respectively	Charu et al. (2024)

increased from 3 to 12 hours. Sheikh et al. (2021) also studied the impact of hydrothermal treatment, noting a 68.72% reduction in amygdalin, a 79.59% reduction in hydrocyanic acid, and a 72.64% decrease in phytate content. Kushwaha et al. (2019) reported that hydrothermal treatment significantly reduced phytates in finger millet, with a 50%

reduction after pearling, and an additional 10.3% reduction compared to untreated samples. In a study on three varieties of finger millet (KMR-344, GE 6834-1, and DHWFM 11-3), soaking for 12, 24, and 48 hours led to a decrease in phytic acid content by 5.90%, 9.59%, and 16.97%, respectively, with the reduction varying based on the variety (Shigihalli et al., 2018). Zhang et al. (2017) reported a 4.49% and 4.51% reduction in phytate content in foxtail millet when cooked or steamed, respectively, attributing the reduction to forming salt complexes with calcium or magnesium (Liu et al., 2019). Kataria et al. (2022) found that heat treatments of teff millet ensued a 26–37% reduction in phytic acid and tannin content. The decrease in phytate levels could be attributed to the breakdown of phytate-protein and phytate-mineral complexes or the cleavage of the phytic acid ring, as phosphorus dissociates from the phytate structure (Ojo et al., 2018). Nour et al. (2015) reported that cooking pearl millet flour in water reduced its phytate content by 14.77%, likely due to the formation of low molecular weight compounds such as penta-phosphate during heating. Onyango et al. (2020) reported that hydrothermal treatment reduced tannin content in finger millet by 76.31%, which they attributed to the development of hydrophobic protein interactions that reduce tannin extractability and leaching.

3.4.3 Heating

Heating methods encompass various techniques such as roasting, boiling, cooking, and autoclaving. Roasting has been known to decrease tannin, phytate, trypsin, and protease inhibitor levels. Sade's research indicated that roasting effectively reduced tannin content by 43.13% and phytate by 47.61% (Sade, 2009). Conversely, due to cooking, Abdelrahman et al. (2005) observed a 6–10% reduction in phytic acid and a 5–8% reduction in polyphenols after cooking. Pushparaj and Urooj (2014) reported a 20–25% reduction in tannin and phytic acid levels, while Mohapatra et al. (2019) noted a 17% decrease in tannins following pressure cooking. Sihag et al. (2015) found that malting achieved the highest reduction of phytic acid at 38.23%, while pressure cooking caused a 49.28% reduction in polyphenols. As reported by Hithamani and Srinivasan (2014), different thermal processing methods led to a substantial decrease in tannin content in pearl millet, with pressure cooking reducing it by 38%, boiling in an open pan for five minutes by 55%, and microwave treatment by 31% and 36% at 450 and 600 W, respectively. Moreover, the tannin content decreased by 25% upon roasting the barnyard millet. This reduction was primarily attributed to thermal degradation and the evaporation of these compounds during the roasting process (Shinde et al., 2023).

3.4.4 Soaking

The process of soaking millets at an ambient temperature of approximately 70 °C for a duration of 10–24 hours enhances water absorption. Prolonged soaking of whole grains can reduce phytic acid levels by 20–30%. Upon cooling the soaked grains to room temperature, the tendency of amylose molecules to reassociate results in a compact structure that renders the kernels firmer and gives them a glossy finish. Before steaming, soaked grains may undergo predrying either in direct sunlight or using dryers. This soaking process is a preliminary step before further processing techniques, such as cooking, fermentation, and malting, and is recognized as a straightforward method for diminishing antinutritional factors (Kumari et al., 2015). Research indicates that soaking grains in water for 12–18 hours effectively lowers the concentrations of both partially and fully soluble phytates and inhibitors of proteolytic enzyme (Handa et al., 2017). According to Nithya et al. (2007), soaking followed by the discarding of water can significantly reduce antinutritional factors through the leaching of polyphenols. Further studies have demonstrated a significant reduction in tannins, oligosaccharides, and phytic acid in beans that were soaked and cooked without soaking water. It is advisable to remove the soaking water, as this method has proven effective in eliminating water-soluble antinutrients. Hithamani and Srinivasan (2014) reported a decrease in tannin content by 74.26% following soaking. The polyphenol oxidase enzyme was activated during this process, contributing to a reduction in tannin content. Singh et al. (2017) stated that finger millet experienced a 78% reduction in phytic acid and a 39% decrease in tannin content upon soaking. Also, there was a 26% and 51% drop in phytates and tannin content, respectively, when pearl millet was soaked for 24 hours in aerated fresh water.

3.4.5 Germination

Germination represents a dynamic metabolic phase where the levels of antinutrients are significantly lowered (Kumari et al., 2015). This process can improve the nutritional profile of legumes and cereals by modifying their chemical composition and reducing the levels of ANFs (Samtiya et al., 2020). According to Sokrab et al. (2012), germination decreases phytic acid levels and boosts polyphenols' content. The phytate levels in various pearl millet (PM) types were in the range of 588–1,382 mg/100 g. When grains were subjected to germination for 24 hours at 30 °C, phytate levels dropped by over 50% due to the activity of an endogenous enzyme called phytase, which breaks down phytate during the process of germination. According to Krishnan et al. (2020), when pearl millet was subjected to combined germination-hydrothermal treatment, there was a 49% reduction in phytate content. Similar findings were stated by Krishnan et al. (2011) when finger millet was germinated; treatment reduced the phytic acid content in the millet by 30.76%. After germination, the phytic acid decreased by 41%, 60%, and 36%, whereas there was a 48%, 52%, and 44% drop in tannin content in foxtail millet, pearl, and finger millet, respectively. The decrease in tannins during germination may be attributed to the leaching of polyphenols into the soaking water and the increased enzymatic activity throughout the germination process. The reduction in the amount of phytic acid due to germination can be linked to enhanced action of the phytase, which breaks down phytate into phosphate and myoinositol phosphates (Mane et al., 2023). Comparable results were observed by Usman (2018) when various varieties of sorghum were germinated for 48 and 96 hours. The phytate content in all sorghum varieties decreased by 32.7–56.3% after 48 hours of germination. Extending the germination period to 96 hours increased the reduction, ranging from 52.1% to 79.2%. The tannin content also decreased significantly after germination.

3.4.6 Fermentation

Fermentation is extensively acknowledged as one of the most primordial methods for food preservation that is currently utilized in the processing of cereal and millet grains. Fermented food items occupy a significant position in human nutrition due to their associated health advantages (Jan et al., 2022a). Adebiyi et al. (2016) defined fermentation as a process that uses microorganisms to convert substrates into new products. Fermentation enhances human nutrition by generating diverse scents, flavors, and textures in food, while also preserving substantial quantities of food through the formation of alcohol, acetic, and lactic acid, reduction in cooking time, etc. (Samtiya et al., 2021). Fermentation provides key benefits owing to the decrease in antinutrients, protein digestibility, an increase in protein bioavailability, and improved nutritional composition (Adebo et al., 2022). Several investigations reveal engrossing results with effect on various antinutrients by fermentation. Osman (2011) conducted a thorough study on fermentation of pearl millet and examined that during 24 hours of fermentation, the activities of amylase inhibitor, trypsin inhibitor, and phytate content were lowered to 9%, 50.8%, and 51.9%, respectively, of initial concentration. Finger millet flour underwent lactic acid fermentation with two *Lactobacillus* strains (*Lactobacillus brevis* and *Lactobacillus plantarum*) and yeast (*Saccharomyces cerevisiae*) at an interval of 12, 24, and 36 hours by Jan et al. (2022b). It was analyzed that there was a 17.11% reduction in phytic acid after yeast fermentation for 36 hours. Sorghum and pearl millet were individually fermented at a temperature of 32±2 °C for 72 hours with the addition of 0.2% ascorbic acid, followed by drying in a hot air oven at 60 °C for 10 hours. Fermented sorghum and pearl millet flour were analyzed and depicted the decrease in phytic acid levels in sorghum and pearl millet after 72 hours from 0.66% to 0.41% (37%) and 1.32% to 0.74% (44%), respectively. Fermentation resulted in a significant reduction in tannin content, decreasing from 312 to 150 mg/g in sorghum and from 402 to 121 mg/g in pearl millet (Gwekwe et al., 2024). Foxtail millet flour was subjected to fermentation by Jan et al. (2022a), which involves an inoculation of sample with the 250 μL of *L. brevis*, 500 μL of *L. plantarum*, and 125 μL of *S. cerevisiae* species. Samples were incubated at 37 °C at 12, 24, and 36 hours of interval and then dried at 50 °C. Fermented samples were analyzed, which showed a reduction in phytic acid by 29.97% after 36 hours of fermentation. The phytic acid content in foxtail millet flour decreased by 46.80% and 29.5% after 36 hours of fermentation with *L. brevis* and *L. plantarum*, respectively. Finger millet was fermented at 28±2 °C for 72 hours and dried at 50 °C for 48 hours by Abioye et al. (2021). Fermented finger millet was analyzed after 0, 24, 48, and 72 hours, which reported the significant ($p < 0.05$) reduction in oxalate content from 1.590 to 1.325, 1.410 and 0.875 mg/100 g, trypsin content from 6.230 to 5.338, 3.121 and 2.731 mg/100 g, phytic acid from 0.850 to 0.594, 0.204 and 0.195 mg/100 g, and tannins from 3.230 to 2.431, 1.896 and 1.537 mg/100 g after 24, 48, and 72 hours of the fermentation treatment. Aksum finger millet was first fermented at 28 °C for 24, 36, and 48 hours at pH 5.2–3.6 using *S. cerevisiae*, then dried at 50 °C for 8 hours. Fermented finger millet flour was used to prepare biscuits by baking in a preheated oven at 210 °C for 18 minutes. Biscuits made from raw grain and fermented grain (24, 36, and 48 hours) were analyzed, revealing a significant ($p < 0.05$) reduction in phytic acid content from 435.50 to 434.03, 431.54, and 430.78 mg/100 g, respectively, and tannin content reduced from

1.21 to 1.04, 0.97, and 0.78 mg/100 g subsequently after 24, 36, and 48 hours of the fermentation process (Adugna et al., 2024).

3.4.7 Other Nonthermal Processing Techniques

These methods involve cold plasma, ultrasound treatment, high-pressure processing, and irradiation. Cold plasma encompasses the production of chemical intermediates, including charged molecules and free radicals, which engage in the functional groups of the antinutritional compounds, thereby facilitating the degradation of these compounds and subsequently diminishing their bioavailability (Jaddu et al., 2024). In the ultrasound technique, the shear forces generated by cavitation provoke the mechanical fragmentation of ANFs. Additionally, cavitation promotes the release and separation of phytochemicals; the localized thermal and pressure variations disrupt the cellular structure, resulting in the breakdown of these compounds, whereas ultrasound enhances the breakdown of the antinutritional compounds by enzymes and increases the bioaccessibility of nutrients (Meena et al., 2024). Ionizing radiation treatment induces the formation of free radicals and reactive species capable of degrading bioactive compounds, thereby diminishing their concentrations. The structural modifications resulting from irradiation lead to a decrease in the chelation ability of antinutrients with minerals and proteins, consequently improving their bioavailability (Nasab et al., 2024). When sorghum millet was treated with irradiation (IR), the IR treatment resulted in an optimal phytic acid reduction of 91% within a duration of 120 seconds, whereas a 90% decrease was attained with a gamma radiation dose of 30 kGy. These treatment modalities disrupt the phytate ring or facilitate the decomposition of phytic acid into inositol and their phosphates, attributed to the activity of free radicals, thereby efficiently diminishing phytate concentrations (Rousta et al., 2014). Charu et al. (2024) investigated the impact of cold plasma on two millets (Barnyard and Pearl) by treating the grains at 2 kV. The treatment resulted in a notable reduction in levels of antinutrients, i.e., phytic acid and tannins. In pearl millet, the tannin content experienced an 18% decrease, while the phytate content underwent a reduction of 57%. In barnyard millet, tannin content was reduced by 24%, while phytic acid levels decreased by 61%. Cold plasma treatment reduces phytates by breaking down the phytic acid into other compounds, a process driven by radicals produced throughout the treatment. This breakdown may lead to the formation of insoluble complexes, thereby enhancing nutrient bioavailability (Bheemaiah Balyatanda et al., 2024).

3.5 Conclusions

Millets, known for their nutritional value and environmental adaptability, are increasingly seen as a sustainable answer to food insecurity and malnutrition. However, their widespread consumption is often hindered by ANFs such as phytic acid, tannins, oxalates, saponins, lectins, and protease inhibitors, which impair nutrient absorption and digestion. These natural plant defenses intervene with the uptake of essential minerals like Fe, Zn, Ca, and Mg, while also reducing protein digestibility, presenting challenges for populations who rely on millet-based diets. To counter these effects, various processing methods have been developed, including traditional techniques like soaking, germination, and fermentation, as well as more advanced methods such as enzymatic treatments and extrusion cooking. Soaking is a straightforward and effective approach that removes water-soluble ANFs like tannins and oxalates, while also activating phytase, an enzyme that breaks down phytates. Germination boosts enzymatic activity, reducing phytates and polyphenols while enhancing certain vitamins and improving both nutrition and sensory attributes. Fermentation, commonly used in millet-based porridges, utilizes microbial activity to decrease phytates and tannins, improving mineral bioavailability and providing probiotic benefits. Heat treatments, such as boiling, steaming, and roasting, are commonly used to deactivate heat-sensitive ANFs like protease inhibitors, though care must be taken to avoid nutrient loss from overheating. Advanced methods like enzymatic treatments specifically target ANFs like phytates, while extrusion cooking—a high-temperature, short-duration process—effectively reduces tannins and protease inhibitors, preserving nutrients and making it suitable for large-scale processing. Despite these advancements, challenges remain in balancing the reduction of harmful ANFs with the retention of beneficial bioactive compounds like polyphenols, which possess antioxidant and anti-inflammatory effects. Additionally, traditional methods are time-consuming, while modern techniques require significant investment and infrastructure, limiting their accessibility in resource-limited settings. To overcome these challenges, a combination of traditional and modern techniques, along with consumer education and the development of biofortified millet varieties with lower ANF content, is recommended. These efforts are crucial for unlocking the full nutritional benefits of millets, ensuring their role in sustainable, healthy diets. With optimized processing and ongoing research, millets can significantly contribute to combating malnutrition, enhancing food security, and promoting global health.

References

Abdelrahman, S. M., El-Maki, H. B., Idris, W. H., Babiker, E. E., & El-Tinay, A. H. (2005). Antinutritional factors content and minerals availability of pearl millet (Pennisetum glaucum) as influenced by domestic processing methods and cultivar. *Journal of Food Technology*, *3*(3), 397–403.

Abioye, V. F., Adejuyitan, J. A., Adeoye, A. O., & Gbadegesin, I. O. (2021). Effect of natural fermentation period on nutritional, anti nutritional, I total phenols, flavonoids and antioxidant contents of finger millet flour. *European Journal of Nutrition & Food Safety*, *13*(3), 74–82.

Adebiyi, J. A., Obadina, A. O., Mulaba-Bafubiandi, A. F., Adebo, O. A., & Kayitesi, E. (2016). Effect of fermentation and malting on the microstructure and selected physicochemical properties of pearl millet (*Pennisetum glaucum*) flour and biscuit. *Journal of Cereal Science*, *70*, 132–139.

Adebo, J. A., Njobeh, P. B., Gbashi, S., Oyedeji, A. B., Ogundele, O. M., Oyeyinka, S. A., & Adebo, O. A. (2022). Fermentation of cereals and legumes: Impact on nutritional constituents and nutrient bioavailability. *Fermentation*, *8*(2), 63.

Adeyemo, S. M., & Onilude, A. A. (2013). Enzymatic reduction of anti-nutritional factors in fermenting soybeans by *Lactobacillus plantarum* isolates from fermenting cereals. *Nigerian Food Journal*, *31*(2), 84–90.

Adugna, A. A., Tolesa, G. N., Abera, S., & Deme, G. D. (2024). Effect of grain fermentation and malting time on nutrient and anti-nutrient composition of biscuits from Aksum finger millet (*Eleusine coracana*). *Cogent Food & Agriculture*, *10*(1), 2336690.

Alarcón, F. J., García-Carreño, F. L., & Navarrete del Toro, M. A. (2001). Effect of plant protease inhibitors on digestive proteases in two fish species, *Lutjanus argentiventris* and *L. novemfasciatus*. *Fish Physiology and Biochemistry*, *24*(3), 179–189.

Amalraj, A., & Pius, A. (2015). Bioavailability of calcium and its absorption inhibitors in raw and cooked green leafy vegetables commonly consumed in India–an in vitro study. *Food Chemistry*, *170*, 430–436.

Anitha, S., Govindaraj, M., & Kane-Potaka, J. (2020). Balanced amino acid and higher micronutrients in millets complements legumes for improved human dietary nutrition. *Cereal Chemistry*, *97*(1), 74–84.

Babiker, E., Abdelseed, B., Hassan, H., & Adiamo, O. (2018). Effect of decortication methods on the chemical composition, antinutrients, Ca, P and Fe contents of two pearl millet cultivars during storage. *World Journal of Science, Technology and Sustainable Development*, *15*(3), 278–286.

Balli, D., Bellumori, M., Pucci, L., Gabriele, M., Longo, V., Paoli, P., Melani, F., Mulinacci, N., & Innocenti, M. (2020). Does fermentation really increase the phenolic content in cereals? A study on millet. *Foods*, *9*(3), 303.

Bheemaiah Balyatanda, S., Gowda, N. N., Subbiah, J., Chakraborty, S., Prasad, P. V., & Siliveru, K. (2024). Physiochemical, bio, thermal, and non-thermal processing of major and minor millets: A comprehensive review on antinutritional and antioxidant properties. *Food*, *13*(22), 3684.

Chakraborty, N., Ghosh, R., Ghosh, S., Narula, K., Tayal, R., Datta, A., & Chakraborty, S. (2013). Reduction of oxalate levels in tomato fruit and consequent metabolic remodeling following overexpression of a fungal oxalate decarboxylase. *Plant Physiology*, *162*(1), 364–378.

Charu, C., Vignesh, S., Chidanand, D. V., Mahendran, R., & Baskaran, N. (2024). Impact of cold plasma on pearl and barnyard millets' microbial quality, antioxidant status, and nutritional composition. *Food and Humanity*, *2*, 100238.

Chung, K. T., Wong, T. Y., Wei, C. I., Huang, Y. W., & Lin, Y. (1998). Tannins and human health: A review. *Critical Reviews in Food Science and Nutrition*, *38*(6), 421–464.

Clemente, M., Corigliano, M. G., Pariani, S. A., Sánchez-López, E. F., Sander, V. A., & Ramos-Duarte, V. A. (2019). Plant serine protease inhibitors: Biotechnology application in agriculture and molecular farming. *International Journal of Molecular Sciences*, *20*(6), 1345.

Coulibaly, A., Kouakou, B., & Chen Jie, C. J. (2011). Phytic acid in cereal grains: Structure, healthy or harmful ways to reduce phytic acid in cereal grains and their effects on nutritional quality. *American Journal of Plant Nutrition and Fertilization Technology*, *1*, 1–22.

D'Imperio, M., Renna, M., Cardinali, A., Buttaro, D., Serio, F., & Santamaria, P. (2016). Calcium biofortification and bioaccessibility in soilless "baby leaf" vegetable production. *Food Chemistry*, *213*, 149–156.

Dayakar Rao, B., Vishala, A. D., Arlene Christina, G. D., & Tonapi, V. A. (2016). *Technologies of millet value added products* (pp. 48). Centre of Excellence on Sorghum, ICAR-IIMR.

Devi, P. B., Vijayabharathi, R., Sathyabama, S., Malleshi, N. G., & Priyadarisini, V. B. (2014). Health benefits of finger millet (*Eleusine coracana* L.) polyphenols and dietary fiber: A review. *Journal of Food Science and Technology*, *51*, 1021–1040.

Gemede, H. F., & Ratta, N. (2014). Antinutritional factors in plant foods: Potential health benefits and adverse effects. *International Jjournal of Nutrition and Food Sciences*, *3*(4), 284–289.

Goudar, G., Manne, M., Sathisha, G. J., Sharma, P., Mokalla, T. R., Kumar, S. B., & Ziouzenkova, O. (2023). Phenolic, nutritional and molecular interaction study among different millet varieties. *Food Chemistry Advances*, *2*, 100150.

Gull, A., Prasad, K., & Kumar, P. (2016). Evaluation of functional, antinutritional, pasting and microstructural properties of Millet flours. *Journal of Food Measurement and Characterization*, *10*, 96–102.

Gupta, R. K., Gangoliya, S. S., & Singh, N. K. (2015). Reduction of phytic acid and enhancement of bioavailable micronutrients in food grains. *Journal of Food Science and Technology*, *52*, 676–684.

Gwekwe, B. N., Chopera, P., Matsungo, T. M., Chidewe, C., Mukanganyama, S., Nyakudya, E., Mtambanengwe, F., Mapfumo, P., & Nyanga, L. K. (2024). Effect of dehulling, fermentation, and roasting on the nutrient and anti-nutrient content of sorghum and pearl millet flour. *International Journal on Food, Agriculture and Natural Resources*, *5*(1), 1–7.

Habib, H., & Fazili, K. M. (2007). Plant protease inhibitors: A defense strategy in plants. *Biotechnology and Molecular Biology Reviews*, *2*(3), 68–85.

Handa, V., Kumar, V., Panghal, A., Suri, S., & Kaur, J. (2017). Effect of soaking and germination on physicochemical and functional attributes of horsegram flour. *Journal of Food Science and Technology*, *54*, 4229–4239.

Hithamani, G., & Srinivasan, K. (2014). Effect of domestic processing on the polyphenol content and bioaccessibility in finger millet (*Eleusine coracana*) and pearl millet (*Pennisetum glaucum*). *Food Chemistry*, *164*, 55–62.

Jaddu, S., Sahoo, S., Sonkar, S., Alzahrani, K., Dwivedi, M., Misra, N. N., & Pradhan, R. C. (2024). Cold plasma treatment of little millet flour: Impact on bioactives, antinutritional factors and functional properties. *Plant Foods for Human Nutrition*, *79*, 503–510.

Jaiswal, H., Singh, O. J., Chauhan, A., Sahu, M. K., & Dv, S. P. (2018). A review on tannins. *European Journal of Biotechnology and Bioscience*, *6*(3), 16–17.

Jan, S., Kumar, K., Ahmed, N., Thakur, P., Chauhan, D., Rizvi, Q. E. H., & Vyas, P. (2022a). Beneficial effect of diverse fermentation treatments on nutritional composition, bioactive components, and anti-nutritional factors of foxtail millet (*Setaria italica* L.). *Journal of Postharvest Technology*, *10*(2), 35–47.

Jan, S., Kumar, K., Yadav, A. N., Ahmed, N., Thakur, P., Chauhan, D., & Dhaliwal, H. S. (2022b). Effect of diverse fermentation treatments on nutritional composition, bioactive components, and anti-nutritional factors of finger millet (*Eleusine coracana* L.). *Journal of Applied Biology and Biotechnology*, *10*(1), 46–52.

Kataria, A., Sharma, S., & Dar, B. N. (2022). Changes in phenolic compounds, antioxidant potential and antinutritional factors of teff (*Eragrostis tef*) during different thermal processing methods. *International Journal of Food Science and Technology*, *57*(11), 6893–6902.

Kaur, T., Panesar, P. S., & Riar, C. S. (2023). Combined effect of hydrothermal treatments and drying temperature on the physicochemical, milling, and nutritional characteristics of foxtail millet. *Journal of Cereal Science*, *114*, 103788.

Khanbabaee, K., & Van Ree, T. (2001). Tannins: Classification and definition. *Natural Product Reports*, *18*(6), 641–649.

Kheya, S. A., Talukder, S. K., Datta, P., Yeasmin, S., Rashid, M. H., Hasan, A. K., Anwar, M. P., Islam, A. A., & Islam, A. M. (2023). Millets: The future crops for the tropics-status, challenges and future prospects. *Heliyon*, *9*, e22123.

Krishnan, R., Dharmaraj, U., Manohar, R. S., & Malleshi, N. G. (2011). Quality characteristics of biscuits prepared from finger millet seed coat based composite flour. *Food Chemistry*, *129*(2), 499–506.

Krishnan, R., Sakhare, S. D., & Meera, M. S. (2020). Effect of germination-hydrothermal treatments followed by fractionation on inhibitory factors, β-carotene and iron-zinc bioaccessibility in pearl millet fractions. *Journal of Cereal Science*, *93*, 102930.

Kumar, A., Tomer, V., Kaur, A., Kumar, V., & Gupta, K. (2018). Millets: A solution to agrarian and nutritional challenges. *Agriculture and Food Security*, *7*(1), 1–15.

Kumari, S., Krishnan, V., & Sachdev, A. (2015). Impact of soaking and germination durations on antioxidants and anti-nutrients of black and yellow soybean (*Glycine max* L.) varieties. *Journal of Plant Biochemistry and Biotechnology*, *24*, 355–358.

Kushwaha, A., Singh, A., Sirohi, R., & Tarafdar, A. (2019). Effect of hydrothermal treatment and milling parameters on milling and nutritional qualities of finger millet (*Eleusine coracana*). *Journal of Agricultural Engineering*, *55*(4), 34–46.

Liu, K., Zheng, J., Wang, X., & Chen, F. (2019). Effects of household cooking processes on mineral, vitamin B, and phytic acid contents and mineral bio-accessibility in rice. *Food Chemistry*, *280*, 59–64.

Mane, R. P., Kshirsagar, R. B., Agarkar, B. S., Patil, B. M., & Katke, S. D. (2023). Effects of germination on antinutritional, colour, functional and physicochemical properties of millets. *Pharma Innovations*, *12*(1), 350–355.

Manzoor, M., Singh, D., Aseri, G. K., Sohal, J. S., Vij, S., & Sharma, D. (2021). Role of lacto-fermentation in reduction of antinutrients in plant-based foods. *Journal of Applied Biology and Biotechnology*, *9*, 7–16.

Martinez, K. B., Mackert, J. D., & McIntosh, M. K. (2017). Polyphenols and intestinal health. In R. R. Watson (Ed.), *Nutrition and functional foods for healthy aging* (pp. 191–210). Academic Press.

Mather, M. (1997). Migraines and tannins—Any relationship? *Headache*, *37*(8), 529.

Meena, L., Gowda, N. N., Sunil, C. K., Rawson, A., & Janghu, S. (2024). Effect of ultrasonication on food bioactive compounds and their bio-accessibility: A review. *Journal of Food Composition and Analysis*, *126*, 105899.

Mehrabadi, M., Franco, O. L., & Bandani, A. R. (2012). Plant proteinaceous alpha-amylase and proteinase inhibitors and their use in insect pest control. In A. R. Bandani (Ed.), *New perspective in plant protection* (pp. 229–246). IntechOpen.

Mohapatra, D., Patel, A. S., Kar, A., Deshpande, S. S., & Tripathi, M. K. (2019). Effect of different processing conditions on proximate composition, anti-oxidants, anti-nutrients and amino acid profile of grain sorghum. *Food Chemistry*, *271*, 129–135.

Nadeem, M., Anjum, F. M., Amir, R. M., Khan, M. R., Hussain, S., & Javed, M. S. (2010). An overview of anti-nutritional factors in cereal grains with special reference to wheat—A review. *Pakistan Journal of Food Sciences*, *20*(1–4), 54–61.

Nakata, P. A., & McConn, M. M. (2000). Isolation of Medicago truncatula mutants defective in calcium oxalate crystal formation. *Plant Physiology*, *124*(3), 1097–1104.

Nasab, S. S., Tahmouzi, S., Feizollahi, E., & Mollakhalili-Meybodi, N. (2024). Impacts of novel non-thermal processing (NTP) on anti-nutritional compounds of food grains and seeds. *Food Control*, *162*, 110469.

Nikmaram, N., Leong, S. Y., Koubaa, M., Zhu, Z., Barba, F. J., Greiner, R., Oey, I., & Roohinejad, S. (2017). Effect of extrusion on the anti-nutritional factors of food products: An Overview. *Food Control*, *79*, 62–73.

Nithya, K. S., Ramachandramurty, B., & Krishnamoorthy, V. V. (2007). Effect of processing methods on nutritional and anti-nutritional qualities of hybrid (COHCU-8) and traditional (CO7) pearl millet varieties of India. *Journal of Biological Sciences*, *7*(4), 643–647.

Nour, M. A. A., Ahmed, M. I. A., Babiker, E. E., Ibrahim, M. A. E., & Ahmed, M. W. A. (2015). Effect of supplementation and cooking on in vitro protein digestibility and antinutrients of pearl millet flour. *American Journal of Food Science and Health*, *1*(3), 69–75.

Ojo, M. A., Ade-Omowaye, B. I. O., & Ngoddy, P. O. (2018). Processing effects of soaking and hydrothermal methods on the components and in vitro protein digestibility of *Canavalia ensiformis*. *International Food Research Journal*, *25*, 720–729.

Onyango, C., Luvitaa, S. K., Unbehend, G., & Haase, N. (2020). Physico-chemical properties of flour, dough and bread from wheat and hydrothermally treated finger millet. *Journal of Cereal Science*, *93*, 102954.

Osman, M. A. (2011). Effect of traditional fermentation process on the nutrient and antinutrient contents of pearl millet during preparation of Lohoh. *Journal of the Saudi Society of Agricultural Sciences*, *10*(1), 1–6.

Paiva, P. M., Pontual, E. V., Coelho, L. C. B. B., & Napoleão, T. H. (2013). Protease inhibitors from plants: Biotechnological insights with emphasis on their effects on microbial pathogens. In A. Méndez-Vilas (Ed.), *Microbial pathogens and strategies for combating them: Science, technology and education*, Edition: Microbiology Book Series (pp. 641–649). Formatex Research Centre.

Popova, A., & Mihaylova, D. (2019). Antinutrients in plant-based foods: A review. *The Open Biotechnology Journal*, *13*(1), 68–76.

Prashanth, P., Joshi, T. J., Singh, S. M., & Rao, P. S. (2024). *Hydrothermal treatment of pearl millet grains: Effects on nutritional composition, antinutrients and flour properties* (Vol. 7, pp. 87–95). Grain & Oil Science and Technology.

Pushparaj, F. S., & Urooj, A. (2011). Influence of processing on dietary fiber, tannin and in vitro protein digestibility of pearl millet. *Food and Nutrition Sciences*, *2*(8), 895–900.

Pushparaj, F. S., & Urooj, A. (2014). Antioxidant activity in two pearl millet (*Pennisetum typhoideum*) cultivars as influenced by processing. *Antioxidants*, *3*(1), 55–66.

Rani, S., Singh, R., Sehrawat, R., Kaur, B. P., & Upadhyay, A. (2018). Pearl millet processing: A review. *Nutrition and Food Science*, *48*(1), 30–44.

Rao, B. S. N., & Prabhavathi, T. (1982). Tannin content of foods commonly consumed in India and its influence on ionisable iron. *Journal of the Science of Food and Agriculture*, *33*(1), 89–96.

Reddy, N. R., & Pierson, M. D. (1994). Reduction in antinutritional and toxic components in plant foods by fermentation. *Food Research International, 27*(3), 281–290.

Ricardo-da-Silva, J. M., Cheynier, V., Souquet, J. M., Moutounet, M., Cabanis, J. C., & Bourzeix, M. (1991). Interaction of grape seed procyanidins with various proteins in relation to wine fining. *Journal of the Science of Food and Agriculture, 57*(1), 111–125.

Rousta, M., Sadeghi, A. A., Shawrang, P., Afshar, M. A., & Chamani, M. (2014). Effect of gamma, electron beam and infrared radiation treatment on the nutritional value and anti-nutritional factors of sorghum grain. *Iranian Journal of Applied Animal Science, 4*(4).

Sade, F. O. (2009). Proximate, antinutritional factors and functional properties of processed pearl millet (*Pennisetum glaucum*). *Journal of Food Technology, 7*(3), 92–97.

Samtiya, M., Aluko, R. E., & Dhewa, T. (2020). Plant food anti-nutritional factors and their reduction strategies: An overview. *Food Production, Processing, and Nutrition, 2*, 1–14.

Samtiya, M., Aluko, R. E., Puniya, A. K., & Dhewa, T. (2021). Enhancing micronutrients bioavailability through fermentation of plant-based foods: A concise review. *Fermentation, 7*(2), 63.

Sene, S., Talla Gueye, M., Sarr, F., Diallo, Y., Salif Sow, M., Faye, A., Min, D., & Gaye, M. L. (2018). Parboiling effect on phytate contains of two Senegalese pearl millet varieties (*Pennisetum glaucum* [L.] R. br.) GB 87-35 and ICRITABI. *Journal of Nutrition, Health and Food Science, 6*, 1–6.

Sharma, K., Kumar, V., Kaur, J., Tanwar, B., Goyal, A., Sharma, R., Gat, Y., & Kumar, A. (2021). Health effects, sources, utilization and safety of tannins: A critical review. *Toxin Reviews, 40*(4), 432–444.

Sheikh, M. A., Saini, C. S., & Sharma, H. K. (2021). Analyzing the effects of hydrothermal treatment on anti-nutritional factor content of plum kernel grits by using response surface methodology. *Applied Food Research, 1*(1), 100010.

Shigihalli, S., Ravindra, U., & Ravishankar, P. (2018). Effect of processing methods on phytic acid content in selected white finger millet varieties. *International Journal of Current Microbiology and Applied Sciences, 7*(2), 1829–1835.

Shinde, E. M., Pawar, V. S., Mundhe, S. S., & Khapre, A. P. (2023). Effect of processing treatment on chemical and nutritional properties of millets. *The Pharma Innovation Journal, 12*, 135–138.

Sihag, M. K., Sharma, V., Goyal, A., Arora, S., & Singh, A. K. (2015). Effect of domestic processing treatments on iron, β-carotene, phytic acid and polyphenols of pearl millet. *Cogent Food & Agriculture, 1*(1), 1109171.

Singh, A. P., & Kumar, S. (2020). Applications of tannins in industry. In A. Aires (Ed.), *Tannins-structural properties, biological properties and current knowledge* (pp. 1–13). IntechOpen. https://doi.org/10.5772/intechopen.85984

Singh, A., Gupta, S., Kaur, R., & Gupta, H. R. (2017). Process optimization for anti-nutrient minimization of millets. *Asian Journal of Dairy and Food Research, 36*(4), 322–326.

Soetan, K. O., & Oyewole, O. E. (2009). The need for adequate processing to reduce the anti-nutritional factors in plants used as human foods and animal feeds: A review. *African Journal of Food Science, 3*(9), 223–232.

Sokrab, A. M., Ahmed, I. A. M., & Babiker, E. E. (2012). Effect of germination on antinutritional factors, total, and extractable minerals of high and low phytate corn (*Zea mays* L.) genotypes. *Journal of the Saudi Society of Agricultural Sciences, 11*(2), 123–128.

Sowbaghya, A. Y., Shanthi, M., Murugan, M., Kokiladevi, E., PuShPAM, A. K., & Chinniah, C. (2019). Proteins and trypsin inhibitors in seeds of various plants. *Indian Journal of Entomology*, *81*(1), 177–181.

Sweta Kumari, S. K., Veda Krishnan, V. K., & Archana Sachdev, A. S. (2015). Impact of soaking and germination durations on antioxidants and anti-nutrients of black and yellow soybean (*Glycine max* L.) varieties. *Journal of Plant Biochemistry and Biotechnology*, *24*(3), 355–358.

Usman, M. A. (2018). Influence of germination time, during malting, on selected anti-nutritional factors of some sorghum [*Sorghum bicolor* (L.) Moench] types. *International Journal of Agriculture Research and Food Production*, *3*, 2536–2734.

Weaver, C. M., Martin, B. R., Ebner, J. S., & Krueger, C. A. (1987). Oxalic acid decreases calcium absorption in rats. *The Journal of Nutrition*, *117*(11), 1903–1906.

Wisessing, A., & Choowongkomon, K. (2012). Amylase inhibitors in plants: Structures, functions and applications. *Functional Plant Science and Biotechnology*, *6*, 31–41.

Yang, J., Punshon, T., Guerinot, M. L., & Hirschi, K. D. (2012). Plant calcium content: Ready to remodel. *Nutrients*, *4*(8), 1120–1136.

Zhang, H., Qiao, Q., Zhao, Y., Zhang, L., Shi, J., Wang, N., Li, Z., & Shan, S. (2024). Expression and purification of recombinant Bowman–Birk trypsin inhibitor from foxtail millet bran and its anticolorectal cancer effect in vitro and in vivo. *Journal of Agricultural and Food Chemistry*, *72*(18), 10439–10450.

Zhang, L., Li, J., Han, F., Ding, Z., & Fan, L. (2017). Effects of different processing methods on the antioxidant activity of 6 cultivars of foxtail millet. *Journal of Food Quality*, *2017*, 1–9.

Section II

Millet Bioactive Compounds and Characterization

4

Millet Grain Phenolics: Classification, Chemical Profiling

Gurjeet Kaur[1], Kanchan Suri[2], Anu Sharma[3], and Zakir Showkat Khan[4]

[1] *Department of Processing and Food Engineering, Punjab Agricultural University, Ludhiana, India*
[2] *PG Department of Agriculture, Khalsa College, Amritsar, India*
[3] *Department of Food Science and Technology, Khalsa College, Amritsar, India*
[4] *Department of Food Technology, Islamic University of Science and Technology, Awantipora, India*

4.1 Introduction

Millets encompass a diverse group of small-seeded grasses that provide a staple food source for millions of people, particularly in arid and semi-arid regions worldwide, and are broadly categorized into major and minor millets. Major millets, such as finger millet (*Eleusine coracana*), pearl millet (*Pennisetum glaucum*), and sorghum (*Sorghum bicolor*), are extensively cultivated and consumed on a larger scale. Millets are staple foods across many regions of Asia and Africa, prized for their rich nutritional value, resilience to drought, and ability to flourish in arid environments. While minor millets—such as foxtail millet (*Setaria italica*), little millet or kutki (*Panicum sumatrense*), barnyard millet (*Echinochloa* spp.), and kodo millet (*Paspalum scrobiculatum*)—are grown on a smaller scale but continue to play a vital role in the diets of specific communities (Khan et al., 2025). Millets are an abundant source of bioactive compounds, especially phenolic compounds, with potential human health-promoting attributes. These phenolic compounds are a class of secondary metabolites found in plants and are further divided into phenolic acids and polyphenols (Giada, 2013), as presented in Table 4.1. Furthermore, phenolic acids are divided into two classes: hydroxybenzoic and hydroxycinnamic acids (Dykes & Rooney, 2006). The main dietary phenolic compounds include phenolic acids, flavonoids, and tannins. Phenolics in the insoluble forms are covalently bound to cell wall structural components such as cellulose, hemicellulose, lignin, and rod-shaped structural proteins (Acosta-Estrada et al., 2014).

Nutraceutical Potential of Millet-Based Food Products: Chemistry and Applications, First Edition.
Edited by Parmjit S. Panesar, Charanjit S. Riar, and Sushil Dhital.

Published 2026 by John Wiley & Sons Ltd.

Table 4.1 Classification of phenolic compounds in millets

Class	Representative compounds	Structural features	Biological roles	Key studies
Phenolic acids	Protocatechuic, *p*-hydroxybenzoic, gallic, ferulic, and caffeic	C_6–C_1 (benzoic), C_6–C_3 (cinnamic); often hydroxylated and methylated	Antioxidant, antimicrobial, and anti-inflammatory	Srinivasan et al. (2020) Yuan et al. (2022) Goudar et al. (2023) Nguyen et al. (2023)
Flavonoids	Catechin, epicatechin, kaempferol glycoside, quercetin, and apigenin	C_6–C_3–C_6 backbone; subclasses: flavanols, flavonols, and flavones	Antioxidant, antidiabetic, and neuroprotective	Xiang et al. (2019) Tang et al. (2022) Abioye et al. (2022)
Proanthocyanidins (condensed tannins)	Procyanidin B1 and B2	Oligomeric flavan-3-ol units	Antioxidant and cardiovascular protection	Xiang et al. (2019) Hassan et al. (2020)
Tannins	Condensed tannins (polymeric flavanol derivatives)	High-molecular-weight polymers of flavanols	Antimicrobial, antioxidant	Ahmad et al. (2018) Dey et al. (2022)
Lignans	Secoisolariciresinol	Dibenzylbutane skeleton—C_6–C_3–C_3–C_6	Antioxidant, estrogenic effects	Ambati and Sucharitha (2019) Kalinová et al. (2022)
Stilbenoids	Resveratrol	C_6–C_2–C_6 structure (stilbene backbone)	Antioxidant, anti-inflammatory	Mir et al. (2024) Wang et al. (2025)
Simple phenols and benzoquinones	Catechol, hydroquinone, dimethoxybenzoquinone	Monophenolic/benzoquinone structures	Antioxidant and defense mechanisms	Ali et al. (2022) Varshini et al. (2024)
Phenolic aldehydes	Vanillin	Benzaldehyde derivatives from lignin or caffeic precursors	Antioxidant and flavor attribute	Li et al. (2021) Zhou et al. (2023)

4.2 Major Millet Grain Phenolics

4.2.1 Finger Millet Phenolics

Finger millet (*Eleusine coracana*) has gained considerable attention due to its high nutritional value, encompassing macro- and micronutrients, as well as its rich profile of bioactive phytochemicals, particularly phenolic compounds. These constituents have been associated with potential protective effects against various chronic diseases, including diabetes, cancer, and cardiovascular disorders (Xiang et al., 2019). Various analytical tools used for phenolic profiling are listed in Table 4.2. Ofosu et al. (2020) evaluated phenolic compounds in ethanolic extracts of Italian finger millet by using ultra performance liquid chromatography coupled with time of flight and mass spectrometry (UPLC-Q-TOF-MS) and identified eight phenolic compounds including (+)-catechin 3-O-glucose, catechin, dihydromyricetin 3-O-rhamnoside, (−)-epicatechin, dihydroquercetin, formononetin, kaempferol, and apigenin. Hithamani and Srinivasan (2014) evaluated the effect of domestic processing of native finger millet on the phenolic compounds by using high-performance liquid chromatography (HPLC) and reported the presence of 22 compounds. Among these 22 compounds, 11 compounds were unidentified, while the other known compounds were gallic acid (10.69 μg/g), protocatechuic acid (14.60 μg/g), *p*-hydroxybenzoic acid (3.62 μg/g), caffeic acid (10.40 μg/g), gentisic acid (4.54 μg/g), *p*-coumaric acid (19.07 μg/g), syringic

Table 4.2 Analytical techniques for profiling phenolics in millets

Technique	Principle	Analytes detected	Advantages	Limitations
HPLC-DAD (± UV–Vis)	Chromatographic separation with diode-array UV–Vis detection	Individual phenolic acids and flavonoids	High sensitivity; quantifies known compounds accurately	Requires standards; limited structural info
HPLC-DAD-ESI-QTOF-MS/MS	HPLC coupled with high-resolution MS via electrospray ionization	Free/bound phenolics and flavonoid derivatives	High-resolution identification; supports untargeted analysis	High cost; complex data analysis
LC–MS/MS (triple quadrupole/ion-trap)	Mass spectrometry tandem analysis following LC	Broad phenolic spectrum	High selectivity/sensitivity for targeted compounds	Requires reference standards
GC–MS	Volatilized compounds separated by gas chromatography before MS detection	Volatile phenolics and some derivatized acids	Good for volatiles; high reproducibility	Requires derivatization; limited to volatile forms
UV–Vis spectroscopy (Folin–Ciocalteu, DPPH)	Colorimetric assays measuring absorbance changes	Total phenolic/flavonoid content; antioxidant activity	Simple; rapid	Nonspecific; only total content measured

acid (10.17 µg/g), ferulic acid (12.88 µg/g), sinapic acid (24.78 µg/g), salicylic acid (413.0 µg/g), and *trans*-cinnamic acid (31.87 µg/g). While, vanillic acid (126.9 µg/g) was found only in roasted finger millet. Hassan et al. (2020) evaluated finger millet varieties by liquid chromatography (LCMS-QTOF) and reported that catechin was identified as the predominant flavonoid, with epicatechin being the next most abundant. The content of catechin varied from 610.4 to 675.1 mg/kg, while epicatechin content varied from 99.1 to 139.5 mg/kg. Moreover, the content of phenolic compounds such as protocatechuic acid and *p*-hydroxybenzoic acid varied from 20.9 to 23.7 mg/kg and 13.5 to 16.8 mg/kg, respectively. The flavonoids including procyanidin B1 and procyanidin B2 were present. Goudar et al. (2023) evaluated finger millet by HPLC and reported that it contains free phenolic acids such as gallic acid (11.93 µg/g), ferulic acid (57.04 µg/g), caffeic acid (38.91 µg/g), syringic acid (30.05 µg/g), and chlorogenic acid (2.29 µg/g). While the bound phenolic acids were ferulic acid (362.12 µg/g), gallic acid (56.12 µg/g), caffeic acid (31.86 µg/g), and chlorogenic acid (5.41 µg/g).

Xiang et al. (2019) evaluated various finger millet varieties by using HPLC and revealed that catechin and epicatechin were the principal polyphenols detected in their free forms. While, in their bound forms, hydroxybenzoic and hydroxycinnamic acid derivatives, with ferulic acid, were most abundant phenolic compounds. In comparison to flavan-3-ols, the concentrations of protocatechuic acid and protocatechuic aldehyde in their free forms were relatively low, ranging from 0.19 to 0.97 mg/100 g and 0.71 to 1.11 mg/100 g, respectively. Caffeic acid and *p*-coumaric acid, two common hydroxycinnamic acid derivatives in cereals (Van Hung, 2016), were detected exclusively in bound forms in finger millet, with concentrations ranging from 0.83 to 2.43 mg/100 g and 1.88 to 3.94 mg/100 g, respectively. As anticipated, ferulic acid was the most abundant polyphenol in the bound fraction, with levels ranging from 59.58 to 67.36 mg/100 g and 35.19 to 40.12 mg/100 g dry weight across millet varieties. Syringic acid, a hydroxybenzoic acid derivative, was also present in bound form but at low concentrations. In Makowe millet samples, syringic acid was detected only in trace amounts, falling below the detection limit.

4.2.2 Pearl Millet Phenolics

Pearl millet (*Cenchrus americanus*) is a vital staple crop cultivated in the semi-arid regions of Asia and Africa, valued for its high nutritional content and exceptional adaptability to diverse abiotic stress conditions (Pattanashetti et al., 2016). Hassan et al. (2020) evaluated pearl millet varieties by LCMS-QTOF and reported that catechin content was 2.50–12.6 mg/kg and epicatechin content was 1.2–1.8 mg/kg. Pearl millet showed the presence of flavonoid like kaempferol glucoside, and its content varied from 196.0 to 213.6 mg/kg. Conversely, certain phenolic compounds, including benzoic acid derivatives such as protocatechuic acid and *p*-hydroxybenzoic acid, as well as flavonoids like procyanidin B1 and procyanidin B2, were not detected. Bouajila et al. (2020) evaluated various pearl millet cultivars and reported 11 phenolic compounds were present and their contents were gallic acid (9.40–13.60 µg/g), ferulic acid (23.60–30.20 µg/g), *p*-coumaric acid (0.60–1.30 µg/g), protocatechuic acid (127.60–452.40 µg/g), hydroxybenzoic acid (253.50–437.60 µg/g), quercitin (0.90–2.40 µg/g), catechin (1.20–23.20 µg/g), *trans*-cinnamic acid (486.30–731.20 µg/g), gentisic acid (19.70–35.40 µg/g), and caffeic acid (1.30–1.70 µg/g), while vanillic acid content was 48.30 μg/g. However, five phenolic compounds such as gallic acid, *trans*-cinnamic acid, protocatechuic acid, quercitin, and catechin were present in all cultivars, while other phenolic compounds like gentisic acid, caffeic acid, vanillic acid, ferulic acid, *p*-coumaric acid, and vanillic acid were present only in few cultivars. Triki et al. (2022) studied the variation in phenolic compounds of southern Tunisian pearl millet germplasm by analyzing methanolic extracts of pearl millet grains and documented the variation of eight phenolic compounds as quinic acid (461.65–1,202.88 μg/g), protocatechuic acid (2.37–3.33 μg/g), caffeic acid (3.87–14.30 μg/g), syringic acid (1.20–2.66 μg/g), 1,3 di-O-caffeoylquinic acid (0.62–0.96 μg/g), *trans*-ferulic acid (7.76–13.91 μg/g), salviolinic acid (1.08–1.87 μg/g), and *p*-coumaric acid (5.28–51.44 μg/g).

Further, Mawouma et al. (2022) evaluated the phytochemical compounds of two pearl millet cultivars by using HPLC and identified gallic acid (165.59–185.79 μg/mL), catechin (12.65–75.73 μg/mL), chlorogenic acid (19.06–34.61 μg/mL), vanillic acid (20.27 μg/mL), and caffeic acid (4.32 μg/mL) as major phenolic compounds. Goudar et al. (2023) assessed pearl millet by HPLC and reported that it contains free phenolic acids such as ferulic acid (47.03 μg/g), caffeic acid (43.88 μg/g), gallic acid (3.87 μg/g), syringic acid (2.44 μg/g), and chlorogenic acid (1.30 μg/g). While, bound phenolic acids were ferulic acid (988.78 μg/g), caffeic acid (45.12 μg/g), syringic acid (10.33 μg/g), and chlorogenic acid (2.42 μg/g). Slama et al. (2020) evaluated the phytochemical composition of pearl millet oil and reported that pearl millet oil contains seven phenolic acids and among them *trans*-cinnamic acid (633 μg/g), protocatechuic acid (387.1 μg/g), and hydroxybenzoic acid (352.0 μg/g) were the most abundant phenolic acids, while gentisic acid (19.7 μg/g), gallic acid (9.4 μg/g), caffeic acid (1.3 μg/g), and *p*-coumaric acid (0.9 μg/g) were phenolic acids present in lower amounts. Authors also reported the presence of two flavonoids including quercitin (1.7 μg/g) and catechin (1.2 μg/g) in pearl millet oil.

4.2.3 Sorghum Phenolics

Sorghum grain is rich in various phenolic acids and flavonoids, with the flavonoid group further subdivided into flavanones, flavonols, anthocyanins, and condensed tannins, also known as proanthocyanidins (Xu et al., 2021). The phenolic acids in sorghum primarily occur as derivatives of benzoic acid or cinnamic acid. Various benzoic acid derivatives present were gallic acid, salicylic acid, *p*-hydroxybenzoic acid, gentisic acid, syringic acid, and protocatechuic acid. On the other hand, various cinnamic acid derivatives were cinnamic acid, sinapic acid, ferulic acid, *o*-coumaric acid, and *p*-coumaric acid (Awika & Rooney, 2004; Xu et al., 2021). The majority of phenolic acids in sorghum—ranging approximately from 70% to 95%—are present in bound form. Among these, ferulic acid is the most prevalent polyphenol, varied from 100 to 500 μg/g, and can constitute up to 90% of the total bound phenolic acids. However, the bioavailability of these bound phenolic acids is low due to their strong covalent bonds, which resist enzymatic digestion (Xiang et al., 2019). Similarly, Hahn and Rooney (1986) stated that sorghum phenolic acids predominantly exist in bound forms, with ferulic acid being the most prominent, accounting for 24–47% of the total. Gallic acid was detected exclusively in the bound fraction, ranging from 12.9 to 46.0 mg/g (dry weight). In contrast, cinnamic acid was generally present only in the free form (2.0–10.7 mg/g, dry weight), except in one variety (SC0719, characterized by a red pericarp and pigmented testa), where it was found solely in the bound form at 19.7 mg/g (dry weight). Further, few researchers also mentioned that the total phenolic acid content in various sorghum varieties has been found to range from 135.5 to 479.4 mg/g. Among these, protocatechuic acid (150.3–178.2 mg/g) and ferulic acid (120.5–173.5 mg/g) were present in the highest concentrations. Smaller quantities were observed for *p*-coumaric (41.9–71.9 mg/g), syringic (15.7–17.5 mg/g), vanillic (15.4–23.4 mg/g), gallic (14.8–21.5 mg/g), caffeic (13.6–20.8 mg/g), cinnamic (9.8–15.0 mg/g), and *p*-hydroxybenzoic acids (6.1–16.4 mg/g) (de Morais Cardoso et al., 2017). Further, Mawouma et al. (2022) evaluated the phytochemical compounds of four sorghum cultivars by using HPLC and reported the presence of phenolic compounds such as gallic acid (104.73–170.01 μg/mL), chlorogenic acid (10.11–59.36 μg/mL), caffeic acid (4.72–7.42 μg/mL), ferulic acid (2.73–55.96 μg/mL), and vanillic acid (18.24–548.65 μg/mL). While catechin (87.41 μg/mL), quercetin (53.85 μg/mL), and kaempferol (11.49 μg/mL) was observed only in red sorghum cultivar, epicatechin (114.80 μg/mL) was detected only in yellow sorghum cultivar. Rao et al. (2018) did characterization of acetone extracts of six sorghum varieties using UHPLC and reported 56 compounds from sorghum grains. However, among them 17 compounds were present in most of varieties and they were protocatechuic acid (0.03–0.06 mg/100 g gallic acid equivalents [GAE]), quercetin-3,4′-O-di-beta-glucopyranoside (0.05–0.13 mg/100 g GAE), caffeic acid (0.92–1.08 mg/100 g GAE), catechin (0.14–9.18 mg/100 g GAE), 1-O-caffeoylglycerol-O-glucoside (0.17–0.65 mg/100 g GAE), ferulic acid (0.91–1.22 mg/100 g GAE), *N′*,*N′*-dicafferoyl spermidine (0.22–0.78 mg/100 g GAE), taxafolin derivative (0.06–0.16 mg/100 g GAE), taxafolin (0.03–0.95 mg/100 g GAE), luteolin-7-o-glucside (0.19–0.37 mg/100 g GAE), pyrano-eriodictyol-(3->4)-catechin-7-o-glucoside isomer (0.15–0.57 mg/100 g GAE), pyrano-eriodictyol-(3->4)-catechin-7-o-glucoside (0.12–0.65 mg/100 g GAE), pyrano-eriodictyol-(3->4)-catechin-7-o-glucoside isomer (0.11–0.47 mg/100 g GAE), procyanidin b1 isomer (0.10–0.28 mg/100 g GAE), eriodictyol-deoxyhexoside (0.10–0.53 mg/100 g GAE), luteolin isomer (0.05–0.72 mg/100 g GAE), and luteolin derivative (0.06–0.73 mg/100 g GAE).

4.3 Minor Millet Grain Phenolics

HPLC, LC–MS, and gas chromatography–mass spectrometry (GC–MS) are widely utilized analytical techniques for phytochemical profiling in minor millets. Methanolic extracts from these grains have been shown to contain a wide range of bioactive compounds, including flavonoids, terpenoids, sterols, phenolics, xanthophylls, xanthinins, and carotenoids. These phytochemicals are associated with several health-promoting properties, such as antioxidant, antidiabetic, and antihyperlipidemic activities (Pujari & Hoskeri, 2022).

4.3.1 Barnyard Millet

Barnyard millet (*Echinochloa frumentacea*) is an underutilized cereal predominantly cultivated and consumed in parts of South Asia, particularly India, Nepal, and Pakistan. (Pujari & Hoskeri, 2022). It has attracted scientific interest due to its adaptability to a wide range of agroclimatic conditions, including tolerance to both biotic and abiotic stresses, as well as its capacity for multiple annual harvests. Nutritionally, barnyard millet is valued for its high levels of micronutrients, dietary fiber, and diverse phytochemicals (Bhatt et al., 2023). Among these bioactive compounds, phenolic acids such as *p*-coumaric, chlorogenic, caffeic, and ferulic acids have been identified as the principal constituents, contributing to its potential health-promoting properties (Anis & Sreerama, 2020). Ofosu et al. (2020) evaluated polyphenols in ethanolic extracts of barnyard millet by using UPLC-Q-TOF-MS and reported the presence of seven compounds including catechin, formononetin, kaempferol, luteolin, apigenin, isorhamnetin, and 3,7-dimethylquercetin. Sharma et al. (2016b) evaluated the effect of germination on phenolic acid of barnyard millet and reported the presence of 10 phenolic acids, i.e., phthalic acid hept4yl isobutyl ester (25.32%), 12-cisoctadecadienoate (16.19%), phthalic acid butyl hex3yl ester (5.96%), hexadecanoic acid methyl ester (5.59%), lupeol (5.16%), ascorbic acid 2,6 dihexadecanoate (4.54%), methyl,10trans, 9,12-octadecadienoic acid (14.59%), 9,12 octadecadienoic acid ethyl ester (11.00%), c-sitosterol (4.03%), and campesterol (7.59%). Goudar et al. (2023) evaluated barnyard millet by HPTLC and reported that it contains free phenolic acids as gallic acid (10.56 μg/g), ferulic acid (27.88 μg/g), caffeic acid (23.65 μg/g), syringic acid (21.56 μg/g), and chlorogenic acid (3.01 μg/g). While ferulic acid (231.84 μg/g), gallic acid (13.91 μg/g), caffeic acid (10.66 μg/g), syringic acid (5.74 μg/g), and chlorogenic acid (2.68 μg/g) were bound phenolic acids present in barnyard millet.

4.3.2 Proso Millet

Proso millet (*Panicum miliaceum* L.) is a widely cultivated minor millet, particularly in tropical regions across several countries, and is considered a staple food for low-income populations in certain areas (Pujari & Hoskeri, 2022). Goudar et al. (2023) did HPLC profiling of proso millet and reported that it contains free phenolic acids such as ferulic acid (41.13 μg/g), gallic acid (4.23 μg/g), syringic acid (15.72 μg/g), caffeic acid (8.69 μg/g), and chlorogenic acid (1.51 μg/g). While, bound phenolic acids constitute majorly ferulic acid (288.57 μg/g) and gallic acid (23.44 μg/g). Further, Kalam Azad et al. (2019) evaluated the phenolic compounds of dehulled and whole proso millet flours by using HPLC and identified hydroxybenzoic acid derivatives, i.e., gallic acid (9.37–11.1 μg/100 g), syringic acid (5.34–11.27 μg/100 g), 4-hydroxybenzoic acid (6.87–9.11 μg/100 g), and hydroxycinnamic acid derivatives, i.e., ferulic acid (38.86–68.63 μg/100 g) and sinapic acid (9.32–12.87 μg/100 g) as major phenolic compounds. Yuan et al. (2022) analyzed methanolic extracts of proso millet by using UPLC-MS/MS and showed the presence of 12 free phenolic compounds, and among them, 6 compounds were identified as *trans-p*-coumaric acid, protocatechualdehyde, *trans*-ferulic acid, chlorogenic acid, *p*-hydroxybenzoic acid, and *p*-hydroxybenzaldehyde, While, other compounds were neochlorogenic acid (3-O-caffeoylquinic acid), *N*′-caffeoyl-*N*″-feruloylspermidine, *N*-(*p*-coumaroyl) serotonin, *N*′,*N*″-dicaffeoylspermidine, and cryptochlorogenic acid. The authors also identified several bound phenolic compounds, including protocatechuic aldehyde, *p*-hydroxybenzoic acid, *p*-hydroxybenzaldehyde, vanillic acid, *trans-p*-coumaric acid, *trans*-ferulic acid, *cis*-ferulic acid, and *cis-p*-coumaric acid.

Furthermore, Xiang et al. (2023) evaluated the effect of germination on the phenolic composition of three proso millet varieties using UPLC-ESI-MS/MS. The analysis revealed the presence of 17 free phenolic compounds, including neochlorogenic acid, protocatechuic aldehyde, *p*-hydroxybenzoic acid, a cinnamic acid derivative, *p*-hydroxybenzaldehyde, chlorogenic acid, feruloylquinic acid, cryptochlorogenic acid, *N,N′*-bis(*p*-coumaroyl) putrescine, various hydroxycinnamic acid derivatives, *trans-p*-coumaric acid, *N′,N″*-dicaffeoylspermidine, *trans*-ferulic acid, *N′*-caffeoyl-*N″*-feruloylspermidine, *N*-(*p*-coumaroyl) serotonin, *N*-feruloylserotonin, and homoeriodictyol, and 18 bound phenolic compounds (protocatechuic aldehyde, *p*-hydroxybenzoic acid, cinnamic acid derivative, *p*-hydroxybenzaldehyde, vanillic acid, *trans-p*-coumaric acid, *cis-p*-coumaric acid, *trans*-ferulic acid, 8,8′-aryltetralin-diferulic acid, *cis*-ferulic acid, 8,8′-diferulic acid, ferulic truxilic acid, triferulic acid, 8,5′-diferulic acid, *trans-trans*-8-O-4′-diferulic acid, 5–5′-diferulic acid, 8–5′-diferulic acid benzofuran, and *trans-cis*-8-O-4′-diferulic acid).

4.3.3 Foxtail Millet

Foxtail millet (*Setaria italica* L.) is an ancient cereal crop widely grown across various regions, particularly in Asia and Africa, where it serves as a staple food for millions. In addition to this, foxtail millet is gaining increasing attention for its rich nutritional and phytochemical profile. It is considered a valuable source of carotenoids, vitamins, phenolic compounds, minerals, and other bioactive substances that contribute to its health-promoting properties (Gao et al., 2024). The phenolic compositions of various milling fractions from two foxtail millet varieties were analyzed using UPLC-Q-TOF-MS, and nine free phenolic compounds from millet grain without husk and bran were identified, i.e., protocatechuic aldehyde (0.47–0.61 mg/kg dw), syringic acid (2.12–4.45 mg/kg dw), *p*-coumaric acid (3.84–6.32 mg/kg dw), ferulic acid (4.59–5.33 mg/kg dw), *N′,N″*-dicaffeoylspermidine (0.12–0.17 mg/kg dw), *N′-p*-coumaroyl-*N′′*-caffeoylspermidine (1.17–2.01 mg/kg dw), kaempferol-C, O-dihexoside (0.19–0.45 mg/kg dw), *N′,N′′*-di-*p*-coumaroylspermidine (0.21–0.42 mg/kg dw), and *N′-p*-coumaroyl-*N′′*-feruloylspermidine (0.36–0.70 mg/kg dw) (Zhang et al., 2021). Authors also reported the presence of nine bound phenolic compounds, i.e., *p*-hydroxybenzaldehyde (12.01–124.75 mg/kg dw), syringic acid (2.30–3.82 mg/kg dw), *trans-p*-coumaric acid (30.24–138.14 mg/kg dw), *trans*-ferulic truxilic (3.64–4.82 mg/kg dw), ferulic acid (174.66–220.23 mg/kg dw), sinapic acid (2.98–4.18 mg/kg dw), 8,8′ aryltetralin DFA (32.11–32.37 mg/kg dw), 8-5′DFA (3.69–5.43 mg/kg dw), and 8-8′DFA (3.83–4.86 mg/kg) from foxtail millet grains without husk and bran. Goudar et al. (2023) evaluated foxtail millet by HPLC and reported that it contains free phenolic acids such as gallic acid (6.23 μg/g), ferulic acid (54.65 μg/g), caffeic acid (13.59 μg/g), syringic acid (8.59 μg/g), and chlorogenic acid (1.42 μg/g). While, bound phenolic acids were ferulic acid (254.20 μg/g), gallic acid (7.35 μg/g), caffeic acid (36.27 μg/g), syringic acid (6.22 μg/g), and chlorogenic acid (2.36 μg/g).

4.3.4 Little Millet

Little millet (*Panicum sumatrense*), commonly referred to as saamai or kutki, is a short-duration crop known for its resilience to both drought and waterlogging conditions (Dey et al., 2022). Chandrasekara and Shahidi (2011) did characterization of free and hydrolyzed fractions of kodo millet grains by using HPLC-DAD-ESI-MS and reported the presence of free phenolic compounds, i.e., hydroxybenzoic acid and its derivatives such as gallic acid

(2.13 μg/g), methyl vanillate (5.11 μg/g), *p*-hydroxybenzoic acid (3.73 μg/g), vanillic acid (15.1 μg/g), syringic acid (9.34 μg/g), vanillin (2.23 μg/g), and hydroxycinnamic acid and its derivatives including sinapaldehyde (2.23 μg/g), *p*-coumaryl pentose (21.8 μg/g), *p*-coumaric acid (63.4 μg/g), sinapic acid (18.1 μg/g), *trans*-ferulic acid (27.8 μg/g), *cis*-ferulic acid (5.24 μg/g), CAD1 (27.8 μg/g), and CAD2 (3.25 μg/g) in the soluble fraction of little millet grains. Authors also characterized the soluble fraction obtained from alkaline hydrolysis of little millet grain and reported the presence of phenolic compounds, i.e., hydroxybenzoic acid and its derivatives such as methyl vanillate (13.9 μg/g), *p*-hydroxybenzoic acid (5.47 μg/g), vanillic acid (45.6 μg/g), syringic acid (12.9 μg/g), and hydroxycinnamic acid and its derivatives such as sinapaldehyde (9.79 μg/g), *p*-coumaric acid (95.4 μg/g), sinapic acid (37.4 μg/g), *trans*-ferulic acid (126 μg/g), and *cis*-ferulic acid (16.7 μg/g). Goudar et al. (2023) also evaluated little millet by HPLC and reported that it contains free phenolic acids such as gallic acid (5.21 μg/g), ferulic acid (29.94 μg/g), caffeic acid (17.59 μg/g), and chlorogenic acid (1.15 μg/g). While, bound phenolic acids constitute majorly ferulic acid (133.58 μg/g), caffeic acid (23.77 μg/g), syringic acid (5.48 μg/g), and chlorogenic acid (3.24 μg/g).

4.3.5 Kodo Millet

Kodo millet (*Paspalum scrobiculatum*), one of the coarsest yet easily digestible ancient grains, is believed to have originated in tropical Africa and was domesticated in India around 3,000 years ago (Dey et al., 2022). Chandrasekara and Shahidi (2011) did characterization of both hydrolyzed and free fractions of kodo millet grains by using HPLC-DAD-ESI-MS and reported free phenolic compounds such as hydroxybenzoic acid and its derivatives including gallic acid (1.80 μg/g), protocatechuic acid (6.84 μg/g), protocatechuic aldehyde (2.82 μg/g), *p*-hydroxybenzoic acid (1.41 μg/g), vanillic acid (25.4 μg/g), vanillin (11.2 μg/g), and hydroxycinnamic acid and its derivatives such as chlorogenic acid (3.56 μg/g), caffeic acid (4.11 μg/g), *p*-coumaryl pentose (10.1 μg/g), cinnamic acid (37.4 μg/g), *p*-coumaric acid (8.65 μg/g), and *trans*-ferulic acid (10.3 μg/g) in soluble fraction of millet grains. Authors also reported the presence of phenolic compounds, i.e., hydroxybenzoic acid and its derivatives such as protocatechuic acid (15.3 μg/g), protocatechuic aldehyde (1.72 μg/g), *p*-hydroxybenzoic acid (12.7 μg/g), vanillic acid (2.97 μg/g), syringic acid (136 μg/g), and hydroxycinnamic acid and its derivatives such as sinapaldehyde (23.9 μg/g), 5-hydroxyl feruloyl hexose (36.6 μg/g), caffeic acid (42.2 μg/g), caffeoyl hexose (39.2 μg/g), *p*-coumaric acid (25.6 μg/g), sinapic acid (53.2 μg/g), *trans*-ferulic acid (350 μg/g), and *cis*-ferulic acid (26.7 μg/g) insoluble fraction yielded from alkaline hydrolysis (esterified) of kodo millet grains. Recently, Goudar et al. (2023) did HPLC profiling of kodo millet and mentioned that it contains free phenolic acids like gallic acid (24.81 μg/g), ferulic acid (99.35 μg/g), caffeic acid (12.45 μg/g), syringic acid (28.22 μg/g), and chlorogenic acid (7.53 μg/g). While, bound phenolic acids were ferulic acid (1,445.06 μg/g), gallic acid (20.52 μg/g), and caffeic acid (20.79 μg/g). Furthermore, Dey et al. (2022) reviewed the phenolic compounds of kodo millet and reported the presence of catechin (1.10 mg/kg), *p*-coumaric acid (1.38 mg/kg), ferulic acid (20.45 mg/kg), sinapic acid (7.9 mg/kg), and naringin (11.97 mg/kg).

4.4 Phytochemical Profiling of Millets

Phenolic compounds not only make millet healthier to eat but also help us to understand the different natural chemicals in the plant. That's why it is important to measure and study these phenolics carefully. Utilizing advanced analytical techniques ensures a thorough understanding of these bioactive compounds in millet (Verma et al., 2023). GC–MS, LC–MS, and HPLC are commonly used techniques for the phytochemical profiling of millets (Table 4.2). Phytochemicals in millet and other plant-based foods are significantly influenced by genetic factors, agroclimatic conditions, and postharvest processing. The role of genetics, particularly cultivar or variety, in determining phytochemical composition is well established (Belobrajdic & Bird, 2013; Shao & Bao, 2015). They are mainly located in the bran and embryo cell walls of the cereal kernel (Belobrajdic & Bird, 2013; Ragaee et al., 2014). Millet is an excellent source of phytochemicals, particularly phenolic compounds (Table 4.3), which have been linked to the prevention of chronic diseases such as diabetes, cancer, and cardiovascular disorders (Xiang et al., 2018). Chandrasekara and Shahidi (2011) reported that the free phenolic content in kodo, finger, foxtail, proso, and pearl millet was 16.2, 9.67, 4.49, 0.55, and 1.27 μmol FAE/g, respectively. The bound phenolic content in these millets was found to be 81.6, 3.83, 0.32, 2.48, and 3.94 μmol FAE/g, respectively. The polyphenolic profile of millet includes hydroxybenzoic acids (protocatechuic, *p*-hydroxybenzoic), hydroxycinnamic acids (*p*-coumaric, ferulic, syringic), flavonoids (quercetin, apigenin, catechin, epicatechin), and proanthocyanidins (Devi et al., 2014).

Phenolic acids are classified as derivatives of cinnamic and benzoic acids. Hydroxycinnamic acids possess a C_6–C_3 structure, while hydroxybenzoic acids have a C_6–C_1 framework. In millet, commonly occurring hydroxycinnamic acids include *p*-coumaric (*p*-CA), caffeic (CA), ferulic (FA), and sinapic (SA) acids, whereas hydroxybenzoic acids such as *p*-hydroxybenzoic (*p*-OHBA), protocatechuic (PA), gallic (GA), vanillic (VA), and syringic (SyA) acids are also present (Manach et al., 2004). Among these, hydroxycinnamic acids are more abundant than hydroxybenzoic acids in plants. Xiang et al. (2018) reported that phenolic compounds in millet exist in both free and bound forms. The predominant phenolic acids found in millet include ferulic, vanillic, caffeic, syringic, and *p*-coumaric acids (Sosulski et al., 1982). Analysis using LC–MS has identified multiple phenolic acids in millet grains, with quinic acid, *p*-coumaric acid, and caffeic acid being the most prominent (Triki et al., 2025). Phytochemicals in millet are primarily extracted using methanol and ethanol solvents, and further identification of specific compounds is achieved through chromatographic techniques such as HPLC, LC–MS, and GC–MS (Sharma et al., 2017a).

Hydroxycinnamic acids are a class of phenolic compounds synthesized in various plant-based foods, including cereal grains (Clifford, 2000). These bioactive compounds exhibit multiple health-promoting properties, such as antioxidant, anticancer, antimicrobial, antifungal, antimalarial, and antiatherogenic effects (Manach et al., 2004). Among the different isoforms of cinnamic acid (CA), *trans*-cinnamic acid (*trans*-3-phenyl-2-propenoic acid, t-CA) is the most prevalent form. Important derivatives of cinnamic acid include *p*-coumaric acid (*p*-CA), caffeic acid (CA), ferulic acid (FA), and sinapic acid (SA) (Manach et al., 2004). Millets are rich sources of hydroxycinnamic acid derivatives, which primarily

Table 4.3 Phytochemical profiling of different millets

Millet type	Major phytochemicals identified	Phytochemical class	Analytical method used	Key studies
Finger millet	Total phenol, tannins, flavonoids	Phenolic acids and polyphenols	UV–Vis spectroscopy and GC–MS	Nakarani et al. (2021) Trivedi and Gupta (2023)
Pearl millet	*Trans*-cinnamic, protocatechuic, hydroxybenzoic, gentisic, gallic, caffeic, *p*-coumaric, quercetin, and catechin	Phenolic acids and flavonoids	UV–Vis spectrophotometry and HPLC-PDA	Slama et al. (2020)
Sorghum	Gallic acid, ferulic acid, catechin, and quercetin	Phenolic acids	HPLC and UV–Vis spectroscopy	Hassan et al. (2020)
Foxtail millet	Total phenolic content, total flavonoid content, and tannins	Phenolic acids and flavonoids	UV–Vis spectroscopy	Sharma et al. (2022)
Barnyard millet	Benzoic acid, 4-ethoxyethyl ester, xanthone, and total phenol, tannins, saponins, and ortho-dihydroxy phenol	Phenolics/ phenylpropanoids	UV–Vis spectroscopy and GC–MS	Sharma et al. (2016a), Panwar et al. (2016)
Kodo millet	Flavonoids-glycosides, caffeic acid, and cinnamic acid	Polyphenols	TLC and HPLC	Sharma et al. (2017b)
Proso millet	Gallic acid, catechin, chlorogenic acid, phenolic, and flavonoid content	Phenolic acids and tannins	LC–MS/MS, Folin–Ciocalteu assay, and HPLC	Park et al. (2014), KabbarePuttaswamy et al. (2024)
Little millet	Alkaloids, flavonoids, tannins, phenols, glycosides, anthocyanins, and coumarins	Phenolic acids and flavonoids	GC–MS and UHPLC-QTOF-MS	Sharma et al. (2024)
Brown top millet	Quercetin, apigenin, and chlorogenic acid	Flavonoids and phenolic acids	LC–MS and UV–Vis	Pranusha et al. (2025)

exist in bound form rather than in free form (Chandrasekara & Shahidi, 2011). The bound phenolics in millet are often linked to the cell wall components, contributing to their stability and bioavailability upon digestion. Hydroxycinnamic acids are characterized by a basic C_6–C_3 structure, featuring a three-carbon side chain with a double bond that can exist in either cis or trans configuration. Xiang et al. (2018) reported that finger millet varieties contain phenolic compounds in both free and bound forms, with the bound fraction being more predominant. The phenolic content in the free form ranges from 114.43 to 179.19 mg/100 g, while the bound form varies between 58.27 and 123.23 mg/100 g.

Ferulic acid, a widely occurring hydroxycinnamic acid derivative, is the predominant phenolic compound in finger millet, with its content ranging from 36.64 to 40.00 mg/100 g (Xiang et al., 2018). Subba Rao and Muralikrishna (2002) reported that native ragi (finger millet in India) contains 18.6 mg per 100 g of ferulic acid. However, ferulic acid serves as the primary bound phenolic compound in finger millet. Xiang et al. (2019) reported that ferulic acid was the predominant compound in Malawian millets, with concentrations ranging from 59.58 to 67.36 mg/100 g DW in central Malawi millets and 35.19 to 40.12 mg/100 g DW in northern Malawi millets. Bouajila et al. (2020) cultivated five pearl millet populations in different agroecological zones of Tunisia and reported that the ferulic acid content ranged from 23.60 to 30.20 µg/g. Triki et al. (2022) recorded *trans*-ferulic acid content in pearl millet varieties from southern Tunisia was reported to range between 7.76 and 13.91 µg/g. Xiang et al. (2023) recorded *trans*-ferulic acid in the proso millet variety at a concentration of 6.67 mg/kg in the free form and 353.21 mg/kg in the bound form. Goudar et al. (2023) reported ferulic acid is present in both free and bound forms in various kinds of millet varieties. The amount of ferulic acid in foxtail millet was determined to be 254.20 µg/g in the bound form and 54.65 µg/g in the free form. Ferulic acid in little millet was found to be 29.94 µg/g in its free form and 133.58 µg/g in its bound form. Ferulic acid content in barnyard millet was 231.84 µg/g (bound form) and 27.88 µg/g (free form). Ferulic acid levels in the kodo millet variety were 99.35 µg/g in the free form and 1,445.06 µg/g in the bound form. Pearl millet had 47.03 µg/g in the free form, whereas proso millet had 41.13 µg/g in the free form and 288.57 µg/g in the bound form. The pearl variety has ferulic acid in 988.78 µg/g in bound form and 47.03 µg/g in free form. Finger millet found ferulic acid in 762.12 µg/g in bound form and 57.04 µg/g free form. Yu et al. (2024) reported that the foxtail variety of millet contains *trans*-ferulic acid at a concentration of 18.17 mg/kg in the free form and 2,325.40 mg/kg in the bound form. Zhang et al. (2025) analyzed *trans*-ferulic acid in whole foxtail millet using UPLC-MS, reporting a concentration of 6.30 mg/kg DW. After milling, the concentration decreased, with polished millet containing 2.33 mg/kg DW in the free form. In the bound form, whole foxtail millet contained 377.86 mg/kg, which reduced to 120.23 mg/kg after polishing.

p-Coumaric acid is the second most abundant phenolic acid in finger millet, with concentrations ranging from 2.23 to 3.24 mg/100 g (Xiang et al., 2018). Subba Rao and Muralikrishna (2002) reported that native finger millet contains approximately 1.20 mg/100 g of *p*-coumaric acid, primarily in its bound form. Chandrasekara and Shahidi (2011) reported that the *p*-coumaric acid content varies among millet types, with kodo millet containing 8.65 µg/g, finger millet 2.10 µg/g, foxtail millet 3.88 µg/g, proso millet 63.4 µg/g, and pearl millet 4.04 µg/g. Hithamani and Srinivasan (2014) reported that the concentration of *p*-coumaric acid in native finger millet is 19.07 µg/g. Xiang et al. (2019)

reported *p*-coumaric acid levels ranging from 1.88 to 3.94 mg/100 g. Bouajila et al. (2020) reported that several pearl millet cultivars contained *p*-coumaric acid in the range of 0.60 to 1.30 μg/g. Triki et al. (2022) reported that *p*-coumaric acid was present in 10 genotypes of pearl millet, with concentrations ranging from 5.28 to 51.44 μg/g. Yu et al. (2024) reported that *trans-p*-coumaric acid was present exclusively in the bound form in the foxtail variety of millet, with a concentration of 1,167.33 mg/kg. Xiang et al. (2023) reported that the proso variety of millet contains *trans-p*-coumaric acid at concentrations of 37.07 mg/kg in the free form and 397.65 mg/kg in the bound form. Zhang et al. (2025) reported that *trans-p*-coumaric acid in whole millet was present at 7.66 mg/kg in the free form, which decreased to 1.32 mg/kg after polishing. In the bound form, it was found at 1,050.4 mg/kg in whole millet and reduced to 138.14 mg/kg after polishing.

Caffeic acid (CA), a common hydroxycinnamic acid derivative found in millets, is typically present only in the bound form, with concentrations ranging from 0.49 to 1.37 mg/100 g (Van Hung, 2016). Chandrasekara and Shahidi (2011) reported that kodo, proso, and pearl millet varieties contain 4.11, 37.3, and 4.88 μg/g, respectively. Hithamani and Srinivasan (2014) analyzed the concentration of caffeic acid in native finger millet and was found to be 10.40 μg/g. Xiang et al. (2018) reported that millet varieties with different colors such as red, brown, white, and reddish contained caffeic acid in the range of 0.89–1.57 mg/100 g dry weight (DW). Xiang et al. (2019) observed caffeic acid concentrations in finger millet ranging from 0.83 to 2.43 mg/100 g. Bouajila et al. (2020) analyzed caffeic acid in five pearl millet varieties cultivated in different agroecological zones of Tunisia. The concentrations ranged from 1.30 to 1.70 μg/g, as determined by RP-HPLC. Triki et al. (2022) recorded caffeic acid was detected in pearl millet varieties within the range of 3.87–14.30 μg/g. Methanol was used for the extraction of phenolic compounds, which were subsequently identified using HPLC–MS. Mawouma et al. (2022) reported the presence of caffeic acid in the mouri pearl millet cultivar from the Far North region of Cameroon, with a concentration of 4.32 μg/mL. Goudar et al. (2023) reported the presence of caffeic acid in various types of millets, both in free and bound forms. In foxtail millet, caffeic acid was found at 13.59 μg/g in the free form and 36.27 μg/g in the bound form. Little millet contained 17.59 μg/g in the free form and 23.77 μg/g in the bound form. Barnyard millet had 23.65 μg/g of caffeic acid in the free form and 10.66 μg/g in the bound form. Kodo millet contained 12.45 μg/g in the free form and 20.79 μg/g in the bound form. Proso millet contained 8.69 μg/g of caffeic acid, detected only in the free form. Pearl millet contained 43.88 μg/g in the free form and 45.12 μg/g in the bound form, while finger millet contained 38.91 μg/g of caffeic acid in the free form and 31.86 μg/g in the bound form. Triki et al. (2025) found caffeic acid in the pearl variety of millet, which ranges from 0.94 to 1.65 mg/100 g.

Mawouma et al. (2022) reported that chlorogenic acid concentrations ranged from 19.06 to 34.61 μg/mL in two pearl millet cultivars grown in the Far North Region of Cameroon. Goudar et al. (2023) reported that various millet varieties contain chlorogenic acid in concentrations ranging from 1.15 to 7.53 μg/g in the free form, with the highest level observed in kodo millet and the lowest in little millet. In the bound form, chlorogenic acid ranged from 2.36 to 5.41 μg/g, with the highest concentration found in finger millet and the lowest in foxtail millet. Hydroxybenzoic acids are the major free phenolic acids in finger millet, contributing about 70–71% of the total grain phenolic content. This group includes *p*-hydroxybenzoic, protocatechuic, vanillic, and syringic acids, with protocatechuic acid being the most abundant at 45 mg/100 g (Mattila et al., 2005). Hydroxybenzoic acids are characterized by a C_6–C_1 structural backbone, with various hydroxyl and methoxyl groups attached to an aromatic ring. Yuan et al. (2022) further identified several hydroxybenzoic acids derivatives in proso millet, including protocatechuic aldehyde, vanillic acid, *p*-hydroxybenzaldehyde, and *p*-hydroxybenzoic acid, detected at a wavelength of 280 nm.

Hithamani and Srinivasan (2014) reported the gallic acid in native finger millet was 10.69 μg/g. Kalam Azad et al. (2019) detected gallic acid in both dehulled and whole proso millet flour using HPLC, with concentrations ranging from 9.37 to 11.1 μg/100 g. Bouajila et al. (2020) reported that gallic acid content ranged from 9.40 to 13.60 μg/g in five pearl millet cultivars grown in different agroecological zones of Tunisia. Similarly, Mawouma et al. (2022) detected gallic acid in two pearl millet grain cultivars grown in the Far North Region of Cameroon, with concentrations ranging from 165.59 to 185.79 μg/mL, as determined by HPLC techniques. Goudar et al. (2023) observed gallic acid content in various millet types, with concentrations as foxtail millet (6.23 μg/g), little millet (5.21 μg/g), barnyard millet (10.56 μg/g), kodo millet (24.81 μg/g), proso millet (4.23 μg/g), pearl millet (3.87 μg/g), and finger millet (11.23 μg/g).

p-Hydroxybenzoic acid has been identified in millets at varying concentrations. Hithamani and Srinivasan (2014) reported that finger millet contains 3.62 μg/g *p*-hydroxybenzoic acid. Oliver-Chen et al. (2019) reported that *p*-hydroxybenzoic acid content is 3.0 μg/g in millet. However, Xiang et al. (2019) found that *p*-hydroxybenzoic acid levels varied across millet varieties, with kodo millet containing 1.41 μg/g, foxtail 4.62 μg/g, proso 15.5 μg/g, little millet 3.73 μg/g, and pearl millet 2.12 μg/g. Kalam Azad et al. (2019) reported that *p*-hydroxybenzoic acid was present in dehulled and whole proso millet flour, with concentrations ranging from 6.87 to 9.11 μg/100 g. Bouajila et al. (2020) analyzed the hydroxybenzoic acid content, which ranged from 253.50 to 437.60 μg/g in five pearl millet cultivars grown along the North African coastline of Tunisia (northeast, center-east, and southeast regions). Xiang et al. (2023) recorded *p*-hydroxybenzoic acid in the proso millet variety at concentrations of 42.74 mg/kg in the free form and 69.93 mg/kg in the bound form.

Xiang et al. (2018) reported that both protocatechuic acid and protocatechuic aldehyde are present in millets in free and bound forms. These compounds were identified in four millet varieties distinguished by color—brown, red, white, and reddish. Protocatechuic acid was detected in the free form at concentrations ranging from 0.16 to 0.79 mg/100 g and in the bound form from 0.31 to 1.37 mg/100 g. Protocatechuic aldehyde was found exclusively in the bound fraction, with concentrations ranging between 0.66 and 2.87 mg/100 g. Xiang et al. (2019) reported the presence of both protocatechuic acid and protocatechuic aldehyde in finger millet from different regions. The results indicated that both compounds were present in free and bound fractions. The concentration of protocatechuic acid ranged from 0.37 to 0.97 mg/100 g DW in the free form and from 0.62 to 1.32 mg/100 g DW in the bound form. On the other hand, protocatechuic aldehyde was detected at concentrations ranging from 0.71 to 1.11 mg/100 g in the free form and from 0.40 to 2.79 mg/100 g in the bound form. Bouajila et al. (2020) reported that pearl millet cultivars grown along the North African coastline of Tunisia were analyzed for protocatechuic acid and catechin using RP-HPLC, with concentrations ranging from 127.60 to 452.40 μg/g and 1.20 to 23.20 μg/g, respectively. Triki et al. (2022) employed LC–MS to quantify protocatechuic acid, with concentrations ranging from 2.37 to 3.33 μg/g. Xiang et al. (2023) reported that the proso millet variety contains protocatechuic aldehyde at concentrations of 41.30 mg/kg in the free form and 31.86 mg/kg in the bound form. Yu et al. (2024) reported that the foxtail variety of millet contains bound protocatechuic acid and protocatechuic aldehyde at concentrations of 18.23 and 1.17 mg/kg, respectively. Zhang et al. (2025) further reported that protocatechuic aldehyde in the foxtail variety is present at 10.33 mg/kg in the free form and 1.07 mg/kg in the bound form.

Vanillic acid, a phenolic compound, has been identified in various millet grains. Shahidi and Chandrasekara (2013) reported that vanillic acid concentration varies among millet types, with kodo millet containing 98.1 μg/g, foxtail millet 118.7 μg/g, proso millet 168.6 μg/g, little millet 162.4 μg/g, and pearl millet 16.0 μg/g. Hithamani and Srinivasan (2014) found the vanillic acid content in roasted finger millet to be 126.9 μg/g, indicating that the compound is formed during the roasting of millet grains, and syringic acid was also detected in finger millet at a concentration of 10.17 μg/g. Oliver-Chen et al. (2019) reported the vanillic acid content as 22.2 μg/g, while syringic acid was found at 13.1 μg/g. Bouajila et al. (2020) reported that vanillic acid was found in the Hawarice variety of pearl millet at a concentration of 48.3 μg/g. Mawouma et al. (2022) reported vanillic acid at a concentration of 20.27 μg/mL in pearl millet from the Far North Region of Cameroon, identified using an HPLC system. Xiang et al. (2023) reported that the proso variety of millet contains vanillic acid at a concentration of 80.50 mg/kg. Zhang et al. (2025) reported that the foxtail variety of millet contains vanillic acid at 12.25 mg/kg in the free form and 22.27 mg/kg in the bound form.

Xiang et al. (2019) also detected syringic acid in millet, with concentrations ranging from 0.30 to 0.63 mg/100 g. Furthermore, Chandrasekara and Shahidi (2011) observed syringic acid levels of 3.54 and 9.34 μg/g in foxtail and little millet varieties, respectively. Kalam Azad et al. (2019) analyzed syringic acid in proso millet using HPLC, and the concentrations were found to range from 5.34 to 11.27 μg/100 g. Triki et al. (2022) also identified quinic acid and syringic acid in 10 genotypes of pearl millet using HPLC–MS. The concentrations ranged from 461.65 to 1,202.88 μg/g for quinic acid and from 1.20 to 2.66 μg/g for syringic acid. Goudar et al. (2023) reported the presence of syringic acid in both bound and free forms across different millet types, including foxtail, little, pearl, proso, and finger millets. The concentrations ranged from 5.48 to 10.33 μg/g in the bound form and from 2.44 to 28.22 μg/g in the free form. Yu et al. (2024) reported that the foxtail variety of millet contains syringic acid at a concentration of 30.40 mg/kg. Zhang et al. (2025) reported that in the foxtail variety of whole millet from Henan, China, the syringic acid content was 3.99 mg/kg in the free form and 50.72 mg/kg in the bound form. Hithamani and Srinivasan (2014) reported that native finger millet contains gentisic acid at a concentration of 4.54 μg/g. Similarly, Bouajila et al. (2020) found gentisic acid in five pearl millet cultivars, with concentrations ranging from 19.70 to 35.40 μg/g across different agroecological zones of Tunisia.

Catechin and epicatechin are phenolic compounds belonging to the flavan-3-ol subclass of flavonoids. Bouajila et al. (2020) analyzed catechin content in pearl millet cultivars using RP-HPLC and reported concentrations of 4.30 μg/g in Hawaria, 23.20 μg/g in Kelibia, 3.90 μg/g in Rejiche, 3.60 μg/g in Zeafrana, and 1.20 μg/g in Gergis. Similarly, Mawouma et al. (2022), using HPLC analysis, found catechin in pearl millet at 87.41 µg/mL in the red cultivar, 12.65 µg/mL in the gawane cultivar, and 75.73 µg/mL in the mouri cultivar. Epicatechin was detected exclusively in yellow sorghum cultivars, with a concentration of 114.80 µg/mL.

4.5 Conclusion

Millets serve as a significant source of naturally occurring phenolic compounds, which contribute to their nutritional value and functional properties. These compounds, including various types of phenolic acids and flavonoids, exist in both soluble and bound forms and demonstrate considerable variation depending on species, genetics, and postharvest processing. The use of modern analytical tools has enhanced the ability to characterize and quantify these phenolics, enabling deeper insights into their bioactivity. However, compared to major cereals, the phenolic profile of millets remains relatively underinvestigated. Further research should focus on optimizing extraction and profiling techniques, improving understanding of bioavailability, and exploring the synergistic effects of phenolics. A more thorough understanding of millet phenolics will support their incorporation into health-promoting food products and dietary interventions aimed at reducing the risk of noncommunicable diseases.

Abbreviations

CAD	chlorogenic acid derivative
dw	dry weight
GAE	gallic acid equivalents
GC–MS	gas chromatography–mass spectrometry
HPLC	high performance liquid chromatography
HPLC-DAD (± UV–Vis)	high performance liquid chromatography with diode array detector (± ultraviolet–visible detection)
HPLC-DAD-ESI-QTOF-MS/MS	high performance liquid chromatography with diode array detector, electrospray ionization, quadrupole time-of-flight tandem mass spectrometry
HPLC-PDA	high-performance liquid chromatography with photodiode array detector
LCMS-QTOF	liquid chromatography
LC–MS/MS (Triple Quadrupole/Ion Trap)	liquid chromatography with tandem mass spectrometry (triple quadrupole or ion trap analyzer)
UHPLC	ultra high-performance liquid chromatography
UPLC-Q-TOF-MS	ultra performance liquid chromatography coupled with time of flight and mass spectrometry

References

Abioye, V. F., Olatunde, S. J., Ogunlakin, G. O., & Abioye, O. A. (2022). Effect of soaking conditions on chemical composition, antioxidant activity, total phenols, flavonoids and antinutritional contents of finger millet. *African Journal of Food, Agriculture, Nutrition and Development*, *22*(7), 20942–20956.

Acosta-Estrada, B. A., Gutiérrez-Uribe, J. A., & Serna-Saldívar, S. O. (2014). Bound phenolics in foods, a review. *Food Chemistry*, *152*, 46–55.

Ahmad, F., Pasha, I., Saeed, M., & Asgher, M. (2018). Biochemical profiling of Pakistani sorghum and millet varieties with special reference to anthocyanins and condensed tannins. *International Journal of Food Properties*, *21*(1), 1586–1597. https://doi.org/10.1080/10942912.2018.1502198

Ali, A., Kumar, R. R., Vinutha, T., Singh, T., Singh, S. P., Satyavathi, C. T., Praveen, S., & Goswami, S. (2022). Grain phenolics: Critical role in quality, storage stability and effects of processing in major grain crops—A concise review. *European Food Research and Technology*, *248*(8), 2197–2213. https://doi.org/10.1007/s00217-022-04026-7

Ambati, K., & Sucharitha, K. V. (2019). Millets-review on nutritional profiles and health benefits. *International Journal of Recent Scientific Research*, *10*(7), 33943–33948.

Anis, M. A., & Sreerama, Y. N. (2020). Inhibition of protein glycoxidation and advanced glycation end-product formation by barnyard millet (*Echinochloa frumentacea*) phenolics. *Food Chemistry*, *315*, 126265.

Awika, J. M., & Rooney, L. W. (2004). Sorghum phytochemicals and their potential impact on human health. *Phytochemistry*, *65*(9), 1199–1221. https://doi.org/10.1016/j.phytochem.2004.04.001

Belobrajdic, D. P., & Bird, A. R. (2013). The potential role of phytochemicals in wholegrain cereals for the prevention of type-2 diabetes. *Nutrition Journal*, *12*, 62.

Bhatt, D., Rasane, P., Singh, J., Kaur, S., Fairos, M., Kaur, J., & Sharma, N. (2023). Nutritional advantages of barnyard millet and opportunities for its processing as value-added foods. *Journal of Food Science and Technology*, *60*(11), 2748–2760.

Bouajila, A., Lamine, M., Rahali, F., Melki, I., Prakash, G., & Ghorbel, A. (2020). Pearl millet populations characterized by Fusarium prevalence, morphological traits, phenolic content, and antioxidant potential. *Journal of the Science of Food and Agriculture*, *100*(11), 4172–4181.

Chandrasekara, A., & Shahidi, F. (2011). Determination of antioxidant activity in free and hydrolyzed fractions of millet grains and characterization of their phenolic profiles by HPLC-DAD-ESI-MS n. *Journal of Functional Foods*, *3*(3), 144–158.

Clifford, M. N. (2000). Chlorogenic acids and other cinnamates – Nature, occurrence and dietary burden. *Journal of the Science of Food and Agriculture*, *80*(7), 1033–1043.

de Morais Cardoso, L., Pinheiro, S. S., Martino, H. S. D., & Pinheiro-Sant'Ana, H. M. (2017). Sorghum (*Sorghum bicolor* L.): Nutrients, bioactive compounds, and potential impact on human health. *Critical Reviews in Food Science and Nutrition*, *57*(2), 372–390.

Devi, P. B., Vijayabharathi, R., Sathyabama, S., Malleshi, N. G., & Priyadarisini, V. B. (2014). Health benefits of finger millet (*Eleusine coracana* L.) polyphenols and dietary fiber: A review. *Journal of Food Composition and Analysis*, *38*, 20–31. https://doi.org/10.1016/j.jfca.2014.05.003

Dey, S., Saxena, A., Kumar, Y., Maity, T., & Tarafdar, A. (2022). Understanding the antinutritional factors and bioactive compounds of kodo millet (*Paspalum scrobiculatum*) and little millet (*Panicum sumatrense*). *Journal of Food Quality*, *2022*(1), 1578448.

Dykes, L., & Rooney, L. W. (2006). Sorghum and millet phenols and antioxidants. *Journal of Cereal Science*, *44*(3), 236–251. https://doi.org/10.1016/j.jcs.2005.12.001

Gao, Y., Ping, H., He, Z., Liu, J., Zhao, M., & Ma, Z. (2024). Characterization of the active components and bioaccessibility of phenolics in differently colored foxtail millets. *Food Chemistry*, *452*, 139355.

Giada, M. D. L. R. (2013). Food phenolic compounds: main classes, sources and their antioxidant power. In *Oxidative stress and chronic degenerative diseases-A role for antioxidants*.

Goudar, G., Manne, M., Sathisha, G. J., Sharma, P., Mokalla, T. R., Kumar, S. B., & Ziouzenkova, O. (2023). Phenolic, nutritional and molecular interaction study among different millet varieties. *Food Chemistry Advances*, *2*, 100150.

Hahn, D. H., & Rooney, L. W. (1986). Effect of genotype on tannins and phenols of sorghum. *Cereal Chemistry*, *63*(1), 4–8.

Hassan, Z. M., Sebola, N. A., & Mabelebele, M. (2020). Assessment of the phenolic compounds of pearl and finger millets obtained from South Africa and Zimbabwe. *Food Science & Nutrition*, *8*(9), 4888–4896.

Hithamani, G., & Srinivasan, K. (2014). Effect of domestic processing on the polyphenol content and bioaccessibility in finger millet (*Eleusine coracana*) and pearl millet (*Pennisetum glaucum*). *Food Chemistry*, *164*, 55–62.

KabbarePuttaswamy, C., Jayadeep, A., & Gurusiddaiah, S. K. (2024). Phytochemical characterization and evaluation of antioxidant activity in differently polished proso millet. *ACS Food Science & Technology*, *4*(12), 2958–2966.

Kalam Azad, M. O., Jeong, D. I., Adnan, M., Salitxay, T., Heo, J. W., Naznin, M. T., & Park, C. H. (2019). Effect of different processing methods on the accumulation of the phenolic compounds and antioxidant profile of broomcorn millet (*Panicum miliaceum* L.) flour. *Food*, *8*(7), 230.

Kalinová, J. P., Marešová, I., Tříska, J., & Vrchotová, N. (2022). Distribution of lignans in *Panicum miliaceum*, *Fagopyrum esculentum*, *Fagopyrum tataricum*, and *Amaranthus hypochondriacus*. *Journal of Food Composition and Analysis*, *106*, 104283. https://doi.org/10.1016/j.jfca.2021.104283

Khan, T., Azad, A. A., & Islam, R. U. (2025). Millets: A comprehensive review of nutritional, antinutritional, glycemic, bioactive, and processing aspects. *Journal of Food Composition and Analysis*, *141*, 107364.

Li, S., Zhao, W., Liu, S., Li, P., Zhang, A., Zhang, J., & Liu, J. (2021). Characterization of nutritional properties and aroma compounds in different colored kernel varieties of foxtail millet (*Setaria italica*). *Journal of Cereal Science*, *100*, 103248. https://doi.org/10.1016/j.jcs.2021.103248

Manach, C., Scalbert, A., Morand, C., Remesy, C., & Jimenez, L. (2004). Polyphenols: Food sources and bioavailability. *American Journal of Clinical Nutrition*, *79*(5), 727–747.

Mattila, P., Pihlava, J. M., & Hellstrom, J. (2005). Contents of phenolic acids, alkyl-and alkenylresorcinols, and avenanthramides in commercial grain products. *Journal of Agricultural and Food Chemistry*, *53*(21), 8290–8295.

Mawouma, S., Condurache, N. N., Turturică, M., Constantin, O. E., Croitoru, C., & Rapeanu, G. (2022). Chemical composition and antioxidant profile of sorghum (*Sorghum bicolor* (L.) Moench) and pearl millet (*Pennisetum glaucum* (L.) R. Br.) grains cultivated in the far-North region of Cameroon. *Foods*, *11*(14), 2026.

Mir, N. A., Yousuf, B., Gul, K., Sheikh, M. A., Riar, C. S., & Singh, S. (2024). Technological and analytical aspects of bioactive compounds and nutraceuticals from coarse grain cereal sources. In C. S. Riar & P. S. Panesar (Eds.), *Bioactive compounds and nutraceuticals from plant sources* (pp. 253–284). Apple Academic Press.

Nakarani, U. M., Singh, D., Suthar, K. P., Karmakar, N., Faldu, P., & Patil, H. E. (2021). Nutritional and phytochemical profiling of nutracereal finger millet (*Eleusine coracana* L.) genotypes. *Food Chemistry*, *341*, 128271. https://doi.org/10.1016/j.foodchem.2020.128271

Nguyen, T. H., Nguyen, L. L., & Baranyai, L. (2023). Influence of sprouting on phenolic acids, carotenoids, and antioxidant activity of millet varieties. *Journal of Agriculture and Food Research*, *14*, 100810. https://doi.org/10.1016/j.jafr.2023.100810

Ofosu, F. K., Elahi, F., Daliri, E. B. M., Chelliah, R., Ham, H. J., Kim, J. H., & Oh, D. H. (2020). Phenolic profile, antioxidant, and antidiabetic potential exerted by millet grain varieties. *Antioxidants*, *9*(3), 254.

Oliver-Chen, C. Y., Costa, S., & Carolo, K. (2019). Phenolic acid. In J. Johnson & T. C. Wallace (Eds.), *Whole grains and their bioactives: Composition and health* (1st ed.). Wiley.

Panwar, P., Dubey, A., & Verma, A. K. (2016). Evaluation of nutraceutical and antinutritional properties in barnyard and finger millet varieties grown in Himalayan region. *Journal of Food Science and Technology*, *53*, 2779–2787. https://doi.org/10.1007/s13197-016-2250-8

Park, H. Y., Kim, Y. J., Kim, K. M., & Lee, J. Y. (2014). Comparative analysis of phytochemicals in several millet grains and their antioxidant activity. *LWT-Food Science and Technology*, *59*(1), 349–355. https://doi.org/10.1016/j.lwt.2014.04.041

Pattanashetti, S. K., Upadhyaya, H. D., Dwivedi, S. L., Vetriventhan, M., & Reddy, K. N. (2016). Pearl millet. In H. D. Upadhyaya & M. Singh (Eds.), *Genetic and genomic resources for grain cereals improvement* (pp. 253–289). Academic Press.

Pranusha, P., Kumar, M. R., Sahi, V. P., & Durai, S. C. (2025). Brown top millet. In *Plant Genebank utilization for trait discovery in millets* (pp. 231–246). Springer. https://doi.org/10.1007/978-981-96-4043-0_9

Pujari, N., & Hoskeri, J. H. (2022). Minor millet phytochemicals and their pharmacological potentials. *Pharmacognosy Reviews*, *16*(32), 101–106.

Ragaee, S., Seetharaman, K., & Abdel-Aal, E. S. M. (2014). The impact of milling and thermal processing on phenolic compounds in cereal grains. *Critical Reviews in Food Science and Nutrition*, *54*(7), 837–849.

Rao, S., Santhakumar, A. B., Chinkwo, K. A., Wu, G., Johnson, S. K., & Blanchard, C. L. (2018). Characterization of phenolic compounds and antioxidant activity in sorghum grains. *Journal of Cereal Science*, *84*, 103–111. https://doi.org/10.1016/j.jcs.2018.07.013

Shahidi, F., & Chandrasekara, A. (2013). Millet grain phenolics and their role in disease risk reduction and health promotion: A review. *Journal of Functional Foods*, *5*, 570–581.

Shao, Y., & Bao, J. (2015). Polyphenols in whole rice grain: Genetic diversity and health benefits. *Food Chemistry*, *180*(8), 86–97.

Sharma, A., Sood, S., Agrawal, P. K., Kant, L., Bhatt, J. C., & Pattanayak, A. (2016a). Detection and assessment of nutraceuticals in methanolic extract of finger (*Eleusine coracana*) and barnyard millet (*Echinochloa frumentacea*). *Asian Journal of Chemistry*, *28*(7), 1633–1637.

Sharma, R., Bhandari, M., Kaur, K., Singh, A., Sharma, S., & Kaur, P. (2022). Molecular interactome and starch–protein matrix, functional properties, phytochemical constituents, and antioxidant activity of foxtail millet (*Setaria italica*) flour as influenced during gaseous ozonation. *Cereal Chemistry*, *99*(5), 1101–1111. https://doi.org/10.1002/cche.10559

Sharma, R., Kumar, D., & Gupta, R. K. (2024). Phytochemical profiling of little millet (*Panicum sumatrense* Roth.). *Indian Journal of Natural Products and Resources*, *15*(4), 555–564. https://doi.org/10.56042/ijnpr.v15i4.9518

Sharma, S., Saxena, D. C., & Riar, C. S. (2016b). Analysing the effect of germination on phenolics, dietary fibres, minerals and γ-amino butyric acid contents of barnyard millet (*Echinochloa frumentaceae*). *Food Bioscience*, *13*, 60–68.

Sharma, S., Saxena, D. C., & Riar, C. S. (2017a). Using combined optimization, GC–MS and analytical technique to analyze the germination effect on phenolics, dietaryfibers, minerals and GABA contents of Kodo millet (*Paspalum scrobiculatum*). *Food Chemistry*, *233*, 20–28.

Sharma, S., Sharma, N., Handa, S., & Pathania, S. (2017b). Evaluation of health potential of nutritionally enriched kodo millet (*Eleusine coracana*) grown in Himachal Pradesh, India. *Food Chemistry*, *214*, 162–168. https://doi.org/10.1016/j.foodchem.2016.07.086

Slama, A., Cherif, A., Sakouhi, F., Boukhchina, S., & Radhouane, L. (2020). Fatty acids, phytochemical composition and antioxidant potential of pearl millet oil. *Journal of Consumer Protection and Food Safety*, *15*, 145–151.

Sosulski, F., Krygier, K., & Hogge, L. (1982). Free, esterified, and insoluble-bound phenolic acids. Composition of phenolic acids in cereal and potato flours. *Journal of Agricultural Food. Chemistry*, *30*(2), 337–340.

Srinivasan, A., Ekambaram, S. P., Perumal, S. S., Aruldhas, J., & Erusappan, T. (2020). Chemical characterization and immunostimulatory activity of phenolic acid bound arabinoxylans derived from foxtail and barnyard millets. *Journal of Food Biochemistry*, *44*(2), e13116. https://doi.org/10.1111/jfbc.13116

Subba Rao, M. V. S. S. T., & Muralikrishna, G. (2002). Evaluation of the antioxidant properties of free and bound phenolic acids from native and malted finger millet (Ragi, *Eleusine coracana* Indaf-15). *Journal of Agricultural and Food Chemistry*, *50*(4), 889–892.

Tang, J., Li, X., Zhang, Y., Yang, Y., Sun, R., Li, Y., Gao, J., & Han, Y. (2022). Differential flavonoids and carotenoids profiles in grains of six poaceae crops. *Food*, *11*(14), 2068. https://doi.org/10.3390/foods11142068

Triki, T., Boussora, F., Gasmi, A., & Tlahig, S. (2025). Phenolic and nutrient profiling of pearl millet seeds from Southern Tunisia: Insights into a nutritious staple crop. *Euro-Mediterranean Journal for Environmental Integration*, *10*, 1919–1929.

Triki, T., Tlahig, S., Benabderrahim, M. A., Elfalleh, W., Mabrouk, M., Bagues, M., & Loumerem, M. (2022). Variation in phenolic, mineral, dietary fiber, and antioxidant activity across southern Tunisian pearl millet germplasm. *Journal of Food Quality*, *2022*(1), 1437306.

Trivedi, P. B., & Gupta, N. (2023). Gas chromatography-mass spectrometry (GC-MS) analysis and phytochemical screening of finger millet and foxtail millet. *Millets*, *2023*, 415–425.

Van Hung, P. (2016). Phenolic compounds of cereals and their antioxidant capacity. *Critical Reviews in Food Science and Nutrition*, *56*(1), 25–35.

Varshini, P. A., Arthanari, P. M., & Paul, R. A. I. (2024). Allelopathic effects of trees leaf extracts on weeds in little millet (*Panicum sumatrense* L.). *Allelopathy Journal*, *63*(2), 187–196.

Verma, R. C., Jitendrakumar, P. H., Ashoka, P., Shekhar, S., Pal, A., Mondal, K., & Singh, B. V. (2023). A review of millet grain phenolics, their health promotion and disease risk reduction. *International Journal of Plant Soil Science*, *35*(20), 863–874.

Wang, S., Mao, J., Xu, Y., Liu, S., Qiao, Z., & Cao, X. (2025). Optimal selenium fertilizer affects the formation of foxtail millet (*Setaria italica* L.) quality by regulating flavonoid metabolism and amino acid metabolism. *Food Science & Nutrition*, *13*(6), e70362. https://doi.org/10.1002/fsn3.70362

Xiang, J., Apea-Bah, F. B., Ndolo, V. U., Katundu, M. C., & Beta, T. (2018). Profile of phenolic compounds and antioxidant activity of finger millet varieties. *Food Chemistry*, *275*, 361–368.

Xiang, J., Li, W., Ndolo, V. U., & Beta, T. (2019). A comparative study of the phenolic compounds and in vitro antioxidant capacity of finger millets from different growing regions in Malawi. *Journal of Cereal Science*, *87*, 143–149.

Xiang, J., Yuan, Y., Du, L., Zhang, Y., Li, C., & Beta, T. (2023). Modification on phenolic profiles and enhancement of antioxidant activity of proso millets during germination. *Food Chemistry: X*, *18*, 100628.

Xu, J., Wang, W., & Zhao, Y. (2021). Phenolic compounds in whole grain sorghum and their health benefits. *Food*, *10*(8), 1921.

Yu, S., Li, C., Wang, X., Herrera-Balandrano, D. D., Johnson, J., & Xiang, J. (2024). Optimization of germination conditions for enriched γ-aminobutyric acid and phenolic compounds of foxtail millet sprouts by response surface methodology. *Food*, *13*, 3340.

Yuan, Y., Xiang, J., Zheng, B., Sun, J., Luo, D., Li, P., & Fan, J. (2022). Diversity of phenolics including hydroxycinnamic acid amide derivatives, phenolic acids contribute to antioxidant properties of proso millet. *LWT – Food Science and Technology*, *154*, 112611.

Zhang, M., Xu, Y., Xiang, J., Zheng, B., Yuan, Y., Luo, D., & Fan, J. (2021). Comparative evaluation on phenolic profiles, antioxidant properties and α-glucosidase inhibitory effects of different milling fractions of foxtail millet. *Journal of Cereal Science*, *99*, 103217.

Zhang, Y., Jiao, J., Li, M., Wei, Z., He, X., Herrera-balandrano, D. D., & Xaing, J. (2025). Effect of milling degree on proximate composition, functional components and antioxidant capacity of foxtail millet. *Food Chemistry*, *27*, 102438.

Zhou, C., Zhou, Y., Liu, T., Li, B., Hu, Y., Zhai, X., & Yang, Z. (2023). Effects of protein components on the chemical composition and sensory properties of millet Huangjiu (Chinese Millet Wine). *Food*, *12*(7), 1458. https://doi.org/10.3390/foods12071458

5

Millet Polyphenols: Extraction, Quality Assessment, and Therapeutic Benefits

José de Jesús Lira-Ricárdez and Lucía Ortega Cabello

Departamento de Sistemas Biológicos, Universidad Autónoma Metropolitana-Unidad Xochimilco, Mexico City, Mexico

5.1 Introduction

Millets (Poaceae) are a group of small-seeded grasses primarily cultivated in Old World regions like Africa, the Indian subcontinent, and China. Their tough adaptation to harsh climatic conditions and nutritional richness makes them a fundamental dietary staple, particularly in underprivileged populations. Beyond their role as a food source, millets are well-known for their important content of bioactive compounds, particularly mono- and polyaromatic compounds known as polyphenols, which contribute to a range of health benefits. These include antioxidative, anti-inflammatory, antidiabetic, and antineoplastic properties (Dekka et al., 2023; Jacob et al., 2024).

The diverse phytochemical profile of millets is key to understanding their therapeutic potential. Compounds such as hydroxybenzoic acids, hydroxycinnamic acids, flavonoids, and tannins have been identified in various millet species. For instance, finger millet is rich in gallic acid, which is a hydrolyzable tannin precursor, ferulic acid, and flavan-3-ols (catechins), while foxtail millet contains flavonoids like rutin and tannins. These compounds not only contribute to health benefits but also enhance the sensory qualities of foods made from millet (Kumari et al., 2017; Kalsi et al., 2023). Recent research has also highlighted the role of processing methods—such as roasting, fermentation, and germination—in modifying the polyphenol content of millets. These techniques can enhance or degrade the bioavailability of polyphenols, thus impacting their antioxidant and therapeutic activities. For example, fermentation has been shown to significantly increase the total phenolic content in pearl millet (Salar et al., 2016). Consequently, understanding the interplay between millet varieties, processing techniques, and their polyphenol profiles is pivotal for harnessing their full therapeutic potential.

Nutraceutical Potential of Millet-Based Food Products: Chemistry and Applications, First Edition.
Edited by Parmjit S. Panesar, Charanjit S. Riar, and Sushil Dhital.

Published 2026 by John Wiley & Sons Ltd.

5.2 Nutritional and Therapeutic Uses of Millet

Millet is a plant that stands out from other plants due to its high yield in biomass and grains as an alternative source of energy and proteins (Popović et al., 2020). There are several species of millet; however, only a few species are cultivated mainly in Africa, India, and China (Figure 5.1). These varieties are classified based on the size of the grain as major and minor millets (Dekka et al., 2023) (Figure 5.2).

These varieties vary in nutritional content regarding proteins which ranges from 6.2 to 12.3 g/100 g, fat from 1.1 to 4.8 g/100 g, and fiber from 2 to 13.6 g/100 g, which can be divided into soluble fiber (1.40–5.65 g/10 g) and insoluble fiber (8.19–15.70 g/100 g) (Jacob et al., 2024; C. Kaur, 2023), apart from carbohydrates, minerals, and vitamins, and have a sweet and astringent taste according to ayurvedic medicine. It has been reported to present different therapeutic activities such as hypoglycemic and antioxidant effects, which make

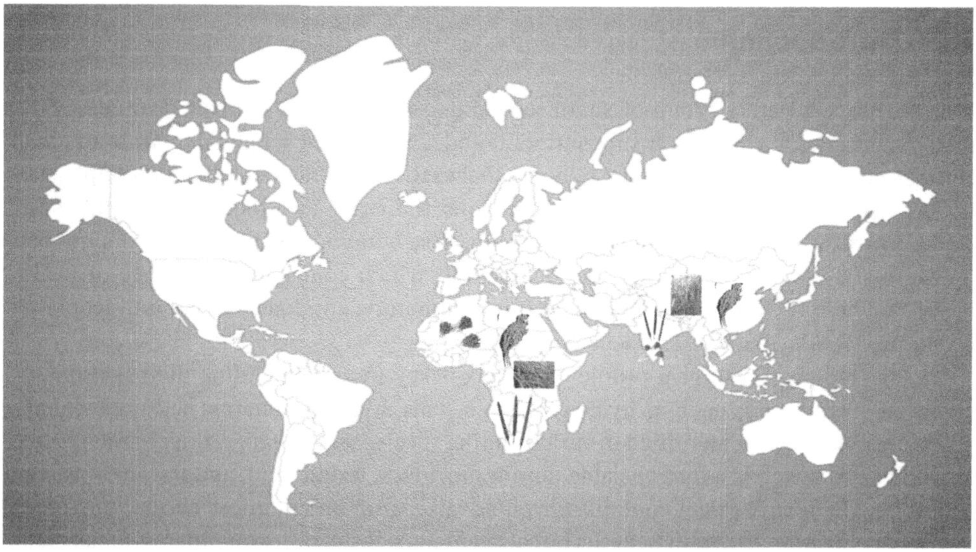

Figure 5.1 Cultivation of some varieties of millet in Africa, India, and China

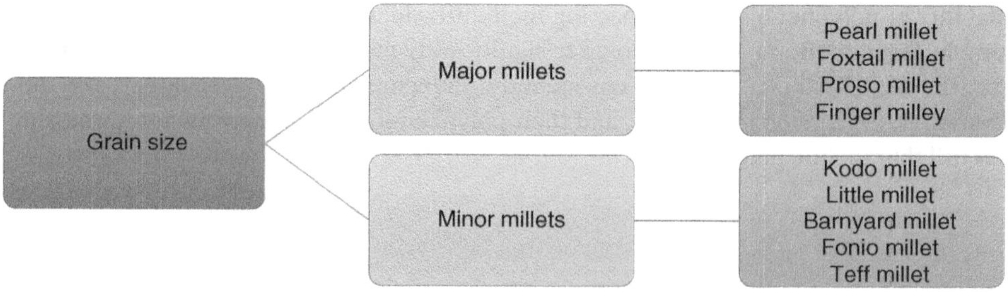

Figure 5.2 Classification of millets based on the grain size. *Source:* Dekka et al. (2023)/John Wiley & Sons

it ideal for treating diabetes and its complications, regulating blood pressure, regulating thyroid function, and acting as an antineoplastic, among others (Jacob et al., 2024). In Table 5.1 the therapeutic activity and phytochemical composition of some millet varieties are found.

Finger millet is considered one of the most important millet varieties in the world because of its grain that grows in Asia and Africa and is a meal in India and Ethiopia in the form of cakes, pancakes, and porridge, among other food, and also has a high nutritional value in vulnerable populations such as children and pregnant women. This millet may appear in different colors due to the levels of calcium and protein content, and its pericarp is easily removed. This seed has been used mainly to treat diabetes, osteoporosis, anemia, management of anxiety, depression, insomnia, weight reduction, lipid-lowering, management of ulcers, and tissue repairs (Erharuyi et al., 2017; Owheruo et al., 2019). It is known that the seed of finger millet has polyphenols such as gallic acid, tannic acid, vanillic acid, ferulic acid, caffeic acid, chlorogenic acid, *trans*-cinnamic acid, *p*-coumaric acid, syringic acid, protocatechuic acid, *p*-hydroxybenzoic acid, quercetin, kaempferol, naringenin, phloroglucinol, apigenin, (+)-catechin, epicatechin, orientin, *trans*-feruloylmalic acid, and dimers and trimers from some of these polyphenols (Figure 5.3). Other metabolites found on finger millet are alkaloids, terpenoids, saponins, cardiac glycosides, and tannins, among others (Erharuyi et al., 2017; D. Mounika et al., 2022; Thagunna et al., 2022); however, some of these metabolites are removed during the roasting process and also depend on the extracting solvent mainly in aqueous and ethanolic extractions (Jadhav et al., 2022; Jagati et al., 2021).

Foxtail millet is produced in different regions of Europe, South America, Africa, and Asia due to its ability to adapt to different weather conditions and has a higher nutritional value compared to rice due to its micronutrient, protein, fiber, and carbohydrate content, and has been used in different foods such as noodles, soups, pancakes, beer, bread, and porridge. It is also known to have ferulic, chlorogenic, salicylic, hydroxycinnamic, and *p*-coumaric acids, pyrogallol, catechin, rutin, tannins, *trans*-lutein, and *trans*-zeaxanthin (Figure 5.4), where the presence and number of compounds depend on the solvent and pH of the extraction used to carry out the extraction (Abedin et al., 2022; John Calvien Hutabarat & Aditya Bowie, 2022; Kalsi & Bhasin, 2023), among the therapeutic activities that have been attributed to foxtail millet are antineoplastic, antidiabetic, antioxidant, anti-inflammatory, antifungal, antibacterial, antiviral, cardioprotector, and lipid-lowering (Abedin et al., 2022; John Calvien Hutabarat & Aditya Bowie, 2022).

Pearl millet has become important due to its nutritional value which could make it equivalent to different cereals and, as a potential source of bioethanol, it may protect against different degenerative diseases as well as aid in bone remodeling (Biradar et al., 2024; Nani et al., 2020), among the polyphenols that have been identified in Pearl millet are protocatechuic acid, gallic acid, *p*-coumaric acid, *p*-hydroxybenzoic acid, vanillic acid, syringic acid, ferulic acid, vanillin, quercetin, catechin, epicatechin, kaempferol glycoside, and myricetin (Bello et al., 2022; Hassan et al., 2020) (Figure 5.5).

Barnyard millet is also known to regulate blood sugar concentration, as well as blood pressure, anti-diabetic, and antineoplastic, among other activities (Chandrkar et al., 2024; Kumaraguru et al., 2024). It is known to possess caffeic acid, ferulic acid, kaempferol, catechins, coumaric acid, hydroxycinnamic acid, and hydroxybenzoic acid (Bala Durairajan et al., 2024) (Figure 5.6).

Table 5.1 Therapeutic activity and phytochemical composition of some millet varieties

Millet variety	Therapeutic activity	Phytochemical composition
Finger millet (*Eleusine coracana*)	Nephroprotective, antidiabetic, antioxidant, wound healing, antimicrobial, lipid-lowering, and calcium regulation (Kalsi et al., 2023)	Tannins and phenolic compounds (Kalsi et al., 2023)
Foxtail millet (*Setaria italica*)	Antioxidant, anti-inflammatory, antineoplastic, lipid-lowering, and metal chelating (R. Sharma & Sharma, 2022)	Polyphenols, flavonoids, tannins, triterpenoids, phytosterols, and saponins (R. Sharma & Sharma, 2022)
Pearl millet (*Pennisetum glaucum*)	Antidiabetic, antioxidant, antimicrobial, blood pressure regulator, anti-inflammatory, and lipid-lowering (El Kourchi et al., 2024)	Polyphenols, flavonoids, carotenoids, alkaloids, tannins, and anthraquinones (El Kourchi et al., 2024; Mawouma et al., 2022)
Barnyard millet (*Echinochloa frumentacea*)	Treatment of celiac disease, obesity, cancer, diabetes, immunomodulator, antioxidant, and anti-inflammatory (Dutta et al., 2023; Tamil Selvan et al., 2023)	Polyphenols, isorhamnetin, carotenoids, and tannins (Dutta et al., 2023; Tamil Selvan et al., 2023)
Kodo millet (*Paspalum scrobiculatum*)	Diabetes, lipid-lowering, antiulcer, cancer, celiac disease, and antioxidant (Jyothi Reddy et al., 2020; Rajesh & Suku, 2022; Ranjan et al., 2023)	Polyphenols, flavonoids, tannins, saponin, phytic acids, alkyl resorcinol, and coumarins (Dey et al., 2022; Kharsan et al., 2023; Yadav, 2022)
Little millet (*Panicum sumatrense*)	Antioxidant, antimicrobial, anti-hemolytic, antifungal, antimicrobial, diabetes, cardioprotector, and prevention of cataracts (Dey et al., 2022; Indirani & Devanesa, 2021; N. A. Dogar et al., 2023)	Phytic acids, tannins, phenolic compounds, flavonoids, cardiac glycosides, saponins, terpenoids, and phlorotannins (Dey et al., 2022; N. A. Dogar et al., 2023)
Proso millet (*Panicum miliaceum*)	Antidiarrheal, spasmolytic, bronchodilator, hypotensive, bone healing, neuroprotector, celiac disease, lipid-lowering, diabetes, antioxidant, and bone strengthening (Panda & Yadav, 2021; Saqib et al., 2022; Y. Zhang et al., 2024b)	Alkaloids, phenols, terpenoids, flavonoids, glycosides, phytates, and tannins (Pujari & Hoskeri, 2022; Saqib et al., 2022)
Brown top millet (*Urochloa ramosa*)	Antioxidant, anti-inflammatory, antineoplastic, cardioprotector, and hypoglycemic (M Singh et al., 2022)	Tannins, saponins, flavonoids, glycosides, quinones, phenols, terpenoids, cardiac glycosides, coumarins, phlorotannins, and anthocyanins (Singh et al., 2022)
Teff (*Eragrostis tef*)	Celiac disease, antidiabetic, anemia (Gebru et al., 2020), and antineoplastic (Seo et al., 2024)	Phenolic acids, flavonoids, C-and glycoside of flavones (Gebru et al., 2020),
Fonio millet/acha (*Digitaria exilis*)	Anti-inflammatory, antidiabetic, and immunomodulatory (Traore, 2011)	Flavonoids in free, C- and O-glycosylated forms (Sartelet et al., 1996), phytic acid, proanthocyanidins, alkaloids, saponins, anthraquinones, terpenoids, phytosterols, and phlorotannins (Zhu, 2020)

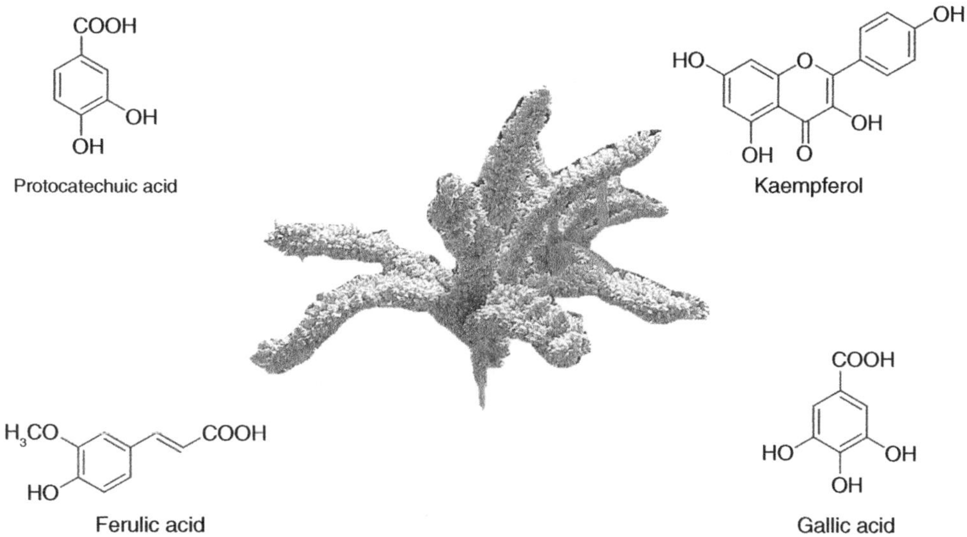

Figure 5.3 Polyphenols present on finger millet. *Source:* Created using ChemDraw.

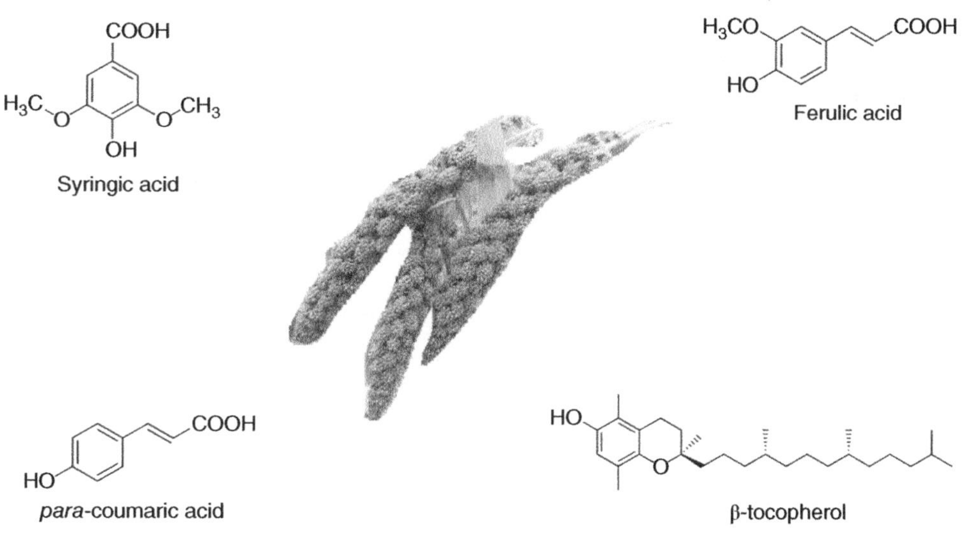

Figure 5.4 Polyphenols are present in foxtail millet. *Source:* Created using ChemDraw

Kodo millet grows mainly in India with antineoplastic and antidiabetic activities as well as a coadjuvant in the treatment of arthritis, in reducing body weight, as well as in neurological disorders, and can modulate the immune response. It is known that kodo millet contains phytochemicals, which are attributed to its biological activity such as catechin, naringin, *p*-coumaric acid, taxifolin, ferulic acid, sinapic acid, and campesterol (Dey et al., 2022) (Figure 5.7).

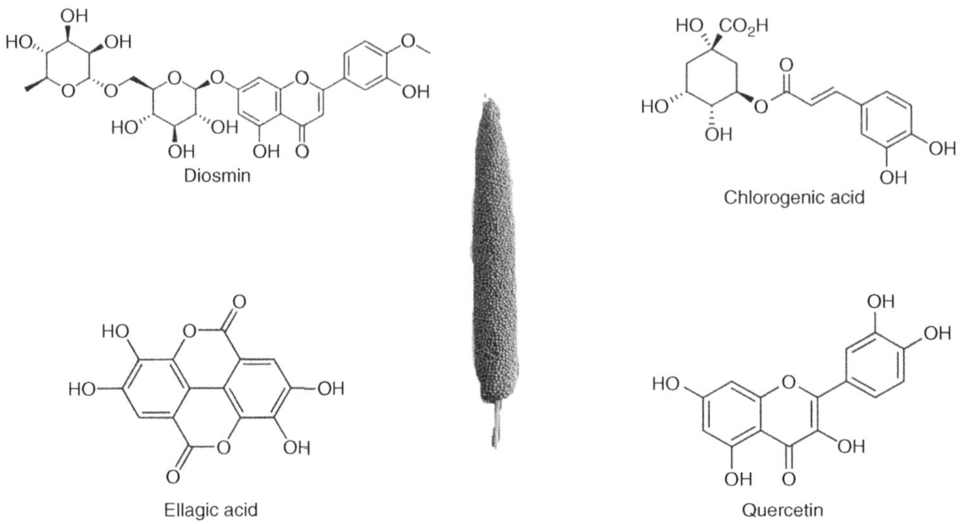

Figure 5.5 Polyphenols are present in pearl millet. *Source:* Created using ChemDraw

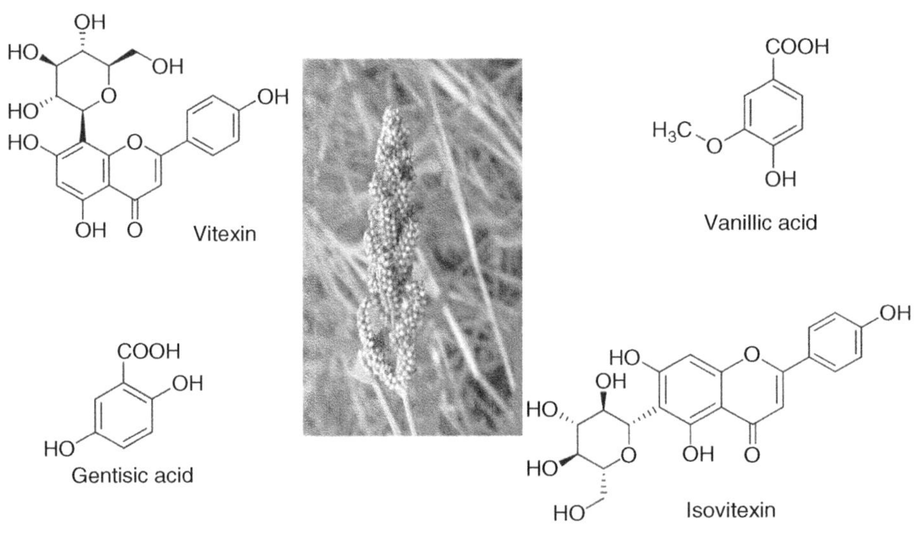

Figure 5.6 Polyphenols are present in barnyard millet. *Source:* Created using ChemDraw.

In the case of little millet, it is rich in phytates, phenolics (such as kaempferol, apigenin, and luteolin), and sterols, among other compounds; these compounds give little millet therapeutic activities such as antidiabetic and antineoplastic (Dey et al., 2022) (Figure 5.8).

5.3 Role of Polyphenols in the Therapeutic Activity of Millets

Polyphenols are natural products found in several sources such as plants, cereals, and fruits, among others that may contribute to different organoleptic features of different foods such as taste (bitterness, astringency, flavor), odor, and oxidative stability; according

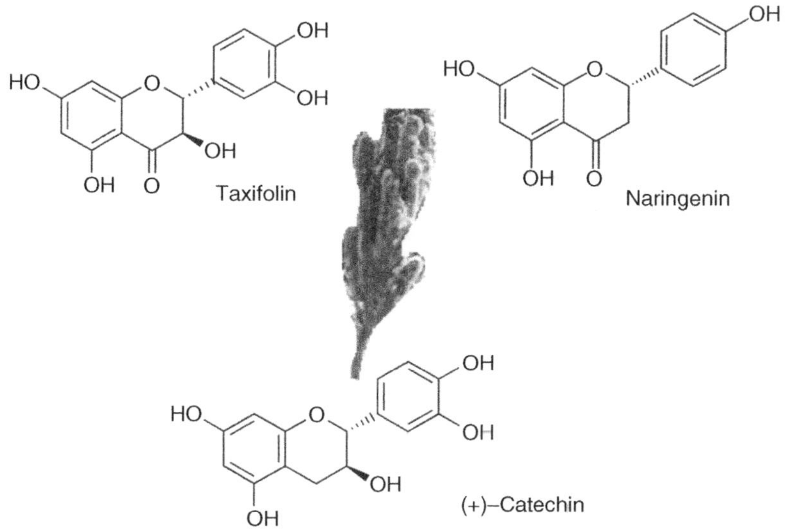

Figure 5.7 Polyphenols present on kodo millet. *Source:* Created using ChemDraw.

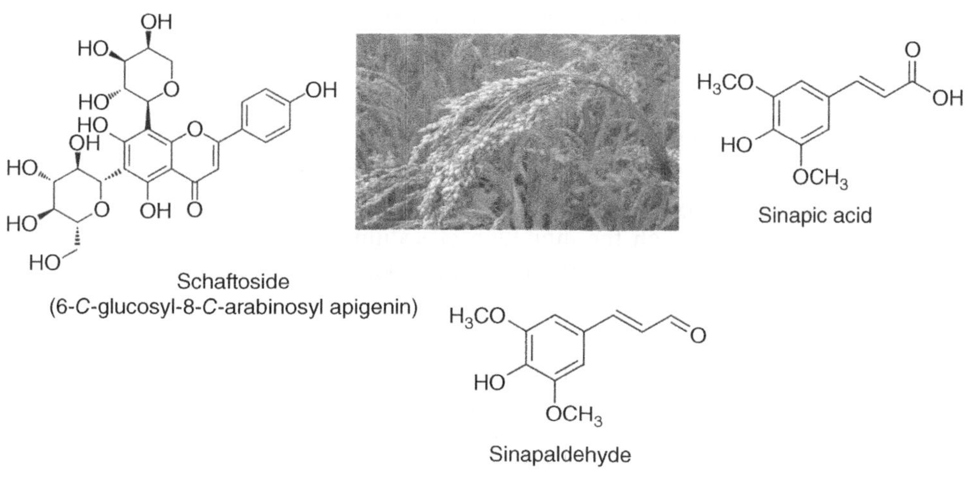

Figure 5.8 Polyphenols are present in little millet. *Source:* Created using ChemDraw.

to Pandey and Rizvi, there are four main classes of phenolic compounds (Figure 5.9) (Pandey & Rizvi, 2009).

The antioxidant activity is important because it is the beginning of many activities of polyphenols such as anti-inflammatory, hepatoprotection, and neuroprotection by regulating pathways such as Nrf2 and Sirt1 and the reduction of NF-κβ, thus preventing neuronal cell death; while in the case of hepatoprotection, polyphenols may act by restoring serum total protein, TNF-α, and caspase 3 (Leri et al., 2020; Yan et al., 2020). Another organ in which polyphenols exert their protective activity is the heart by reducing inflammation, decreasing NF-κβ, and preventing the phosphorylation of MAPK; polyphenols also reduce

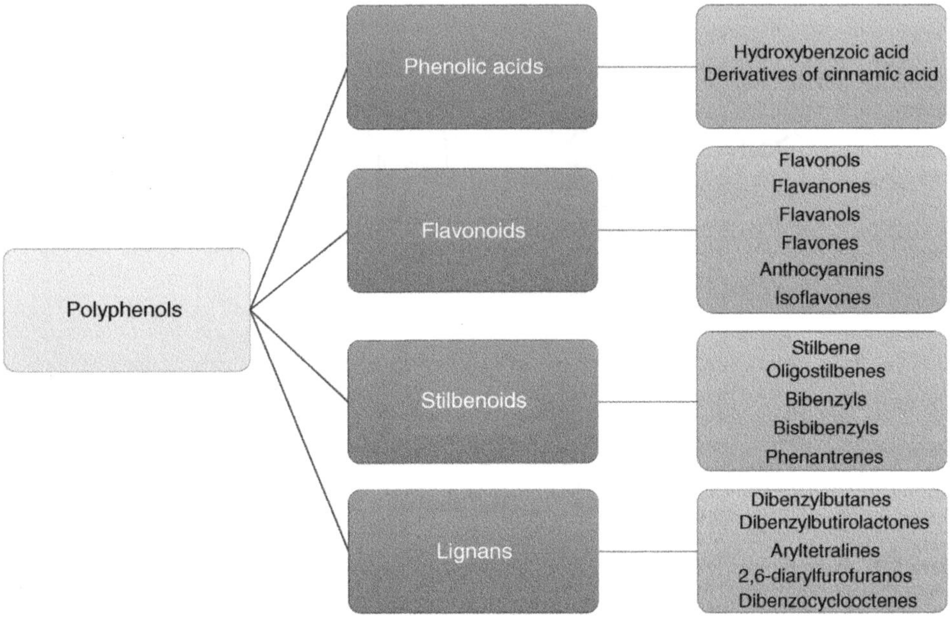

Figure 5.9 Classification of polyphenols based on their chemical structures. *Source:* Cui et al. (2020), Gupta and Pandey (2020), and Pandey and Rizvi (2009)

the autophagy and apoptotic processes, modulating Ca^{2+}-dependent protein kinases and normalizing Ca^{2+} efflux, among other mechanisms (Najjar & Feresin, 2021).

Polyphenols, as previously mentioned, may regulate blood sugar concentration by reducing glucose absorption through the inhibition of sodium-independent transporters and stimulation of insulin secretion with the activation of kinases and insulin-sensitive receptors by regulating membrane channels sensitive to potassium (Leri et al., 2020). At the same time, polyphenols may contribute to modulating the lipid concentration in blood by inhibiting enzymes from the lipase family and by impairing the formation of micelles as well as the modification of hepatic enzymes responsible for xenobiotic metabolism, such as aromatic hydrocarbon receptors Keap1 and AhR (Nakamura, 2023).

It has been assessed that the antineoplastic activities of polyphenols are due to the interaction with many therapeutic targets that are involved in a variety of processes such as signaling, catalysis, and DNA binding. An example of these mechanisms is the binding of polyphenols to aromatic hydrocarbon receptors; these receptors are responsible for activating the metabolism of different hydrocarbons, causing a conformational change and exposing a nuclear translocation signal which transfers this complex to the nucleus and the receptor forms a heterodimer activating xenobiotic metabolism. Polyphenols act as agonists of this receptor through competitive binding, thus inhibiting the activation of xenobiotic metabolism and interrupting, in the case of polyphenols containing five hydroxyl residues, while those with a different number of hydroxyl residues act as antagonists (Lyubitelev & Studitsky, 2023). Other mechanisms involved in the antineoplastic activity of polyphenols are the activation of apoptotic pathways by caspases via p53 (Chimento et al., 2023; E. Sharma et al., 2022). Other mechanisms involved in this activity are cell cycle arrest in different phases depending on the cell type by regulating cyclins, autophagy, regulation of mRNA, and inhibition of other receptors such as P-glycoprotein (E. Sharma et al., 2022).

Due to their capacity to modulate inflammatory pathways, it can be said that polyphenols also have immunomodulatory activity and therefore have a role in autoimmune diseases such as osteoarthritis by regulating the expression through suppressing the IRAK1/NLRP3 and caspase-3 pathways with a subsequent decrease in proinflammatory cytokines and by reducing the apoptosis of chondrocytes (Sirše, 2022).

5.4 Pretreatment of Millets and Their Polyphenol Extraction

Phenolic compounds represent an important group of bioactive compounds with multiple health benefits, as has been previously described, but also these compounds pose a crucial processing challenge related to the complexity and diversity of chemical structures, including phenolic acids, flavonoids, tannins, some lignans, and their respective glycosides, along with their chemical stability that can be affected by some outdoor conditions (temperature, storage time, pH) and industrial procedures for food production (Albuquerque et al., 2021; Ali et al., 2022; L. Shi et al., 2022).

For this consideration, pretreatment procedures in millet grains are crucial for preserving the bioactive compounds that exhibit beneficial properties. Decortication or coat dehulling is the most common postharvest grain treatment, which is necessary for adequate access to the millet kernel for food processing and the elimination of some antinutrients such as phytates or saponins, but also millet grains can be submitted to other industrial techniques such as hydrothermal (parboiling), mechanical (milling, roasting, puffing), and chemical treatments (soaking). Polyphenols can be affected by these procedures, diminishing the amount of some antinutrient polyphenols, but also the amount and quality of some important phenolic compounds such as flavonoids and phenolic acids (Mahajan et al., 2024; Mittal et al., 2024). Some studies showed that water soaking treatment can proportionate a less-decreased polyphenols amount in some grains such as finger millet and foxtail millet, allowing a reasonable amount of beneficial polyphenols, and additionally, some novel techniques such as soak-steam thermal treatment have exhibited polyphenol-increased amount in foxtail millet, related to the liberation and migration of phenolic compounds from the outer seed layers to the endosperm, enhancing their content (Kaur et al., 2024; Vf et al., 2022; Wang et al., 2022).

When dealing with natural products, the pretreatment of the vegetable material is important because, depending on the pretreatment technique, there can be an optimal extraction yield, since physical methods such as mechanical disruption, high-pressure-assisted processes, pulsed electric field application, ultrasound irradiation, freeze–thaw process, continuing with biological methods such as enzymatic to chemical methods, and the hybrid between different techniques (Zhao et al., 2014). Regarding millets, there are common pretreatment techniques such as heating, germination, and fermentation. Wang et al. (2022) mentioned these techniques; in the case of fermentation, researchers have used lactobacilli and yeast, reaching the maximum total phenolic content within 72 hours where vitexin and its glycoside increased significantly. Similar research obtained flavonoids and cinnamic acid derivatives increasing their content after acidic hydrolysis (Balli et al., 2020); Salar et al. (2016) extracted polyphenols from pearl millet with fermentation using *Aspergillus awamori* MTCC-548 through solid-state fermentation in which the polyphenol content increased fivefold after eight days of fermentation compared to unfermented millet. There may be other techniques that may reduce the polyphenol content such as malting and blanching (Archana et al., 1998).

Heating helps to improve the quality of polyphenols depending on the method, for example, roasting of proso millet where the amount of phenolic acids such as syringic, gallic, *p*-hydroxybenzoic, ferulic, and synaptic acids were increased while catechin decreased after millets were roasted at 110 °C for 10 minutes, while with finger millet decreased (Wang et al., 2022). Other used techniques to treat millets are extrusion, boiling, and microwave (Li et al., 2023); in the research by Li et al. (2023), high-pressure boiling, extrusion, baking, steam boiling, and microwave were evaluated at different temperature, times, and potency regarding microwave pretreatment, in which steam boiling yielded the highest polyphenol content. In general, the authors noticed three changes in antioxidant activity during the pretreatment in all techniques: in the first stage, superficial polyphenols are degraded having consequences in the antioxidant activity; in the second stage, due to cellular damage, some polyphenols are released, and the antioxidant activity increases, followed by their immediate degradation decreasing again the antioxidant activity. In the case of steam boiling due to the presence of water, the influence of heat is decreased, thus allowing the protection of free polyphenols before and after cellular destruction.

In another research (Kalam Azad et al., 2019), different treating methods were used on kodo millet (roasting, steaming, puffing, and extrusion); the authors demonstrated significant differences in the phenolic and flavonoid content, in which roasting at 110 °C for 10 minutes presented the highest content of both components where gallic acid was the phenolic acid with the highest content. The authors believe that heat aids in the hydrolysis of *C*-glycosylflavones as well as the degradation of cellular components. Coupling of pretreatment methods is important as well, as the research reported by Balasubramaniam et al. (2019), in which the authors compared enzymatic assisted ultrasound using xylanase and conventional ultrasound pretreatment of finger millet with heat reflux; the authors found that the enzymatic process enhanced phenolic content, while conventional ultrasound presented similar results with heat reflux.

Various nonthermal pretreatment techniques are applied in the mullet processing under different circumstances (Table 5.2). Ultrasonication, a nonthermal pretreatment technique, utilizes high-frequency sound waves to induce cavitation, a process that forms microscopic bubbles that implode and create localized high pressure and temperature. This mechanical disruption breaks down cell walls, facilitating the release of intracellular bioactive compounds such as phenolics. Ultrasonication has been widely applied to finger millet, where studies have shown a significant increase in phenolic content—up to 2.3 times higher than that obtained through conventional extraction methods. This enhancement is attributed to the improved mass transfer and the breakdown of bonds holding phenolic compounds in their bound forms, resulting in higher antioxidant activity. The technique is particularly effective in preserving heat-sensitive compounds and offers a green, solvent-saving

Table 5.2 Nonthermal pre-treatments for millet grains

Technique name	Millet species	Information on phenolic compounds	References
Cold plasma	Finger millet, foxtail millet, and pearl millet	Increase in phenolic content due to the release of bound compounds through RONS-media and reduction of antinutrients	(Balendran et al., 2024; Jaddu et al., 2024; Monica et al., 2023)
Germination	Finger millet, foxtail millet, and proso millet	Significant increase in phenolic content due to enzymatic hydrolysis releasing bound polyphenols	(Nguyen et al., 2023; Xiang et al., 2023; Zhang et al., 2024a)
Pulsed electric field (PEF)	Pearl millet and foxtail millet	Increase in phenolic compounds through electroporation and enhanced mass transfer	(Bheemaiah Balyatanda et al., 2024; Ramakrishnan et al., 2023)
Ultrasonication	Barnyard millet, browntop millet, and finger millet	2.3-fold increase in phenolic compounds compared to conventional methods	(Balasubramaniam et al., 2019; Bhandari et al., 2025; Mounika et al., 2024)

alternative for functional food applications (Balasubramaniam et al., 2019; Bheemaiah Balyatanda et al., 2024; Ramakrishnan et al., 2023).

Cold plasma is an innovative nonthermal technology that has shown potential in processing millets by enhancing their nutritional and functional properties. Ionized gas at low temperatures is used to induce chemical modifications on the grain's outer layers without damaging its overall integrity. Cold plasma has been applied to finger millet and foxtail millet, and the polyphenol content has revealed that the treatment enhanced total polyphenolic content (TPC) by breaking down the cell wall matrix and increasing the availability of bound phenolic compounds. This enhancement is attributed to the oxidative stress generated by reactive oxygen and nitrogen species (RONS), which facilitate the release of phenolic compounds from their bound forms, improving the antioxidant properties of the treated millet grains (Jaddu et al., 2024; Ramakrishnan et al., 2023).

During the extraction of polyphenols, one important variable that influences the yield is solvent selection, where common solvents used in polyphenol extraction can be either alcoholic media such as methanol or ethanol, acetone, or acidic or aqueous solutions of these solvents, in which it is necessary to consider as well that polyphenols may be complexed with other molecules such as sugars and proteins, thus affecting their polarity. The use of acidic media is important because at alkaline pH some hydroxycinnamic acid derivatives may be degraded (Ajila et al., 2011). An example of solvent selection in millets is the one reported by Sunagar and Sreerama (2023), in which they tested different solvents for polyphenol extraction from browntop millet in which 80% acetone and 70% methanol were the best solvents for the extraction of polyphenols. Gallic, caffeic, and ferulic acids were identified as phenolic acids, while myricetin and kaempferol were the main flavonoids.

Even though there are classic natural product extraction methods, such as percolation, maceration, and Soxhlet extraction, among other techniques, alternative methods have been noticed due to their use of less solvent and energy, thus becoming ecofriendly techniques such as supercritical fluids, enzymatic extraction, microwave, ultrasound, and the use of eutectic solvents (Nayak et al., 2024; H. Zhang et al., 2022). In the research by H. Zhang et al. (2022), in which they use switchable hydrophilicity solvents by changing pH, thus simplifying operations such as evaporation and minimizing loss of product. Such solvents were prepared by using mixtures of octanoic, nonanoic, or decanoic acid with linalool in a 1 : 1 ratio as a hydrophobic solvent; then 4% NaOH solution was added to convert it into a hydrophilic solvent. Afterward, millet powder was mixed with the hydrophilic mixture and was vortexed for 80 seconds, followed by centrifugation at 3,600 g for 2 minutes; once centrifuged, the mixture and the supernatant were separated, and a 1 M HCl solution was added to make the solvent hydrophobic and thus separate the mixture in two phases. Then, the aqueous phase was separated and lyiophilized. This extraction was compared with ethanol showing good opportunities with this alternative solvent, in which the extraction yield was higher than that of ethanol, allowing its recycle in multiple extractions. The influence of the pH in polyphenol extraction from millets is supported by Chethan and Malleshi (2007b), in which the authors reported that basic media results in precipitation decrease polyphenol content, while at acidic and near neutral pH, polyphenols remain stable because it can prevent their oxidation (Ajila et al., 2011; Chethan & Malleshi, 2007b).

Zheng et al. (2022) coupled the use of different eutectic solvents with ultrasound for polyphenol extraction from foxtail millet and compared it with conventional solvent extraction using 80% methanol solution. The eutectic solvent that yielded the highest polyphenol content was a mixture of betaine and glycerol in a 1 : 2 ratio compared to methanolic extraction because of its viscosity, followed by the addition of water in order to decrease the viscosity due to breakage of hydrogen bonds for 30 minutes at 60 °C at 250 W; these extraction conditions yielded more than the double the soluble phenolic content compared to methanolic extraction, while insoluble phenolic content was similar in both extraction methods. The highest soluble polyphenols were *p*-coumaric acid, followed by apigenin-*C*-dihexoside, while ferulic acid followed by *p*-coumaric acid were the polyphenols with the highest content as insoluble polyphenols.

Subcritical water extraction (SWE) and supercritical carbon dioxide ($SCCO_2$) extraction have been applied to millets for efficient phenolic recovery. In *foxtail millet*, SWE at 200 °C enhanced the release of hydroxybenzoic acids, while $SCCO_2$ extraction with ethanol as a co-solvent improved the recovery of apigenin derivatives. These methods preserve the phenolic profile by avoiding extreme thermal degradation, and $SCCO_2$, in particular, has shown high selectivity for lipophilic phenolic compounds, such as tocopherols from millet oils (Essien et al., 2020; Y. Shi et al., 2015; J. Zhang et al., 2020).

In the review carried out by Nayak et al. (2024), a microwave extraction method in kodo millet is reported, where approximately six minutes at 100 °C using 50% ethanol solution was higher than solvent extraction.

5.5 Characterization of Phenolic Extracts from Millets

Due to their nutritional and therapeutic properties, millets have become important, where the use of analytical techniques may help not only for polyphenol quantification but also for the characterization of these polyphenols either by HPLC-DAD or HPLC-MS techniques and the elucidation of their therapeutic activities, as reported in the research of Goudar et al. (2023), where they compared little, barnyard, foxtail, kodo, proso, finger, and pearl millets by carrying out methanolic extractions to obtain free phenolic compounds, followed by a basic extraction and solvent extraction in the remaining material after methanolic extractions to obtain the bound phenolic compounds. Table 5.3 shows, from major to minor, free and bound phenolic compounds in all the analyzed millets.

In this regard, kodo millet presented the highest free phenolic compound, followed by finger > pearl > barnyard > foxtail > proso > little millets. With regard to bound phenolic compounds, from highest to lowest were kodo > pearl > finger > proso > foxtail > barnyard millets. In the research of Bello et al. (2022), the authors carried out the polyphenol extraction with an ethanol:water mixture (1 : 1) using percolation technique, and UV–Vis and FT-IR spectra were obtained. The UV–Vis spectra of ethanolic extracts from fonio, finger, and pearl millets were different in shape and intensity in a wavelength range from 325 to 425 nm, where fonio and finger millets present a common maximum absorption wavelength (λmax) at approximately 350 nm, while pearl millet presented a λmax at approximately 400 nm. After λmax, in the case of fonio millet, a gradual decrease in the absorption spectra occurred, while with finger millet, a first decrease was observed at 400 nm, followed by a second decrease at 550 nm; regarding pearl millet, three absorbance decreases were observed at 475, 525, and 625 nm. With respect to FT-IR spectra, more signals in common were found in the case of fonio and finger millets at 3,279, 3,021, 2,921, 2,851, and 1,544 cm^{-1}, while only one signal was common between finger and pearl millets (1,640 cm^{-1}) and between fonio and pearl millets (1,462 cm^{-1}) attributed to phenolic acids (protocatechuic, gallic, *p*-coumaric, *p*-hydroxybenzoic, vanillic, syringic, and ferulic acids) and flavonoids (quercetin, catechin, myricetin, and epicatechin).

In another research (Pradeep & Sreerama, 2015), the authors analyzed the phenolic acids and flavonoids profiles of barnyard, proso, and foxtail millets with different processing treatments (raw, germination, steam, and microwave), followed by an extraction with acidic methanol (HCl 1% w/v) and the mixture was refluxed for 30 minutes; afterward, the mixture was centrifuged and concentrated under reduced pressure, polyphenols were quantified with UV–Vis spectrometry, and the profiles were analyzed with RP-HPLC using as mobile phase a mixture of acidic water (water adjusted to pH 2.8 with formic acid) and acetonitrile in gradient elution, different wavelengths were used to detect flavonoids (342 nm), phenolic acids (320 nm), and benzoic acids derivatives (254 nm). In the case of phenolic content, the germ of barnyard millet presented the highest phenolic content (246.89 mg of ferulic acid equivalents/100 g), while the germ of foxtail millet had the highest flavonoid content (157.73 mg of catechin equivalents/100 g). Among the benzoic acid derivatives that were identified were gallic acid, *p*-hydroxybenzoic acid, and vanillic acid, where the germ of foxtail millet, the raw material of proso millet, and the germ of barnyard millet presented the highest amount of gallic acid (27.44 μg/g), *p*-hydroxybenzoic acid (34.18 μg/g), and vanillic acid (150.11 μg/g), respectively. Identified cinnamic acid

Table 5.3 Free and bound phenolic compounds detected from mayor to minor content in the analyzed millets

Free phenolic acid compounds						
Foxtail	**Little**	**Barnyard**	**Kodo**	**Proso**	**Pearl**	**Finger**
Ferulic	Ferulic	Ferulic	Ferulic	Ferulic	Ferulic	Ferulic
Caffeic	Caffeic	Caffeic	Syringic	Syringic	Caffeic	Caffeic
Syringic	Gallic	Syringic	Gallic	Caffeic	Gallic	Syringic
Gallic	Chlorogenic	Gallic	Caffeic	Gallic	Syringic	Gallic
Chlorogenic		Chlorogenic	Chlorogenic	Chlorogenic	Chlorogenic	Chlorogenic
Bound phenolic acid compounds						
Foxtail	**Little**	**Barnyard**	**Kodo**	**Proso**	**Pearl**	**Finger**
Ferulic	Ferulic	Ferulic	Ferulic	Ferulic	Ferulic	Ferulic
Caffeic	Caffeic	Gallic	Caffeic	Gallic	Caffeic	Gallic
Gallic	Syringic	Caffeic	Gallic		Syringic	Caffeic
Syringic	Chlorogenic	Syringic			Chlorogenic	Chlorogenic
Chlorogenic		Chlorogenic				

Source: Adapted from Goudar et al. (2023).

derivatives were caffeic, chlorogenic, ferulic, cinnamic, and *p*-coumaric acids, where the raw material of proso millet, the germ of barnyard millet, and the raw material of foxtail millet presented the highest content of caffeic acid (15.97 μg/g), sinapic acid (52.62 μg/g), and *p*-coumaric acid (45.41 μg/g) respectively; it is to be noticed that the germ of foxtail millet also presented the highest amount of chlorogenic (45.68 μg/g) and ferulic (353.52 μg/mL) acids. Regarding flavonoids, apigenin, naringenin, kaempferol, myrecetin, and rutin were identified, where the raw material of barnyard millet presented the highest content of naringenin and kaempferol (85.81 and 118.92 μg/g, respectively), while the germ and the raw material of foxtail millet, as well as the germ of proso millet, presented the highest content of apigenin (146.17 μg/g), rutin (17.14 μg/g), and myrecetin (72.64 μg/g), respectively. The authors discuss that the variation in the polyphenol content could be either due to the activation of different biosynthetic pathways (mainly during the germination process by activating *p*-coumarate 3-hydroxylase, *O*-methyl transferase, and ferulate 5-hydroxylase) or to the modification of the seed structure in the different procedures that improved the penetration of the solvent and therefore enhance the extraction, while steam and microwave yielded the lowest phenolic content probably due to oxidative stress.

The influence of the germination state on the polyphenol profile of millets is supported by the research of Xiang et al. (2023), in which the authors used proso millet varieties (black, red, and yellow) and let germinate for six days at 25 °C with 90% relative humidity, followed by freeze-drying and grinding and extracted with methanol 80% and centrifuged, phenolic and flavonoid content were quantified with UV–Vis spectrometry, and polyphenols were identified by UPLC (ultra performance liquid chromatography) coupled to ESI-MS with a mixture of acidic water (0.1% formic acid) and acidic methanol (0.1% formic acid) as mobile phase with gradient elution. In the case of total phenolics, it was observed that the content increased as days passed by, where on the sixth day of germination, the samples presented the highest phenolic content with a gradual change in the free:bound phenolics ratio; with regard to flavonoid content, black variety presented the highest flavonoid content at the beginning of the experiment and decreased as days passed by, while the other varieties remained constant for two days, beginning their increase at the fourth day of the experiment. Table 5.4 shows the identified free and bound phenolic compounds in the proso millet.

In Table 5.5, it can be seen which compounds increased or decreased their content during the germination process.

The variation of individual phenolic compounds in this research may support the hypothesis of the activation of different enzymes responsible for the biosynthesis of polyphenols during the germination process, as reported by Pradeep and Sreerama (2015).

As it was observed in the previous research, the variety may influence the polyphenol profile as well; this can also be observed in the review presented by Chethan and Malleshi (2007a), in which the authors report different phenolic content on different finger millet varieties (Indian brown, African brown, and Indian white) with values oscillating between 0.003% and 2.44%, with the African brown variety being the one that presented the highest range of polyphenol content (0.54–2.44%). The authors also noticed that the ideal conditions for extracting polyphenols were in acidified medium (1% HCl) under reflux, with methanol being the best solvent to extract polyphenols. The authors report that among the compounds identified in finger millet were gallic, protocatechuic, *p*-hydroxybenzoic,

Table 5.4 Identified free and bound phenolic compounds of the analyzed proso millet samples

Phenolic compound type	Free	Bound
Hydroxybenzoic acids	Protocatechualdehyde *p*-Hydroxybenzoic acid *p*-Hydroxybenzaldehyde	Protocatechualdehyde *p*-Hydroxybenzoic acid *p*-Hydroxybenzaldehyde Vanillic acid
Hydroxycinnamic acids	*Trans*-*p*-coumaric acid *Trans*-ferulic acid Feruloylquinic acid *N,N′*-bis-(*p*-coumaroyl)-putrescine Hydroxycinnamic acid derivatives	*Trans*-*p*-coumaric acid *Cis*-*p*-coumaric acid *Trans*-ferulic acid 8,8′-aryltetralin-DFA *Cis*-ferulic acid 8,8′-DFA 8,5′-DFA *Trans*-trans-8-O-4′–DFA 5–5′-DFA 8–5′-DFA benzofuran *Trans*-cis-8-O-4′–DFA
Hydroxycinnamic acid spermidines	*N′,N′′*-dicaffeoylspermidine *N′*-caffeoyl-*N′′*-feruloyl Spermidine	NA
Hydroxycinnamic acid serotonins	*N*-(*p*-coumaroyl) serotonin *N*-feruloylserotonin	NA

DFA, diferulic acid; TFA, triferulic acid.
Source: Adapted from Xiang et al. (2023).

gentisic, vanillic, and syringic acids as hydroxybenzoic acid derivatives; ferulic, caffeic, *cis*- and *trans*-*p*-coumaric, cinnamic, and sinapic acids as hydroxycinnamic acid derivatives; quercetin, orientin, isoorientin, vitexin, saporin, violanthin, leucine, and tricin as flavonoids; some of these compounds were identified with RP-HPLC with a mixture of 0.1% trifluoroacetic acid: methanol as mobile phase with isocratic elution. In another research (Xiang et al., 2019), other finger millet varieties were analyzed and more phenolic compounds were identified such as catechin-*O*-dihexoside, protocatechuic aldehyde, procyanidin dimer, catechin, epicatechin, apigenin-*C*-pentosyl-*O,C*-dihexoside, rutin, quercetin-*O*-dihexoside, methyl vanillate, ferulic truxilic acid, 8,8′-aryltetralyn-DFA, 5,5′-DFA, 8,5′-DFA, 8,5′-DFA benzofuran, 8-*O*-4'-DFA, and triferulic acid within methanolic extracts analyzed with HPLC coupled to quadrupole time of flight mass spectrometry using a mixture of acidic methanol (0.1% formic acid) and acidic water (0.1% formic acid). Other identified compounds in finger millet are salicylic and chlorogenic acids, gallocatechin, epigallocatechin, taxifolin, kempherol, narigenin, diadzein, procyanidin B1, and isovitexin (Udeh et al., 2017). It was also observed in another research (Karki et al., 2024) that the polyphenol profile changed according to the germination state; however, the behavior in phenolic, flavonoids, and tannin content was different than that reported by Xiang et al. (2023) with

Table 5.5 Changes in individual phenolic content during the germination process

Modification of phenolic content	Bound compound	Free compound
Increased	8-5′-DFA 8-5′-DFA benzofuran *Trans-cis*-8-*O*-4′–DFA	Feruloylquinic acid *N,N*′-bis-(*p*-coumaroyl) putrescine Hydroxycinnamic acid derivatives
Decreased	Cinnamic acid derivative Ferulic truxilic acid TFA	Neochlorogenic acid Protocatechualdehyde *p*-Hydroxybenzoic acid Cinnamic acid derivatives Chlorogenic acid Cryptochlorogenic acid
Constant	*p*-Hydroxybenzaldehyde Vanillic acid *Trans-p*-coumaric acid *Cis-p*-coumaric acid *Trans*-ferulic acid 8,8′-aryltetralin-DFA *Cis*-ferulic acid 8,8′-DFA Protocatechualdehyde *p*-Hydroxybenzoic acid *Trans-trans*-8-*O*-4′–DFA 5-5′-DFA	*p*-Hydroxybenzaldehyde *Trans-p*-coumaric acid a *N*′,*N*′′-dicaffeoylspermidine *Trans*-ferulic acid a *N*′-caffeoyl-*N*′′-feruloylspermidine *N*-(*p*-coumaroyl) serotonin *N*-feruloylserotonin Homoeriodictyol

DFA, diferulic acid; TFA, triferulic acid.
Source: Adapted from Xiang et al. (2023).

proso millet; in the case of finger millet, the three contents decreased gradually within two days of germination.

The influence of millet variety on the polyphenol profile is also supported by the research reported by Ofosu et al. (2020), in which the authors compared the phenolic content and profile from four different varieties (millet, Italian millet, barnyard millet, and finger Italian millet) cultivated in Korea, and ethanolic extractions were carried out; total phenolic, flavonoids, and condensed tannins were quantified with UV–Vis spectrometry, and their identification was carried out with UHPLC (ultra high performance liquid chromatography) coupled to quadrupole time of flight mass spectrometry using a mixture of 0.1% formic acid and methanol as mobile phase with gradient elution. Finger Italian millet presented the highest phenolic and flavonoid content (136.4 and 115.8 g/g, respectively), while barnyard millet presented the highest condensed tannin content (59.54 g/g). In Table 5.6, it can be seen the polyphenols identified in both varieties.

As can be seen in literature, millets have common compounds, although they differ in their quantities. In Table 5.7, it can be observed the most common polyphenols found in different millets that match the reviewed literature.

Table 5.6 Identified polyphenol in the Italian finger and barnyard millets

Barnyard millet	Finger Italian millet
(+)-Catechin	(+)-Catechin 3-*O*-glucose
Formomentin	(+)-Catechin
Kaempferol	Dihydromyricetin 3-*O*-rhamnoside
Luteolin	(−)-Epicatechin
Apigenin	Dihydroquercetin
Isorhamnetin	Formononetin
3,7-Dimethylquercetin	Kaempferol
	Apigenin

Source: Adapted from Ofosu et al. (2020).

Table 5.7 Phenolic compounds found on different millets

Hydroxybenzoic acid derivatives	Hydroxycinnamic acid derivatives	Flavonoids
Gallic acid	Sinapaldehyde	Catechin
Methyl vanillate	Sinapyl glycoside	Taxifolin-*O*-hexoside
Protocatechuic acid	Ferulyl hexose	Gallocatechin
Protocatechuic aldehyde	Chlorogenic acid	Taxifolin-*O*-pentoside
p-Hydroxybenzoic acid	5-Hydroxy feruloyl hexose	Procyanidin dimer B1
Gentisic acid	Caffeic acid	Epigallocatechin
Vanilllic acid	Caffeoyl hexose	Epicatechin
Syringic acid	*p*-Coumaryl pentose	6-*C*-glucosyl-8-*C*-rhamnosyl Apigenin
p-Hydroxybenzaldehyde	Cinnamic acid isomer	Taxifolin
Vanillin	*p*-Coumaric acid	Vitexin (apigenin-8-*C*-glu)
	Sinapic acid	Luteolin 7-*O*-glucoside
	Trans-ferulic acid	Isovitexin (apigenin-6-*C*-glu)
	Cis-ferulic acid	Tricin
	p-Coumaric truxilic acid	Myricetin hexoside
	Ferulic truxilic acid	Myricetin pentoside
	Tri FA1	Myricetin
	8,5′-Benzo DiFA	Kaempferol hexoside
	Tri FA 2	Quercetin-*O*-pentoside
	8,8′-Aryl DiFA	Procyanidin dimer B2
	8,5′-DiFA	Quercetin
	CAD 1	6-*C*-glucosyl-8-*C*-arabinosyl apigenin
	Di FA 4	6,8-Di-*C*-glucosyl apigenin
	5,5′-DiFA	Apigenin
	Tri FA 3	
	CAD 2	
	Tri FA 4	

Source: Adapted from Chandrasekara and Shahidi (2011).

5.6 Conclusion

The growing interest in millets as nutraceuticals and phytopharmaceuticals is sustained by the relevance of their polyphenol-related health-promoting properties. This review emphasizes the significance of polyphenols in the antioxidant, anti-inflammatory, and antidiabetic activities of the millet group, alongside their potential in mitigating chronic diseases such as cancer and cardiovascular disorders. For instance, hydroxyphenolic compounds such as gallic acid, ferulic acid, and flavonoids have demonstrated roles in oxidative stress reduction, anti-inflammatory modulation, and metabolic regulation (Leri et al., 2020; Lyubitelev & Studitsky, 2023; Najjar & Feresin, 2021). Additionally, the impact of processing techniques on polyphenols' bioavailability and therapeutic efficacy highlights the importance of optimizing methods like fermentation, roasting, and germination to maximize health benefits (Salar et al., 2016; Xiang et al., 2023). Advances in extraction and characterization techniques, such as ultrasound-assisted methods and the use of deep eutectic solvents, have shown promise in improving polyphenol recovery from millets (Zhang et al., 2022; Zheng et al., 2022). Future research should prioritize elucidating the molecular mechanisms of millet-derived polyphenols, their synergistic interactions with other bioactive compounds, and their long-term health effects in human populations. By integrating sustainable processing methods and advanced extraction technologies, the nutraceutical and functional food potential of millets can be fully harnessed, contributing to both health and food security globally.

References

Abedin, M. J., Abdullah, A. T. M., Satter, M. A., & Farzana, T. (2022). Physical, functional, nutritional and antioxidant properties of foxtail millet in Bangladesh. *Heliyon*, *8*(10), e11186. https://doi.org/10.1016/j.heliyon.2022.e11186

Ajila, C. M., Brar, S. K., Verma, M., Tyagi, R. D., Godbout, S., & Valéro, J. R. (2011). Extraction and analysis of polyphenols: Recent trends. *Critical Reviews in Biotechnology*, *31*(3), 227–249. https://doi.org/10.3109/07388551.2010.513677

Albuquerque, B. R., Heleno, S. A., Oliveira, M. B. P. P., Barros, L., & Ferreira, I. C. F. R. (2021). Phenolic compounds: Current industrial applications, limitations and future challenges. *Food & Function*, *12*(1), 14–29. 10.1039/D0FO02324H

Ali, A., Kumar, R. R., Vinutha, T., Singh, T., Singh, S. P., Satyavathi, C. T., Praveen, S., & Goswami, S. (2022). Grain phenolics: Critical role in quality, storage stability and effects of processing in major grain crops—A concise review. *European Food Research and Technology*, *248*(8), 2197–2213. https://doi.org/10.1007/S00217-022-04026-7

Archana, Sehgal, S., & Kawatra, A. (1998). Reduction of polyphenol and phytic acid content of pearl millet grains by malting and blanching. *Plant Foods for Human Nutrition*, *53*(2), 93–98. https://doi.org/10.1023/A:1008060604880

Bala Durairajan, M., Velavan Sundararajan, V., Kannan, G., Mathews Paul, B., Muniyandi, K., & Thangaraj, P. (2024). Elicitation of nutritional, antioxidant, and antidiabetic potential of barnyard millet (*Echinochloa esculenta* (A. Braun) H. Scholz) sprouts and microgreens through in vitro bio-accessibility assessment. *Food Chemistry*, *441*, 138282. https://doi.org/10.1016/J.FOODCHEM.2023.138282

Balasubramaniam, V. G., Ayyappan, P., Sathvika, S., & Antony, U. (2019). Effect of enzyme pretreatment in the ultrasound assisted extraction of finger millet polyphenols. *Journal of Food Science and Technology*, *56*(3), 1583–1594. https://doi.org/10.1007/s13197-019-03672-2

Balendran, S. D., Kalaivendan, R. G. T., & Theagarajan, R. (2024). Recent advances with cold plasma technology for millet processing: A brief review. *Journal of Food Process Engineering*, *47*(1), e14409. https://doi.org/10.1111/JFPE.14409

Balli, D., Bellumori, M., Pucci, L., Gabriele, M., Longo, V., Paoli, P., Melani, F., Mulinacci, N., & Innocenti, M. (2020). Does fermentation really increase the phenolic content in cereals? A study on millet. *Foods*, *9*(3), 303. https://doi.org/10.3390/foods9030303

Bello, O. M., Ogbesejana, A. B., Balkisu, A., Osibemhe, M., Musa, B., & Oguntoye, S. O. (2022). Polyphenolic fractions from three millet types (Fonio, finger millet, and pearl millet): Their characterization and biological importance. *Clinical Complementary Medicine and Pharmacology*, *2*(1), 100020. https://doi.org/10.1016/j.ccmp.2022.100020

Bhandari, M., Sharma, R., Bobade, H., Sharma, S., & Pathania, S. (2025). Ultrasonication-induced modification of finger millet and barnyard millet flours: Phytoconstituents and antioxidant capacity, functional properties, and microstructural arrangement. *Journal of Food Processing and Preservation*, *2025*(1), 6640001. https://doi.org/10.1155/JFPP/6640001

Bheemaiah Balyatanda, S., Gowda, N. A. N., Subbiah, J., Chakraborty, S., Prasad, P. V. V., & Siliveru, K. (2024). Physiochemical, bio, thermal, and non-thermal processing of major and minor millets: A comprehensive review on antinutritional and antioxidant properties. *Foods*, *13*(22), 3684. https://doi.org/10.3390/FOODS13223684

Biradar, V. M., Kumar, P., Ramya, M. Y., Nayak, P., Sekhar, M., Bhushan, S., & Babu, S. (2024). A sustainable nutricereal: A review on nutrient and bio-active composition and its potential health benefits of pearl millet. *International Journal of Advanced Biochemistry Research*, *8*(3S), 30–35. https://doi.org/10.33545/26174693.2024.v8.i3Sa.684

Chandrasekara, A., & Shahidi, F. (2011). Determination of antioxidant activity in free and hydrolyzed fractions of millet grains and characterization of their phenolic profiles by HPLC-DAD-ESI-MSn. *Journal of Functional Foods*, *3*(3), 144–158. https://doi.org/10.1016/J.JFF.2011.03.007

Chandrkar, L., Sahu, T., & Shahi, S. (2024). A systematic review of barnyard millet. *African Journal of Biological Sciences*, *6*(9), 2673–2680.

Chethan, S., & Malleshi, N. G. (2007a). Finger millet polyphenols: Characterization and their nutraceutical potential. *American Journal of Food Technology*, *2*(7), 582–592. https://doi.org/10.3923/ajft.2007.582.592

Chethan, S., & Malleshi, N. G. (2007b). Finger millet polyphenols: Optimization of extraction and the effect of pH on their stability. *Food Chemistry*, *105*(2), 862–870. https://doi.org/10.1016/J.FOODCHEM.2007.02.012

Chimento, A., De Luca, A., D'Amico, M., De Amicis, F., & Pezzi, V. (2023). The involvement of natural polyphenols in molecular mechanisms inducing apoptosis in tumor cells: A promising adjuvant in cancer therapy. *International Journal of Molecular Sciences*, *24*(2), 1680. https://doi.org/10.3390/ijms24021680

Cui, Q., Du, R., Liu, M., & Rong, L. (2020). Lignans and their derivatives from plants as antivirals. *Molecules*, *25*(1), 183. https://doi.org/10.3390/molecules25010183

Dekka, S., Paul, A., Vidyalakshmi, R., & Mahendran, R. (2023). Potential processing technologies for utilization of millets: An updated comprehensive review. *Journal of Food Process Engineering*, *46*(10), e14279. https://doi.org/10.1111/jfpe.14279

Dey, S., Saxena, A., Kumar, Y., Maity, T., & Tarafdar, A. (2022). Understanding the antinutritional factors and bioactive compounds of kodo millet (*Paspalum scrobiculatum*) and little millet (*Panicum sumatrense*). *Journal of Food Quality*, *2022*, 1–19. https://doi.org/10.1155/2022/1578448

Dogar, N. A., Ghous, M., Hanif, A., Ali, Z., Jabbar, M., & Mehmood, T. (2023). Phytochemical screening and study of anti-microbial, anti-haemolytic, anti-fungal and anti-oxidant activities of ethanol extract of *Panicum sumatrense* leaves. *Pakistan Journal of Science*, *75*(3), 488–495. https://doi.org/10.57041/pjs.v75i03.1006

Dutta, A., Joshi, D., Suri, S., Trivedi, N., Acharya, R., Malik, R., Bose, S., & Dobhal, A. (2023). Bioactive and techno-functional properties of barnyard millet (*Echinochloa frumentacea*). In R. Sharma, V. Nanda, & S. Sharma (Eds.), *Nutri-cereals* (pp. 213–231). CRC Press. https://doi.org/10.1201/9781003251279-9

El Kourchi, C., Belhoussaıne, O., Elhrech, H., Harhar, H., Ullah, R., Bari, A., Maggi, F., Caprioli, G., Bouyahya, A., & Tabyaouı, M. (2024). Antidiabetic, antioxidant, and phytochemical profile of *Pennisetum glaucum* cultivated in central-southern Morocco and imported from India. *Journal of Agriculture and Food Research*, *16*, 101197. https://doi.org/10.1016/j.jafr.2024.101197

Erharuyi, O., Oseghale, I., Imieje, V., Falodun, A., & Iheanacho, C. (2017). A review of the phytochemistry and pharmacology of *Eleusine coracana* Linn (Poaceae): A popular Nigerian edible grain. *Tropical Journal of Natural Product Research*, *1*(6), 227–235. https://doi.org/10.26538/tjnpr/v1i6.1

Essien, S. O., Young, B., & Baroutian, S. (2020). Recent advances in subcritical water and supercritical carbon dioxide extraction of bioactive compounds from plant materials. *Trends in Food Science & Technology*, *97*, 156–169. https://doi.org/10.1016/J.TIFS.2020.01.014

Gebru, Y. A., Sbhatu, D. B., & Kim, K.-P. (2020). Nutritional composition and health benefits of Teff (*Eragrostis tef* (Zucc.) Trotter). *Journal of Food Quality*, *2020*, 1–6. https://doi.org/10.1155/2020/9595086

Goudar, G., Manne, M., Sathisha, G. J., Sharma, P., Mokalla, T. R., Kumar, S. B., & Ziouzenkova, O. (2023). Phenolic, nutritional and molecular interaction study among different millet varieties. *Food Chemistry Advances*, *2*, 100150. https://doi.org/10.1016/J.FOCHA.2022.100150

Gupta, A., & Pandey, A. K. (2020). Antibacterial lead compounds and their targets for drug development. In C. Egbuna, S. Kumar, J. C. Ifemeje, S. M. Ezzat, & S. Kaliyaperumal (Eds.), *Phytochemicals as lead compounds for new drug discovery* (pp. 275–292). Elsevier. https://doi.org/10.1016/B978-0-12-817890-4.00018-4

Hassan, Z. M., Sebola, N. A., & Mabelebele, M. (2020). Assessment of the phenolic compounds of pearl and finger millets obtained from South Africa and Zimbabwe. *Food Science & Nutrition*, *8*(9), 4888–4896. https://doi.org/10.1002/fsn3.1778

Indirani, K., & Devanesa, M. (2021). Review on nutritional profiles and health benefits of little millets – India. *International Journal of Research in Engineering and Science*, *9*(11), 7–11.

Jacob, J., Krishnan, V., Antony, C., Bhavyasri, M., Aruna, C., Mishra, K., Nepolean, T., Satyavathi, C. T., & Visarada, K. B. R. S. (2024). The nutrition and therapeutic potential of millets: An updated narrative review. *Frontiers in Nutrition*, *11*, 1346869. https://doi.org/10.3389/fnut.2024.1346869

Jaddu, S., Sahoo, S., Sonkar, S., Alzahrani, K., Dwivedi, M., Misra, N. N., & Pradhan, R. C. (2024). Cold plasma treatment of little millet flour: Impact on bioactives, antinutritional factors and functional properties. *Plant Foods for Human Nutrition*, *79*(2), 503–510. https://doi.org/10.1007/S11130-024-01171-0

Jadhav, B. N., Nikam, S., Salunke, S., Jagdale, G., Jagtap, P., & Kharak, J. (2022). Phytochemical analysis of selective millets. *Journal of Emerging Technologies and Innovative Research*, *9*(8), e131–e146.

Jagati, P., Mahapatra, I., & Dash, D. (2021). Finger millet (Ragi) as an essential dietary supplement with key health benefits: A review. *International Journal of Home Science*, *7*(2), 94–100. https://doi.org/10.22271/23957476.2021.v7.i2b.1152

John Calvien Hutabarat, D., & Aditya Bowie, V. (2022). Bioactive compounds in foxtail millet (*Setaria italica*)-extraction, biochemical activity, and health functional: A review. *IOP Conference Series: Earth and Environmental Science*, *998*(1), 012060. https://doi.org/10.1088/1755-1315/998/1/012060

Jyothi Reddy, G., Bhaskar Reddy, K., & Subba Reddy, G. V. (2020). Antidiabetic and antidyslipidemic activities of hexane/ethyl acetate/methanol fractions of *Paspalum scrobiculatum* Linn. grains in high-fat diet and streptozotocin-induced diabetic rats. *Pharmacognosy Research*, *12*(3), 230–235.

Kalam Azad, M. O., Jeong, D. I., Adnan, M., Salitxay, T., Heo, J. W., Naznin, M. T., Lim, J. D., Cho, D. H., Park, B. J., & Park, C. H. (2019). Effect of different processing methods on the accumulation of the phenolic compounds and antioxidant profile of broomcorn millet (*Panicum miliaceum* L.) Flour. *Foods*, *8*(7), 230. https://doi.org/10.3390/foods8070230

Kalsi, R., Bhasin, J., Goksen, G., & Kashyap, P. (2023). Exploration of nutritional, pharmacological, and the processing trends for valorization of finger millet (*Eleusine coracana*): A review. *Food Science & Nutrition*, *11*(11), 6802–6819. https://doi.org/10.1002/fsn3.3659

Kalsi, R., & Bhasin, J. K. (2023). Nutritional exploration of foxtail millet (*Setaria italica*) in addressing food security and its utilization trends in food system. *eFood*, *4*(5), e111. https://doi.org/10.1002/efd2.111

Karki, S., Dangal, A., Pokharel, A., Dhakal, L., & Timsina, P. (2024). The effect of germination on the nutritional quality and sensory evaluation of finger millet (*Elusine coracana* L., var. *Kabre-1)* malt incorporated biscuit. *Cogent Food & Agriculture*, *10*(1), 2419149. https://doi.org/10.1080/23311932.2024.2419149

Kaur, C. (2023). Millet: A therapeutic grain. *The Pharma Innovation*, *12*(5), 2289–2293. https://doi.org/10.22271/tpi.2023.v12.i5ab.20202

Kaur, T., Panesar, P. S., & Riar, C. S. (2024). Optimization of hydrothermal (soak-boil and soak-steam) treatments of foxtail (*Setaria italica* L.) millet and their effect on milling, antioxidant, pasting, and drying characteristics. *Journal of Food Measurement and Characterization*, *18*(1), 500–515. https://doi.org/10.1007/S11694-023-02168-0

Kharsan, R., Khokhar, D., & Panjwani, M. (2023). Phytochemical evaluation of kodo (*Paspalum scrobiculatum* L.) bran. *Journal of Research in Chemistry*, *4*(1), 26–30.

Kumaraguru, G., Ramalingam, A., Shanmugam, G., Krishna Kumar, A. K., Muthusamy, S., & Malaiyandi, J. (2024). Investigating antioxidant, anti-obesity, and anticancer potential in raw and processed flour extracts of barnyard millets. *Journal of Food Measurement and Characterization*, *18*(7), 6124–6144. https://doi.org/10.1007/s11694-024-02633-4

Kumari, D., Madhujith, T., & Chandrasekara, A. (2017). Comparison of phenolic content and antioxidant activities of millet varieties grown in different locations in Sri Lanka. *Food Science and Nutrition*, *5*(3), 474–485. https://doi.org/10.1002/fsn3.415

Leri, M., Scuto, M., Ontario, M. L., Calabrese, V., Calabrese, E. J., Bucciantini, M., & Stefani, M. (2020). Healthy effects of plant polyphenols: Molecular mechanisms. *International Journal of Molecular Sciences*, *21*(4), 1250. https://doi.org/10.3390/ijms21041250

Li, L., Ling, Y., He, T., Li, C., & Cao, R. (2023). Analysis of polyphenols in millet and processing effects on antioxidant activity. *Food Science and Technology*, *43*, 1–8. https://doi.org/10.5327/fst.129722

Lyubitelev, A., & Studitsky, V. (2023). Inhibition of cancer development by natural plant polyphenols: Molecular mechanisms. *International Journal of Molecular Sciences*, *24*(13), 10663. https://doi.org/10.3390/ijms241310663

Mahajan, M., Singla, P., & Sharma, S. (2024). Sustainable postharvest processing methods for millets: A review on its value-added products. *Journal of Food Process Engineering*, *47*(1), e14313. https://doi.org/10.1111/JFPE.14313

Mawouma, S., Condurache, N. N., Turturică, M., Constantin, O. E., Croitoru, C., & Rapeanu, G. (2022). Chemical composition and antioxidant profile of sorghum (*Sorghum bicolor* (L.) Moench) and pearl millet (*Pennisetum glaucum* (L.) R.Br.) grains cultivated in the far-north region of Cameroon. *Foods*, *11*(14), 2026. https://doi.org/10.3390/foods11142026

Mittal, V., Kumar, S., Gupta, R., Panghal, A., & Kumar, N. (2024). Review on millets dehulling and pearling technologies and machinery. *Journal of Food Process Engineering*, *47*(5), e14611. https://doi.org/10.1111/JFPE.14611

Monica, V., Anbarasan, R., & Mahendran, R. (2023). Influence of cold plasma in accelerating the germination and nutrient composition of foxtail millet (*Setaria italica* L.). *Plasma Chemistry and Plasma Processing*, *43*(6), 1843–1861. https://doi.org/10.1007/S11090-023-10368-1

Mounika, A., Shanmugam, A., Rao, T. J., Sinija, V. R., Rawson, A., & Baskaran, N. (2024). Evaluation of quality characteristics of ultrasound–treated browntop millet grains. *Chemical Engineering and Processing – Process Intensification*, *203*, 109919. https://doi.org/10.1016/J.CEP.2024.109919

Mounika, D., Sangeetha, U., & Sireesha, G. (2022). Estimation of phytochemicals in millets and selected millet products. *Indian Journal of Applied and Pure Biology*, *37*(3), 810–820.

Najjar, R. S., & Feresin, R. G. (2021). Protective role of polyphenols in heart failure: Molecular targets and cellular mechanisms underlying their therapeutic potential. *International Journal of Molecular Sciences*, *22*(4), 1668. https://doi.org/10.3390/ijms22041668

Nakamura, Y. (2023). Molecular mechanisms for polyphenol functionality. *Journal for the Integrated Study of Dietary Habits*, *33*(4), 165–169. https://doi.org/10.2740/jisdh.33.4_165

Nani, A., Belarbi, M., Khan, N. A., & Hichami, A. (2020). Nutritional properties and plausible benefits of Pearl millet (*Pennisetum glaucum*) on bone metabolism and osteoimmunology: A mini-review. *The North African Journal of Food and Nutrition Research*, *4*(8), 336–342. https://doi.org/10.51745/najfnr.4.8.336-342

Nayak, N., Bhujle, R. R., Nanje-Gowda, N. A., Chakraborty, S., Siliveru, K., Subbiah, J., & Brennan, C. (2024). Advances in the novel and green-assisted techniques for extraction of bioactive compounds from millets: A comprehensive review. *Heliyon*, *10*(10), e30921. https://doi.org/10.1016/j.heliyon.2024.e30921

Nguyen, T. H. D., Nguyen, L. L. P., & Baranyai, L. (2023). Influence of sprouting on phenolic acids, carotenoids, and antioxidant activity of millet varieties. *Journal of Agriculture and Food Research*, *14*, 100810. https://doi.org/10.1016/J.JAFR.2023.100810

Ofosu, F. K., Elahi, F., Daliri, E. B.-M., Chelliah, R., Ham, H. J., Kim, J.-H., Han, S.-I., Hur, J. H., & Oh, D.-H. (2020). Phenolic profile, antioxidant, and antidiabetic potential exerted by millet grain varieties. *Antioxidants*, *9*(3), 254. https://doi.org/10.3390/antiox9030254

Owheruo, J. O., Ifesan, B. O. T., & Kolawole, A. O. (2019). Physicochemical properties of malted finger millet (*Eleusine coracana*) and pearl millet (*Pennisetum glaucum*). *Food Science & Nutrition*, *7*(2), 476–482. https://doi.org/10.1002/fsn3.816

Panda, J., & Yadav, A. (2021). Proso millet: Therapeutic effects and nutritional health security. *Agriculture & Food: E-Newsletter*, *3*(11), 1–3.

Pandey, K. B., & Rizvi, S. I. (2009). Plant polyphenols as dietary antioxidants in human health and disease. *Oxidative Medicine and Cellular Longevity*, *2*(5), 270–278. https://doi.org/10.4161/oxim.2.5.9498

Popović, V., Ikanović, J., Rajičić, V., Ksenija, M., Ljubičić, N., Kostic, M., Radovic, M., & Šarčević-Todosijević, L. (2020). Millet – Panicum miliaceum L. production trend in the world. Importance of millet in nutrition and for bioenergy. In N. Aleksić & V. Karajović (Eds.), *XXIV International Eco-Conference XI Safe Food* (pp. 297–306). Ecological Movement of Novi Sad.

Pradeep, P. M., & Sreerama, Y. N. (2015). Impact of processing on the phenolic profiles of small millets: Evaluation of their antioxidant and enzyme inhibitory properties associated with hyperglycemia. *Food Chemistry*, *169*, 455–463. https://doi.org/10.1016/J.FOODCHEM.2014.08.010

Pujari, S. N., & Hoskeri, J. H. (2022). Minor millet phytochemicals and their pharmacological potentials. *Pharmacognosy Reviews*, *16*(32), 100–106. https://doi.org/10.5530/phrev.2022.16.15

Rajesh, C., & Suku, J. (2022). Evaluation of antiulcer activity of hydro-alcoholic extract of seeds of *Paspalum scrobiculatum* Linn. (Poaceae) in pylorus ligated wistar albino rats. *World Journal of Pharmaceutical Research*, *11*(5), 1760–1772.

Ramakrishnan, S. R., Antony, U., & Kim, S. J. (2023). Non-thermal process technologies: Influences on nutritional and storage characteristics of millets. *Journal of Food Process Engineering*, *46*(10), e14215. https://doi.org/10.1111/JFPE.14215

Ranjan, R., Singh, S., Dhua, S., Mishra, P., Chauhan, A. K., & Gupta, A. K. (2023). Kodo millet (*Paspalum scrobiculatum*). In R. Sharma, V. Nanda, & S. Sharma (Eds.), *Nutri-cereals* (pp. 193–211). CRC Press. https://doi.org/10.1201/9781003251279-8

Salar, R. K., Purewal, S. S., & Bhatti, M. S. (2016). Optimization of extraction conditions and enhancement of phenolic content and antioxidant activity of pearl millet fermented with *Aspergillus awamori* MTCC-548. *Resource-Efficient Technologies*, *2*(3), 148–157. https://doi.org/10.1016/J.REFFIT.2016.08.002

Saqib, F., Al-Huqail, A. A., Asma, M., Chicea, L., Hogea, M., Irimie, M., & Gavris, C. (2022). Dose-dependent spasmolytic, bronchodilator, and hypotensive activities of *Panicum miliaceum* L. *Dose-Response*, *20*(1), 155932582210795. https://doi.org/10.1177/15593258221079592

Sartelet, H., Serghat, S., Lobstein, A., Ingenbleek, Y., Anton, R., Petitfrère, E., Aguie-Aguie, G., Martiny, L., & Haye, B. (1996). Flavonoids extracted from fonio millet (*Digitaria exilis*) reveal potent antithyroid properties. *Nutrition*, *12*(2), 100–106. https://doi.org/10.1016/0899-9007(96)90707-8

Seo, J., Lee, H. J., & Ju, J. (2024). Anti-growth and anti-metastatic potential of raw and thermally treated Eragrostis tef extract in human cancer cells. *Nutrients*, *16*(16), 2612. https://doi.org/10.3390/nu16162612

Sharma, E., Attri, D. C., Sati, P., Dhyani, P., Szopa, A., Sharifi-Rad, J., Hano, C., Calina, D., & Cho, W. C. (2022). Recent updates on anticancer mechanisms of polyphenols. *Frontiers in Cell and Developmental Biology*, *10*, 1005910. https://doi.org/10.3389/fcell.2022.1005910

Sharma, R., & Sharma, S. (2022). Anti-nutrient & bioactive profile, in vitro nutrient digestibility, techno-functionality, molecular and structural interactions of foxtail millet (Setaria italica L.) as influenced by biological processing techniques. *Food Chemistry*, *368*, 130815. https://doi.org/10.1016/j.foodchem.2021.130815

Shi, L., Zhao, W., Yang, Z., Subbiah, V., & Suleria, H. A. R. (2022). Extraction and characterization of phenolic compounds and their potential antioxidant activities. *Environmental Science and Pollution Research*, *29*(54), 81112–81129. https://doi.org/10.1007/S11356-022-23337-6

Shi, Y., Ma, Y., Zhang, R., Ma, H., & Liu, B. (2015). Preparation and characterization of foxtail millet bran oil using subcritical propane and supercritical carbon dioxide extraction. *Journal of Food Science and Technology*, *52*(5), 3099–3104. https://doi.org/10.1007/S13197-014-1311-0

Singh, A. M., Rekha, N. D., Udayashankar, A. C., & Sumana, K. (2022). Phytochemical analysis and evaluation of antimicrobial, antioxidant, anti-inflammatory and antiangiogenic activities of methanol extract of *Urochloa ramosa*. *Research Journal of Pharmacy and Technology*, *15*(8), 3571–3579. https://doi.org/10.52711/0974-360X.2022.00599

Sirše, M. (2022). Effect of dietary polyphenols on osteoarthritis—Molecular mechanisms. *Life*, *12*(3), 436. https://doi.org/10.3390/life12030436

Sunagar, R. R., & Sreerama, Y. N. (2023). Implication of solvent polarities on browntop millet (*Urochloa ramosa*) phenolic antioxidants and their ability to protect oxidative DNA damage and inhibit α-amylase and α-glucosidase enzymes. *Food Chemistry*, *411*, 135474. https://doi.org/10.1016/J.FOODCHEM.2023.135474

Tamil Selvan, J., Kalaiselvi, C., Jothimanivannan, C., Sakthivel, M., Kalaiyarasan, V., & Subashvarma, S. (2023). Pharmacological activities of barnyard millets – A review. *World Journal of Pharmaceutical Research*, *12*(3), 59–65.

Thagunna, B., Rimal, A., Kaur, J., Dhakal, Y., & Paudel, B. (2022). Finger millet: A powerhouse of nutrients its amino acid, micronutrient profile, bioactive compounds, health benefits, and value-added products. *Journal of Research in Agriculture and Animal Science*, *9*(8), 36–44.

Traore, D. (2011). *Effects of Digitaria exilis (Fonio) on inflammation and diabetes pathogenesis*. Oklahoma State University.

Udeh, H. O., Duodu, K. G., & Jideani, A. I. O. (2017). Finger millet bioactive compounds, bioaccessibility, and potential health effects – A review. *Czech Journal of Food Sciences*, *35*(1), 7–17. https://doi.org/10.17221/206/2016-CJFS

Vf, A., Sj, O., Go, O., Abioye, O. A., & Abioye, V. (2022). Effect of soaking conditions on chemical composition, antioxidant activity, total phenols, flavonoids and antinutritional contents of finger millet. *African Journal of Food, Agriculture, Nutrition and Development, 22*(7), 20942–20956. https://doi.org/10.18697/ajfand.112.20480

Wang, H., Fu, Y., Zhao, Q., Hou, D., Yang, X., Bai, S., Diao, X., Xue, Y., & Shen, Q. (2022). Effect of different processing methods on the millet polyphenols and their anti-diabetic potential. *Frontiers in Nutrition, 9*, 780499. https://doi.org/10.3389/FNUT.2022.780499

Xiang, J., Apea-Bah, F. B., Ndolo, V. U., Katundu, M. C., & Beta, T. (2019). Profile of phenolic compounds and antioxidant activity of finger millet varieties. *Food Chemistry, 275*, 361–368. https://doi.org/10.1016/J.FOODCHEM.2018.09.120

Xiang, J., Yuan, Y., Du, L., Zhang, Y., Li, C., & Beta, T. (2023). Modification on phenolic profiles and enhancement of antioxidant activity of proso millets during germination. *Food Chemistry: X, 18*, 100628. https://doi.org/10.1016/J.FOCHX.2023.100628

Yadav, S. (2022). Efficacy of kodo millets *Paspalum scrobiculatum*: A systematic review. *Bioscience Biotechnology Research Communications, 15*(3), 379–385. https://doi.org/10.21786/bbrc/15.3.2

Yan, Z., Zhong, Y., Duan, Y., Chen, Q., & Li, F. (2020). Antioxidant mechanism of tea polyphenols and its impact on health benefits. *Animal Nutrition, 6*(2), 115–123. https://doi.org/10.1016/j.aninu.2020.01.001

Zhang, H., Zhao, W., Bai, T., Fu, L., Chen, Z., Jing, X., & Wang, X. (2022). Sustainable extraction of polyphenols from millet using switchable deep eutectic solvents. *LWT - Food Science and Technology, 170*, 114082. https://doi.org/10.1016/J.LWT.2022.114082

Zhang, J., Wen, C., Zhang, H., Duan, Y., & Ma, H. (2020). Recent advances in the extraction of bioactive compounds with subcritical water: A review. *Trends in Food Science & Technology, 95*, 183–195. https://doi.org/10.1016/J.TIFS.2019.11.018

Zhang, J., Yang, J., & Yin, Y. (2024a). Germination promotes flavonoid accumulation of finger millet (*Eleusine coracana* L.): Response surface optimization and investigation of accumulation mechanism. *Plants, 13*(16), 2191. https://doi.org/10.3390/plants13162191

Zhang, Y., Wang, Q., Pan, H., Xu, H., Deng, W., & Sun, X. (2024b). *Panicum miliaceum* L. and bone healing properties in rat model of femur fracture by activating phosphate stimulating macrophages via BMP2 and RANK signaling pathway. *Pharmacognosy Magazine, 20*(1), 16–29. https://doi.org/10.1177/09731296231188459

Zhao, S., Baik, O.-D., Choi, Y. J., & Kim, S.-M. (2014). Pretreatments for the efficient extraction of bioactive compounds from plant-based biomaterials. *Critical Reviews in Food Science and Nutrition, 54*(10), 1283–1297. https://doi.org/10.1080/10408398.2011.632698

Zheng, B., Yuan, Y., Xiang, J., Jin, W., Johnson, J. B., Li, Z., Wang, C., & Luo, D. (2022). Green extraction of phenolic compounds from foxtail millet bran by ultrasonic-assisted deep eutectic solvent extraction: Optimization, comparison and bioactivities. *LWT - Food Science and Technology, 154*, 112740. https://doi.org/10.1016/J.LWT.2021.112740

Zhu, F. (2020). Fonio grains: Physicochemical properties, nutritional potential, and food applications. *Comprehensive Reviews in Food Science and Food Safety, 19*(6), 3365–3389. https://doi.org/10.1111/1541-4337.12608

6

Millets Dietary Fibers: Extraction, Quality Characteristics, and Health Benefits

Farhan M. Bhat[1] *and Charanjit S. Riar*[2]

[1] *Department of Food Science, Nutrition and Technology, CSKHPKV-Palampur, Palampur, India*
[2] *Department of Food Engineering and Technology, Sant Longowal Institute of Engineering and Technology,, Longowal, Punjab, India*

6.1 Introduction

Among the many health advantages of millet is its ability to reduce blood sugar and cholesterol levels. Those who have celiac disease or gluten sensitivity can enjoy them because they are gluten-free. Millet has a dietary fiber content that is equivalent to whole wheat and nearly double that of rice (Table 6.1). Millets are low in glycemic index, gluten-free, and contain phytochemicals that help prevent diseases like cancer and rectify lifestyle issues. Treatments of anemia and calcium deficiencies, particularly in pregnant women and young children, are additional advantages. According to Gupta et al. (2015), millets include both soluble and insoluble fractions of dietary fiber. The soluble fibers include pectins, arabinoxylans, and β-glucans, while the insoluble fibers include cellulose and hemicelluloses. Two important dietary fiber sources are finger millet and pearl millet. Millet consumption aids in the prevention of metabolic problems and the treatment of lifestyle disorders. These are easier to consume and have a higher bioavailability because they are given as food. Millet supplementation in diet-based trials has had positive outcomes for performance and health. Studies on adding millets to the diet have shown encouraging results, including improvements in anemia and overall health and performance.

Indian primary school students' anthropometric indices increased when they were regularly supplemented with a multi-millet health mix that included finger millet, foxtail millet, tiny millet, kodo millet, and wheat with pulses (Durairaj et al., 2019). Because of their high soluble fiber content, millets can help decrease blood cholesterol levels and trap fat in the gut. This can lessen your risk of developing heart disease, or atherosclerosis. According to Chen et al. (2017), millets are a potential diet for people with hypertension since they are a good supply of hypotensive components such as dietary fiber, protein, minerals, and other phytochemicals. Including millet in the diet enhances nutrition in general and

Nutraceutical Potential of Millet-Based Food Products: Chemistry and Applications, First Edition.
Edited by Parmjit S. Panesar, Charanjit S. Riar, and Sushil Dhital.

Published 2026 by John Wiley & Sons Ltd.

Table 6.1 Functional characteristics of dietary fiber of different types of millets

Millet	Functional characteristics	References
Foxtail millet (kakum/kangni)	Foxtail millet, like the majority of millets, is also an excellent source of crude fiber, aids in digestion, and promotes bowl movement, all of which contribute to a laxative impact that is good for stomach health	Sharma and Niranjan (2017)
Proso millet (chena/barri)	Proso millet brans have drawn a lot of study interest because of their high dietary fiber content, minerals, and bioactive substances. Proso millet husks are a possible source of natural antioxidants due to their excellent antioxidant qualities	Bagdi et al. (2011)
Sorghum millet (jowar)	Dietary fiber concentration in sorghum flour ranges from 6–8%, while β-glucan fiber level is 3–6%. With about 6.7 gs of dietary fiber per 100 gs, sorghum helps with digestion by giving meals more volume and avoiding constipation. Weight management can benefit from its soluble fiber's ability to control blood sugar, reduce cholesterol, and increase satiety	Khalid et al. (2022)
Finger millet (ragi)	The amount of nutritious fiber in finger millet is larger than that of other grains. Due to its high fiber content, ragi atta has been utilized for decades in newborn porridges. Finger millet's insoluble fiber supports improved digestive health, reduced blood lipids, fermentability, and regular bowel movements	Rathore et al. (2019)
Pearl millet (bajra)	Pearl millet fiber is a blend of carbohydrates, oligosaccharides, and polysaccharides, including gums, resistant starch, cellulose, hemicelluloses, pectin compounds, and inulin. Insoluble dietary fibers found in pearl millet (5 g/100 g) and soluble dietary fibers (3 g/100 g) create insoluble complexes with divalent cations like iron and zinc, which prevent the intestines from absorbing them. The fibers and phytates combine to produce fiber–phytate–mineral complexes, which chelate minerals	Nandini and Salimath (2001)
Browntopmillet (korle)	By clearing the waste from the colon, brown top millet, which has 12.5 g of fiber per 100 g, aids in the body's detoxification. As the colon ferments, fiber serves as a prebiotic	Lei et al. (2006)
Barnyard millet (sanwa)	With 2.4 g of fiber per serve, the grain has the most fiber of any grain or millets produced. Barnyard millet's soluble and insoluble dietary fibers support digestive health by reducing diarrhea, gas, bloating, and cramps	Kulkarni et al. (1992)
Little millet (moraiyo)	A gluten-free option for managing weight is little millet. Its high dietary fiber content helps with digestion and supports regularity in the bowels for efficient detoxification, reducing overeating and promoting extended fullness. Its rich dietary fiber helps to reduce blood sugar swings by delaying the absorption and digestion of carbohydrates. Fiber also helps reduce appetite, improve bowel movements, and lower cholesterol	Sushmita et al. (2020)
Buckwheat millet (kuttu)	A significant amount of fiber, which your body cannot digest, is present in buckwheat. The health of the colon benefits from this vitamin. Fiber, which is mostly made up of cellulose and lignin, accounts up 2.7% of the weight of boiling groats. Buckwheat has 4.8% soluble and 2.2% insoluble dietary fiber, which promotes health by lowering cholesterol and obesity while also causing negative side effects like protein and mineral deficiency	Zhu (2020)
Kodo millet	The dietary fiber included in kodo millet is both soluble and insoluble, with the insoluble fiber acting as a prebiotic to promote beneficial bacteria in the digestive tract. Kodo millets' high fiber content helps to eradicate conditions like cramping, bloating, constipation, and excessive gas. It promotes the development of beneficial bacteria, which are necessary for a healthy digestive system	Hadimani and Malleshi (1993)

intestinal health in particular. Between 15% and 20% of millet is dietary fiber, which includes β-glucan, arabinoxylan, and nonstarchy polysaccharides. Cellulose and hemicellulose, two types of insoluble fibers, give stool volume and facilitate regular bowel motions. Some soluble fibers, such as pectins, arabinoxylans, and β-glucans, have the ability to gel and have a variety of physiological consequences (Mutshinyani et al., 2020).

Inhibiting the absorption of macronutrients, decreasing dyslipidemia, the postprandial glucose response, and increasing food viscosity, they demonstrated gel-forming capabilities by absorbing water. The gut microbiota (*Bacteroides* and *Bifidobacterium*) may break it down during colonic fermentation to create short-chain fatty acids (SCFAs) (Gupta et al., 2015). Hepatic bile synthesis using intracellular cholesterol is increased and blood cholesterol levels are decreased as a result of soluble dietary fiber reduction of bile acid absorption in the small intestine (ileum) and increase of bile acid excretion. Cellulose and hemicellulose are examples of insoluble dietary fibers that are more prone to colonic fermentation by the gut microbiota and improve insulin sensitivity. Insoluble fibers increase intestinal transit and fecal volume by increasing water-holding capacity and eliminating toxins from the digestive system (Basen & Kurrer, 2021). Constructed from a small number of embryos, pericarps, testa, aleurone layer, and endosperm, millet bran is a very cheap and easily accessible by-product that provides dietary fiber, protein, fat, minerals, and vitamins.

Dietary fibers may naturally be prebiotics if the gut bacterium breaks them down in the colon to create various SCFA, including butyric, propionic, and acetic acids (Singh et al., 2021). Consuming millets guarantees a healthy intake of dietary fiber, which is critical for general health. Millet fiber encourages appropriate laxation, which wards off diverticulosis, diverticulitis, and constipation because of its sluggish or partial fermentation by the colon's microbiota. Children, teenagers, adults, and even the elderly are frequently experiencing constipation these days as a result of consuming too many refined foods. Foods made from millet can help lessen the issue of constipation. High in dietary fiber that is accessible to the microbiota, finger millet has been linked to antiobesity and antidiabetic effects. A considerable quantity of indigestible dietary fiber can be found in the millet bran fraction. As a result, the fiber component is significantly reduced when the bran section is removed during decortication or dehulling. Dehulling millet grains more than 30% is said to result in a considerable loss of nutritious fiber.

Most of the millets are eaten in their decorticated state, so controlling the amount of dehulling is essential to maximizing the fiber content (Sharma et al., 2015). The impact of grinding on the fiber components of foxtail millet revealed that while the fiber content of foxtail millet grew dramatically with increasing germination time, the insoluble dietary fiber was lower in the milled fraction than in whole millet flour. Millet fiber digestibility differs depending on the fiber type, processing techniques, and individual gut flora differences. Fiber structure is changed during processing, which also affects its bioavailability and digestibility. Soluble fibers are broken down into smaller molecules during the fermentation process, which the colon cells subsequently absorb and use as fuel (Barber et al., 2020). Frequent intake of millet fibers has been linked to better gastrointestinal function, increased absorption of nutrients, and decreased risk of a number of digestive disorders, high-fiber millets are linked to a lower risk of colon rectal cancer, and consumption of foxtail millet has been found to reduce inflammation in the colon (Aune et al., 2011). In rats, eating foxtail millet was observed to reduce colonic inflammation and the risk of colitis-associated colorectal cancer caused by azoxymethane (AOM)/dextran sodium sulfate (DSS) mouse model.

Foxtail millet microbial metabolites stimulated aryl hydrocarbon receptors and G-protein-coupled receptors while blocking STAT3 phosphorylation, hence mediating the regulatory actions (Zhang et al., 2020). The bound polyphenol from foxtail millet bran induced apoptosis in human colorectal cancer HCT-116 cells and in nude mice with HCT-116 by increasing the generation of reactive oxygen species (ROS). Bound polyphenol from foxtail millet bran caused ROS to be generated in HCT-116 cells, which was followed by ROS-induced apoptosis. By blocking the epithelial-mesenchymal transition via STAT3 signaling pathways, a peroxidase enzyme from foxtail millet bran was discovered to prevent cell migration in human colon cancer cells (Ramadoss & Sivalingam, 2021).

Certain sorghum lines, particularly black sorghum, contain up to 10 mg/g of 3-deoxyanthocyanidins in their bran. When tested against different human cancer cell lines, this chemical showed substantial antiproliferative activity, and its potential was on par with quercetin, one of the most powerful antioxidant compounds (Yang et al., 2009). Dehulling has been shown to lessen the millet grains' antioxidant qualities, and hulls are healthier sources of total phenolic content. Protein glycation is inhibited, protein aggregation formation is decreased, and oxidative DNA damage and hydroxyl radical-induced protein fragmentation are prevented by millet phenolics. Among millets, phenolic extracts of finger and kodo millets were 97% effective against peroxyl-mediated DNA scission (Chandrasekara & Shahidi, 2011). According to Hadimani and Malleshi (1993), millet grain's nutritional composition shows that it is a good source of energy, protein, vitamins, minerals, and trace elements. The edible part of the millet kernel is a rich source of phytochemicals, including dietary fiber and polyphenols (0.2–0.3%).

Millets have antioxidant properties, and the phytates, polyphenols, and tannins they contain play a significant role in aging and metabolic diseases (Bravo, 1998). Grain sorghum has a 1.6% mineral and fiber content that is practically identical. About 349 Kcal and 72.6% of calories come from it, making it a good source of energy (Gopalan et al., 1996). The grain's main carbohydrate is starch. Cellulose, hemicellulose, and simple sugars make up the remaining carbs. Starch's amylose concentration ranges from 21.28%. Additionally high in dietary fiber (14.3%) is sorghum. According to reports, kodo millet and small millet also contain 37–38% of dietary fiber, which was formerly regarded as an"antinutrient" but is now regarded as a nutraceutical and the highest among grains (Hegde & Chandra, 2005). Millet's bran, also known as complex unavailable polysaccharides, is a major source of dietary fiber. Dietary fibers are important for lowering blood glucose levels and insulin response because of their increased viscosity, glycemic index, and water-holding ability (Easwaran et al., 1991). Additionally, it lowers the incidence of gastrointestinal diseases and cholesterol levels (Potdar et al., 1994). Its excellent cost-economics and negligible impact on sensory qualities make it a viable alternative to standard dietary fibers including cellulose, β-glucan, and inulin (Charalampopoulos et al., 2002).

6.2 Methods of Extraction of Dietary Fibers from Millets

6.2.1 Hydrothermal Treatment

A technique called hydrothermal extraction uses hot water to remove nutritional fiber from millet. At high temperatures and pressures (Figure 6.1), water is utilized in this process for hydrolysis, extraction, and structural material modification. The grains are soaked, steamed, and then dried during the process. Because it uses less harmful solvents, produces

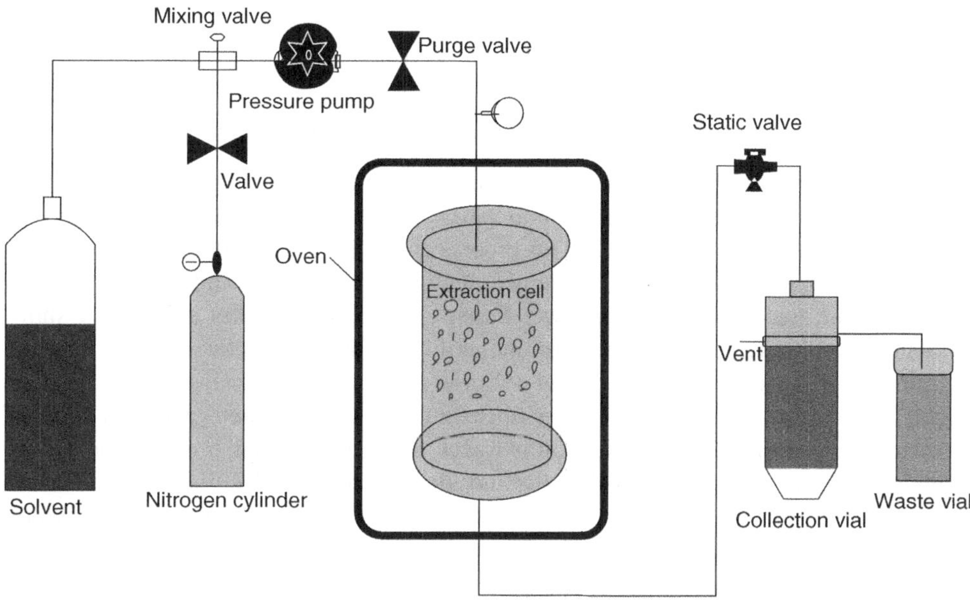

Figure 6.1 A schematic set-up for pressure solvent extraction of fiber from millet

a higher extraction yield, is highly pure, and is profitable, hydrothermal extraction of dietary fiber from millet bran is a straightforward and environmentally benign method. According to a study on pearl millet bran, the ideal parameters for this process were 20 : 1 water-to-solid ratio, 90 °C, and 15 minutes. Antinutrient substances like tannins, phytic acid, phenols, and trypsin inhibitors can be reduced via hydrothermal processing.

Prior to extraction in the water extraction procedures covered below, the grain samples were ground, inspected for appropriate particle size, and refluxed with ethanol. Besides aiding in the deactivation of β-glucanase, refluxing in ethanol also aids in the elimination of the majority of lipids. The solubility and subsequent gelatinization of starch can be avoided by extracting the starch at 47–50 °C. The extracted molecular weight of β-Glucan was influenced by the cultivar type and extraction temperature. Since endogenous β-glucanases are responsible for the breakdown of the polysaccharide backbone, conventional extraction methods result in a low molecular weight β-glucan product. Because it can provide both an effective extraction solvent and the medium for β-glucanases to be inactivated, the use of pressurized hot water as an extraction solvent may be a novel, adaptable, and promising method for this process. By applying pressure, water in a liquid condition between the boiling point (100 °C) and the critical point (374 °C) is referred to as pressurized hot water. Water's density, surface tension, viscosity, and diffusion all drastically alter with changes in pressure and, particularly, temperature, which impacts the extraction process's mass transfer. The diffusion into the vegetal matrix and the release of chemicals are facilitated by pressurized hot water's increased diffusivity and lower viscosity compared to room temperature water. The hydrogen bonds that bind the carbohydrates to the solid matrix are also weakened by rising temperatures, which speeds up the desorption of the compounds. High temperatures can also change the structure of the natural matrix or start the hydrolysis of chemicals that have already dissolved. To promote complete penetration of the matrix particles and to allow the flow of the extraction solvent, dry, finely divided solid samples are desired. By increasing the solvent-matrix contact surface, grinding particles larger extraction than 1 mm can improve efficiency.

6.2.2 Ultrasound-Assisted Extraction of Dietary Fibers from Millet

In contrast to conventional methods, ultrasound-assisted extraction of dietary fibers from millet involves using high-frequency sound waves, or ultrasound, to break down the cell walls of the millet grains (Figure 6.2). This process allows for a more effective extraction of the dietary fiber within the grain, frequently yielding a higher yield and better quality of extracted fiber. Because it requires fewer harsh chemicals and takes a relatively short extraction time, it is regarded as an environmentally friendly and effective method of isolating millet fiber. By using less solvent and a shorter extraction period, ultrasonication may be able to increase the amount of dietary fiber that is extracted. When cavitation bubbles are formed in the millet tissue by ultrasound waves, they quickly burst, producing localized high shear stresses that degrade cell walls and liberate the digestive fiber. The extracted fiber's structure can be altered by ultrasound treatment, improving its ability to retain water and swell. Additionally, ultrasound-assisted extraction uses lower temperatures than traditional extraction techniques, which reduces the thermal destruction of delicate chemicals and maintains their bioactivity. According to contemporary developments in green extraction technologies, the ultrasound-assisted extraction approach is regarded as somewhat safe and sustainable.

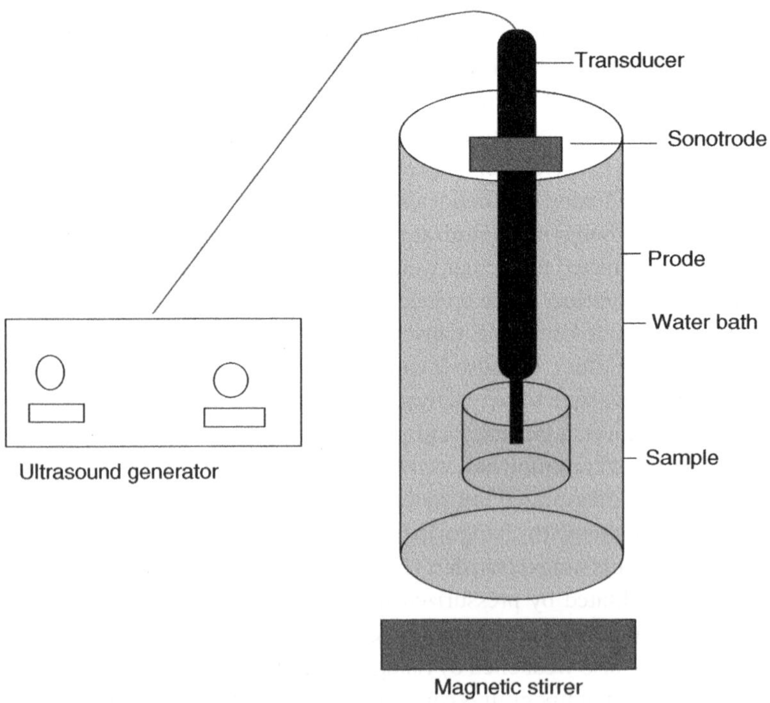

Figure 6.2 Schematic diagram of ultrasound-assisted extraction of dietary fiber from millets

The total dietary fiber extracted from pearl millet bran with the maximum extraction yield (78.8%, w/w) was produced by combining ultrasonic treatment with enzymatic extraction at amplitude (35%), temperature (50 °C), and liquid-to-solid ratio (40 : 1) after treatment duration (10 minutes). The studies were conducted using a probe-type ultrasonicator with a maximum power input of 700 W and a frequency of 20 kHz. The ultrasound treatment/hydration duration was varied between 10 and 30 minutes, and the ultrasound amplitude was varied between 30% and 70% with a pulse on/off period of five seconds. Because of the stronger intensity that is provided across a small surface during the extraction process, a probe system is favored over a bath. By connecting probes to a transducer and operating at 20 kHz, waves are delivered directly to extraction media with less ultrasonic energy loss.

6.2.3 Microwave-Assisted Extraction of Dietary Fiber

An effective method for analyzing and isolating dietary fiber from a variety of food sources, microwave-assisted extraction (MAE) of dietary fiber uses microwave radiation to quickly heat a sample. This breaks down plant cell walls and makes it possible to extract dietary fiber components like cellulose, hemicellulose, and pectin efficiently. The extraction time is significantly reduced compared to traditional methods, and the yield is maximized. Direct heat generation from microwaves breaks down cell structures quickly, allowing the components of dietary fiber to be released into the solvent. It may cause the solvent to dipole rotate and rapidly elevate its temperature, increasing the solubility of the chemical (Ahmad & Langrish, 2012). Furthermore, microwave radiation can pass straight through the wall material to the sample's core, heating it more rapidly and uniformly and saving more time and energy (Jamaluddin et al., 2013). More significantly, a microwave could burst the cell membrane and release active chemicals by raising intracellular pressure in fresh tissue (Wei et al., 2019). A mixture of milled millet and a solvent is heated quickly using microwave radiation in MAE of dietary fiber from millets. This process effectively breaks down the plant cell walls and extracts the dietary fiber in a short amount of time, making it a quicker and more effective method than traditional extraction techniques.

6.2.4 High-Pressure Techniques for FiberExtraction

Using extremely high pressure to break down plant cell walls, "High Pressure Homogenization" is a high-pressure method for extracting dietary fiber (Figure 6.3). Other related methods include supercritical fluid extraction, which uses a supercritical fluid under high pressure to extract fiber components from plant material. Breakdown of plant cell walls, which are made up of complex polysaccharides including cellulose, hemicellulose, and pectin, is the main mechanism underlying high-pressure processing (HPP)-assisted fiber extraction. These cell walls undergo considerable stress under high pressure, which leads to microcracks and ruptures that make it easier for fiber components to be released. By improving the solubility of plant metabolites and the permeability of cell membranes, high pressure can further improve the mass transfer process by facilitating the more effective extraction of fiber components into the surrounding solvent. Using a fluid as the medium and pressure applied isostatically from all directions in a small area, HPP is a nonthermal technique that transfers pressures of

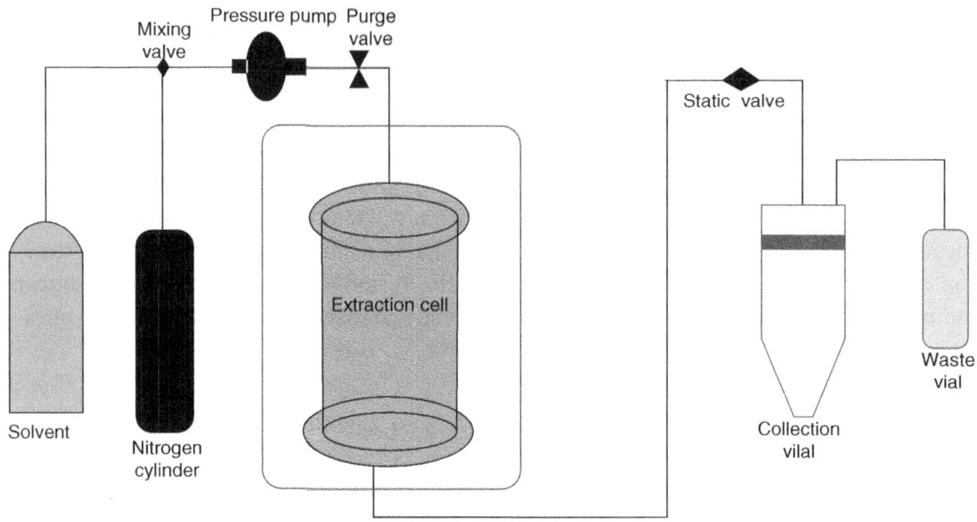

Figure 6.3 Diagrammatic representation of high-pressure extraction of fiber

thousands of atmosphere pressures (100–1000 MPa). Because high pressure can increase the mass transfer rate and some component extraction, as well as cause cell deformation and cell wall breaking, HPP can effectively extract plant-based compounds. Because of these characteristics, HPP is a highly effective and environmentally responsible substitute technique for polysaccharide extraction.

6.2.5 Enzymatic Method

Following the procedures outlined in work (Zheng et al., 2022), the millet bran dietary fiber was prepared by removing the protein (using NaOH) and starch (using α-amylase and diastatic enzyme). A precisely weighed sample (weighing 5.00 g) was mixed with 50 mL of distilled water, which is 10 times the sample's weight. Following that, α-amylase (≥3,700 U/g) and diastatic enzyme (≥100,000 U/g) (1.5% the sample, w/w) were added and thoroughly blended. The suspension was then kept in a water bath at 57 °C for 90 minutes. After the reaction was finished, 3.6% NaOH solution (10 times the sample's weight, w/w) was added, stirred, and then placed in a water bath at 76 °C for 96 minutes.

Prosky et al. (1988) used a procedure wherein 1 g of defatted flours were sequentially digested by heat-stable α-amylase (95–100 °C), amylo-glucosidase (60 °C), and protease. The insoluble dietary fiber was filtered and cleaned using acetone, 95% ethanol, and hot distilled water (70 °C). Overnight, the insoluble dietary fiber is vacuum-dried at 50 °C. 95% ethanol was used to precipitate a mixture of filtrate and hot water washings for soluble dietary fibers. Following precipitate filtering, residues were cleaned with 78% ethanol, 95% ethanol, and acetone before being vacuum-dried for an entire night at 50 °C. To calculate the total dietary fibers, the insoluble dietary fiber and soluble dietary fiber values are combined.

6.3 Quality Characteristics of Dietary Fiber from Millets

Over 60% of the dietary fiber in foxtail millet bran was higher than that of rice (27.04%) and wheat bran (44.46%) (Elleuch et al., 2011). According to these results, foxtail millet bran may prove to be a beneficial dietary fiber source. According to the results, the dietary fiber yield and purity of foxtail millet bran were 72.20% and 79.22%, respectively. According to functional investigations, foxtail millet bran dietary fiber has a decreased binding capacity for water but a higher binding capacity for lipophilic substances, including bile salts, cholesterol, and lard. In comparison to goods made from wheat, rice, and barley, those made from foxtail millet had lower glycemic indexes, higher levels of protein, dietary fiber, and resistant starch, and a slower rate of starch digestion. Foods with a medium glycemic index, such as the created foxtail millet products, can be utilized as diabetics' therapeutic foods. The high fiber and calcium content of kheer made from foxtail millet can help prevent osteoporosis and make it a nutritious snack. Because of its high fiber content, resistant starch, and reduced glycemic index, foxtail millet products can help prevent noncommunicable disorders including diabetes and obesity.

The significant levels of dietary fiber, minerals, and bioactive substances found in the brans that were separated from proso millet have drawn a lot of research interest. Thus, the pharmaceutical and nutritional food businesses might make use of proso millet brans on a commercial basis. The significant antioxidant qualities of proso millet husks make them a possible source of natural antioxidants, according to Devisetti et al. (2014). Water and oil can be bound by capillary and molecular forces, and the reticular structure of the fibers can trap more of these molecules (Cao et al., 2021). Fiber-rich materials' ideal degree of use in a variety of processed foods is determined by their hydration qualities, and high hydration allows for the integration of nutrients into foods (Cao et al., 2021). Consequently, when applied as an ingredient, the high hydration of proso millet brans help prevent water loss and preserve the freshness and moist texture of baked items.

According to Martino et al. (2012), sorghum is a rich source of fibers because its nonstarch carbohydrate is mainly made up of soluble fibers (10% to 25%) and insoluble fibers (75–90%). These fibers are found in the cell walls of the pericarp and endosperm and make up roughly 6–15 g per 100 g of grain. The fiber and antioxidants included in sorghum millet make it a great addition to baked products, cereal, breakfast bowls, and smoothies. Other traits of millet and sorghum grains that significantly affect their consumer acceptance are their high fiber content and poor nutritional digestion.

Although finger millet contains a larger percentage of nutritional fiber than many other cereals, it nonetheless contains carbohydrates like any other cereal. Dietary fiber's beneficial effects are typically ascribed to either slower stomach emptying or the creation of unabsorbable complexes with available carbohydrates in the gut lumen. These two characteristics may cause delayed absorption of carbohydrates and a decrease in the absolute amount absorbed (Rasmussen et al.,1991). Soluble fiber in millets is reduced by 39% when they are hydrothermally processed, while total dietary fiber remains unchanged. The millet's dietary fiber content decreased by roughly 33.2% as a result of decortication; however, the amount of soluble fiber content increased significantly by 170% (Dharmaraj & Malleshi, 2010). The functional fiber fraction found in finger millets, known as resistant starch, prevents enzymatic digestion and offers advantages by lowering several digestive issues. It avoids digestion and instead transports fermentable metabolites and SCFAs to the colon, such as butyrate, which may protect against colon cancer by controlling the growth of colonic cells. Furthermore, resistant starch enhances the texture and looks of fiber and has therapeutic benefits.

Five grams of insoluble and three grams of soluble dietary fibers per 100 g are found in pearl millet (Nandini & Salimath, 2001). These fibers create insoluble complexes with divalent cations like iron and zinc, which prevent the intestines from absorbing them. Fiber–phytate–mineral complexes are created when the fibers and phytates chelate minerals. Scientists' attention was drawn to pearl millet by the growing market demand for nutritious foods with low gluten and high crude fiber.

In terms of fiber content, brown top millet has 12.5 g of fiber per 100 g (Indian Institute of Millets Research [IIMS], Dayakar, 2012). Brown top millet is known for having a high percentage of nutritious fiber in comparison to other millets. As a result, this nutri-cereal can be added to food to improve general health and prevent a number of diseases related to lifestyle, including type 2 diabetes, obesity, and heart disease. Barnyard millet is a nutritionally significant crop as well. According to Veena et al. (2005), it is a good supply of dietary fiber, a reasonable quantity of highly digested protein, and a good amount of both soluble and insoluble components. Little millet has a low glycemic index because of its high dietary fiber content, and a recent study on the grain found that it had a hypoglycemic effect because of this greater the amount of fiber in the (Itagi et al., 2012). The high proportion of carbohydrates to crude fiber helps to maintain blood sugar levels by ensuring a delayed release of blood glucose. It has been demonstrated that the resistant starch in barnyard millet lowers triglycerides, serum cholesterol, and blood glucose in rats.

The amount of dietary fiber in various processed buckwheat fractions varies, with unhusked buckwheat having a dietary fiber content of 23.8%, husked buckwheat having a concentration of 10.3%, and groats having a dietary fiber content of 7% (Qin et al., 2010). The buckwheat's seed coat contains most of its dietary fiber, which is reduced after milling (Wefers & Bunzel, 2015). There are two types of dietary fiber: soluble (found in pectin and gums) and insoluble (found in lignin and cellulose). Buckwheat has 4.8% soluble and 2.2% insoluble dietary fiber, which might help prevent obesity and high cholesterol while also having negative effects including causing protein and mineral deficiency (Zhu, 2020). The husk also contains resistant starch, which is classified as fiber since it is difficult to digest. Buckwheat has a low to medium glycemic index and is a rich source of fiber. This implies that most patients with type 2 diabetes should be able to eat it without risk. According to reports, kodo millet and tiny millet have the greatest dietary fiber content of any cereal at 37–38%. They are also low in fat but high in polyunsaturated fatty acids (Malleshi & Hadimani, 1993).

6.4 Functional Role of Fibers Isolated from Millets

Pectin, a soluble dietary fiber, makes significantly less than 15% of millet cell walls. This substance is composed of galacturonan, rhamnogalacturonan, and glucuronoarabinoxylan. Pectin is widely employed in food, medicine, materials, and other industries because of its good gelling, thickening, stability, and emulsification capabilities, as well as its anti-inflammatory, antioxidant, hypoglycemic, prebiotic, and other biological activities

(Cui et al., 2020). Pectin has been utilized in food applications to make fruit juices and acidified milk drinks because of its good thickening and stability qualities, as well as jams and jellies because of its good gelling property. It is also a biopolymer that can be utilized in packaging and as a molecule that carries antimicrobials and antioxidants to preserve food (Naqash et al., 2017). A form of soluble fiber that is present in millets and other grains, β-glucans serve a variety of purposes in the body and in food. In the intestinal tract, β-glucans generate SCFAs, aid in digestion, and inhibit the absorption of cholesterol. Inhibiting infections, β-glucans promote the proliferation of good gut microorganisms. Numerous health benefits are offered by β-glucans, including the control of blood sugar and insulin levels, the reduction of cholesterol, the prevention of heart disease, the decrease of blood pressure and obesity, and the presence of anticancer and anti-inflammatory properties. In addition to a low-cholesterol diet, the US Food and Drug Administration has advised consuming 3 g of β-glucan daily to lower the risk of cardiovascular illnesses.

A functional component of millets, arabinoxylans are nonstarch polysaccharides that have been associated with several health advantages, such as immunomodulation, anticancer activity, increased fecal bulk, decreased insulin resistance, antioxidant, anti-inflammatory, and prebiotic properties (Kim et al., 2020). A water-soluble polymer, inulin belongs to the class of fructans, which are indigestible carbohydrate polymers. With very low-calorie content, it can be used as a sugar or fat alternative. It functions similarly to dietary fibers and helps to improve conditions of the gastrointestinal tract. Due to its beta structure, which renders it an indigestible fiber that cannot be hydrolyzed by human enzymes generated by the pancreas and salivary glands, inulin has prebiotic benefits. Consequently, the large intestine, the area of the gastrointestinal system that is most extensively colonized, receives undigested inulin. Numerous substances that are created when Bifidobacteria digest inulin have an impact on the intestinal and systemic physiology, including controlling cell division, differentiation, and metabolism (Cherbut, 2022).

6.5 Conclusion

Millet fibers excellent cost-economics and negligible impact on sensory qualities make it a viable alternative to standard dietary fibers, including cellulose, β-glucan, and inulin. A form of soluble fiber that is present in millets and other grains, β-glucans serve a variety of purposes in the body and in food. In the intestinal tract, β-glucans generate SCFAs, aid in digestion, and inhibit the absorption of cholesterol. Inhibiting infections, β-glucans promote the proliferation of good gut microorganisms. A water-soluble polymer, inulin belongs to the class of fructans, which are indigestible carbohydrate polymers. With very low-calorie content, it can be used as a sugar or fat alternative. It functions similarly to dietary fibers and helps to improve conditions of the gastrointestinal tract. Due to its beta structure, which renders it an indigestible fiber that cannot be hydrolyzed by human enzymes generated by the pancreas and salivary glands, inulin has prebiotic benefits. The significant levels of dietary fiber, minerals, and bioactive substances found in the brans that were separated from millet have drawn a lot of research interest. Thus, the pharmaceutical and nutritional food businesses might make use of millet brans on a commercial basis. The significant antioxidant qualities of millet husks make them a possible source of natural antioxidants.

References

Ahmad, J., & Langrish, T. A. G. (2012). Optimisation of total phenolic acids extraction from mandarin peels using microwave energy: The importance of the Maillard reaction. *Journal of Food Engineering*, *109*(1), 162–174.

Aune, D., Chan, D. S. M., Lau, R., Vieira, R., Greenwood, D. C., Kampman, E., & Norat, T. (2011). Dietary fiber, whole grains, and risk of colorectal cancer: Systematic review and dose-response meta-analysis of prospective studies. *BMJ*, *343*, d6617–d6617.

Bagdi, A., Balázs, G., Schmidt, J., Szatmári, M., Schoenlechner, R., Berghofer, E., & Tömösközia, S. (2011). Protein characterization and nutrient composition of Hungarian proso millet varieties and the effect of decortications. *ActaAlimentaria*, *40*(1), 128–141.

Barber, T. M., Kabisch, S., Pfeiffer, A. F. H., & Weickert, M. O. (2020). The health benefits of dietary fiber. *Nutrients*, *12*, 3209.

Basen, M., & Kurrer, S. E. (2021). A close look at pentose metabolism of gut bacteria. *The FEBS Journal*, *288*, 1804–1808.

Bravo, L. (1998). Polyphenols: Chemistry, dietary sources, metabolism and nutritional significance. *Nutrition Reviews*, *56*, 317–333.

Cao, Y., Zhao, J., Tian, Y., Jin, Z., Xu, X., Zhou, X., & Wang, J. (2021). Physicochemical properties of rice bran after ball milling. *Journal of Food Processing and Preservation*, *45*. Article e15785

Chandrasekara, A., & Shahidi, F. (2011). Bioactivities and antiradical properties of millet grains and hulls. *Journal of Agricultural and Food Chemistry*, *59*, 9563–9571.

Charalampopoulos, D., Wang, R., Pandiella, S., & Webb, C. (2002). Application of cereals and cereal components in functional foods: A review. *International Journal of Food Microbiology*, *79*(1–2), 131–141.

Chen, J., Duan, W., Ren, X., Wang, C., Pan, Z., Diao, X., & Shen, Q. (2017). Effect of foxtail millet protein hydrolysates on lowering blood pressure in spontaneously hypertensive rats. *European Journal of Nutrition*, *56*, 2129–2138.

Cherbut, C. (2022). Inulin and oligofructose in the dietary fibre concept. *British Journal of Nutrition*, *87*(2), S159–S162.

Cui, J., Ren, W., Zhao, C., Gao, W., Tian, G., Bao, Y., Lian, Y., & Zheng, J. (2020). The structure–property relationships of acid- and alkali-extracted grapefruit peel pectins. *Carbohydrate Polymers*, *229*, 115524.

Dayakar R.B., Bhaskarachary, K., Arlene Christina, G.D., Sudha Devi G., & Vilas, A.T. (2012). Nutritional and health benefits of millets. ICAR_Indian Institute of Millets Research (IIMR) Rajendranagar, Hyderabad, pp. 112.

Devisetti, R., Yadahally, S. N., & Bhattacharya, S. (2014). Nutrients and antinutrients in foxtail and proso millet milled fractions: Evaluation of their flour functionality. *LWT-Food Science and Technology*, *59*(2), 889–895.

Dharmaraj, U., & Malleshi, N. G. (2011). Changes in carbohydrates, proteins and lipids of finger millet after hydrothermal processing. *LWT-Food Science and Technology*, *44*, 1636–1642. https://doi.org/10.1016/j.lwt.2010.08.014

Durairaj, M., Gurumurthy, G., Nachimuthu, V., Muniappan, K., & Balasubramanian, S. (2019). Dehulled small millets: The promising nutricereals for improving the nutrition of children. *Maternal and Child Nutrition*, *15*, e12791. https://doi.org/10.1111/mcn.12791

Easwaran, P. P., Mageshwari, V. S., & Narayanan, G. (1991). Therapentic use of globe artichoke in non-insulin dependent diabetes mellitus and hypercholesterolemia. *Indian Journal of Nutrition and Dietetics*, *28*(12), 321–327.

Elleuch, M., Bedigian, D., Roiseux, O., Besbes, S., Blecker, C., & Attia, H. (2011). Dietary fibre and fibre-rich by-products of food processing: Characterisation, technological functionality and commercial applications: A review. *Food Chemistry*, *124*(2), 411–421.

Gopalan, C., Ramasastri, B. V., & Balasubramanian, S. C. (1996). *Nutritive value of Indian foods*. National Institute of Nutrition, Indian Council of Medical Research.

Gupta, R. K., Gangoliya, S. S., & Sangh, N. K. (2015). Reduction of phytic acid and enhancement of bioavailable macronutriems in food grains. *Journal of Food Science and Technology*, *52*, 676–684.

Hadimani, N. A., & Malleshi, N. G. (1993). Studies on milling, physicochemical properties, nutrient composition and dietary fiber content of millets. *Journal of Food Science and Technology*, *30*, 193–198.

Hegde, P. S., & Chandra, T. S. (2005). ESR spectroscopic study reveals higher free radical quenching potential in kodo millet (*Paspalum scrobiculatum*) compared to other millets. *Food Chemistry*, *92*, 177–182.

Itagi, S., Naik, R., Bharati, P., & Sharma, P. (2012). Readymade foxtail millet mix for diabetics. *International Journal of Natural Sciences*, *3*(1), 47–50.

Jamaluddin, M. A., Ismail, K., Ishak, M. A. M., Ghani, Z. A., Abdullah, M. F., Safian, M. T., Idris, S. S., Tahiruddin, S., et al. (2013). Microwave-assisted pyrolysis of palm kernel shell: Optimization using response surface methodology (RSM). *Renewable Energy*, *55*, 357–365.

Khalid, W., Ali, A., Arshad, M. S., Afzal, F., Akram, R., Siddeeg, A., Kousar, S., Rahim, M. A., Aziz, A., Maqbool, Z., & Saeed, A. (2022). Nutrients and bioactive compounds of *Sorghum bicolor* L. used to prepare functional foods: A review on the efficacy against different chronic disorders. *International Journal of Food Properties*, *25*(1), 1045–1062.

Kim, C. S., Byeon, S., & Shin, D.-M. (2020). Sources of dietary fiber are differently associated with prevalence of depression. *Nutrition*, *12*, 2813.

Kulkarni, L. R., Naik, R. K., & Katarki, P. A. (1992). Chemical composition of minor millets. *Karnataka Journal of Agricultural Sciences*, *5*(3), 255–258.

Lei, V., Friis, H., & Michaelsen, K. (2006). Spontaneously fermented millet product as a natural probiotic treatment for diarrhoea in young children: An intervention study in Northern Ghana. *International Journal of Food Microbiology*, *110*, 246–253.

Malleshi, N. G., & Hadimani, N. A. (1993). Studies on milling, physicochemical properties, nutrient composition and dietary fiber content of millets. *Journal of Food Science and Technology.*, *30*, 17–20.

Martino, H. S. D., Tomaz, P. A., Moraes, E. A., Conceic'aoI, L. L. d., Oliveira, D. d. S., Queiroz, V. A. V., & Ribeiro, S. M. R. (2012). Chemical characterization and size distribution of sorghum genotypes for human consumption. *Revista do Instituto Adolfo Lutz (Impresso)*, *71*(2), 337–344.

Mutshinyani, M., Mashau, M. E., & Jideani, A. L. O. (2020). Bioactive compounds, antioxidant activity and consumer acceptability of porridges of finger millet (*Eleusine coracana*) flours: Effects of spontaneous fermentation. *International Journal of Food Properties*, *23*, 1692–1710.

Nandini, C. D., & Salimath, P. V. (2001). Carbohydrate composition of wheat, wheat bran, sorghum, and bajra with good chapatti/roti (Indian flatbread) making quality. *Food Chemistry*, *73*, 197–203.

Naqash, F., Masoodi, F. A., Rather, S. A., Wani, S. M., & Gani, A. (2017). Emerging concepts in the nutraceutical and functional properties of pectin—A review. *Carbohydrate Polymers, 168*, 227–239.

Potdar, K., Mistra, M., & Yesikar. (1994). A study on the effect of soyabean, wheat bran and pyrodextrins on the blood sugar and lipid profile in obese diabetics. *Indian Journal of Nutrition and Dietetics, 31*(10), 299–302.

Prosky, L., Asp, N., Schweizer, T., De Vries, W., & Furda, I. (1988). Determination of insoluble, soluble and total dietary fiber in foods and food products: Interlaboratory study. *Journal of the Association of Official Analytical Chemists, 71*, 1017–1023.

Qin, P., Wang, Q., Shan, F., Hou, Z., & Ren, G. (2010). Nutritional composition and flavonoids content of flour from different buckwheat cultivars. *International Journal of Food Science and Technology, 45*, 951–958.

Ramadoss, D. P., & Sivalingam, N. (2021). Vanillin extracted from proso and barnyard millets induces cell cycle inhibition and apoptotic cell death in MCF-7 cell line. *Journal of Cancer Research and Therapeutics, 17*, 1425–1433.

Rasmussen, O., Winther, C., & Hermansen, K. (1991). Glycemic responses to different types of bread in IDDM patients: Studies at constant insulinaemia. *European Journal of Clinical Nutrition, 45*(2), 97–103.

Rathore, T., Singh, R., Kamble, D. B., Upadhyay, A., & Thangalakshmi, S. (2019). Review on finger millet: Processing and value addition. *The Pharma Innovation Journal, 8*, 283–291.

Sharma, N., & Niranjan, K. (2017). Foxtail millet: Properties, processing, health benefits, and uses. *Food Reviews International, 34*(4), 329–363.

Sharma, S., Saxena, D. C., & Riar, C. S. (2015). Antioxidant activity, total phenolics, flavonoids and antinutritional characteristics of germinated foxtail millet (*Setaria italica*). *Cogent Food & Agriculture, 1*, 1081728.

Singh, V., Hwang, N., Ko, G., & Tatsuya, U. (2021). Effects of digested Cheonggukjang on human microbiota assessed by in vitro fecal fermentation. *Journal of Microbiology, 59*, 217–227.

Sushmita, V. P., Lakshmi, J., & Lakshmi, K. (2020). Nutrient composition of little millet varieties. *The Andhra Agricultural Journal, 67*(1), 73–76.

Veena, S., Chimmad, B. V., Naik, R. K., & Shanthakumar, G. (2005). Physico-chemical and nutritional studies in Barnyard millet. Karnataka. *The Journal of Agricultural Science, 18*, 101e105.

Wefers, D., & Bunzel, M. (2015). Characterization of dietary fiber polysaccharides from dehulled common buckwheat (*Fagopyrumesculentum*) seeds. *Cereal Chemistry, 92*(6), 598–603.

Wei, C. H., Allot, A., Leaman, R., & Lu, Z. (2019). PubTator central: Automated concept annotation for biomedical full text articles. *Nucleic Acids Research, 47*, 587–593.

Yang, L., Browning, J. D., & Awika, J. M. (2009). Sorghum 3-deoxyanthocyanins possess strong phase II enzyme inducer activity and cancer cell growth inhibition properties. *Journal of Agricultural and Food Chemistry, 57*, 1797–1804.

Zhang, B., Xu, Y., Liu, S., Lv, H., Hu, Y., Wang, Y., Li, Z., Wang, J., Ji, X., Ma, H., Wang, X., & Wang, S. (2020). Dietary supplementation of foxtail millet ameliorates colitis-associated colorectal cancer in mice via activation of gut receptors and suppression of the STAT3 pathway. *Nutrients, 12*, 2367.

Zheng, Y., Wang, X., Guo, M., Yan, X., Zhuang, Y., Sun, Y., et al. (2022). Two novel antihypertensive peptides identified in millet bran glutelin-2 hydrolysates: Purification, in silico characterization, molecular docking with ace and stability in various food processing conditions. *Foods*, *11*, 1355.

Zhu, F. (2020). Dietary fiber polysaccharides of amaranth, buckwheat and quinoa grains: A review of chemical structure, biological functions and food uses. *Carbohydrate Polymers*, *248*, 116819.

7

Millet Proteins: Extraction and Nutritional Characteristics

Qun Shen, Qingyu Zhao, Liangxing Zhao, and Luman Sang

College of Food Science and Nutritional Engineering, China Agricultural University, Beijing, China

7.1 Cultivation and Distribution of Millet

Millet, belonging to the Poaceae family and having smaller seeds, is usually cultivated in the semi-arid tropical regions of Asia and Africa. As stated in the FAO Production Yearbook, the term "millet" refers to a category of small-seeded, annual plants within the Poaceae family, cultivated for their utility in human diets and as fodder for livestock. Taxonomically, it includes crops such as foxtail millet, proso millet, and barnyard millet, most of which belong to the Paniceae tribe. These grains are often referred to as "millet grains" due to their small size after de-hulling.

Millet crops can be broadly divided into two categories: major millet crops, which include pearl millet (*Pennisetum glaucum*), foxtail millet (*Setaria italica*), proso millet (*Panicum miliaceum*), and finger millet (*Eleusine coracana*); and minor millet crops, such as fonio (*Digitaria exilis*), kodo millet (*Paspalum scrobiculatum*), and barnyard millet (*Echinochloa frumentacea*). These diverse species of millet are cultivated across multiple regions globally. For example, pearl millet is the most widely cultivated variety, especially in India, Nepal, and several African nations. In China, foxtail millet is the predominant species, while proso millet is more commonly cultivated in both North America and China. In Japan, smaller quantities of barnyard millet, Indian barnyard millet, and chicken millet are grown (Kumar et al., 2021) (Table 7.1).

According to FAOSTAT (2024) data, the total cultivated area and production of millet crops have shown a declining trend between 1994 and 2022. In the last decade, the area under millet cultivation has stabilized between 30 and 35 million hectares, with total production ranging from 25 to 30 million tons. The majority of millet cultivation is focused in Asia and Africa, where Asia produces 48.5% of the world's total millet yield, while Africa contributes 47.6%. From 1994 to 2022, the leading millet-producing nations, ranked by average output, were India, Nigeria, Niger, China, and Mali. India, as the global leader in

Nutraceutical Potential of Millet-Based Food Products: Chemistry and Applications, First Edition.
Edited by Parmjit S. Panesar, Charanjit S. Riar, and Sushil Dhital.

Published 2026 by John Wiley & Sons Ltd.

Table 7.1 Major cultivation millets in the world

Generally used name	Other common and vernacular names	Taxonomy subfamily, tribe, genus, species, and subspecies where applicable authority(s)	Major producing regions/countries
Pearl millet	Bulrush millet Cattail millet Babala (southern Africa) Bajra (India)	Subfamily: Panicoideae Tribe: Paniceae *Pennisetum glaucum* (L.) R. Br.	West and central Africa (Niger, Mali, Burkina Faso, Nigeria), India, East and Southern Africa (Sudan, Namibia, Botswana), the United States, Brazil
Foxtail millet	Intalian millet Forxtail bristle grass German millet Hungarian millet	Subfamily: Panicoideae Tribe: Paniceae *Setaria italica* (L.) P. Beauv subsp. *italica*	Eurasia-Eastern and Southern Europe, tropical and subtropical Asia (China, India, Korea), the United States, Australia
Finger millet	Ragi (India) Wimbi (East Africa)	Subfamily: Chloridoideae Tribe: Eragrostideae *Eleusine coracana* (L.) Gaertn	Eastern and Southern Africa (Ethiopia, Uganda, Kenya, Zimbabwe) and Asia-Near East to Far East (India, Nepal, China)
Proso millet	Common millet Broom millet Hog millet Panic millet	Subfamily: Panicoideae Tribe: Paniceae *Panicum miliaceum* (L.) L. subsp. *miliaceum*	Eurasia (China, Kazakhstan, Afghanistan, India, Turkey, Romania), the United States, Australia
Teff			Ethiopia, Eritrea, South Africa, the United States, Australia
White fonio			Africa-Sahel region (Guinea, Nigeria, both species)

millet production, achieved a remarkable yield of 11,849,190 tons in 2022, which was significantly lower than the 13,208,480 tons produced in 2021 and the 12,488,470 tons in 2020. India's millet cultivation area is also gradually shrinking. On the other hand, Nigeria's millet cultivation and production have been steadily increasing, with production figures from 2018 to 2022 standing at 1,879,340, 1,925,075, 1,919,054, 1,926,950, and 1,941,220 tons, respectively. Figure 7.1 illustrates the top 10 millet-producing countries based on their average production between 1994 and 2022 (FAOSTAT, 2024).

7.2 Nutritional Value of Millet

Despite the decline in millet cultivation area and production, millet remains a staple food in many African regions, serving as a crucial source of energy and protein for local populations. Millet is composed of 60–70% carbohydrates, 6–19% protein, 1.5–5% fat, 12–20% dietary fiber, 2–4% minerals, and a variety of phytochemicals (Table 7.2) (McDonough

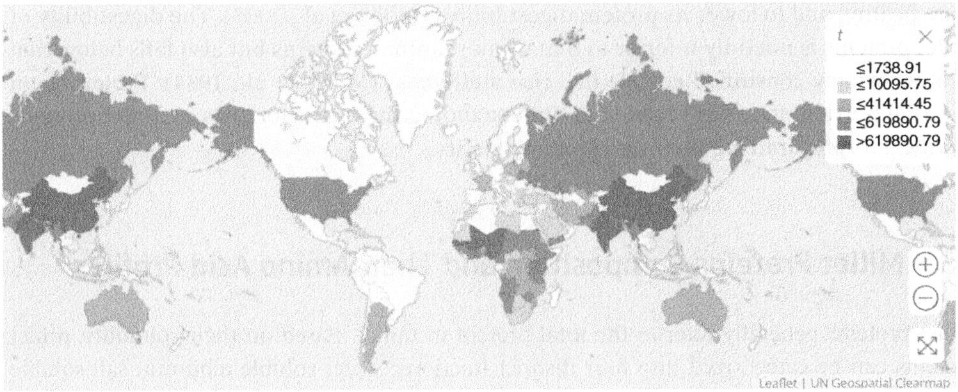

Figure 7.1 Millets planted in the world and top 10 producers of the average millets production from 1994 to 2022

Table 7.2 Nutrients in different millet varieties (%)

Millets	Protein	Fat	Crude fiber	Ash	Starch
Pearl millet	14.5	5.1	2.0	2.0	71.6
	8.6–19.4	1.5–6.8	1.4–7.3	1.6–3.6	63.1–78.5
Foxtail millet	11.7	3.9	7.0	3.0	55.1
	6.0–14.0	1.2–5.2	2.6–8.6	1.5–3.6	51.0–59.1
Finger millet	8.0	1.5	3.0	3.0	59.0
	6.9–10.9	1.0–4.6	2.0–6.8	2.3–3.9	
Proso millet	13.4	6.7	6.3	4.2	57.1
	6.4–15.9	2.9–11.1	4.6–10.7	1.4–8.8	56.1–58.0
Teff	10.9	2.4	2.7	2.5	73
	7.9–12.6	2.3–2.5	2.4–3.0	2.2–2.8	
White fonio	8.1	3.3	5.9	3.5	68
	5.1–11.1	1.3–5.2	0.4–11.3	1.0–6.0	

Source: Adapted from McDonough (2000).

et al., 2000). The fact that millets do not contain gluten, making them suitable for people with coeliac disease, millets can also be exploited in the management of type 2 diabetes due to their hypoglycemic property. The other side of the coin is protein digestibility, which is lower in millets compared to many other grains (Mertz et al., 1984).

The digestibility of plant storage proteins is generally lower than that of most animal proteins, a phenomenon that can be attributed to several factors. These include the presence of protease inhibitors or tannins that hinder digestive enzymes, reduced protein solubility, the structural configuration of proteins within protein bodies, and limited enzyme accessibility due to the rigidity of the cell wall and/or seed coat (Becker & Yu, 2013). In addition, disulfide-mediated protein cross-linking has been shown to occur in foxtail millet upon heating and to lower its protein digestibility (Duodu et al., 2003). The digestibility of millet proteins is not only inferior to that of most animal proteins but also falls below that of other widely consumed cereals like rice and wheat (Mertz et al., 1984). Protein is an essential nutrient in millet crops, and understanding the characteristics of millet proteins is critical for maximizing their nutritional benefits.

7.3 Millet Proteins Composition and Their Amino Acid Profiles

Millet proteins generally refer to the total protein in millet. Based on their solubility, millet proteins can be categorized into four distinct fractions: water-soluble albumin, salt-soluble globulin, alcohol-soluble prolamin, and alkali-soluble glutelin. The distribution of these protein fractions varies across different types of millet. For example, in common millet, prolamins represent the most abundant storage protein, making up roughly 40–65% of the total protein content, similar to the protein profile found in wheat (Kumar & Parameswaran, 1998).

Prolamin and glutelin in pearl millet (*P. glaucum*) and finger millet (*E. coracana*) are present in nearly equal proportions, both serving as the major storage proteins. On the other hand, in species like barnyard millet (*Echinochloa* spp.), little millet (*Panicum sumatrense*), and kodo millet (*P. scrobiculatum*), glutelin is the predominant protein fraction, and these species are notably abundant in sulfur-containing amino acids.

Millet proteins, overall, exhibit a high nutritional value, especially due to their beneficial composition of specific amino acids such as leucine, isoleucine, and lysine. The four protein fractions—albumin, globulin, prolamin, and glutelin—are different in their solubility, structure, functional properties, and nutritional characteristics (Parameswaran & Thayumanavan, 1995).

7.3.1 Globulins and Albumins

Albumins and globulins are primarily found in the millet bran and millet germ. In finger millet (*E. coracana*), the albumin content ranges from 9.6% to 12.2% (Ramachandra et al., 1978). Kodo millet (*P. scrobiculatum*) contains 5.6–11.7 mg/g of albumins and globulins (constituting 55–109 mg/g of total protein) (Sudharshana et al., 1988). The albumin and globulin fractions in foxtail millet (*S. italica*) are abundant in glutamic acid, aspartic acid, glycine, alanine, lysine, and arginine (Danno & Natake, 1980).

7.3.2 Prolamins

Prolamins are an important storage protein in millet, primarily found in the endosperm and generally present as large molecular complexes. The molecular weight of prolamins ranges from 27 to 11 kDa (Parameswaran & Thayumanavan, 1995). Research has shown that over 40% of the amino acids present in prolamins from foxtail millet are essential amino acids, playing a vital role in supporting the optimal functioning of the human body (Parameswaran & Thayumanavan, 1995). Another characteristic of prolamins is their high proline content, which facilitates their interaction with and binding to phytochemicals. This interaction supports the transport and absorption of bioactive compounds in the body (Chen et al., 2022).

Foxtail millet prolamins are composed of 18 different amino acids, with glutamic acid, leucine, and alanine being the most prevalent, making up 23.3%, 15.86%, and 9.5% (w/w) of the total quantified amino acids, respectively Zhang et al., (2022b). However, it is important to note that the absence or insufficient levels of lysine and tryptophan in prolamins of foxtail millet negatively impact their nutritional value (Kumar & Parameswaran, 1998).

As prolamins represent the major protein fraction in foxtail millet, their amino acid composition significantly influences key properties such as solubility, palatability, and overall nutritional quality of the protein. The variation of acidic and basic amino acid content affects the protein's solubility by altering the ionization of surface groups (Ge et al., 2021). The relatively high proportion of nonpolar amino acids further contributes to the unique solubility characteristics of prolamins (Chen et al., 2022). Regarding texture and taste, the palatability of foxtail millet prolamins is closely linked to the proportions of various amino acids present (Zhang et al., 2021).

7.3.3 Glutelins

Glutelins are a class of proteins that are soluble in alkaline solutions and are predominantly found in the endosperm. In foxtail millet (*S. italica*), the glutelin content ranges from 20% to 35%. Glutelins are mainly localized in the endosperm of millet and contain significantly higher levels of lysine compared to albumins and globulins. They are also rich in glutamic acid and aspartic acid. The levels of glutelin and lysine in millet can vary significantly depending on the specific variety and the conditions under which it is grown (Taylor & Schussler, 1986).

Additionally, research has shown that finger millet (*E. coracana*) contains a specific type of glutelin known as "Eleusinian glutelin," which represents the second largest storage protein fraction in this species. This glutelin includes key amino acids such as lysine and tryptophan (Virupaksha et al., 1975).

7.4 Properties of Millet Proteins

7.4.1 Extraction of Total Millet Protein

In the realms of food processing and biotechnology, the extraction and application of plant proteins have long been areas of active research. Methods such as alkaline extraction-acid precipitation and enzymatic hydrolysis are widely used due to their efficiency and feasibility.

7.4.1.1 Method of Alkaline Extraction-Acid Precipitation

The method of alkaline extraction-acid precipitation is a classical approach that leverages the solubility differences of proteins at different pH levels. Under alkaline conditions, particularly at pH values exceeding their isoelectric point (pI), proteins acquire a net negative charge, which increases their solubility. Conversely, when the pH is lowered to near the isoelectric point, protein net charge approaches zero, diminishing electrostatic repulsion between molecules and causing proteins to aggregate and precipitate. This method takes advantage of these properties by dissolving millet proteins in an alkaline solution, followed by lowering the pH to near the isoelectric point to precipitate and isolate the proteins.

Several factors significantly impact protein extraction using this method, including pH, temperature, and extraction time. pH is a critical determinant in both the dissolution and precipitation phases. Typically, during the solubilization phase, the pH is adjusted to between 9 and 11 to ensure maximal protein solubility. For optimal precipitation, the pH is usually lowered to a range of 4.5–5.5 (Fu et al., 2020). Experimental results indicate that the extraction efficiency of millet proteins using the alkaline extraction-acid precipitation method is around 44.30%, with a purity of approximately 39% (He, 2016; Liu et al., 2009).

7.4.1.2 Enzymatic Hydrolysis

Enzymatic hydrolysis involves using specific enzymes to catalyze the hydrolysis of protein molecules. During this process, certain peptide bonds are cleaved, resulting in smaller peptides or even free amino acids. This method is highly selective and operates under mild reaction conditions, making it suitable for extracting or isolating target proteins from complex mixtures.

When applied to the extraction of millet proteins, enzymatic hydrolysis effectively enhances protein extraction efficiency while reducing the amount of insoluble material. By selecting appropriate enzymes and optimizing reaction conditions, enzymatic hydrolysis can maximize protein yield and purity while preserving protein functionality and structural integrity. It is important to note that different enzymes exhibit varying effects on protein hydrolysis (Han, 2023). For instance, Gu (2013) used enzyme-assisted extraction to isolate proteins from millet bran, achieving an extraction rate of 38.14% and a purity of 30.26%.

7.5 Isolation of Millet Protein Fractions

The isolation of specific fractions from millet protein largely depends on their solubility across various solvents. The method of traditional Osborne is commonly used for stepwise extraction, effectively separating the four main protein types: albumins, globulins, prolamins, and glutelins. Danno and Natake (1980) employed NaCl, ethanol, and SDS to isolate protein fractions from foxtail millet in Japan. Their results indicated the presence of 7.24–7.71% water-soluble protein, 5.64–6.53% salt-soluble protein (in 2% NaCl), 39.61–44.73% alcohol-soluble protein (in 70% ethanol), 19.92–24.78% SDS-soluble protein (in 1% SDS), and 17.98–20.02%, 1% SDS-ME-soluble protein. Zhao et al. (2011), using the classical Osborne extraction method, analyzed the protein composition of foxtail millet. They found that prolamins accounted for the largest fraction (46.0%), followed by glutelins (21.1%), while albumins and globulins were present in smaller quantities at 2.8% and 2.7%, respectively.

7.6 Properties and Structure of Proteins in Millet

The properties of proteins in millet, including amino acid composition, oil and water absorption capacity, stability, and isoelectric point, are key factors that not only determine their nutritional value but also affect their performance in food processing and industrial applications.

7.6.1 Amino Acid Composition

Researchers often use chromatographic methods to determine the amino acid content of millet proteins (Kamara et al., 2009). Similar to other grains, lysine is the primary limiting amino acid in millet proteins, followed by tryptophan. Table 7.3 systematically summarizes the protein levels and amino acid profiles of different millet species. Compared to wheat and rice, millet is rich in methionine, cysteine, isoleucine, aspartic acid, glutamic acid, alanine, and proline. Foxtail millet flour is particularly high in leucine and alanine content.

7.6.2 Physicochemical Characteristics

7.6.2.1 Foaming Capacity and Foam Stability

Foaming capacity (FC) and foam stability (FS) are critical parameters in evaluating a protein's ability in forming and sustaining foam, a property that plays a significant role in food processing applications. Sharma et al. (2023) reported that foxtail millet prolamins have foaming capacity and foam stability values of 7.32% and 57.89%,

Table 7.3 Comparison of amino acids among different kinds of corn, wheat, and rice

	Name								
Protein composition	**Pearl millet**	**Finger millet**	**Broomcorn millet**	**Small millet**	**Kodo millet**	**Millet**	**Mitte**	**Wheat**	**Rice**
Protein content	10.96	7.16	12.5	10.13	10.52	12.26	8.92	10.57	9.93
Histidine	2.15	2.37	2.1	2.35	2.23	2.42	2.14	2.56	2.45
Isoleucine	3.45	3.7	4.1	4.14	4.44	6.04	4.55	3.78	4.29
Leucine	8.52	8.86	12.2	8.08	10.84	12.46	11.96	6.13	8.09
Lysine	3.42	2.83	1.5	2.42	1.91	2.13	1.42	2.42	3.7
Methionine	1.86	2.74	2.2	2.21	2.73	3.08	2.69	1.77	2.6
Cysteine	2.2	1.48	1	1.85	–	0.45	1.92	2.24	1.84
Phenylalanine	4.88	5.7	5.5	6.14	9.56	6.86	6.27	5.03	5.36
Threonine	3.55	3.84	3	4.24	3.85	4.72	3.89	2.58	3.28
Tryptophan	1.33	0.91	0.8	1.35	1.32	1.31	1.32	0.99	1.27
Valine	4.79	5.65	5.4	5.31	6.78	6.59	5.49	5.12	6.06
Alanine	7.68	6.71	3.2	8.6	2.73	13.77	11	3.64	5.51
Arginine	4.54	4.33	3.2	5.96	3.68	4.68	3.18	5.13	7.72
Aspartic acid	7.77	6.4	6.2	5.51	7.7	7.52	5.61	5.44	8.73
Glutamic acid	17.42	20.22	21.3	20.22	25.73	24.1	18.25	27.06	18.92
Glycine	3.52	3.59	2.1	3.43	2.83	2.84	3.12	4.19	4.18
Proline	5.3	5.42	7.3	6.04	10.34	10.14	7.33	10.25	4.31
Serine	4.44	4.81	6.3	5.6	4.97	6.54	5.5	4.8	4.95
Tyrosine	2.67	3.37	4	3.43	5.64	5.23	3.87	3.12	4.36

respectively, indicating that foxtail millet prolamins can both produce and maintain foam. The foaming performance of prolamins is strongly correlated with both surface hydrophobicity and solubility (Kaur & Singh, 2005). Following ultrasonic treatment, both the FC and FS of the prolamins showed significant improvement relative to untreated samples, with FC increasing by 20% and FS by 40%. This improvement is attributed to the augmented exposure of hydrophobic protein regions, which promotes their absorption at the air-water interface, thereby augmenting foam formation and sustainability. These proteins manifest optimal foaming capacity at pH 10, suggesting that an alkaline milieu is conducive to the development of stable foam architectures. The propensity of proteins to form enduring foam is profoundly connected to their solubility, given that proteins with elevated solubility engage more extensively with water, thereby facilitating bubble generation (Li et al., 2020).

7.6.2.2 Isoelectric Point

Isoelectric Point (pI): pI represents the specific pH value at which a protein exhibits a net charge of zero in solution. At this point, proteins typically have the lowest solubility and are prone to precipitation due to the absence of net charge. The determination of the isoelectric point (pI) is essential for protein purification, separation processes, and investigations into protein functionality. Current methods for determining pI include electrophoresis, titration, and predictive approaches.

Isoelectric Focusing (IEF): This is a high-resolution technique that separates proteins in a gradient of pH values. In an electric field, proteins undergo electrophoretic migration until they reach a position where the ambient pH matches their isoelectric point (pI), at which point they stop moving. This method is suitable for analyzing complex protein mixtures and can accurately determine the pI of individual proteins. However, no studies have yet utilized IEF to measure the isoelectric point of millet proteins.

Titration: This method involves gradually changing the pH of a protein solution and recording the corresponding potential or absorbance to plot a titration curve. The isoelectric point is usually located outside the buffer region of the titration curve, at the inflection point. For example, by preparing a solution of isolated millet proteins at a given concentration, adjusting the pH with low-concentration acid or base solutions across a gradient from pH 2 to 12 (in 0.5 increments), and measuring the effect of pH on protein solubility, the isoelectric point can be calculated. Gu (2013) preliminarily determined the millet bran protein PI to be in the range of pH 3–4 through solubility experiments in different buffer solutions, later refining the value to pH 3.5

Predictive Methods: By analyzing the protein amino acid sequence and their pKa values, bioinformatics algorithms can be used to predict the isoelectric point. Commonly used online tools include Expasy's pI/Mw calculator and databases like Proteome-pI 2.0.

7.6.3 Millet Protein Structure

Millet proteins are functional proteins; their structure determines their application properties in food and feed, such as solubility, emulsification, and foaming capacity. However, in practical production processes, millet proteins often undergo a series of processing steps such as heating, extrusion, and acid–base treatment. These operations may alter the proteins' molecular structure which leads to affecting their functional characteristics and overall quality. Understanding how these processing methods impact the structure and quality of millet proteins is of vital importance for optimizing production processes, enhancing product quality, and applying millet proteins in a wider range of industries.

Determining the protein structure is critical across several scientific disciplines (Varadi & Velankar, 2023). In terms of secondary structure, Guo et al. (2019) studied that the millet protein's secondary structure includes α-helices, β-sheets, and β-turns, but no random coil structures were found. Among the four major protein fractions, except for albumin, the other three are primarily composed of β-sheets, followed by α-helices and β-turns, with the lowest content being β-antiparallel sheets. Prolamins were particularly rich in β-turn and β-antiparallel sheet structures.

Protein three-dimensional structures can display fundamental information regarding their functions (Floudas et al., 2006). In cereals, there are more than 100 crystal structures experimentally characterized and cataloged in the Protein Data Bank (PDB) for rice, wheat, and maize. Similarly, for sorghum, which is closely related to millet, 40 experimentally determined structures have been deposited. These are including hydroxynitrile lyase (2.3 Å resolution), dhurrinase 1 (1.8 Å resolution) (Lauble et al., 2002), and hydroxycinnamoyltransferase (2.012 Å resolution) (Walker et al., 2013). However, progress in the experimental determination of millet protein crystal structures has been notably limited. Currently, seven protein structures have been determined for pearl millet, four for foxtail millet, three for finger millet, and one for barnyard millet and another for proso millet.

7.6.3.1 Impact of Processing on Millet Protein Structure

Millet must undergo various cooking processes before it can be consumed. Common processing methods include boiling, extrusion, high temperature and pressure treatments, fermentation, and enzymatic hydrolysis. These processing techniques significantly affect the quality, nutrition, and structure of millet proteins. Studies have shown that cooked millet proteins exhibit lower digestibility than uncooked millet proteins (Fu et al., 2021), primarily due to the disulfide bond-mediated protein crosslinking that occurs during heating (Duodu et al., 2003). In contrast, pearl millet protein digestibility improves after extrusion processing (Cisse et al., 1998). Thus, it is crucial to investigate the structural modifications that millet proteins undergo during different processing methods.

7.6.3.2 Effects of Boiling on Millet Protein Structure

Boiling increases the carbohydrate level of millet from 78.51 to 78.91 g/100 g, reduces fat from 3.24 to 2.3 g/100 g, and increases protein content from 8.28 to 9.07 g/100 g. While boiling does not significantly alter the protein bodies' size or shape, it induces microstructural changes. Raw protein bodies feature large cavities, whereas cooked protein bodies display only small pores. Boiling also exposes tryptophan residues and unfolds proteins, promoting protein aggregation through intermolecular hydrophobic interactions, as indicated by fluorescence quenching studies (Zhang et al., 2021). Zhang et al. (2021) found that boiling increased the β-sheet content of millet prolamins while decreasing the proportion of α-helices and random coils. Heat-induced denaturation and unfolding of prolamin molecules result from the generation of disulfide linkages and the accumulation of antinutritional compounds, which impede the entry of digestive enzymes. Microscopic observations show that pepsin cannot digest the protein bodies in cooked millet (Gulati et al., 2018).

7.6.3.3 Influence of Extrusion on Millet Protein Structure

Extrusion is characterized as a high-temperature, short-time (HTST) processing technique. Research on Kodo millet found that extrusion at 230 °C for three minutes increased protein content from 7.92% to 8.12%. This increase in protein is due to the breakdown of proteins into low-molecular-weight fragments. Guo et al. observed that during extrusion, α-helix content increased while β-sheet content decreased across all four protein fractions. Additionally, the β-turns and β-antiparallel sheets in albumin increased, while they decreased in the other three protein fractions. The millet prolamin subunit structure remained relatively unchanged before and after extrusion, but the tertiary structure of albumin, globulin, and glutelin underwent significant changes, with large subunits breaking down into smaller ones due to heat and shear stress (Guo et al., 2019).

7.6.3.4 Influence of High-Pressure Processing on Millet Protein Structure

High-pressure processing (HPP) is an innovative nonthermal approach that applies substantial pressure in a specialized chamber, serving as an effective replacement for heat processing (Sruthi & Rao, 2021). Zhang et al. (2022b) assessed the impact of HPP at 100–500 MPa for 10 min on millet prolamins and found that their thermal stability increased with higher pressures. Su (2020) reported that HPP at 100–500 MPa disrupted hydrophobic and ionic interactions in millet proteins, reducing the stability of their secondary structure. Treatment of millet protein at 200 and 600 MPa for 5 and 20 min resulted in molecular bond breakage within the protein (Gulati et al., 2018). HPP reduced the α-helix structures and a corresponding rise in the β-sheet formations in millet prolamins, indicating changes in secondary structure under ultra-high pressure (Zhang, 2023).

7.6.3.5 Millet Protein Structure Modeling

The workflow for protein structure modeling begins with the extraction of proteins from particular tissues of plants cultivated under controlled environmental conditions. The main steps include employing specialized extraction techniques to selectively isolate and concentrate the specific protein group, followed by digestion with specific proteases. Then, techniques such as gel electrophoresis or liquid chromatography are used to separate the complex protein mixtures. Next, the separated proteins are identified by mass spectrometry, and bioinformatics tools are utilized to align the identified peptides with the predicted protein sequences derived from the plant genome, generating proteomic data. Most proteomic data are stored and freely accessible through online databases such as NCBI (https://www.ncbi.nlm.nih.gov/), UniProt (https://www.uniprot.org/), Phytozome (https://phytozome-next.jgi.doe.gov/), and PPDB (http://ppdb.tc.cornell.edu/).

For these databases, UniProt addresses the exponential growth in proteome data by constructing a curated collection of reference proteomes for accurate comparison and analysis. Researchers can retrieve related data from the UniProt database and computer-aided tools are employed to predict the three-dimensional protein structures. Notably, a newly introduced artificial intelligence-based tool has been introduced for predicting protein structures, AlphaFold, and has been used for the rapid and accurate prediction of protein structures (Jumper et al., 2021). AlphaFold has already predicted thousands of protein models for rice and maize.

Research on millet protein structure modeling lags that of other major cereal proteins. To date, researchers have predicted protein models for key proteins involved in abiotic stress tolerance in millet using AlphaFold (Figure 7.2) (Kosová et al., 2018). AlphaFold has modeled key stress tolerance proteins such as glutathione peroxidase (GPx), early dehydration response protein (ERD4), *S*-adenosyl-L-methionine-dependent methyltransferases (SAM-Mtases) superfamily, heat shock protein 90 (HSP90), Cu-Zn superoxide dismutase (SOD-CuZn), and others. The active site residues of these proteins were identified using the Sitemap tool integrated into the Maestro software platform (Figure 7.2). Currently, AlphaFold has created 179 protein models for *Cenchrus americanus* (pearl millet) and *S. italica* (foxtail millet), while other millet species have only a few models (Ceasar et al., 2024).

7.7 Health Benefits of Millet Proteins

Recent studies have underscored the extensive health benefits associated with millet proteins, including antidiabetic effects, alleviation of lipid metabolism disorders, blood pressure reduction, antioxidant, anti-inflammatory, and antimicrobial activities. Millet protein peptides, derived from millet proteins through microbial fermentation or enzymatic hydrolysis, play a pivotal role in these bioactivities. These small peptides have attracted considerable attention owing to their wide range of physiological functions.

7.7.1 Antidiabetic Effects

Insulin sensitivity refers to how effectively insulin facilitates glucose metabolism. Millet proteins have been shown to significantly improve insulin sensitivity (Choi et al., 2005). Fu et al. (2021) discovered that both uncooked and cooked millet proteins could reduce insulin resistance, alleviate disorder of intestinal flora caused by diabetes, and modulate the GLP-1R/PI3K/AKT pathway to reduce the blood glucose levels of diabetic mice. Further studies indicate that the primary antidiabetic component in millet is prolamin, supported by Nishizawa et al. (2009) and Fu (2021). Additionally, their study demonstrated that both uncooked and cooked millet prolamins obviously lowered fasting blood glucose of diabetic mice while increasing insulin secretion. However, only the cooked prolamin improved glucose tolerance and pancreatic β-cell function (Fu, 2021). Some reports suggest that heat-treated millet prolamin benefits endothelial cell metabolism, which is closely related to diabetes (Wei et al., 2013).

Key enzymes in glucose metabolism include α-glucosidase, DPP-IV, and α-amylase, which, when inhibited, can slow carbohydrate absorption, stimulate insulin secretion, and reduce postprandial blood glucose spikes. These enzymes are important targets for screening antidiabetic peptides. Karas et al. (2019) found that heat-treated millet

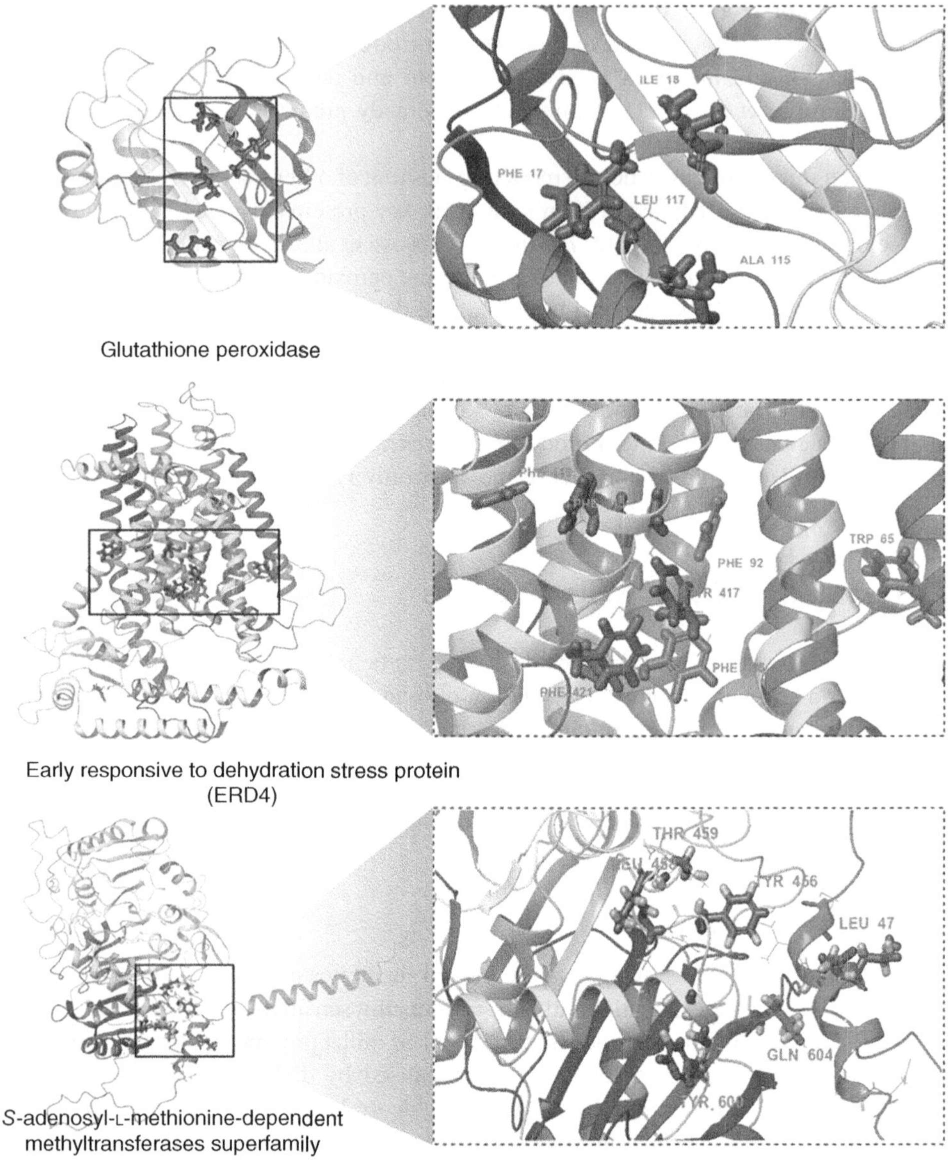

Figure 7.2 Structure diagram of glutathione peroxidase, early responsive to dehydration stress protein (ERD4), and *S*-adenosyl-L-methionine-dependent methyltransferases superfamily. *Source:* Ceasar et al. (2024)/Springer Nature

protein hydrolysates inhibited α-glucosidase activity, while Gu et al. (2021) identified peptides such as NDWHTGPLS and TYPHQQPPIT from millet proteins with strong DPP-IV inhibitory activity. Additionally, Shen et al. (2022a) isolated millet prolamin peptides like QQLRPF, AGAGPQGRP, and FALQGAAFLGSA that exhibit dual inhibition of α-glucosidase and α-amylase.

7.7.2 Regulation of Lipid Metabolism

Dysregulated lipid metabolism is marked by increased levels of triglycerides, total cholesterol, and low-density lipoprotein (LDL-C), alongside decreased high-density lipoprotein (HDL-C). Choi et al. (2005) discovered that supplemented millet protein can improve cholesterol metabolism in KK-Ay diabetic mice, significantly increasing HDL-C levels while stabilizing LDL-C. Similar results were gotten by Trivedi et al. (2015), who noted increased plasma HDL-C levels. Additionally, millet bran protein hydrolysates prepared by Shan et al. (2023) activated PPARγ signaling pathways, alleviating fatty degeneration of liver and reducing fat accumulation in nonalcoholic fatty liver disease (NAFLD) mice.

Pancreatic lipase, an endogenous enzyme in the gastrointestinal tract, degrades fats in collaboration with bile acids. Cholesterol esterase, primarily found in the small intestine, hydrolyzes dietary cholesterol esters. Inhibiting these enzymes can reduce fat absorption, potentially improving conditions like hyperlipidemia. Mudgil et al. (2023) reported that millet protein hydrolysates prepared with alkaline protease, bromelain, and chymotrypsin inhibited pancreatic lipase and cholesterol esterase, showing potential in lipid regulation. Further, millet prolamin peptides exhibited strong lipase inhibitory activity (Jakubczyk et al., 2019), with WQHQY and WQHMMP being identified as potent inhibitors (Shen et al., 2022b).

7.7.3 Blood Pressure Reduction

Angiotensin-converting enzyme (ACE) plays a crucial role as a key regulator within the renin-angiotensin and kallikrein-kinin systems, critical for blood pressure regulation. Hou et al. (2020) found that extrusion and fermentation treatments significantly enhanced ACE inhibitory activity in millet peptides compared to raw millet flour. Chen et al. (2017) also demonstrated that protein hydrolysates from raw and extruded millet reduced spontaneously hypertensive rats' blood pressure by significantly suppressing serum ACE activity.

Ultrafiltration is a method used for enriching bioactive peptides from protein hydrolysates. Su (2019) found that ultrafiltered millet peptides smaller than 1 kDa exhibited the best ACE inhibitory activity compared to 1–3 kDa and larger fractions. Similarly, Karas et al. (2019) noted that <3 kDa peptides from millet proteins had lower ACE IC_{50} values after in vitro digestion. Further, Zhao (2022) identified FNPMFNPM from millet prolamins as a potent ACE inhibitor.

7.7.4 Antioxidant Properties

Oxidative damage triggered by reactive oxygen species (ROS) can lead to aging, tissue damage, and cancer development. Wang et al. (2018) demonstrated that millet peptides, prepared with alkaline protease, effectively scavenged superoxide anion, hydroxyl, and 1,1-diphenyl-2-picrylhydrazyl (DPPH) radicals in a concentration-dependent manner.

Hu et al. (2020) employed a pepsin-trypsin digestion system to isolate seven new peptides from sprouted millet proteins, with EDDQMDPMAK and QNWDFCEAWEPCF significantly reducing ROS production in Caco-2 cells.

Ji (2020) isolated two novel antioxidant peptides, PFLF and IALLIPF, from <1 kDa alkaline protease hydrolysates, with DPPH scavenging activity reaching 149 μMTE/g protein and ORAC values of 1180 MTE/g protein. Millet bran peptides CFMTY, CTGTPYC, and SELE exhibited excellent ferrous ion chelation abilities (Xu et al., 2022). Agrawal et al. (2019) discovered that the antioxidant properties of millet peptides are strongly associated with the presence of serine and threonine residues. Agrawal et al. (2016) concluded that millet antioxidant peptides often contain hydrophobic amino acids like leucine, proline, and glycine, which enhance their solubility in fats, facilitating better interaction with free radicals.

7.7.5 Anti-inflammatory Effects

Hu et al. (2020) found that small peptides (EDDQMDPMAK and QNWDFCEAWEPCF) from germinated and heat-treated millet significantly reduced levels of TNF-α and IL-6 in RAW264.7 cells. Similarly, Ji et al. (2020) reported that prolamin peptides PFLF and IALLIPF alleviated inflammation in LPS-induced RAW264.7 cells by inhibiting the production of nitric oxide and reducing the levels of inflammatory cytokine. Zhang et al. (2022a) prepared hydrolysates from millet protein (featuring LPF, ANP, PY, YW, and IPP sequences) that significantly lowered TNF-α and IL-6 levels in DSS-induced colitis mice. Notably, glycine-rich millet peptides exhibited the highest anti-inflammatory activity (Jakubczyk et al., 2019).

7.7.6 Antimicrobial Activity

Proteins and peptides isolated from finger millet, barnyard millet, and proso millet exhibited a wide spectrum of antimicrobial activity against *Pseudomonas aeruginosa* (MTCC 424) and *Salmonella enterica* (MTCC 739). Bisht et al. (2016) discovered that millet peptides effectively inhibited pathogenic Gram-positive bacteria, suggesting potential applications as novel antimicrobial agents in the medical field. Amadou et al. (2013) utilized *Lactobacillus paracasei* Fn032 to ferment millet flour, producing peptides that inhibited *Escherichia coli* activity. Joshi et al. (1998) isolated a cysteine proteinase inhibitor from millet, rich in serine, glycine, and tryptophan, which showed antimicrobial activity by inhibiting *Trichoderma reesei* and important plant pathogens like fusarium, aspergillus, and others. Proteins and peptides derived from finger millet, barnyard millet, and proso millet also demonstrated antimicrobial activity against *P. aeruginosa* and *S. enterica* (Bisht et al., 2016).

7.7.7 Others

Yu et al. (2014) found that millet bran peptides, hydrolyzed with bromelain, inhibited liver injury and precancerous lesions. Millet proteins have been shown to prevent acute liver damage induced by D-galactosamine, with steamed millet proteins being less effective than raw (Fu et al., 2022). Additionally, millet peptides prepared with alkaline protease displayed significant antifatigue effects in mice (Cao et al., 2011). Liu et al.'s (2012) research demonstrated that millet peptides enhance the immune response by promoting phagocytosis in macrophages and increasing spleen function index.

7.8 Conclusion and Future Perspectives

Millets are gluten-free, rich in carbs, protein, fiber, and minerals, which are usually cultivated in the semi-arid tropical regions of Asia and Africa. Millet proteins can be divided into albumin, globulin, prolamin, and glutelin; prolamin is the most abundant in millets. Alkaline extraction-acid precipitation and enzymatic hydrolysis are two methods mostly used to extract millet protein, while enzymatic methods can improve yield and purity.

Although millet protein digestibility is lower than other grains due to factors like protease inhibitors and structural issues, millet protein has alleviation on lipid metabolism disorders, reduction of blood pressure and blood glucose, antioxidant, anti-inflammatory, and antimicrobial activities. Treatments such as boiling, extrusion, HPP, etc., will alter protein content, structure, digestibility, and stability, and also alters its health benefits.

Nowadays, artificial intelligence-based technology is applied in the research of major cereal protein, while research on millet protein structure modeling lags far behind that of other cereal proteins.

Future research can focus on optimizing the extraction process to enhance the extraction rate and purity of millet protein, explore the health benefits of millet protein, and gain a deeper understanding of the structure of millet protein by establishing a visualization model. These studies are of vital importance for the development of functional foods based on millet protein.

References

Agrawal, H., Joshi, R., & Gupta, M. (2016). Isolation, purification and characterization of antioxidative peptide of pearl millet (*Pennisetum glaucum*) protein hydrolysate. *Food Chemistry*, *204*, 365–372.

Agrawal, H., Joshi, R., & Gupta, M. (2019). Purification, identification and characterization of two novel antioxidant peptides from finger millet (*Eleusine coracana*) protein hydrolysate. *Food Research International*, *120*, 697–707. https://doi.org/10.1016/j.foodres.2018.11.028

Amadou, I., Le, G. W., Amza, T., Sun, J., & Shi, Y. H. (2013). Purification and characterization of foxtail millet-derived peptides with antioxidant and antimicrobial activities. *Food Research International*, *51*, 422–428. https://doi.org/10.1016/j.foodres.2012.12.045

Becker, P. M., & Yu, P. Q. (2013). What makes protein indigestible from tissue-related, cellular, and molecular aspects? *Molecular Nutrition & Food Research*, *57*, 1695–1707. https://doi.org/10.1002/mnfr.201200592

Bisht, A., Thapliyal, M., & Singh, A. (2016). Screening and isolation of antibacterial proteins/peptides from seeds of millets. *International Journal of Current Pharmaceutical Review and Research*, *8*, 96–99.

Cao, X. Y., Zhu, Y., Liu, J. L., Lu, X. L., & Li, Q. J. (2011). Anti-fatigue effect of millet polypeptide on mice (in Chinese). *Chinese Public Health*, *27*, 1153–1154.

Ceasar, S. A., Prabhu, S., & Ebeed, H. T. (2024). Protein research in millets: Current status and way forward. *Planta*, *260*(2), 43. https://doi.org/10.1007/s00425-024-04478-z

Chen, J., Duan, W., Ren, X., Wang, C., Pan, Z. L., Diao, X. M., & Shen, Q. (2017). Effect of foxtail millet protein hydrolysates on lowering blood pressure in spontaneously hypertensive rats. *European Journal of Nutrition*, *56*(6), 2129–2138. https://doi.org/10.1007/s00394-016-1252-7

Chen, X., Wu, Y. C., Liu, Y., Qian, L. H., Zhang, Y. H., & Li, H. J. (2022). Single/co-encapsulation capacity and physicochemical stability of zein and foxtail millet prolamin nanoparticles. *Colloids and Surfaces B-Biointerfaces*, *217*, 112685. https://doi.org/10.1016/j.colsurfb.2022.112685

Choi, Y. Y., Osada, K., Ito, K., Nagasawa, T., Choi, M. R., & Nishizawa, N. (2005). Effects of dietary protein of Korean foxtail millet on plasma adiponectin, HDL-cholesterol, and insulin levels in genetically type 2 diabetic mice. *Bioscience Biotechnology and Biochemistry*, *69*(1), 31–37. https://doi.org/10.1271/bbb.69.31

Cisse, C. T., Dotou, C. R., Kpekpede, F., & Diadhiou, F. (1998). Spontaneous and complete uterine rupture, on scarred gravid uterus. *Contraception Fertilite Sexualite*, *26*, 876–880.

Danno, G., & Natake, M. (1980). Isolation of foxtail millet proteins and their subunit structure. *Agricultural and Biological Chemistry*, *44*, 913–918. https://doi.org/10.1080/00021369.1980.10864039

Duodu, K. G., Taylor, J. R. N., Belton, P. S., & Hamaker, B. R. (2003). Factors affecting sorghum protein digestibility. *Journal of Cereal Science*, *38*(2), 117–131. https://doi.org/10.1016/S0733-5210(03)00016-X

FAOSTAT. (2024). Crops and livestock products, (1994 to 2022). Retrieved October 16, from https://www.fao.org/faostat/en/#data/QCL.

Floudas, C. A., Fung, H. K., McAllister, S. R., Mönnigmann, M., & Rajgaria, R. (2006). Advances in protein structure prediction and de novo protein design: A review. *Chemical Engineering Science*, *61*(3), 966–988. https://doi.org/10.1016/j.ces.2005.04.009

Fu, Y. X. (2021). Virtual screening of millet protein to improve blood glucose metabolism and its hypoglycemic peptides in diabetic mice [Doctor's thesis in Chinese, China Agricultural University].

Fu, Y. X., Liu, Z. Y., Zhang, F., & Shen, Q. (2022). Comparison of preventive effects of millet protein on acute liver injury in mice before and after cooking (in Chinese). *Journal of the Chinese Institute of Food Science and Technology*, *22*, 165–172. https://doi.org/10.16429/j.1009-7848.2022.11.018

Fu, Y. X., Yin, R. Y., Guo, E. H., Cheng, R. H., Diao, X. M., Xue, Y., & Shen, Q. (2021). Protein isolates from raw and cooked foxtail millet attenuate development of type 2 diabetes in streptozotocin-induced diabetic mice. *Molecular Nutrition & Food Research*, *65*(6), 2000365. https://doi.org/10.1002/mnfr.202000365

Fu, Y. X., Yin, R. Y., Liu, Z. Y., Niu, Y., Guo, E. H., Cheng, R. H., Diao, X. M., Xue, Y., & Shen, Q. (2020). Hypoglycemic effect of prolamin from cooked foxtail millet (*Setaria italic*) on streptozotocin-induced diabetic mice. *Nutrients*, *12*(11), 3452. https://doi.org/10.3390/nu12113452

Ge, J., Sun, C. X., Mata, A., Corke, H., Gan, R. Y., & Fang, Y. P. (2021). Physicochemical and pH-dependent functional properties of proteins isolated from eight traditional Chinese beans. *Food Hydrocolloids*, *112*, 106288. https://doi.org/10.1016/j.foodhyd.2020.106288

Gu, H. F., Gao, J., Shen, Q., Gao, D. X., Wang, Q., Tangyu, M. Z., & Mao, X. Y. (2021). Dipeptidyl peptidase-IV inhibitory activity of millet protein peptides and the related mechanisms revealed by molecular docking. *LWT-Food Science and Technology*, *138*, 110587. https://doi.org/10.1016/j.lwt.2020.110587

Gu, N. (2013). Enzymatic extraction and characterization of foxtail millet bran protein [Master's thesis in Chinese), anjing Agriculutral University].

Gulati, P., Sabillon, L., & Rose, D. J. (2018). Effects of processing method and solute interactions on pepsin digestibility of cooked proso millet flour. *Food Research International*, *109*, 583–588. https://doi.org/10.1016/j.foodres.2018.05.005

Guo, L. D., Xu, L., Ou, C. Z., Ding, Y. Y., Zhang, G. P., Ni, C. L., & Cheng, J. J. (2019). Molecular compositions and structural properties of protein in millet. *Food Science*, *40*(24), 201–206. https://doi.org/10.7506/spkx1002-6630-20181225-293

Han, Z. T. (2023). Physical method assisted enzymatic extraction of high temperature soybean meal protein and its influence on its properties [Master's thesis in Chinese, Jilin University].

He, S. J. (2016). Preparation of active peptide by fermentation of millet protein and biological function of millet bran oil [Master's thesis in Chinese, Hefei University of Technology].

Hou, D. Z., Chen, J., & Shen, Q. (2020). Effects of extrusion and fermentation on ACE inhibitory activity and antioxidant activity of millet polypeptide (in Chinese). *Journal of the Chinese Institute of Food Science and Technology*, *20*(5), 174–180.

Hu, S., Yuan, J. L., Gao, J. Y., Wu, Y., Meng, X. Y., Tong, P., & Chen, H. B. (2020). Antioxidant and anti-inflammatory potential of peptides derived from in vitro gastrointestinal digestion of germinated and heat-treated foxtail millet (*Setaria italica*) proteins. *Journal of Agricultural and Food Chemistry*, *68*(35), 9415–9426. https://doi.org/10.1021/acs.jafc.0c03732

Jakubczyk, A., Szymanowska, U., Karas, M., Zlotek, U., & Kowalczyk, D. (2019). Potential anti-inflammatory and lipase inhibitory peptides generated by *in vitro* gastrointestinal hydrolysis of heat treated millet grains. *Cyta-Journal of Food*, *17*, 324–333. https://doi.org/10.1080/19476337.2019.1580317

Ji, Z. W. (2020). Study on the preparation of foxtail millet prolamins peptide and its antioxidant and anti-inflammatory activities [Doctor's thesis in Chinese, Jiangnan University].

Ji, Z. W., Mao, J. Q., Chen, S. G., & Mao, J. (2020). Antioxidant and anti-inflammatory activity of peptides from foxtail millet (*Setaria italica*) prolamins in HaCaT cells and RAW264.7 murine macrophages. Food. *Bioscience*, *36*, 100636. https://doi.org/10.1016/j.fbio.2020.100636

Joshi, B. N., Sainani, M. N., Bastawade, K. B., Gupta, V. S., & Ranjekar, P. K. (1998). Cysteine protease inhibitor from pearl millet: A new class of antifungal protein. *Biochemical and Biophysical Research Communications*, *246*, 382–387.

Jumper, J., Evans, R., Pritzel, A., Green, T., Figurnov, M., Ronneberger, O., Tunyasuvunakool, K., Bates, R., Žídek, A., Potapenko, A., Alex Bridgland, A., Meyer, C., Kohl, S. A. A., Ballard, A. J., Cowie, A., Romera-Paredes, B., Nikolov, S., Jain, R., Adler, J., ... & Hassabis, D. (2021). Highly accurate protein structure prediction with AlphaFold. *Nature*, *596*, 583–589. https://doi.org/10.1038/s41586-021-03819-2

Kamara, M. T., Huiming, Z., Kexue, Z., Amadou, I., & Tarawalie, F. (2009). Comparative study of chemical composition and physicochemical properties of two varieties of defatted foxtail millet flour grown in China. *American Journal of Food Technology*, *4*, 255–267. https://doi.org/10.3923/ajft.2009.255.267

Karas, M., Jakubczyk, A., Szymanowska, U., Jederka, K., Lewicki, S., & Złotek, U. (2019). Different temperature treatments of millet grains affect the biological activity of protein hydrolyzates and peptide fractions. *Nutrients*, *11*(3), 550. https://doi.org/10.3390/nu11030550

Kaur, M., & Singh, N. (2005). Studies on functional, thermal and pasting properties of flours from different chickpea (*Cicer arietinum* L.) cultivars. *Food Chemistry*, *91*, 403–411. https://doi.org/10.1016/j.foodchem.2004.06.015

Kosová, K., Vítámvás, P., Urban, M. O., Prášil, I. T., & Renaut, J. (2018). Plant abiotic stress proteomics: The major factors determining alterations in cellular proteome. *Frontiers in Plant Science*, *9*, 122. https://doi.org/10.3389/fpls.2018.00122

Kumar, A., Tripathi, M. K., Joshi, D., & Kumar, V. (2021). *Millet and millet technology*. Springer. (Chapter 2)

Kumar, K. K., & Parameswaran, K. P. (1998). Characterisation of storage protein from selected varieties of foxtail millet (*Setaria italica* [L] Beauv). *Journal of the Science of Food and Agriculture*, *77*, 535–542. https://doi.org/10.1002/(sici)1097-0010.199808)77:4<535::Aid-jsfa77>3.3.Co;2-G

Lauble, H., Miehlich, B., Förster, S., Wajant, H., & Effenberger, F. (2002). Crystal structure of hydroxynitrile lyase from Sorghum bicolor in complex with the inhibitor benzoic acid: A novel cyanogenic enzyme. *Biochemistry*, *41*, 12043–12050. https://doi.org/10.1021/bi020300o

Li, C., Yang, J. Z., Yao, L., Qin, F. L., Hou, G. F., Chen, B., Jin, L. H., Deng, J. J., & Shen, Y. H. (2020). Characterisation, physicochemical and functional properties of protein isolates from *Amygdalus pedunculata* Pall seeds. *Food Chemistry*, *311*, 125888. https://doi.org/10.1016/j.foodchem.2019.125888

Liu, J. L., Cao, X. Y., & Li, Q. J. (2009). Comparative study on extraction methods of millet protein (in Chinese). *Food Industry*, *30*(03), 30–31.

Liu, J. L., Cao, X. Y., Li, Q. J., & Hou, X. (2012). Effects of millet polypeptide on immune regulation in mice (in Chinese). *Chinese Public Health*, *28*, 44–45.

McDonough, C. M., Rooney, L. W., & Serna-Saldivar, S. O. (2000). The millets. In K. Kulp (Ed.), *Food science and technology: Handbook of cereal science and technology* (2nd ed., pp. 177–210). CRC Press.

Mertz, E. T., Hassen, M. M., Carins-Whittern, C., Kirleis, A. W., Tu, L., & Axtell, J. D. (1984). Pepsin digestibility of proteins in sorghum and other major cereals. *Proceedings of the National Academy of Sciences of the United States of America-Biological Sciences*, *81*, 1–2. https://doi.org/10.1073/pnas.81.1.1

Mudgil, P., Alblooshi, M., Singh, B. P., Devarajan, A. R., & Maqsood, S. (2023). Pearl millet protein hydrolysates exhibiting effective *in-vitro* antioxidant, antidiabetic and anti-lipidemic properties as potential functional food ingredient. *International Journal of Food Science and Technology*, *58*, 3264–3272. https://doi.org/10.1111/ijfs.16151

Nishizawa, N., Togawa, T., Park, K. O., Sato, D., Miyakoshi, Y., Inagaki, K., Ohmori, N., Ito, Y., & Nagasawa, T. (2009). Dietary Japanese millet protein ameliorates plasma levels of adiponectin, glucose, and lipids in type 2 diabetic mice. *Bioscience Biotechnology and Biochemistry*, *73*(2), 351–360. https://doi.org/10.1271/bbb.80589

Parameswaran, K. P., & Thayumanavan, B. (1995). Homologies between prolamins of different minor millets. *Plant Foods for Human Nutrition, 48*, 119–126. https://doi.org/10.1007/bf01088307

Ramachandra, G., Virupaksha, T. K., & Shadaksharaswamy, M. (1978). Comparison of protein-fractions of finger millet. *Phytochemistry, 17*, 1487–1490. https://doi.org/10.1016/s0031-9422(00)94626-0

Shan, S. H., Zhou, J. Q., Yin, R. P., Zhang, L. Z., Shi, J. Y., Qiao, Q. Q., & Li, Z. Y. (2023). Millet bran protein hydrolysate displays the anti-non-alcoholic fatty liver disease effect via activating peroxisome proliferator- activated receptor γ to restrain fatty acid uptake. *Journal of Agricultural and Food Chemistry, 71*(3), 1628–1642. https://doi.org/10.1021/acs.jafc.2c08169

Sharma, N., Sahu, J. K., Choudhary, A., Meenu, M., & Bansal, V. (2023). High intensity ultrasound (HIU)-induced functionalization of foxtail millet protein and its fractions. *Food Hydrocolloids, 134*, 108083. https://doi.org/10.1016/j.foodhyd.2022.108083

Shen, Q., Fu, Y. X., Zhao, Q. Y., Wang, C., Xue, Y., Zhao, L. X., & Niu, Y. (2022a). Cooked millet olysin peptides that inhibit α-amylase and α-glucosidase. Chinese Invention Patent. CN114716524B.

Shen, Q., Zhao, Q. Y., Hao, Z. H., Zhang, X. Y., Zhao, L. X., & Wang, C. (2022b). Two millet olysin peptides that inhibit the activity of pancreatic lipase and cholesterol esterase. Chinese Invention Patent. CN115925854B.

Sruthi, N. U., & Rao, P. S. (2021). Effect of processing on storage stability of millet flour: A review. *Trends in Food Science & Technology, 112*, 58–74. https://doi.org/10.1016/j.tifs.2021.03.043

Su, S. Y. (2019). Study on the preparation of foxtail millet ACE inhibitory peptides and their antihypertensive activity [Master's thesis in Chinese, Dalian Polytechnic University].

Su, X. (2020). Effect of ultra-high pressure treatment on functional properties of millet protein [Master's thesis in Chinese, Shanxi Agricultural University].

Sudharshana, L., Monteiro, P. V., & Ramachandra, G. (1988). Studies on the proteins of dodu millet (*Paspalum scrobiculatum*). *Journal of the Science of Food and Agriculture, 42*, 315–323. https://doi.org/10.1002/jsfa.2740420405

Taylor, J. R. N., & Schussler, L. (1986). The protein composition of the different anatomical parts of sorghum grain. *Journal of Cereal Science, 4*, 361–369. https://doi.org/10.1016/s0733-5210(86)80040-6

Trivedi, A. K., Arya, L., Verma, M., Verma, S. K., Tyagi, R. K., & Hemantaranjan, A. (2015). Genetic variability in proso millet *Panicum miliaceum* germplasm of central Himalayan region based on morpho-physiological traits and molecular markers. *Acta Physiologiae Plantarum, 37*, 23. https://doi.org/10.1007/s11738-014-1770-y

Varadi, M., & Velankar, S. (2023). The impact of AlphaFold protein structure database on the fields of life sciences. *Proteomics, 23*(17), 2200128. https://doi.org/10.1002/pmic.202200128

Virupaksha, T. K., Ramachandra, G., & Nagaraju, D. (1975). Seed proteins of finger millet and their amino-acid composition. *Journal of the Science of Food and Agriculture, 26*, 1237–1246. https://doi.org/10.1002/jsfa.2740260823

Walker, A. M., Hayes, R. P., Youn, B., Vermerris, W., Sattler, S. E., & Kang, C. (2013). Elucidation of the structure and reaction mechanism of sorghum hydroxycinnamoyltransferase and its structural relationship to other coenzyme a-dependent transferases and synthases. *Plant Physiology, 162*(2), 640–651. https://doi.org/10.1104/pp.113.217836

Wang, S., Liu, J. L., Huo, Y. P., Yan, X., Tian, S. Q., He, Y., & Chao, X. Y. (2018). Purification of antioxidant peptides from millet and itsprotective effect against oxidative stress induced by H_2O_2 in INS-1 cells (in Chinese). *Food Science, 39*(19), 169–174.

Wei, Y., Liu, G. L., Yang, J. Y., Zheng, R. X., Jiang, L. H., & Bao, P. J. (2013). The association between metabolic syndrome and vascular endothelial dysfunction in adolescents. *Experimental and Therapeutic Medicine, 5*(6), 1663–1666. https://doi.org/10.3892/etm.2013.1055

Xu, B. F., Wang, X. Y., Zheng, Y. J., Li, Y., Guo, M., & Yan, Z. Y. (2022). Novel antioxidant peptides identified in millet bran glutelin-2 hydrolysates: Purification, *in silico* characterization and security prediction, and stability profiles under different food processing conditions. *LWT-Food Science and Technology, 164*, 113634. https://doi.org/10.1016/j.lwt.2022.113634

Yu, S. J., Wang, C. Q., Yuan, M., Lian, W. S., Zi, Y., Chen, X. M., & Hao, Z. P. (2014). Inhibitory effect of millet bran polypeptide on liver injury and precancerous lesions. *Food Science, 35*, 213–217.

Zhang, B. W., Xu, Y. C., Zhao, C. Y., Zhang, Y. H., Lv, H., Ji, X. M., Wang, J., Pang, W. W., Wang, X. W., & Wang, S. (2022a). Protective effects of bioactive peptides in foxtail millet protein hydrolysates against experimental colitis in mice. *Food & Function, 13*(5), 2594–2605. https://doi.org/10.1039/d1fo02482e

Zhang, F., Fu, Y., Liu, Z. Y., & Shen, Q. (2021). Comparison of the characteristics of prolamins among foxtail millet varieties with different palatability: Structural, morphological, and physicochemical properties. *International Journal of Biological Macromolecules, 186*, 194–205. https://doi.org/10.1016/j.ij.iomac.2021.07.051

Zhang, Y. D. (2023). Effects of ultra-high pressure treatment on structural properties and functional properties of millet gliadin [Master's thesis in Chinese, Shanxi Agricultural University].

Zhang, Y. D., Zhang, X. B., Zhang, Z. H., Chen, Z. J., Jing, X., & Wang, X. W. (2022b). Effect of high hydrostatic pressure treatment on the structure and physicochemical properties of millet gliadin. *LWT-Food Science and Technology, 154*(15), 112755. https://doi.org/10.1016/j.lwt.2021.112755

Zhao, Q. Y., Shen, Q., Hao, Z. H., Chen, C. Y., & Wang, C. (2022) ACE inhibitory peptides from two multi-grain sources, preparation method and application thereof. Chinese Invention Patent. CN116041427B.

Zhao, X. W., Wei, Y. M., Wang, Z. C., & Zhang, B. (2011). Study on physicochemical properties of millet protein (in Chinese). *Food and Feed Industry, 4*, 34–37.

8

Millet Starch: Extraction, Structure, and Functional Properties

Chuanjie Chen and Fan Zhu

School of Chemical Sciences, The University of Auckland, Auckland, New Zealand

8.1 Introduction

Starchy cereal grains are the major carbohydrate source of human diet. As introduced in the beginning of this book, millets are a group of important cereals in many regions of the world. The protein, starch, lipids, vitamins, and others in millet grains provide nutrients for the human body. Many plants store energy by producing starch in granular form. During the germination of cereal grains, starch becomes the nutritional source of the biological process. When the grains are served as a whole, starch, as the major component, significantly affects the quality of cereal grains (Li et al., 2017b). When flour (or starch) produced from starchy grains is used individually or incorporated into processed foods (e.g., noodles, breads, biscuits, and cookies), the properties of starch significantly affect the processed foods. The sensory properties (including texture, flavor, and appearance) of cooked whole grains and related foods are affected by the composition and properties of the starch therein. Great diversities are found in starches of different sources, between which genetic and environmental factors are the known determinants. Modifications on starch overcome some limitations that are associated with native starch and increase the versatility of starch application. This chapter started with an introduction of starch isolation from millet grains and then summarized the structures and properties of millet starch. The modifications and applications of millet starch were summarized as well.

8.2 Isolation and Composition of Millet Starch

Isolation of starch from millet kernels follows procedures that are common to starch extraction from cereals (Halal et al., 2019). As starch granules are encapsulated within cells, breakdown of wall materials should be done before the releasing of starch granules (Halal

Nutraceutical Potential of Millet-Based Food Products: Chemistry and Applications, First Edition.
Edited by Parmjit S. Panesar, Charanjit S. Riar, and Sushil Dhital.

Published 2026 by John Wiley & Sons Ltd.

et al., 2019). Washing prior to mechanical disruption (e.g., milling or grinding) of kernels can reduce the amount of introduced soluble and insoluble impurities. Using dehulled cereals for starch isolation is recommended, as grain hulls are likely to introduce more impurities and thus affect the color, flavor, and rheological properties of the bulk starch (Singh et al., 2024). Steeping the kernels (or flour) in aqueous solution(s) assists the separation of starch and nonstarch components. After milling, the diluted flour slurry is filtered through cheesecloth and/or sieved to separate fibrous components and proteins. The residue is discarded or resuspended for repeated extraction, which increases the yield. Starch granules in the filtrate are recovered using centrifugation (commonly seen in laboratory-scale practice) or under gravity. Some water-soluble components are readily dissolved in the steeping solution. Repeated washing of the sediments using aqueous solution or water is conducted to pursue a high starch purity.

As far as the foremost mechanical disruption of millet grains is concerned, dry- and wet-milling methods are the most used. Dry-milling grinds dried millet kernels into flour, followed by additional steps to separate starch from nonstarch components in aqueous solutions (Wang et al., 2020a). Compared to wet-milling, dry-milling has been less used for millet starch isolation partly due to the higher degree of granular damage caused by dry-milling (Wang et al., 2020a). Intensive milling might lead to structural changes and property modification (Wang et al., 2020a). In the wet-milling method, millet kernels are soaked in solutions to soften kernels and to some extent weaken the physical and/or chemical barriers that may hinder the releasing of starch granules (Heena et al., 2024). The conditions of steeping varied significantly between studies. Three systems, namely neutral/near neutral, acidic, and alkaline solution, have been used for the wet-milling-based isolation of starch from millet kernel (Heena et al., 2024; Singh et al., 2023; Yanez & Walker, 1986). Being rich in nutrients, the slurry prepared by milling kernels in aqueous solutions may be subjected to biological contamination and starch degradation. Thus, adding sodium azide, mercury (II) chloride, or sodium metabisulfite into neutral or near-neutral solution has been a common practice to inhibit bacteria growth/amylase activity in the slurry (Aruna & Parimalavalli, 2022). Millet starches isolated based on steeping in neutral sodium azide solution, neutral mercuric chloride solution, and acidic sodium metabisulfite solution (with lactic acid added) were compared (Aruna & Parimalavalli, 2022). Steeping in neutral sodium azide solution resulted in a higher starch yield but higher protein content. Starch obtained from this steeping condition had lower swelling power, water-holding, and oil-holding capacities compared to those based on other conditions. Millet starch isolated based on steeping in acidic sodium metabisulfite solution had much higher gelatinization temperatures than those obtained based on other conditions (Aruna & Parimalavalli, 2022). Protein-starch interactions have been of primary concern. Diluted sodium hydroxide has been the most used soaking solution, which assists the denaturation and solubilization of proteins on one hand and inhibits the bacteria growth/amylase activity on the other hand. Using acidic solution is less common practice in starch isolation. Adding dilute acidic/alkaline solution or repeated washing with water was sometimes applied to neutralize the slurry when an alkaline/acidic steeping was used. Extraction method affected the chemical composition (e.g., protein) and properties (e.g., pasting properties) of the isolated millet starch (Singh et al., 2023; Yanez & Walker, 1986). The differences might be because of minor components (e.g., lipids and proteins) (Singh et al., 2023). The presence of minor components not only reduced the starch content but also may affect starch properties via interaction with starch at granular and molecular levels (Wang et al., 2020b). A higher starch yield (by 20%) was achieved using an alkaline method for millet starch extraction than that using the neutral and acidic ones. The impact of steeping temperature on the isolation of millet starch was less studied specifically but should not be neglected. In a previous study (Sandhu & Siroha, 2017), millet seeds were soaked in 0.1% (w/w) sodium metabisulfite solution at 50 °C for 18–20 hours, where starch annealing might occur.

8.3 Chemical Composition of Millet Starch

The proportions of amylose and amylopectin are the most important parameters leading to the variation in starch properties (Jane et al., 1999). Most starches had amylose contents (ACs) ranging between 15% and 30%. Waxy and high-amylose species can have AC out of this range. Trace amount of amylose was detected in waxy-type millet starches (Chao et al., 2014). For most nonwaxy millet starches, the ACs distributed widely between 10% and 35% (Table 8.1). Notably, the methods for the quantification of AC could amplify the difference between millet starches in AC. Amylose molecules are known to have high iodine binding capacity, and the amylose-iodine complex develops blue color. However, iodine binding can also occur to amylopectin molecules, leading to an overestimation of AC. The AC as estimated by iodine-binding-based methods is thus clarified as apparent AC. The extent of overestimation depends on the amylopectin structure. Such an overestimation was also found in the studies of millet starch. The amyloses contents of millet starches measured by fractionation-based method were lower than those measured by iodine-binding-based colorimetric methods (Wankhede et al., 1979).

The diversity in the AC of millet starches stems from variations in genetics and growth conditions. Li et al. (2020) collected proso millet grains of 93 accessions from China and reported that the apparent ACs of the starch collections ranged between 0% and 32.3% with a mean value of 16.4%. Qi et al. (2020) reported that the apparent ACs of eight foxtail millet starches ranged between 16.8% and 26.8%. Growing conditions (temperature in particular) had a significant effect on the apparent AC of a nonwaxy proso millet starch, but the effect was not significant on waxy proso millet (Wang et al., 2023a). The variations in the composition of the nonstarch components were related to the species, growth conditions, and the abovementioned isolation procedures (Aruna & Parimalavalli, 2022; Halal et al., 2019; Wang et al., 2023a).

Except for the molecular composition, isolated starches contain nonstarch compounds such as protein, lipid, and mineral substance. The contents of lipid and protein in millet starches were in ranges of 0.1–0.9% and 0.2–4.3%, respectively (Table 8.1). In a comparative study, millet starch had a higher protein content but lower ash content than those of other cereal starches including wheat, rice, maize, barley, oats, and sorghum starches (Kaur et al., 2018). The contents of lipid and protein in starch are greatly dependent on the methods used for starch isolation (Wankhede et al., 1979). Madhusudhan and Tharanathan (1995) analyzed the composition of bound lipid fractions and that of granule-bound proteins in an isolated finger millet starch. The ratio of saturated to unsaturated fatty acids was 1:0.81 in the finger millet starch with C16 and C18:2 fatty acids being the major fatty acids (Madhusudhan & Tharanathan, 1995). The granule-bound lipids and proteins are

Table 8.1 Starch yield and chemical composition of millet starch

Source	No.	Starch yield (%)	Amylose (%)	Lipid (%)	Protein (%)	Ash (%)	References
Pearl (*Pennisetum glaucum*)	1	70.4	32.5				Annor et al. (2014b)
	1		6	0.2	0.2	1.4	Choi et al. (2004)
	3	53.1–56.5	24–31.9			0.02–0.03	Hoover et al. (1996)
	1	60.2	22.8	0.92	0.68	0.78	Wankhede et al. (1990)
	5		20–22	0.09–0.19	0.55–0.77	0.10–0.13	Beleia et al. (1980)
	1		17.6	0.03	0.90	0.14	Dey et al. (2024)
	1		28.14		0.31	0.86	Wu et al. (2014)
Finger (*Eleusine coracana*)	1	63.4	32.4				Annor et al. (2014b)
	1		28.4		0.58	1.60	Wu et al. (2014)
	1		17.6	0.04	0.81	0.14	Dey et al. (2024)
	3	57.5–59.2	14.4–16.2	0.78–0.85	1.1–1.24	0.08–0.14	Wankhede et al. (1979)
Foxtail (*Setaria italica*)	1	69.1	28.6				Annor et al. (2014b)
	1	56.2	16.9–17.5	1.2	1.35	0.12	Wankhede et al. (1979)
	1		25.7		0.56	0.39	Wu et al. (2014)
	10		11.2–25.2				Liu et al. (2011)
	53		11.4–27.1				Fujita et al. (1996)
	1		18.3	0.06	0.52	0.14	Dey et al. (2024)
	8		16.8–26.8	0.21–0.47	0.10–0.26		Qi et al. (2020)
	34		18.3–25.3				Li et al. (2019)
Little (*Panicum miliare*)	1		18.8				Kumari and Thayumanavan (1998)
Barnyard (*Echinocloa frumentacea*)	1		20.0				Kumari and Thayumanavan (1998)
	1		17.6		0.58	1.06	Wu et al. (2014)

Kodo (*Paspalum scrobiculatum*)	1		19.6				Kumari and Thayumanavan (1998)
	1			0.16	0.56		Annor et al. (2013)
Proso (*Panicum miliaceum*)	1	93.7	33.9				Annor et al. (2014b)
	1		0.38		0.45	0.58	Wu et al. (2014)
	1[a]		27.2–29.1	0.59–0.6	0.69–4.31	0.51–0.63	Yanez and Walker (1986)
	4	61.8–68.2	29.2–32.6		1.1–2.1	0.48–0.83	Yanez et al. (1991)
	95		0–32.3				Li et al. (2020)
	1		17.2				Kumari and Thayumanavan (1998)
Hybrid barnyard (*Echinochloa frumentacea*)	1		27.68		0.44	0.81	Wu et al. (2014)
White fonio (*Digitaria exilis*)	2		22.6–26.1	2.6–2.9	0.3–0.7	0.4–0.8	Carcea and Acquistucci (1997)

Note: No., number of genotypes analyzed in specific study.
[a] Starch extracted with different methods.

associated with the structure of starch granules and physicochemical properties (Xu et al., 2024). The relationship between millet starch and the protein and lipid therein remains to be studied.

8.4 Structures of Millet Starch

8.4.1 At Granular Level

Starch granules from different botanical sources had extensively diverse morphology. Starch granules of oval/elongated, round/spherical, polygonal, triangular, lenticular, disk-like, and irregular shapes have been reported (Jane et al., 1994). The diversity could be much larger when the detailed differences in the long axis, short axis, thickness, and others of starch granules are compared. Most millet starch granules have polygonal shapes (Wu et al., 2014), which are commonly found in cereal starches (Jane et al., 1994). Some spherical granules were found as well. Most starch granules have the size of between submicrometer to ~200 μm. Depending on the species, the size of millet starches ranged from several micrometers to ~20 μm with a peak located at the range of 5–8 μm (Annor et al., 2014b). Small granules with the size smaller than 1 μm had been reported in foxtail millet starch (Kumari & Thayumanavan, 1998). Spherical granules had smaller size and in much lower proportion than polygonal granules of millet starch (Wu et al., 2014). Yang et al. (2023a) observed the change of millet starch granules in developing broomcorn millet grains. The size of starch granules increased with developing grains in the first 12 days after pollination, and starch granules changed from having more round shapes to having more polygonal ones from 12 to 21 days after the pollination. It was also noted that broomcorn millet grains of nonwaxy type had larger granules than waxy type (Yang et al., 2023a). In a case study, the granule sizes of starches of waxy proso millet but not of nonwaxy type altered with changing environmental conditions (Wang et al., 2023a).

In most cases, starch granules presented individually in separate amyloplasts, while some amyloplasts contain more than one granule, and they tend to be in a compound form (Matsushima & Hisano, 2019). Starches isolated from certain sources like wheat and barley grains could have bimodal sizes (Jane et al., 1994). Unimodal size distributions of millet starches were reported in most studies (Zhu, 2014), whereas a compound form of starch granules was observed in finger millet (McDonough et al., 1986).

Channels, pores, and cavities are reported on starch granules of various sources such as corn, rice, wheat, pea, and potato (Zhu et al., 2024). The position of pores, channels, and cavities may be either on the surface or penetrate the granule interior, and they are likely to pack proteins and lipids (Zhu et al., 2024). Procedures of the starch isolation might result in similar structures (Zhu et al., 2024). Pores were observed on the surface of a pearl millet starch (Choi et al., 2004). Millet starches of other species with such structural characteristics remain to be discovered. It was suggested that channels, pores, and cavities on the starch granules might ease the diffusion of chemicals and enzymes and thus accelerate corresponding reactions (Huber & Bemiller, 2001). Investigating the granular structure of millet starch is helpful for understanding its properties such as granular disintegration and digestion.

8.4.2 At Molecular Level

8.4.2.1 Amylopectin Structure

The molecular weight of amylopectin molecules ranges widely between 2×10^6 and 7×10^8 depending on botanical source (Bertoft, 2018). The molecular weight of millet amylopectins from different varieties was reported (Li et al., 2017a; Madhusudhan & Tharanathan, 1996; Wankhede et al., 1979; Yang et al., 2019). The weight-average molecular weight (M_w) of the amylopectin of a finger millet starch was 1.3×10^7 Da (Madhusudhan & Tharanathan, 1996). In another study, the M_w of amylopectin from a foxtail millet was 1.4×10^8 Da (Li et al., 2017a). A large discrepancy was found in the reported values of different studies, which might be severely affected by the limitations associated with corresponding methodology (e.g., technique and calibration). Size exclusion chromatography, for example, is commonly used for characterizing the size distribution of starch molecules. However, the branched nature of amylopectin makes it fail to create a well-calibrated relationship between the hydrodynamic volume and molecular size (Castro et al., 2005). The number average molecular weight of a finger millet starch starches (AC = 25.3%) was lower than those of other amylose-containing starches (AC = 17–36%) from potato, green gram, wheat, tapioca, chickpea, and maize. The millet starch had lower intrinsic viscosity than the others. The results might indicate that millet amylopectin might have molecular size lower than the others.

The highly branched amylopectin molecules consist of unit chains of various but conservative lengths (6 to ~100 glucosyl units). Amylopectin has polymodal chain length distributions (Hizukuri, 1986). However, limited by the difference in the resolution of methodology, bimodal distributions are commonly reported with the short unit chains being the majority (Jane et al., 1999). The bimodal distribution was explained by a well-known model named "cluster model." In this model, a dense appearance of short amylopectin branches (e.g., $DP \leq 36$) formed a densely branched subunit assigned "cluster," and the long amylopectin chains (e.g., $DP > 36$) connect the clusters to assemble the giant amylopectin molecule (Bertoft, 2018). The molar ratio of amylopectin short to long chains (S_{ap}:L_{ap}) was considered a coarse parameter to indicate the average cluster size of amylopectin. The S_{ap}:L_{ap} of most starches ranged between 6 and 22 (Bertoft et al., 2008). The S_{ap}:L_{ap} of millet amylopectin were 13.4, 12.4, 13.7, and 15.9 for foxtail, proso, finger, and pearl species, respectively (Annor et al., 2014a). Based on the S_{ap}:L_{ap} of millet amylopectin, the cluster size of millet amylopectin is considered high. The average lengths of long chains (LCL) of the four species were 51.0, 50.2, 51.4, and 52.1 glucosyl residues, respectively. Their average chain lengths and average short chain lengths were in the narrow ranges of 18.0–18.1 and 15.5–15.8 glucosyl residues, respectively (Annor et al., 2014a). In general, the average chain length of millet amylopectin was at a low level (Jane et al., 1999). Amylopectins in nonwaxy millet starches had a longer average chain length than those of waxy types (Yang et al., 2019). Notably, the average amylopectin chain lengths obtained based on different techniques and/or different calibration methods lack comparability.

The connectivity of unit chains in amylopectin is a combination of Harworth and Staudinger types (Bertoft et al., 2008). Unit chains of amylopectin are classified into A-chains (unsubstituted) and B-chains (substituted). Whole A-chains are external, while B-chains consist of external and internal parts with the outmost branch points as the boundary. The molar ratio of A- to B-chains in amylopectin is around 1.0 (Bertoft et al., 2008). The molar ratio of A- to B-chains was 1.03, 1.03, 1.02, and 0.98 for the amylopectins of foxtail, proso, finger, and pearl millet starches, respectively (Annor et al., 2014a). β-Amylase hydrolyzes starch from the nonreducing ends at the external part until the reaction is blocked by the outmost branch points. The limit value of β-hydrolysis (β-LV) estimates the limit proportion of external part of starch molecules that could be enzymatically removed. The β-LV of millet amylopectins from pearl (56.0%), foxtail (58.7%), and finger (52.1–56.7%) millet starches were reported (Table 8.2) (Gaffa et al., 2004; Madhusudhan & Tharanathan, 1996; Wankhede et al., 1979). Amylopectin could be externally hydrolyzed by using phosphorylase *a* and β-amylase successively. The resulting limit dextrins (named ϕ,β-limit dextrins) were slightly different from β-limit dextrins in the leaving stub (Bertoft, 2018). The ϕ,β-limit values were 59.7%, 59.8%, 57.7%, and 58.2% for amylopectins from foxtail, proso, finger, and pearl millets, respectively (Annor et al., 2014a). The average external and internal chain lengths of millet amylopectins from these species were estimated to be 11.8–12.3 and 4.7–5.1 glucosyl residues, respectively (Annor et al., 2014a).

Blue value and iodine binding affiliation (%) are used to characterize the ability of starch molecules to form helical complexes with iodine. Formation of iodine-amylose complex develops blue color, while formation of iodine-amylopectin complex develops purple color (Madhusudhan & Tharanathan, 1996). The iodine binding affiliation of a pearl millet amylopectin was 1.08%, which was at a low level compared to those of other amylopectins from twelve botanical sources (0.00–5.31% for normal starches and 6.79–9.30% for high-amylose corn starches) (Jane et al., 1999). The iodine binding affiliations of millet amylopectins were 0.67–0.72% and 0.64% for three foxtail species and one finger species (Wankhede et al., 1979). Gaffa et al. (2004) reported values of 1.37 and 1.47 for two pearl millet amylopectins. The diversity might stem from the differences in amylopectin structure (i.e., the availability of chain segments to form helices with iodine). In general, the iodine binding affiliations of millet starches are at a low level.

Except for the conventional concerns of molecular size and chain lengths of amylopectin in general, other concerns included the nonrandomly branched structure of amylopectin (Bertoft et al., 2024). Amylopectin is described as a highly branched molecule constructed by many densely branched blocks, which are connected by linear inter-block segments (Bertoft et al., 2024). Building block composition of amylopectin from different starches was reported (Bertoft et al., 2012). The degree of branching of waxy proso and foxtail millet starches were 4.2% and 3.9%, respectively (Yang et al., 2019). The nonrandomly branched structure of millet starch has not been studied in detail.

8.4.2.2 Amylose Structure

Amylose molecules are linear or slightly branched (Wang et al., 2019a). Gel permeation chromatography (GPC) is one of the most used techniques for defining amylose structure. The technique separates molecules according to their hydrodynamic volumes at certain conditions. However, diversities in apparatus, calibration method, and other experimental factors of the GPC used in different studies caused inevitably biased results and made the comparison of results less feasible between different studies. For example, Madhusudhan & Tharanathan (1996) reported that the amylose of a finger millet starch had molecular weight of 1.6×10^6 Da, while Wankhede et al. (1979) reported a value range of 2.54×10^5–2.80×10^5 Da for finger and foxtail millet starches. The discrepancy might be related to the sample profile per se and the inconsistent experimental settings. Using a modified Park-Johnson method,

Table 8.2 Structural properties of amylopectin of millet starch

Source	No.	DP_n/molecular weight	LV%	CL	ECL	ICL	References
Pearl (*Pennisetum glaucum*)	1		58.2 (φ,β)	18	12	5	Annor et al. (2014a)
	2	9000–9100 (DP_n)	56 (β)	20–21	13.2–13.8	5.8–6.2	Gaffa et al. (2004)
Finger (*Eleusine coracana*)	1		57.7 (φ,β)	17.9	11.9	5.1	Annor et al. (2014a)
	1	1.3×10^7	52.1(β)	20.7	12.1	6.3	Madhusudhan and Tharanathan (1996)
	3	$>2 \times 10^7$	54.6–56.7 (β)				Wankhede et al. (1979)
Foxtail (*Setaria italica*)	1		59.7 (φ,β)	18	12.3	4.8	Annor et al. (2014a)
	1	$>2 \times 10^7$	58.7(β)				Wankhede et al. (1979)
Proso (*Panicum miliaceum*)	1		59.8 (φ,β)	18.1	12.3	4.8	Annor et al. (2014a)

Note: No., number of genotypes analyzed in specific study; DP_n, number-average degree of polymerization; LV(%), limit value (%); β, β-limit value (%); φ,β, φ,β-limit value (%); CL, average unit chain length; ECL, average external chain length; ICL, average internal chain length.

the number-average DP (DP_n) of amylose from different botanical sources were found in the range of 720–1970 glucosyl residues (Takeda et al., 1987). Using a modified Park-Johnson methods, the DP_n of millet amylose (from pearl millet) was 1060 glucosyl residues (Gaffa et al., 2004). Using calibrated GPC, the DP_n of amylose from eight foxtail varieties were in the range of 878–1128. These results suggested that the size of millet amylose molecule was at a relatively low level in the starch family.

More than 30% (on a molar basis) of amylose are branched for starches from various botanical sources (Takeda et al., 1987). The number of chains per amylose molecule (NC) was between 2.9 and 9.8 for starches from different botanical sources (Takeda et al., 1987). The NC of two pearl millet amylose was reported to be 3.9 and 4.1 glucosyl residues (Gaffa et al., 2004), which is at the lower end of the abovementioned range. It should be noted that the reported NC is the average of total amylose including linear and branched ones. The NC of branched amylose from various botanical sources ranged from 5.3 to 17.1 (Takeda et al., 1987) and that for branched millet amylose has never been reported.

The position of branch points in branched amyloses caused structural variation as well (Wang et al., 2019a). The β-LV of amyloses from various botanical sources ranged between 75% and 91% (Takeda et al., 1987). The β-LV of amyloses from finger millet starches ranged between 82% and 83%, which were slightly higher than that (78%) of amylose from foxtail millet starch (Wankhede et al., 1979). In another study, the β-LV of amyloses from two pearl millet starches (80% and 84%) were reported (Gaffa et al., 2004). A high β-LV (93.7%) was reported for amylose from a finger millet (ragi) starch, suggesting a less branched nature of the amylose. As linear amylose would be hydrolyzed completely by β-amylase, the actual β-LV for branched amylose would be lower than the reported value for whole amylose. Techniques that could be used to separate intact linear and branched amyloses remain absent. Annor et al. (2014b) analyzed the difference in the molecular structure of branched amylose from different millet starches and proposed that the branched amylose molecules from proso and foxtail millets had shorter interbranch chain length than those of finger and pearl species. Previous study indicated that the NC of potato amyloses did not vary with the molecular size of amylose, to which the variations in unit chain length contributed (Wang et al., 2019a). The case for millet amylose remains to be studied.

Unit chains of amylose are divided into long (LC_{AM}) and short chains (SC_{AM}), though the unit chains of amylose are generally longer than those of amylopectin. The weight-based ratios of LC_{AM} to SC_{AM} were 0.60, 0.63, 1.27, and 0.95 for foxtail, proso, finger, and pearl millet starches, respectively (Annor et al., 2014b). The average lengths of unit chains of amylose (CL_{AM}) and their distributions varied widely (270–525 glucosyl units) for starches from different sources (Takeda et al., 1987). The CL_{AM} of two pearl millet starches were 260 and 270 glucosyl units. The amylose chain length distributions of eight foxtail millet starches from different cultivars were diverse, but the diversity was conservative (Xing et al., 2023). The amylose chain length distributions of these foxtail millet starches were parameterized using a biosynthesis-based model (Xing et al., 2023). The obtained parameters had biological implications but should be discussed upon comparison.

8.4.2.3 Intermediate Materials

Intermediate materials were sometimes distinguished structurally from amylose and amylopectin. Precipitation-based methods were used to isolate the three fractions (Klucinec & Thompson, 1998). Madhusudhan and Tharanathan (1996) isolated crude amylose from a finger millet by using Concanavalin A (ConA) to precipitate amylopectin, the supernatant was further fractioned using 1-butanol to precipitate true amylose, the starch materials in the collected supernatant were defined as intermediate fraction between amylose and amylopectin. The molecular weight of the intermediate materials was 4.5×10^6 Da. Its β-LV (75%) was higher than that of amylopectin from the same origin (52.1%) (Madhusudhan & Tharanathan, 1996).

8.4.3 At Supramolecular Level

Starch granules are semi-crystalline (Pérez & Bertoft, 2010). Starches are classified into A-, B-, and C- types according to their crystallographic pattern of X-ray diffraction. A-type starch contains nanocrystals of monoclinic lattice, while B-type starch has hexagonal structure (Pérez & Bertoft, 2010). Some starches have a combined diffraction pattern of A and B types, which are C type. Millet starch granules are A type in different reports (Zhu, 2014). There was great inconsistency in relative crystallinities of millet starch granules in various studies (Annor et al., 2014b; Qi et al., 2020; Yang et al., 2019). This was due to the different methods used to define the amorphous and crystalline regions in the diffractograms. Waxy millet starches had higher relative crystallinity than their nonwaxy counterparts of the same species (Yang et al., 2019). Indeed, amylose was considered largely amorphous in native starch granules.

Starch nanocrystals are not randomly distributed but stack alternatively with amorphous layers in between, which are known as the lamellar structure of starch granules. The thicknesses of the crystalline and amorphous lamellae of foxtail millet starches were ~6.5 and ~2.5 nm, respectively (Shi et al., 2023). A broomcorn millet starch was included in a comparative study on 27 starches from different botanical origins (Sanderson et al., 2006). The proportion of crystalline lamellar thickness in the total repeat distance of the millet starch was at a relatively high level among 13 A-type starches.

8.5 Physicochemical Properties of Millet Starch

8.5.1 Gelatinization

Gelatinization includes the amorphization of crystalline structure, uncoiling of double helices, and the swelling and disintegration of starch granules. In general, gelatinization occurs when starch is heated in the presence of plasticizer (e.g., water, glycerol, ethylene glycol, and 1,4-butanediol) (Ai & Jane, 2015). Alkaline solutions (e.g., NaOH and KOH), neutral salt solutions (e.g., $CaCl_2$ and LiCl), and organic solvents like dimethyl sulfoxide could lead to starch gelatinization at or below room temperatures. In food systems, starch gelatinization mostly refers to the one in the presence of water (Ai & Jane, 2015). The order-to-disorder transition of starch gelatinization could be detected in situ using various techniques. Differential scanning calorimetry (DSC) is the most common technique used for defining starch gelatinization. The onset, peak, and conclusion temperatures of millet starches from various species were reported (Table 8.3). Large diversity was found not only between species but also among varieties of the same species.

Table 8.3 Gelatinization of millet starch measured by differential scanning calorimetry

Source	No.	Starch:water ratio (w:w)	Heating rate (°C/min)	Gelatinization parameters				References
				T_o (°C)	T_p (°C)	T_c (°C)	ΔH (J/g)	
Pearl (*Pennisetum glaucum*)	1	1:3	5	62.8	67.9		12.3	Annor et al. (2014b)
	1	1:3	10	65.3	76.4	81.6	8.5	Choi et al. (2004)
	2	1:2	5	66.2–67.2	69.7–71.4	86.3–91	14.3–14.7	Gaffa et al. (2004)
	3	1:3	10	60.9–64.5	67.5–70	74–78	10.5–14.7	Hoover et al. (1996)
	6	7:10	5	63.4–67.7	69.3–71.6	74.5–76.3	10.6–12.4	Sandu and Siroha (2017)
	1	1:3	10	69.6	73.7	87.2	9.9	Shaikh et al. (2020)
Finger (*Eleusine coracana*)	1	1:3	5	63.9			13.2	Annor et al. (2014b)
Foxtail (*Setaria italica*)	1	1:3	5	66.7			11.8	Annor et al. (2014b)
	51	1:2	5		62.4–75.1		8.2–13.5	Fujita et al. (1993)
	2	1:2	10	65.1–68.3	71.0–73.0	76.37–8.7	6.6–8.5	Yang et al. (2019)
	8	1:3	10	65.4–68.0	70.9–73.2	75.6–80.3	11.8–15.2	Qi et al. (2020)
	34	1:3	10	66.8–69.5	72.1–73.8	76.3–78.6	17.6–25.8	Li et al. (2019)
	21	1:3	10	55.6–72.4	64.5–80.7	71.9–87.8	9.2–20.4	Xu et al. (2022)
	8	1:3	5	66.4–70.1	71.1–74.1	74.6–77.3	7.1–9.3	Xing et al. (2023)

Proso (*Panicum miliaceum*)	1	1:3	5	68.4	72.2		13.1	Annor et al. (2014b)
	1[a]	1:3	10	72.7–73.6	75.8–77.6	84.4–89.5	13.2–14.8	Yanez and Walker (1986)
	19	1:2	5		65.8–80.2		6.4–11.4	Fujita et al. (1993)
	4	1:2.7	10	67.8–69.0	69·0–73.9	75.5–81.8	13.0–14.9	Yanez et al. (1991)
	2	1:2	5	62–69	67–74	77–78	9.6–12.6	Tomita et al. (1981)
	4	1:3	10	66.8–70.0	72.8–76.6	78.3–82.4	10.4–14.5	Wen et al. (2014)
	2	1:2	10	64.6–71.1	70.5–77.9	77.48–2.3	9.6–10.8	Yang et al. (2019)
	95	1:3	10	67.4–75.5	71.5–79.0	76.5–84.0	11.9–17.6	Li et al. (2020)
	4	1:3	10	67.3–73.6	70.8–76.2	80.3–85.4	12.0–16.6	Wang et al. (2023a)
Barnyard (*Echinocloa frumentacea*)	1	1:2	5	60	68	71	5.0	Tomita et al. (1981)

Note: No., number of genotypes analyzed in specific study; T_o, onset temperature; T_p, peak temperature; T_c, conclusion temperature; ΔH, melting enthalpy change.
[a] Two different extraction methods (alkaline and acid).

8.5.2 Swelling Power

Swelling power (SP) of millet starch at certain temperatures varied with the millet variety (Bangar et al., 2021; Wu et al., 2014). Starch heated at a higher temperature tended to have higher SP than that at a lower one. SP of proso millet starch at 90 °C ranked second among various unconventional starches cultivated in China (Yang et al., 2023b). In a comparative study of Kaur et al. (2018), the SP (at 85 and 95 °C) of a pearl millet starch was the highest among seven cereal starches. SP of amylose-containing and waxy millet starches showed differences including the lower SP of waxy types at low temperatures and higher SP at high temperature (Yang et al., 2019). Amylose and amylopectin play different roles in the swelling and solubilizing behaviors of starch granules. For example, waxy starch normally had higher gelatinization temperatures than their amylose-containing counterparts. The former might have a lower degree of gelatinization at low temperature and thus lower swelling power than the latter. By comparison, amylose tended to restrict the swelling of starch granules and thus millet starch with normal AC had lower SP than their waxy counterparts. SP of proso millet starches as affected by cultivation location was reported (Wen et al., 2014), which might be related to various factors including chemical composition and thermal properties as exemplified earlier.

8.5.3 Water Solubility

The water solubility of starch is largely due to the amylose leaching upon starch gelatinization. The water solubility index of millet starches varied with the starch source. In a comparative study, pearl millet starch had lower solubility than sorghum, mango kernel, and water chestnut starches (Sinhmar et al., 2024). In another study of starch samples collected from China, the solubility of proso millet was at a middle level compared to seven starches from other sources (Yang et al., 2023b). A higher solubility of millet starch than those of other cereal starches (wheat, rice, maize, barley, oats, and sorghum) was reported (Kaur et al., 2018). The inconsistency was related to the great diversity presented within millet starches of different sources and the limited number of samples included in case studies. Wu et al. (2014) compared the solubility of millet starches from six species and found that finger and pearl millet starches had the highest and lowest solubility among all. Large variation was also found in the solubility of millet starches from different genotypes of the same species (Li et al., 2019, 2020). Yang et al. (2019) reported higher solubilities of amylose-containing millet starches than their waxy counterparts at low temperatures and an opposite trend at high temperatures. Amylose could leach from partially swollen starch granules more easily than amylopectin, and it had structural stabilizing effect on starch granules upon extensive swelling.

8.5.4 Light Transmittance

Light transmittance (LT) of starch paste is of importance for its impact on the appearance of foods, especially those that require transparent appearance (e.g., jelly). The LT values of millet starch reported in different studies varied dramatically due to the different starch concentrations used. Wu et al. (2014) compared starches from six millet varieties and found a significantly higher LT of a waxy proso variety than those of other varieties. Among amylose-containing cereal starches (AC = 17.7–24.7%), the LT of a millet starch was lower than most cereal starches (wheat, rice, maize, barley, and sorghum) except for oat starches (Kaur et al., 2018). Hoover et al. (1996) studied the effect of pH on the LT of pearl millet starches of three cultivars. The LT of these starches was not dramatically different and showed a similar trend against pH. The LT increased slightly with the increasing pH at a low range (pH < 8) and increased dramatically at a high range (pH = 8–13), which was related to the inhibition of inter-chain hydrogen bonding under alkaline conditions. The LT of millet starch seemed quite stable during storage, slightly decreasing from 97.5% on the first day to 95.5% after one week (Balasubramanian et al., 2014).

8.5.5 Pasting Properties

Pasting properties of starch have been of concern partly because of its connection with the processability of starchy foods. Gelatinization and retrogradation of starch while changing temperature are recorded as the changes in viscosity based on rheological principles (Balet et al., 2019). The pasting properties of starch have been studied by using various instruments (Balet et al., 2019). Rapid visco analyzer (RVA) has become a top choice due to the sensitivity, repeatability, and versatility upon its use (Balet et al., 2019). The pasting properties of millet starches as evaluated using RVA are summarized (Table 8.4). Great diversity in the pasting properties of millet starches has been observed. Large-scale comparative studies have been carried out on foxtail and proso millets individually (Li et al., 2019, 2020). Results of these studies demonstrated a huge variation among starches from different genotypes of the same species. Little millet starch had the lowest peak viscosity (PV) among all, being only half the value of starches from barnyard, finger, and proso millets (Bangar et al., 2021). Li et al. (2020) found that the PV of 95 genotypes of proso millet starches ranged widely between 185 and 316 RVU. Their breakdown viscosity (BD) and setback viscosity (SB) also varied in the ranges of 43–194 RVU and −120 to 63 RVU, respectively. The variation between the samples should take diverse compositional and structural attributes into consideration. Millet starches of waxy type had higher PV and BD but lower SB than their amylose-containing counterparts (Yang et al., 2019). Liu et al. (2011) reported similar results, in which the PV and BD of foxtail millet starches (10 varieties) were negatively correlated with apparent amylose content (AAM), and their SB were positively correlated with AAM. Comparison between results from different reports is impossible due to the differences in experimental setting (e.g., starch concentration and program for temperature changing). Compared to starches from different botanical sources, millet starch, although only a single species was studied, did not present outstanding pasting properties (Jane et al., 1999).

8.5.6 Flow Properties

Steady shear test has been used to measure the flow properties of starch paste (Ahmed et al., 2023). It detects the fluidity of starch paste in response to changing external force. The flow properties of starch may indicate the potential of a starch paste as thickeners and predict its performance during food processing operations such as cooking and mixing (Ahmed et al., 2023). Only a few studies investigated the flow properties of millet starch. Like starch of most botanical sources, millet starch paste was non-Newtonian fluids and shear-thinning (Wang et al., 2019b). Great diversity was found for millet starches of different genotypes of the same species (Sandhu & Siroha, 2017; Wang et al., 2019b).

Table 8.4 Pasting properties of millet starch as estimated by RVA

Source	No.	Starch (%)	PV	BD	SB	PT (°C)	References
Pearl (*Pennisetum glaucum*)	1	9.2	202	107	20.8	75.2	Choi et al. (2004)
	2	9	204–205	56–92	129–156	69.4–74	Gaffa et al. (2004)
	1	9.2	226	132	269	84.2	Balasubramanian et al. (2014)
Finger (*Eleusine coracana*)	1		246	52	111	79.1	Bangar et al. (2021)
Foxtail (*Setaria italica*)	10	10.5	345–425	183–237	142–188	72.0–78.5	Liu et al. (2011)
	8	10.7	232–307	117–177	105–183	77.0–81.0	Qi et al. (2020)
	1		182	81	159	80.0	Bangar et al. (2021)
	8	9.4	241–268	109–151	114–139		Xing et al. (2023)
	34	6.3	65–93	9–24	20–51	79.2–87.1	Li et al. (2019)
Proso (*Panicum miliaceum*)	1	9.2	235	155	42	76.0	Sun et al. (2014)
	1	10.7	196	86	26	77.0	Yang et al. (2019)
	95	6.3	185–316	43–194	−120–63	78.8–82.8	Li et al. (2020)
	1		245	110	74	79.9	Bangar et al. (2021)
Barnyard (*Echinocloa frumentacea*)	1		249	120	159	82.4	Bangar et al. (2021)
Kodo (*Paspalum scrobiculatum*)	1		228	99	141	81.5	Bangar et al. (2021)
Little (*Panicum miliare*)	1		127	17.7	68	84.1	Bangar et al. (2021)

Note: No., number of genotypes analyzed in specific study; the units for viscosities were converted to RVU; PV, peak viscosity; BD, breakdown; SB, setback; PT, pasting temperature.

For example, the yield stress values of six pearl millet starch pastes (starch concentration = 9.09%, w/w), as estimated by fitting with Herschel–Bulkley model, were found to be in the range of 25.6–178.1 Pa (Sandhu & Siroha, 2017). The yield stresses of proso millet starches were negatively correlated with their ACs (Wang et al., 2019b).

8.5.7 Viscoelastic Properties

The viscoelastic properties of millet starches of six varieties during gelatinization and cooling were compared (Wu et al., 2014). Finger millet starch gelatinized at a lower temperature than barnyard, hybrid barnyard, finger, foxtail, and pearl millet starches (Wu et al., 2014). Barnyard millet starch had the highest gelatinization temperature and the highest maximum storage and loss moduli during heating. During cooling, barnyard and proso millet starches developed weaker gels compared to millet starches of other varieties. The variations were closely related to the molecular composition and their structures. Amylose and amylopectin play different roles in determining the rheological behavior of starch gels. In a comparative study, the storage and loss moduli of pearl millet starch were comparable to those of mango kernel starch, and they were lower than those of sorghum and water chestnut starches (Sinhmar et al., 2024). Chemical composition and amylopectin molecular structure should be responsible for the variations. For example, the water chestnut starch in this study had the highest storage moduli which was related to its high AC.

8.5.8 Texture

Texture properties of millet starch gels varied (Li et al., 2019, 2020; Liu et al., 2011; Wang et al., 2019b). Hardness (HD) and adhesiveness (ADH) are commonly obtained to evaluate the texture of starch gels. It was reported that the HD and ADH of proso millet starches from 95 species ranged widely between 1.7 and 61.8 g and −94.4 and −0.8 g/s, respectively. The HD and ADH of 34 Chinese foxtail millets ranged from 7.6 to 14.0 g and from −20.9 to −13.9 g/s, respectively (Li et al., 2019). The HD and ADH of starch gels are highly related to the ratio of amylose and amylopectin and their structural profiles (Wang et al., 2019b). In general, starch with higher AC tends to have higher HD, while amylopectin is responsible for the ADH of starch gels. Amylopectin structure did not vary significantly in the millet starches studied (Wang et al., 2019b).

8.5.9 Retrogradation

Starch retrogradation occurs to cooling or cold gels following starch gelatinization (Wang et al., 2015). DSC is widely used for characterizing starch retrogradation, in which starch chains tend to reassociate and form ordered structure (Wang et al., 2015). Millet starches from different species and genotypes had different retrogradation tendency as estimated by rescanning the stored gels using DSC (Hoover et al., 1996; Li et al., 2019, 2020). Jane et al. (1999) compared the retrogradation of starch from different botanical sources and found that the cattail millet starch had a higher percentage of retrogradation than most A-type starches. Syneresis is an indicator of retrogradation in starch gels. Water may be expelled out of the starch gel during starch retrogradation. Yang et al. (2023a) compared starches from various nontraditional sources in China and found that the proso millet starch had a significantly lower syneresis than the others.

8.5.10 Enzymatic Digestibility

The digestibility of starch is important to the nutritional value of starchy foods or feeds. Granular, gelatinized, and retrograded samples of native or modified starches are studied. The susceptibility of starch to different enzymes such as human salivary α-amylase, pancreatin, glucoamylase, and their mixtures in vitro could be used to estimate the in vivo digestibility of starch (Zhu, 2014). During hydrolysis, starch fractions are hydrolyzed at different rates and are thus classified into three categories (Englyst et al., 1992). The rapid digestible starch is associated with the postprandial blood sugar, which is of particular concern for people with diabetes. Resistant starch has been extensively studied due to their potential promoting effects on gut health (Bojarczuk et al., 2022). The enzymatic digestibility of millet starch has been extensively studied (Annor et al., 2017; Cisse et al., 2018; Kaimal et al., 2021; Wankhede et al., 1990). Millet grains are more resistant than other cereals and had hypoglycemic property (Annor et al., 2017). It was reported that starch from kodo millet flour had more resistant starch and less slow digestible starch (SDS) than rice and wheat starches (Annor et al., 2013). The susceptibility of a proso millet starch to porcine pancreatic α-amylase was similar to that of maize starch (Yanez et al., 1991). However, in another study, the digestibility of starch isolated from pearl millet as estimated by a modified Englyst method was not slow (Kaur et al., 2018). The resistant starch content of the millet starch (either in native form or after cooked) was higher than those of barley and sorghum starches but was lower than those of wheat, rice, oat, and maize starches (Kaur et al., 2018). Comparison of results from different studies was difficult due to the different enzymes and hydrolytic conditions used. Overall, isolated millet starches are not as resistant as they are in grains. Gelatinized millet starch was more susceptible to enzymatic hydrolysis than its granular counterpart as the protection of granular structure was lost (Wankhede et al., 1990). Granule-bound proteins and lipids contributed to a lower digestibility of millet starch (Annor et al., 2013; Kaimal et al., 2021). The ways millet was cooked and served might make a difference as well. The glycemic index (GI) of millet foods were significantly different (GI of millet porridge =93.6; GI of millet steamed bread = 89.6; GI of millet pancakes of different recipes (76.2–83.0); GI of cooked millet = 64.4) (Ren et al., 2016). A recent review summarized the digestibility of millets upon different processes (Vidhyalakshmi & Meera, 2024). The interaction of starch with coexisting food components such as fatty acids affected the digestibility of starch (Annor et al., 2015).

8.6 Modifications of Millet Starch for Improved Functionality and Their Applications

The versatility of native starch is limited. In most cases, starches need modification when targeted functionality is required for a specific application. Modifications of millet starches have been of interest in the past 20 years. Physical, chemical, and enzymatic methods have been conducted to modify millet starches to improve their versatility (Table 8.5). This section summarizes the changed functionality of modified millet starches.

Table 8.5 Modifications of millet starches

Type of modifications	Millet types	Methodology	Key findings	References
Hydroxypropylation	Finger	Starch, water, and $NaSO_4$ were mixed at a weight ratio of 5:10:1 for 30 min after the pH was adjusted to 11.15 with NaOH. Propylene oxide (10, 20, 30, and 40 mL) was added to the suspension and maintained at 40 °C with shaking for 24 h	Hydroxypropylation increased the swelling power, solubility, and PV of starch, and decreased the turbidity and syneresis, SB, gelatinization temperatures, and enthalpy change of starch. The fraction of digestible starch increased	Lawal (2009)
Hydroxypropylation	Pearl	Starch (100 g) was dispersed in 150 mL distill water before addition of 15 g Na_2SO_4, followed by magnetic stirring for 40 min. The pH was adjusted to 11 with 1 N NaOH before 20 mL of propylene oxide was added. The reaction was kept at 35 °C for 24 h and then stopped by neutralized with 0.1 N HCl. The starch was recovered after thorough washing	Hydroxypropylation increased the swelling power and solubility of starch and decreased the gelatinization temperatures and enthalpy. The retrogradation of starch was inhibited as indicated by DSC and syneresis analysis	Shaikh et al. (2020)
Succinylation	Pearl	Starch slurry (40%, w/w) was adjusted to have pH 9.5 using 3% NaOH. Succinic anhydride (2% on starch weight) was added slowly within 2 h and then terminated by adjusting the pH to 6.5 with 0.5 M HCl. The starch was recovered after thoroughly washing	Succinylation increased the swelling power and solubility of starch and decreased the gelatinization temperatures and enthalpy. The retrogradation of starch was inhibited as indicated by DSC and syneresis analysis	Shaikh et al. (2020)
Cross-linking	Pearl	Starch was well dispersed in 0.5% sodium hydroxide solution. Epichlorohydrin was added in different concentrations (0.1%, 0.3%, 0.5%, and 0.8%). The reaction was maintained for 5 h and then stopped by using 1 HCl to adjust the pH to 5. The starch was recovered after thoroughly washing	The AC, swelling power, and solubility of starch decreased with the degree of cross-linking. The PV and HPV of starch increased at low epichlorohydrin concentrations (0.1–0.3) and then decreased at high concentrations. The BD and SB of modified starch were lower than those of native starch	Siroha et al. (2021)
Acetylation	Finger	The pH of a well-dispersed starch slurry (35%, w/w) was adjusted to 8.0–8.5. The reaction was conducted for 90 min before the pH was adjusted to 4.5 with 0.5 M HCl. The starch was washed thoroughly with distilled water before recovery	Acetylation increased bulk density, tapped density, and particle density, swelling power, solubility of starch, and gelatinization temperatures of both granular and retrograded starches. The PV, HPV, and CPV of starch increased, while the SB, gelatinization enthalpy change of native and retrograded starches decreased	Afolabi et al. (2012)

(*Continued*)

Table 8.5 (Continued)

Type of modifications	Millet types	Methodology	Key findings	References
Oxidation	Finger	Starch (100 g) was dispersed in 500 mL water, following by an adjustment of pH to 9.0–9.5 using 3% NaOH. NaOCl (10 g) was slowly added in 30 min and kept for 4 h with constant stirring. The starch was neutralized using 0.5 M HCl and washed thoroughly with distilled water before recovery	Acetylation increased bulk density, tapped density, particle density, swelling power, and solubility of starch, and gelatinization temperatures of both granular and retrograded starches. The PV, HPV, and CPV of starch increased, while the SB, gelatinization enthalpy of native and retrograded starches decreased	Afolabi et al. (2012)
Heat moisture treatment	Foxtail	Wet starch (75%, w/w) was sealed in bag and kept at 4 °C for 8 h to equilibrate moisture. The sample was then heated at 110 °C for 3 h	The swelling power, solubility, PV, HPV, CPV, BD, hardness, consistency, and syneresis of starch decreased, while the SB and cohesiveness of starch increased	Dey and Sit (2017)
Heat moisture treatment	Finger	The moisture content of starch was adjusted to 20%, 25%, and 30%, respectively, before sealing the samples in glass jars. The samples were then heated in an air oven at 100 °C for 16 h	Upon heat moisture treatment, the swelling power, PV, HPV, CPV, BD, SB, and gelatinization enthalpy of starch decreased, while the solubility, water-holding capacity, oil-holding capacity, and gelatinization temperatures of starch increased. Moisture content affected the extents of property changes. Heat moisture treatment at 30% moisture content had less significant impacts to starch compared to those at 20% and 25%	Adebowale et al. (2005)
Heat moisture treatment	Proso	The moisture content of starch was adjusted to 15%, 20%, 25%, and 30%. The wet starch was equilibrated at 4 °C before heated at 100 °C for 2 h	Upon heat moisture treatment, most granules maintained their shapes while some melted. Gelatinization temperatures increased with moisture content, while granular crystallinity and gelatinization enthalpy change of starch decreased with the moisture content. PV, HPV, CPV, DB, and SB increased at moisture content of 15% while those decreased at moisture contents of 20% and 25%. The content of rapid digestible starch decreased with increased moisture content. The content of slowly digestible starch in modified starch was higher than that in native starch. The resistant starch contents in modified starches obtained at 20%, 25%, and 30% were higher than those in native starch	Wang et al. (2023b)

Annealing	Proso	Starch was dispersed in water (five times by weight) in a beaker. The sample was heated at 30, 40, 50, and 60 °C for 24 h	Upon annealing, most granules maintained their shapes at 30 °C, while more melted granules were found at higher incubation temperatures (40–60 °C). Gelatinization temperatures increased with incubation temperature, while crystallinity and gelatinization enthalpy change of starch granules decreased with the incubation temperature. No significant change was found in the resistant starch content upon annealing. Upon annealing, the content of rapid digestible starch decreased, while that of slowly digestible starch increased	Wang et al. (2023b)
Annealing	Finger	Starch was dispersed in excess water and kept at 50 °C for 48 h	Upon annealing, the solubility, gelatinization temperatures, and enthalpy of starch increased, while the swelling power decreased	Adebowale et al. (2005)
Dry heating treatment	Proso	Starch was dried at 40 °C to reach a moisture content of 8%. The sample was then heated at 130 °C for 2 and 4 h	Starch granules largely maintained their shapes but had decreased crystallinity. Little change was found in the thermal properties of starch upon dry heating treatments. The PV and BD of starch decreased, while the HPV and CPV increased. Prolonged treatment had a more significant impact	Sun et al. (2014)
Ultrasound treatment	Foxtail	Starch slurry (30%) was treated with ultrasound for 20 min at 40 kHz	Ultrasound treatment decreased the swelling power, hardness, consistency, and syneresis of starch and increased paste cohesiveness. Little change in solubility and pasting properties was detected	Dey and Sit (2017)
Ultrasound treatment	Not given	Starch slurry (5%) was treated with ultrasound of different conditions (frequencies = 80 KHz along or 40 and 80 KHz combined; temperature = 25 or 60 °C)	The effects of ultrasound treatments were limited and varied slightly with conditions. Modifications at 60 °C caused gelatinization-like change to starch	Hu et al. (2019)
High pressure treatment	Pearl	Starch slurry (10%, w/w) was treated with high pressure of 200, 400, and 600 MPa for 10, 20, and 30 min with temperature being controlled below 30 °C	High-pressure treatment decreased the swelling power, solubility, oil-holding capacity, PV, HPV, CPV, BD, and SB of starch, while increasing the water holing capacity of starch. The effects depended on the pressure and duration of treatment	Mirzababaee et al. (2022)

(*Continued*)

Table 8.5 (Continued)

Type of modifications	Millet types	Methodology	Key findings	References
Microwave treatment	Proso	Starch slurry (20%, w/w) was heated at a microwave power of 500 W for 10 min	Microwave treatment decreased the swelling power and solubility of starch at low temperatures but increased these properties at high temperatures. The molecular weight and crystallinity of starch decreased associated with decreased gelatinization temperatures. The PV, HPV, CPV, BD, and SB of starch paste were decreased	Zheng et al. (2020)
Acid hydrolysis	Foxtail	Starch slurry (40%, w/w) was dispersed in 0.14 N HCl at 50 °C for 8 h and then adjust the pH to 5.5 with 1 N NaOH	The solubility, syneresis, BD, and cohesiveness of starch increased, while the swelling power, PV, HPV, CPV, SB, hardness, and consistency decreased	Dey and Sit (2017)
Acid hydrolysis	Finger	Starch slurry (33%, w/w) was treated with 0.5 N HCl, HNO_3, H_2SO_4, and H_3PO_4 for 1.5 h at 50 °C. The solution was neutralized using dilutes NaOH solution	All acid treatments led to decreased molecular weight, viscosity, and iodine-binding capacity	Singh and Ali (2008)
Enzymatic hydrolysis	Foxtail	Starch/enzyme water solution (pH 4.7) containing 10% starch and 0.1% α-amylase of *Aspergillus oryzae* (20,000 U/g) was incubated at 37 °C for 90 min, followed by heating the suspension to 70 °C for 5 min to inactivate the enzyme	The solubility, syneresis, CPV, BD, SB, hardness, and cohesiveness of starch increased, while the swelling power, PV, HPV, and consistency decreased	Dey and Sit (2017)

Note: PV, HPV, CPV, BD, and SB are peak viscosity, hot paste viscosity, cold paste viscosity, breakdown viscosity, and setback viscosity of starch paste, respectively, as measured by rapid visco analyzer.

8.6.1 Chemical Modifications

Different chemical modifications have been applied to millet starch to pursue desired properties (Afolabi et al., 2012; Lawal, 2009; Shaikh et al., 2017a, 2020; Siroha et al., 2021). Lawal (2009) etherified a finger millet starch with propylene oxide. The swelling power, solubility, and paste clarity of the hydroxypropylated millet starch at certain temperatures increased with the degree of substitution. The pasting properties of the millet starch changed upon the hydroxypropylation with an increased peak viscosity but lower setback viscosity. This suggested inhibited gelation. Succinylation of millet starch, compared to its hydroxypropylation, has a larger extent of increasing past viscosity (Shaikh et al., 2020). However, compared to a decreased setback value upon hydroxypropylation for millet starch, succinylation resulted in an increased setback value and a promoted gelation of the starch.

Cross-linking modification of starch is applied to increase structural stability. A pearl millet starch was modified using epichlorohydrin (Siroha et al., 2021). The swelling power, solubility, and pasting viscosity increased at a low epichlorohydrin concentration and then decreased with the degree of cross-linking at high concentrations. The cross-linking increased resistant starch content, reduced paste viscosity, and lowered yield stress. Cross-linking on barnyard millet starch using sodium trimetaphosphate resulted in similar outcomes as the abovementioned method using epichlorohydrin did (Siroha et al., 2021). Finger millet starch was acetylated and oxidized separately (Afolabi et al., 2012). Compared to the native starch, increased swelling power, solubility, and pasting viscosity were observed for both acetylated and oxidized starches. The increases in swelling power and pasting viscosity were more significant for oxidized millet starch than for acetylated millet starch (Afolabi et al., 2012). Singh and Ali (2008) compared the structure and properties of a finger millet starch upon hydrolysis using different kinds of acids (HCl, HNO_3, H_2SO_4, and H_3PO_4 solutions of 0.5 N). They also compared the results with those of wheat, maize, tapioca, green gram, chickpea, and potato starches. The acid treatment caused the degradation of starch molecules and decreased the intrinsic viscosity (Afolabi et al., 2012). Among these acids, HCl and HNO_3 produced the highest slurry thinning effects on the millet starch which was related to the severe molecular breakdown.

8.6.2 Physical Modifications

Physical modifications may be used to modify starch properties. Both hydrothermal treatments are conducted under certain moisture conditions when starch is subjected to incubation at mid-heating conditions. Annealing is a physical treatment involving incubation of starch with excess or intermediate water contents (>40%) at around 50–70 °C (Fonseca et al., 2021). By comparison, heat moisture treatment is normally conducted at low moisture levels (<35%, w/w) and at high temperatures (e.g., 100–120 °C). Adebowale et al. (2005) performed heat moisture treatment (100 °C for 16 hours at starch concentrations of 70%, 75%, and 80%) and annealing (50 °C for 48 hours in excess water) on finger millet starch and found significantly increased gelatinization temperatures of the modified starch and its retrograded counterpart. Upon the hydrothermal treatments, the finger millet starch exhibited decreased swelling power and increased water solubility.

Similar results were obtained by Wang et al. (2023b) when heat moisture treatment and annealing were conducted on proso millet starch. Sun et al. (2014) modified proso millet starch at even lower moisture level (8%) and higher temperature (130 °C for 2 and 4 hours), which was defined as dry heat treatment. Depending on the conditions, the treatments barely affected or decreased the gelatinization temperatures of the millet starch. The conclusion gelatinization temperature of the millet starch decreased upon the dry moisture treatment.

Except for hydrothermal treatment, ultrasonication (Hu et al., 2019), microwave treatment (Zheng et al., 2020), and high-pressure treatments (Mirzababaee et al., 2022; Sun et al., 2022) were used to modify millet starches. Hu et al. (2019) used ultrasound of single (40 kHz) and dual (40 + 80 kHz) frequency to modify potato and millet starches at two temperatures (25 and 60 °C). The modifications were less effective in changing the properties of millet starch than of potato starch. Limited changes were found in the swelling power, thermal and pasting properties, granular morphology, short and long range ordered structures of millet starch upon ultrasonication at these conditions (Hu et al., 2019). Microwave treatment at a power of 500 W for 10 minutes significantly induced molecular degradation of proso millet starch (20 g in 80 mL water) (Zheng et al., 2020). The results were associated with increased AC and resistant starch content and decreased pasting viscosities (Zheng et al., 2020). The microwave modification on millet starch resulted in the gelatinization of millet starch as autoclave did but induced more severe molecular degradation. High hydrostatic pressure treatment on 10% (w/w) starch slurry significantly changed the structure and properties of pearl millet starch, and the outcome depended on the pressure level and treatment duration (Mirzababaee et al., 2022). Under the majority of the conditions tested, the water-holding capacity was increased and the viscosities of pasting were decreased (Mirzababaee et al., 2022). Millet kernels were subjected to ultrasonication (Yadav et al., 2021) and microwave treatment (Zhi et al., 2022). The ultrasonication improved the hydration of millet kernels and eased the releasing of starch upon isolation. Millet starch extracted with the assistance of ultrasonication had higher crystallinity, reduced size, and homogeneous size distribution than that without.

8.6.3 Combined Modifications

Dey and Sit (2017) modified foxtail millet starch using heat moisture treatment, acid hydrolysis, enzymatic treatment, ultrasound treatment individually, and designed combinations of these modification methods. A wide range of variations was created depending on the type of modifications and their combinations. Swelling power of starch decreased upon modification, and the decrease became more significant with increasing number of modifications. Combined modifications were proved to create properties of various patterns and might be suitable for various applications. However, combining modifications significantly affected the production cost of modified starch, which must be of the top concern for food manufacturers. Ultra-high pressure and cold plasma in combination were used to modify proso millet starch (Sun et al., 2022). The combined modification led to depolymerization of starch molecules and increased thermal stability and resistance toward shearing.

8.6.4 Modified Properties of Millet Starch in Biological Processes

Starches isolated from malted finger, pearl, and foxtail millets had structures and properties different from their native counterparts (Malleshi et al., 1986). Higher AC, smaller granular size, higher gelatinization temperatures, lower swelling power, and higher solubility of the starches in malted millets were detected. The effect of germination on the structures and properties of millet starch was compared with that of other cereal starches (Li et al., 2017a). Such biological process caused pits and holes on the surface of the starch granule and reduced the particle size and crystallinity for all the studied starches. Millet and sorghum grains after the germination had a significantly decreased peak viscosity of starch therein, which was not observed in brown rice and oat grains. Germination caused increased amylose leaching to millet and sorghum starches. The results suggested the affected grain quality related to starch during germination, which might be an undesirable situation during grain storage.

8.6.5 Applications of Millet Starch

Millet starch was used for developing functional films (Gautam et al., 2020; Yang et al., 2018). By incorporating clove leaf oil into millet starch film, the product was used as packaging material with antimicrobial and antioxidant functions (Yang et al., 2018). Native, hydroxypropylated, succinylated, oxidized, and acetylated millet starches were compared in terms of their application in custard formulations (Shaikh et al., 2017b). Among these products, the custard prepared with the hydroxypropylated millet starch had superior sensory preference and storage stability over those prepared using others (Shaikh et al., 2017b). A subsequent study by the same authors investigated the properties of white sauce with the incorporation of native, hydroxypropylated, and succinylated millet starches (Shaikh et al., 2020). The white sauce incorporated with hydroxypropylated millet starch exhibited no detectable retrogradation during 14 days of storage.

8.7 Conclusions

Millet starch exhibited properties that are common to many cereal starches, indicating the potential of being complementary to conventional starches. The large diversity in species and genotypes of millet allows a huge variation in the structure and properties of millet starch. Although native millet starches, like most native starches from other botanical sources, are not fully qualified for certain applications, they can be modified via various methods to have desirable properties. Many aspects of millet starch remain to be studied, including the detailed branched structure of millet amylopectin and the branched profile of millet amylose. More modification methods such as those by using radiation, pulsed electric field, and debranching enzymes can be done.

References

Afolabi, T. A., Olu-Owolabi, B. I., Adebowale, K. O., Lawal, O. S., & Akintayo, C. O. (2012). Functional and tableting properties of acetylated and oxidised finger millet (*Eleusine coracana*) starch. *Starch-Stärke*, *64*, 326–337.

Adebowale, K. O., Afolabi, T. A., & Olu-Owolabi, B. I. (2005). Hydrothermal treatments of Finger millet (*Eleusine coracana*) starch. *Food Hydrocolloids, 19*, 974–983.

Ahmed, J., Dhull, S. B., & Chandak, A. (2023). Rheology and rheological measurements of starch. In J. Ahmed & S. Basu (Eds.), *Advances in food rheology and its applications* (pp. 521–556). Elsevier.

Ai, Y., & Jane, J. (2015). Gelatinization and rheological properties of starch. *Starch-Stärke, 67*, 213–224.

Annor, G. A., Marcone, M., Bertoft, E., & Seetharaman, K. (2013). In vitro starch digestibility and expected glycemic index of kodo millet (*Paspalum scrobiculatum*) as affected by starch-protein-lipid interactions. *Cereal Chemistry, 90*, 211–217.

Annor, G. A., Marcone, M., Bertoft, E., & Seetharaman, K. (2014a). Unit and internal chain profile of millet amylopectin. *Cereal Chemistry, 91*, 29–34.

Annor, G. A., Marcone, M., Bertoft, E., & Seetharaman, K. (2014b). Physical and molecular characterization of millet starches. *Cereal Chemistry, 91*, 286–292.

Annor, G. A., Marcone, M., Corredig, M., Bertoft, E., & Seetharaman, K. (2015). Effects of the amount and type of fatty acids present in millets on their *in vitro* starch digestibility and expected glycemic index (eGI). *Journal of Cereal Science, 64*, 76–81.

Annor, G. A., Tyl, C., Marcone, M., Ragaee, S., & Marti, A. (2017). Why do millets have slower starch and protein digestibility than other cereals? *Trends in Food Science & Technology, 66*, 73–83.

Aruna, M., & Parimalavalli, R. (2022). Influence of acid on isolation of starch from pearl millet (underutilized crop) and its characteristics. *Materials Today: Proceedings, 66*, 920–927.

Bangar, S. P., Siroha, A. K., Nehra, M., Trif, M., Ganwal, V., & Kumar, S. (2021). Structural and film-forming properties of millet starches: A comparative study. *Coatings, 11*, 954.

Balet, S., Guelpa, A., Fox, G., & Manley, M. (2019). Rapid visco analyser (RVA) as a tool for measuring starch-related physiochemical properties in cereals: A review. *Food Analytical Methods, 12*, 2344–2360.

Balasubramanian, S., Sharma, R., Kaur, J., & Bhardwaj, N. (2014). Characterization of modified pearl millet (*Pennisetum typhoides*) starch. *Journal of Food Science and Technology, 51*, 294–300.

Beleia, A., Varriano-Marston, E., & Hoseney, R. C. (1980). Characterization of starch from pearl millets. *Cereal Chemistry, 57*, 300–303.

Bertoft, E., Piyachomkwan, K., Chatakanonda, P., & Sriroth, K. (2008). Internal unit chain composition in amylopectins. *Carbohydrate Polymers, 74*, 527–543.

Bertoft, E., Koch, K., & Åman, P. (2012). Building block organisation of clusters in amylopectin from different structural types. *International Journal of Biological Macromolecules, 50*, 1212–1223.

Bertoft, E. (2018). Analyzing starch molecular structure. In S. Malin & L. Nilsson (Eds.), *Starch in food. Structure, function and applications* (2nd ed., pp. 97–149). Woodhead Publishing.

Bertoft, E., Blennow, A., & Hamaker, B. R. (2024). Perspectives on starch structure, function, and synthesis in relation to the backbone model of amylopectin. *Biomacromolecules, 25*(9), 5389–5401.

Bojarczuk, A., Skąpska, S., Khaneghah, A. M., & Marszałek, K. (2022). Health benefits of resistant starch: A review of the literature. *Journal of Functional Foods, 93*, 105094.

Carcea, M., & Acquistucci, R. (1997). Isolation and physicochemical characterization of fonio (*Digitaria exilis* Stapf) starch. *Starch-Stärke*, *49*, 131–135.

Castro, J. V., Ward, R. M., Gilbert, R. G., & Fitzgerald, M. A. (2005). Measurement of the molecular weight distribution of debranched starch. *Biomacromolecules*, *6*, 2260–2270.

Chao, G., Gao, J., Liu, R., Wang, L., Li, C., Wang, Y., Qu, Y., & Feng, B. (2014). Starch physicochemical properties of waxy proso millet (*Panicum Miliaceum* L.). *Starch-Stärke*, *66*, 1005–1012.

Choi, H., Kim, W., & Shin, M. (2004). Properties of Korean amaranth starch compared to waxy millet and waxy sorghum starches. *Starch-Stärke*, *56*, 469–477.

Cisse, F., Erickson, D. P., Hayes, A. M. R., Opekun, A. R., Nichols, B. L., & Hamaker, B. R. (2018). Traditional Malian solid foods made from sorghum and millet have markedly slower gastric emptying than rice, potato, or pasta. *Nutrients*, *10*, 124.

Dey, A., & Sit, N. (2017). Modification of foxtail millet starch by combining physical, chemical and enzymatic methods. *International Journal of Biological Macromolecules*, *95*, 314–320.

Dey, A., Dutta, D., Singh, A., & Sit, N. (2024). Physicochemical and functional properties of starches isolated from different millets. *Measurement: Food*, *15*, 100188.

Englyst, H. N., Kingman, S. M., & Cummings, J. H. (1992). Classification and measurement of nutritionally important starch fractions. *European Journal of Clinical Nutrition*, *46 Suppl 2*, S33–S50.

Fonseca, L. M., El Halal, S. L. M., Dias, A. R. G., & Zavareze, E. R. (2021). Physical modification of starch by heat-moisture treatment and annealing and their applications: A review. *Carbohydrate Polymers*, *274*, 118665.

Fujita, S., Morita, T., & Fujiyama, G. (1993). The study of melting temperature and enthalpy of starch from rice, barley, wheat, foxtail- and proso-millets. *Starch-Stärke*, *45*, 436–441.

Fujita, S., Sugimoto, Y., Yamashita, Y., & Fuwa, H. (1996). Physicochemical studies of starch from foxtail millet (*Setaria italica* Beauv.). *Food Chemistry*, *55*, 209–213.

Gaffa, T., Yoshimoto, Y., Hanashiro, I., Honda, O., Kawasaki, S., & Takeda, Y. (2004). Physicochemical properties and molecular structures of starches from millet (*Pennisetum typhoides*) and sorghum (*Sorghum bicolor* L. Moench) cultivars in Nigeria. *Cereal Chemistry*, *81*, 255–260.

Gautam, N., Garg, S., & Yadav, S. (2020). Underutilized finger millet crop for starch extraction, characterization, and utilization in the development of flexible thin film. *Journal of Food Science and Technology*, *58*, 4411–4419.

Halal, S. L. M. E., Kringel, D. H., Zavareze, E. R., & Dias, A. R. G. (2019). Methods for extracting cereal starches from different sources: A review. *Starch-Stärke*, *71*, 1900128.

Heena, Kumar, N., Singh, R., Upadhyay, A., & Giri, B. S. (2024). Application and functional properties of millet starch: Wet milling extraction process and different modification approaches. *Heliyon*, *10*, e25330.

Hizukuri, S. (1986). Polymodal distribution of the chain lengths of amylopectins, and its significance. *Carbohydrate Research*, *147*, 342–347.

Hu, A., Li, A., & Zheng, J. (2019). Dual-frequency ultrasonic effect on the structure and properties of starch with different size. *LWT-Food Science and Technology*, *106*, 254–262.

Huber, K. C., & BeMiller, J. N. (2001). Location of sites of reaction within starch granules. *Cereal Chemistry*, *78*, 173–180.

Hoover, R., Swamidas, G., Kok, L. S., & Vasanthan, T. (1996). Composition and physicochemical properties of starch from pearl millet grains. *Food Chemistry*, *56*, 355–367.

Jane, J., Kasemsuwan, T., Leas, S., Zobel, H., & Robyt, J. F. (1994). Anthology of starch granule morphology by scanning electron microscopy. *Starch-Stärke*, *46*, 121–129.

Jane, J., Chen, Y. Y., Lee, L., Mcpherson, A. E., Wong, K., Radosavljević, M., & Kasemsuwan, T. (1999). Effects of amylopectin branch chain length and amylose content on the gelatinization and pasting properties of starch. *Cereal Chemistry*, *76*, 629–637.

Kaimal, A. M., Mujumdar, A. S., & Thorat, B. N. (2021). Resistant starch from millets: Recent developments and applications in food industries. *Trends in Food Science & Technology*, *111*, 563–580.

Kaur, H., Gill, B. S., & Karwasra, B. L. (2018). *In vitro* digestibility, pasting, and structural properties of starches from different cereals. *International Journal of Food Properties*, *21*, 70–85.

Klucinec, J. D., & Thompson, D. B. (1998). Fractionation of high-amylose maize starches by differential alcohol precipitation and chromatography of the fractions. *Cereal Chemistry*, *75*, 887–896.

Kumari, S. K., & Thayumanavan, B. (1998). Characterization of starches of proso, foxtail, barnyard, kodo, and little millets. *Plant Foods for Human Nutrition*, *53*, 47–56.

Lawal, O. S. (2009). Starch hydroxyalkylation: Physicochemical properties and enzymatic digestibility of native and hydroxypropylated finger millet (*Eleusine coracana*) starch. *Food Hydrocolloids*, *23*, 415–425.

Li, C., Oh, S., Lee, D., Baik, H., & Chung, H. (2017a). Effect of germination on the structures and physicochemical properties of starches from brown rice, oat, sorghum, and millet. *International Journal of Biological Macromolecules*, *105*, 931–939.

Li, H., Fitzgerald, M. A., Prakash, S., Nicholson, T. M., & Gilbert, R. G. (2017b). The molecular structural features controlling stickiness in cooked rice, a major palatability determinant. *Scientific Reports*, *7*, 43713.

Li, K., Zhang, T., Sui, Z., Narayanamoorthy, S., Jin, C., Li, S., & Corke, H. (2019). Genetic variation in starch physicochemical properties of Chinese foxtail millet (*Setaria italica* Beauv.). *International Journal of Biological Macromolecules*, *133*, 337–345.

Li, K., Zhang, T., Narayanamoorthy, S., Jin, C., Sui, Z., Li, Z., Li, S., Wu, K., Liu, G., & Corke, H. (2020). Diversity analysis of starch physicochemical properties in 95 proso millet (*Panicum miliaceum* L.) accessions. *Food Chemistry*, *324*, 126863.

Liu, C., Liu, P., Yan, S., Qin, Z., & Shen, Q. (2011). Relationship of physicochemical, pasting properties of millet starches and the texture properties of cooked millet. *Journal of Texture Studies*, *42*, 247–253.

Malleshi, N. G., Desikachar, H. S. R., & Tharanathan, R. N. (1986). Physico-chemical properties of native and malted finger millet, pearl millet and millet starches. *Starch-Stärke*, *38*, 202–205.

Madhusudhan, B., & Tharanathan, R. N. (1995). Legume and cereal starches–Why differences in digestibility? – Part II. Isolation and characterization of starches from rice (*O. sativa*) and ragi (finger millet, *E. coracana*). *Carbohydrate Polymers*, *28*, 153–158.

Madhusudhan, B., & Tharanathan, R. N. (1996). Structural studies of linear and branched fractions of chickpea and finger millet starches. *Carbohydrate Research*, *284*, 101–109.

Matsushima, R., & Hisano, H. (2019). Imaging amyloplasts in the developing endosperm of barley and rice. *Scientific Reports*, *9*, 3745.

McDonough, C. M., Rooney, L. W., & Earp, C. F. (1986). Structural characteristics of *Eleusine corocana* (finger millet) using scanning electron and fluorescence microscopy. *Food Microstructure*, *5*, 247–256.

Mirzababaee, S. M., Ozmen, D., Hesarinejad, M. A., Toker, O. S., & Yeganehzad, S. (2022). A study on the structural, physicochemical, rheological and thermal properties of high hydrostatic pressurized pearl millet starch. *International Journal of Biological Macromolecules, 223*, 511–523.

Pérez, S., & Bertoft, E. (2010). The molecular structures of starch components and their contribution to the architecture of starch granules: A comprehensive review. *Starch-Stärke, 62*, 389–420.

Qi, Y., Wang, N., Yu, J., Wang, S., Wang, S., & Copeland, L. (2020). Insights into structure-function relationships of starch from foxtail millet cultivars grown in China. *International Journal of Biological Macromolecules, 155*, 1176–1183.

Ren, X., Chen, J., Molla, M. M., Wang, C., Diao, X., & Shen, Q. (2016). *In vitro* starch digestibility and *in vivo* glycemic response of foxtail millet and its products. *Food & Function, 7*, 372–379.

Sanderson, J. S., Daniels, R. D., Donald, A., Blennow, A., & Engelsen, S. B. (2006). Exploratory SAXS and HPAEC-PAD studies of starches from diverse plant genotypes. *Carbohydrate Polymers, 64*, 433–443.

Sandhu, K. S., & Siroha, A. K. (2017). Relationships between physicochemical, thermal, rheological and in vitro digestibility properties of starches from pearl millet cultivars. *LWT-Food Science and Technology, 83*, 213–224.

Shaikh, M., Ali, T. M., & Hasnain, A. (2017a). Effects of different modification reagents on functional properties of pearl millet starches. *Starch-Stärke, 69*, 1–10.

Shaikh, M., Ali, T. M., & Hasnain, A. (2017b). Utilization of chemically modified pearl millet starches in preparation of custards with improved cold storage stability. *International Journal of Biological Macromolecules, 104*, 360–366.

Shaikh, M., Ali, T. M., & Hasnain, A. (2020). Comparative study on the application of chemically modified corn and pearl millet starches in white sauce. *Journal of Food Processing and Preservation, 44*, 14393.

Shi, P., Zhao, Y., Qin, F., Liu, K., & Wang, H. (2023). Understanding the multi-scale structure and physicochemical properties of millet starch with varied amylose content. *Food Chemistry, 410*, 135422.

Siroha, A. K., Bangar, P. S., Sandhu, K. S., Trif, M., Kumar, M., & Guleria, P. (2021). Effect of cross-linking modification on structural and film-forming characteristics of pearl millet (*Pennisetum glaucum* L.) starch. *Coatings, 11*, 1163.

Singh, V., & Ali, S. Z. (2008). Properties of starches modified by different acids. *International Journal of Food Properties, 11*, 495–507.

Singh, H., Singh, R., Kumar, N., Kaur, B. P., & Upadhyay, A. (2023). Effects of wet-milling extraction methods on nutritional, functional, and structural properties of barnyard millet starch. *Starch-Stärke, 76*, 2300136.

Singh, S. M., Joshi, T. J., & Rao, P. S. (2024). Technological advancements in millet dehulling and polishing process: An insight into pretreatment methods, machineries and impact on nutritional quality. *Grain & Oil Science and Technology, 7*, 186–195.

Sinhmar, A., Sharma, S., Pathera, A. K., Nehra, M., Thory, R., Nain, V., & Godara, S. K. (2024). Comprehensive characterization of starch from diverse sources: Physicochemical, and functional properties. *Starch-Stärke, 76*, 2300280.

Sun, Q., Gong, M., Li, Y., & Xiong, L. (2014). Effect of dry heat treatment on the physicochemical propertiesand structure of proso millet flour and starch. *Carbohydrate Polymers*, *110*, 128–134.

Sun, X., Saleh, A. S. M., Lu, Y., Sun, Z., Zhang, X., Ge, X., Shen, H., Yu, X., & Li, W. (2022). Effects of ultra-high pressure combined with cold plasma on structural, physicochemical, and digestive properties of proso millet starch. *International Journal of Biological Macromolecules*, *212*, 146–154.

Tomita, Y., Sugimoto, Y., Sakamoto, S., & Fuwa, H. (1981). Some properties of starches of grain amaranths and several Millets. *Journal of Nutritional Science and Vitaminology*, *27*, 471–484.

Takeda, Y., Hizukuri, S., Takeda, C., & Suzuki, A. (1987). Structures of branched molecules of amyloses of various origins, and molar fractions of branched and unbranched molecules. *Carbohydrate Research*, *165*, 139–145.

Vidhyalakshmi, R., & Meera, M. S. (2024). Role of millets in pre-diabetes and diabetes: Effect of processing and product formulation. *Journal of Food Science and Technology*, *61*, 1848–1861.

Wankhede, D. B., Shehnaz, A., & Rao, M. R. R. (1979). Preparation and physicochemical properties of starches and their fractions from finger millet (*Eleusine-coracana*) and foxtail millet (*Setaria italica*). *Starch-Stärke*, *31*, 153–159.

Wankhede, D. B., Rathi, S. S., Gunjal, B. B., Patil, H. B., Walde, S. G., Rodge, A. B., & Sawate, A. R. (1990). Studies on isolation and characterization of starch from pearl millet (*Pennisetum americanum* [L.] Leeke) grains. *Carbohydrate Polymers*, *13*, 17–28.

Wang, S., Li, C., Copeland, L., Niu, Q., & Wang, S. (2015). Starch retrogradation: A comprehensive review. *Comprehensive Reviews in Food Science and Food Safety*, *14*, 568–585.

Wang, K., Vilaplana, F., Wu, A., Hasjim, J., & Gilbert, R. G. (2019a). The size dependence of the average number of branches in amylose. *Carbohydrate Polymers*, *223*, 115134.

Wang, Y., Yu, Z., Chen, H., & Shang, M. (2019b). Effect of amylose content on the physicochemical properties of proso millet starch. *Current Topics in Nutraceutical Research*, *17*, 42–49.

Wang, Q., Li, L., & Zheng, X. (2020a). A review of milling damaged starch: Generation, measurement, functionality and its effect on starch-based food systems. *Food Chemistry*, *315*, 126267.

Wang, S., Chao, C., Cai, J., Niu, B., Copeland, L., & Wang, S. (2020b). Starch-lipid and starch-lipid-protein complexes: A comprehensive review. *Comprehensive Reviews in Food Science and Food Safety*, *19*, 1056–1079.

Wang, H., Wu, E., Ma, Q., Zhang, H., Feng, Y., Yang, P., Gao, J., & Feng, B. (2023a). Comparison of the fine structure and physicochemical properties of proso millet (*Panicum miliaceum* L.) starch from different ecological regions. *International Journal of Biological Macromolecules*, *249*, 126115.

Wang, J., Ye, Y., Chai, Z., Liu, H., Wei, X., Ye, X., Tian, J., & Fang, H. (2023b). Physicochemical properties and *in vitro* digestibility of proso millet starch modified by heat-moisture treatment and annealing processing. *International Journal of Biological Macromolecules*, *235*, 123829.

Wen, Y., Liu, J., Meng, X., Zhang, D., & Zhao, G. (2014). Characterization of proso millet starches from different geographical origins of China. *Food Science and Biotechnology*, *23*, 1371–1377.

Wu, Y., Lin, Q., Cui, T., & Xiao, H. (2014). Structural and physical properties of starches isolated from six varieties of millet grown in China. *International Journal of Food Properties, 17*, 2344–2360.

Xing, B., Yang, X., Zou, L., Liu, J., Liang, Y., Li, M., Zhang, Z., Wang, N., Ren, G., Zhang, L., & Qin, P. (2023). Starch chain-length distributions determine cooked foxtail millet texture and starch physicochemical properties. *Carbohydrate Polymers, 320*, 121240.

Xu, Z., Liu, X., Ma, M., He, J., Sui, Z., & Corke, H. (2024). Reduction of starch granule surface lipids alters the physicochemical properties of crosslinked maize starch. *International Journal of Biological Macromolecules, 259*, 129139.

Yadav, S., Mishra, S., & Pradhan, R. C. (2021). Ultrasound-assisted hydration of finger millet (*Eleusine Coracana*) and its effects on starch isolates and antinutrients. *Ultrasonics Sonochemistry, 73*, 105542.

Yanez, G. A., & Walker, C. E. (1986). Effect of tempering parameters on extinction and ash of proso millet flours, and partial characterization of proso starch. *Cereal Chemistry, 63*, 164–167.

Yanez, G. A., Walker, C. E., & Nelson, L. A. (1991). Some chemical and physical properties of proso millet (*Panicum milliaceum*) starch. *Journal of Cereal Science, 13*, 299–305.

Yang, S., Cao, L., Kim, H., Beak, S., & Song, K. B. (2018). Utilization of foxtail millet starch film incorporated with clove leaf oil for the packaging of Queso Blanco cheese as a model Food. *Starch-Stärke, 70*, 1700171.

Yang, Q., Zhang, W., Li, J., Gong, X., & Feng, B. (2019). Physicochemical properties of starches in proso (non-waxy and waxy) and foxtail millets (non-waxy and waxy). *Molecules, 24*, 1743.

Yang, Q., Yuan, Y., Liu, J., Han, M., Li, J., Jin, F., & Feng, B. (2023a). Transcriptome analysis reveals new insights in the starch biosynthesis of non-waxy and waxy broomcorn millet (*Panicum miliaceum* L.). *International Journal of Biological Macromolecules, 230*, 123155.

Yang, J., Reddy, C. K., Fan, Z., & Xu, B. (2023b). Physicochemical and structural properties of starches from non-traditional sources in China. *Food Science and Human Wellness, 12*, 416–423.

Zheng, M., Xiao, Y., Yang, S., Liu, H., Liu, M., Yaqoob, S., Xu, X. Y., & Liu, J. (2020). Effects of heat–moisture, autoclaving, and microwave treatments on physicochemical properties of proso millet starch. *Food Science & Nutrition, 8*, 735–743.

Zhi, W., Zhou, Y., Wang, R., Wang, M., Wang, W., Hu, A., & Zheng, J. (2022). Effect of microwave treatment on the properties of starch in millet kernels. *Starch-Stärke, 74*, 2200063.

Zhu, F. (2014). Structure, physicochemical properties, and uses of millet starch. *Food Research International, 64*, 200–211.

Zhu, J., Han, L., Wang, M., Yang, J., Fang, Y., Zheng, Q., Zhang, X., Cao, J., & Hu, B. (2024). Formation, influencing factors, and applications of internal channels in starch: A review. *Food Chemistry: X, 21*, 101196.

9

Millet Bioactives: Bioavailability and Enhancement of Their Bioaccessibility

Divyanshi Thakur, Tejinder Kaur, Amandeep Singh, Charanjit S. Riar, and Parmjit S. Panesar

Department of Food Engineering and Technology, Sant Longowal Institute of Engineering and Technology, Longowal, Punjab, India

9.1 Introduction

Millets are climate-adaptive crops, which means they can thrive in poor soil types and inadequate water supply. In a period of climatic uncertainty, these characteristics make millets indispensable for global food structures. They are nutrient-dense crops that require relatively lesser inputs in comparison to the majorly grown cereals like wheat and rice. These crops can provide nutritional security to upcoming generations as they can grow in drylands. The increasing awareness of wholesome living has strengthened the merchandise for the need for more natural and healthier food options. Consumers have focused attention on foods with higher amounts of phytochemicals and quality proteins available from cheaper plant sources such as grain (millets) due to their nutritional and health benefits (Abhilasha et al., 2022). Millets are small-seeded grains that belong to the *Poaceae* family and are majorly grown in semi-arid and tropical regions of Asia and Africa. They are resilient to climate change and possess potential benefits for improved nutrition and health security. These draught-tolerant crops have a short growing season and serve as staple food for the replacement of major cereals. Millets need lesser inputs while growing as they have minimal water requirements, which minimizes their production costs (Sharma et al., 2023) as they are identified as C4 crops (Singh et al., 2021). Their low carbon footprint and excellent photosynthetic efficiency make them climate-smart crops suitable for the future. Millets can be grown along with other crops as they are well suited for various climates while being resistant to certain pests and diseases (Saleh et al., 2013). As reported earlier, rice requires about 1,200 mm of water, whereas millets need only 350–450 mm of water. They have a better nutritional profile compared to major cereals like wheat, rice, and maize, which makes them an alternate for the other starchy cereals. Their gluten-free nature has increased the options for celiac disease patients. For the 1% of the worldwide population

Nutraceutical Potential of Millet-Based Food Products: Chemistry and Applications, First Edition.
Edited by Parmjit S. Panesar, Charanjit S. Riar, and Sushil Dhital.

Published 2026 by John Wiley & Sons Ltd.

affected by celiac disease, millets provide a secure and nourishing alternative to traditional gluten-free choices, which often lack essential nutrients. These grains are potential sources of minerals like calcium and iron. All the millets have different nutritional profiles. They are good sources of carbohydrates, protein, dietary fiber, and antioxidants in varied proportions. Millets contain calcium, iron, zinc, and potassium in significant amounts (Priya et al., 2024). These grains are capable of eradicating hunger, malnutrition, and other metabolic disorders as they possess higher amounts of calcium than any other cereal grain. Millets have made a strong comeback in the daily platter after decades of ignorance (Harish et al., 2024).

Wheat, rice, and maize provide nutritional security, whereas millets give multiple securities, for example, food, fiber, fodder, ecology, health, and livelihood, while maintaining the fertility of soil as well, providing sustainability owing to the rising population and water crises around the globe. This multifaceted advantage unveils millets as unique among staple crops, providing solutions to both agricultural and public health concerns. Interest in consumption of polyphenol is growing as they have health benefits, for example, improved gut health and a reduction in heart diseases. Phenolic compounds present in millet grains are extensively studied due to their antioxidant potential. Millets with pigmented testa possess higher amounts of polyphenols (Serna-Saldivar & Espinosa-Ramírez, 2019). Major phenolic compounds in the millet are ferulic, caffeic, and sinapic acids, while luteolin and kaempferol were the major flavonoids in free fractions of foxtail millet (Xiang et al., 2019). Millets are consumed because of their rich nutritional profile and higher phytochemical content.

Kodo (*Paspalum scrobiculatum*) and finger millet (*Eleusine coracana*) cultivars characterized by dark brown seed coat pigmentation show significantly higher concentrations of phenolic compounds in comparison to white or yellow pericarp varieties such as pearl millet (*Pennisetum glaucum*), proso millet (*Panicum miliaceum*), foxtail millet (*Setaria italica*), and little millets (*Panicum sumatrense*). Grain phenolics are potential natural source of antioxidants. The bioactive compounds not only safeguard the plant from environmental stress but also provide comparable protective advantages to consumers. Millets are the staple food source as they provide dietary security, which is immensely important for assessing the bioaccessibility and bioavailability of the ingested bioactive compounds present in the millets. Bioaccessibility refers to the amount of ingested nutrients or the bioactive compound assimilated in the gastrointestinal tract, whereas bioavailability means the amount of ingested nutrient or bioactive compounds entering the bloodstream. Both these assessments are dependent on the nutritional composition and interactions of these nutritional components in the food.

The paradox of millets lies in their nutrient content versus bioavailability challenges an issue modern food science is uniquely positioned to address. Millets are considered the most efficient staple to meet dietary nutritional security; thus, it is important to assess bioavailability and bioaccessibility of nutrients. Bioavailability and bioaccessibility of the ingested nutrients are dependent on the nutritional composition of food and its molecules' interactions with food components and their interaction. Bioavailability of nutrients signals toward the quality which is basically dependent on the type of food matrix and chemical changes occurring during digestion. Bioavailability of nutritional constituents is affected by various physiological processes such as reduction in solubility and formation of insoluble compounds in the GI tract, decreased release rate of nutrients from the food, reduced permeability of the intestinal mucus layer, and molecular transformations of the nutrients into the bloodstream (Tomar et al., 2022). Emerging technologies show potential in improving millet nutrient bioavailability by 40–60%, addressing the gap between potential and delivered nutrition. Literature reports that various fractions of grains have a positive impact on the health of animals as well as humans through two mechanisms: firstly by breaking down the indigestible fibers to influence the gut microbiota composition and function, and secondly by providing substrates like resistant starch, nonstarch polysaccharides, and phenolic compounds, which are metabolized into beneficial, microbiota metabolites (Călinoiu & Vodnar, 2018; Gong et al., 2018).

This chapter focuses on the bioavailability and bioaccessibility of millet bioactives, as well as techniques for improving their delivery. The phenolic acids, flavonoids, and minerals found in millets can offer significant health benefits when they are released effectively during digestion (bioaccessibility) and absorbed systemically (bioavailability). Although millets contain significant quantities of these compounds; however, their bioavailability is hampered by various processing techniques which are further discussed.

9.2 Bioactive Compounds in Millet

Apart from its role as a dietary staple, millets are being acknowledged as a superfood loaded with bioactive compounds, providing significant health benefits. Millets are abundant sources of various bioactive compounds, including polyphenols, flavonoids, tannins, and phytosterols, which contribute substantially to their health-promoting properties. These bioactive components exhibit a wide range of beneficial characteristics such as antioxidant, anti-inflammation, and anticancer properties (Gowda et al., 2022). Phenolic acids such as vanillic acid and sinapic acid, as well as flavonoids like quercetin and catechin, serve important functions in neutralizing reactive oxygen species, lowering oxidative stress, and modulating metabolic pathways related to diabetes and cardiovascular disease (CVD). Moreover, millets contain several important bioactive compounds such as lignans, which promote hormonal balance, and phytic acid, recognized for its function in mineral absorption and detoxification. The presence of anthocyanins in certain millet varieties enhances their antioxidant capacity, aiding in the prevention of cellular damage. Millets also contain saponins, carotenoids, and dietary fibers, which increase their nutritional value and disease-preventing properties. Their adaptability and scientifically proved benefit make them a vital component in the future of a sustainable and healthy diet (Figure 9.1). Researchers and industry personnel are currently investigating innovative methods to harness millets' potential (Khan et al., 2025).

9.2.1 Polyphenols

Millets are acknowledged for their rich nutritional composition, including carbohydrates, proteins, crude fiber, lipids, calcium, iron, and niacin. Apart from this, they have received attention due to the presence of polyphenols which have been extensively studied for their antioxidant, anti-inflammatory, and disease-preventive properties (Akanbi & Barrow, 2018). Polyphenols are an extensive group of compounds that occur naturally having an aromatic

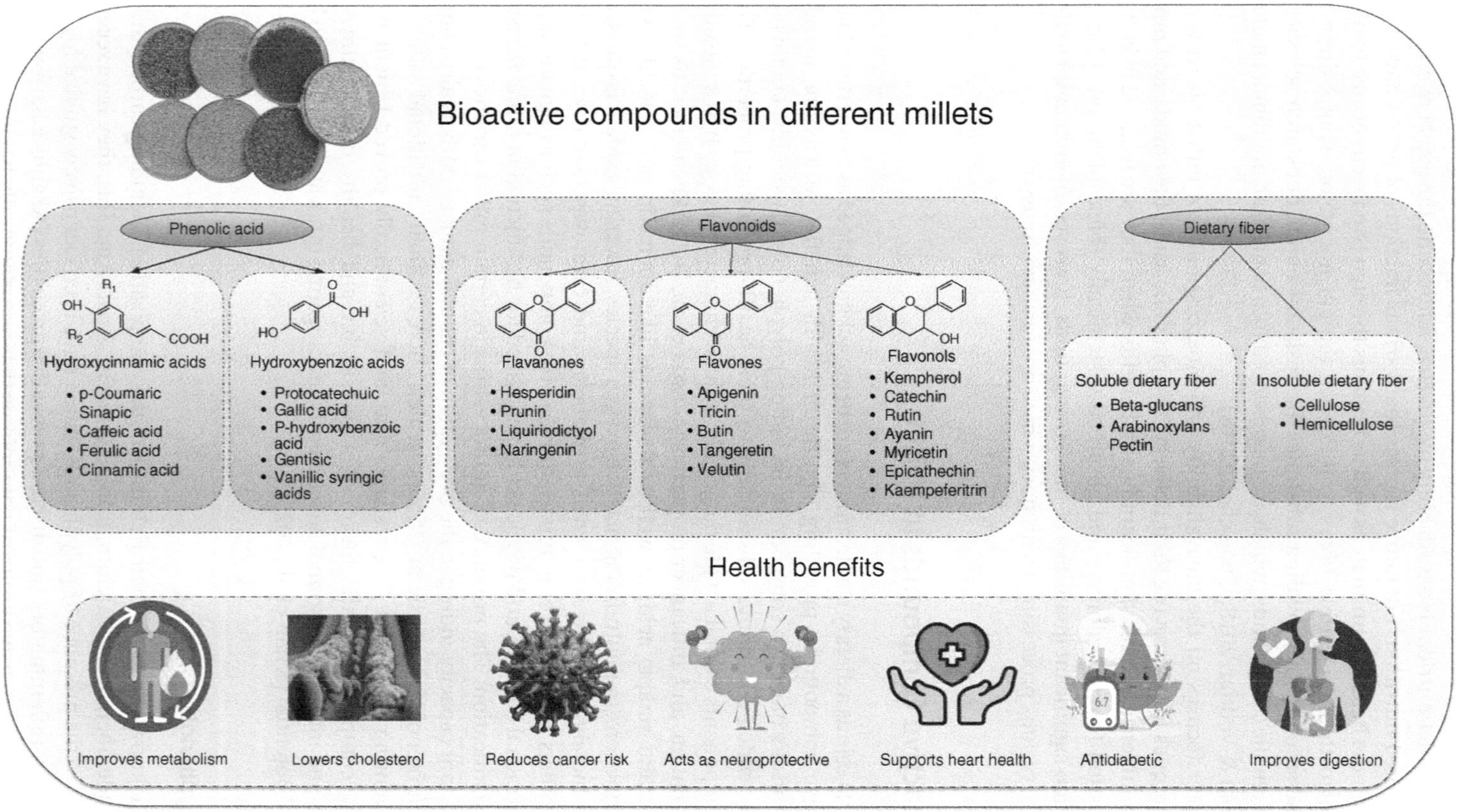

Figure 9.1 Bioactive compounds in millets and their health benefits. *Source:* Nisarg Nutritions, 2023, March 15.

ring structure, one or more hydroxyl groups, and a broad range of substituents. These compounds often appear in the form of glycosides, attached to sugar molecules, or in complexes with organic acids, amines, carbohydrates, lipids, and other phenolic compounds (Chauhan et al., 2018). These compounds are known for strong antioxidant properties and other therapeutic benefits. Polyphenols act as powerful antioxidants in the human body, countering oxidative stress by neutralizing free radicals and decreasing cellular destruction. Their ability to modulate oxidative processes not only aids in cellular integrity but also significantly reduces the risk of noncommunicable diseases such as cardiovascular disorders, diabetic conditions, and certain cancers (Kumari et al., 2017). Polyphenols help to reduce oxidative stress and its related ailments, including neurological diseases, cardiovascular problems, chronic fatigue syndrome, and sickle cell anemia. Additionally, these compounds exhibit bioactive attributes as neuroprotective, antiviral, anti-inflammatory, and anticancer activity, emphasizing their significance in health and disease prevention (Arts & Hollman, 2005; Holland et al., 2017). The specific bioactive compounds present in various millet are detailed in Table 9.1.

Chethan and Malleshi (2007) noted significant differences in polyphenol content within finger millet varieties, with brown varieties having 1.2–2.3% and white varieties having 0.3–0.5% polyphenols. These polyphenols are important antioxidants that help to prevent oxidative disorders associated with stress, including cancer and CVD, emphasizing millet as a functional food with significant health promoting benefits. Chandrasekara and Shahidi (2010) conducted a comprehensive analysis of phenolic compounds in various millet varieties, including kodo, finger, foxtail, proso, pearl, and little millets. Their research showed that kodo millet exhibited significantly higher bound phenolics (81.64 μmol FAE/g), significantly contributing to its antioxidant potential, while proso millet showed the lowest levels. Bound fractions, rich in ferulic and p-coumaric acids, contributed most to antioxidant activity. The study highlights millets' potential as gluten-free, antioxidant-rich functional foods.

9.2.2 Phenolic Acids

Millets and other plant foods contain significant amounts of phenols, which are important antioxidants. Phenolics, a major group of phytochemicals, are vital for human health and are a key part of plant-based diets (Hoda et al., 2019). Phenolic acids are predominantly derived from benzoic acid and divided into two types: hydroxybenzoic acid derivatives and hydroxycinnamic acid derivatives. Hydroxybenzoic acid is made up of several compounds, including protocatechuic acid, vanillic acid, syringic acid, and gallic acid. These chemicals are often found bonded, usually as part of complex structures such as lignins and hydrolysable tannins, which contribute to their stability and functionality in plants. Hydroxycinnamic acid derivatives include substances like p-coumaric acid, caffeic acid, ferulic acid, and sinapic acid. These phenolic acids mostly occur in conjugated forms, forming ester bonds with structural plant components such as cellulose, lignin, and various proteins. In finger millet, the primary phenolic acid is ferulic acid, followed by caffeic acid, syringic acid, vanillic acid, and p-coumaric acid (Hegde et al., 2002). These compounds contribute significantly to the grain's structural integrity and antioxidant properties, which increase its nutritional as well as health benefits. Furthermore, their presence in bound forms ensures

Table 9.1 The content of specific bioactive compounds present in various millets

Millets	Phenolic acids (μg/g)					Flavonoids (μg/g)			Dietary fiber (%)		References
	Ferulic acid	p-Coumaric acid	Gallic acid	p-Hydroxybenzoic acid	Caffeic acid	Catechin	Quercetin	Kaempferol	Soluble dietary fiber	Insoluble dietary fiber	
Pearl millet	1,035.8	196.6	5.4	47.9	89	2.5–12.6	53.85	11.49	1.91	9.85	Goudar et al. (2023), Pradeep and Sreerama (2017). Hassan et al. (2020), Shahidi and Chandrasekara (2013), and Mawouma et al. (2022)
Finger millet	41–405	1.8–41.1	68.1	13.5–16.8	70.8	58.81	12.75	1.35	1.4	15.7	Udeh et al. (2017), Hithamani and Srinivasan (2014), Chethan et al., (2008a), Chandrasekara and Shahidi (2012b), Devi et al. (2014), Hassan et al. (2020), and Khare et al. (2020)
Barnyard millet	27.9–231.8	797.4	24.5	12.26	34.3	1.46	-	6.3	4.15	8.19	Goudar et al. (2023), Pradeep and Sreerama (2017), Dayakar Rao et al. (2018), Joshi and Srivastava (2020), Ofosu et al. (2020), and Kumari et al. (2025)
Kodo millet	99.4–1,445.1	802	45.3	31.2	33.2	0.11	3.45	60.6	3.85	13.13	Shahidi and Chandrasekara (2013), Goudar et al. (2023), Khare et al. (2020), and Pradeep and Sreerama (2017)
Proso millet	358.9	1,155.4	27.7	126	8.7	—	—	—	2.05	9.48	Goudar et al. (2023) and Pradeep and Sreerama (2017)
Foxtail millet	308.9	200.3	13.6	21.8	49.9	20.97	75.89	51.61	5.87	11.06	Goudar et al. (2023) and Pradeep and Sreerama (2017), Zhu et al. (2018), Pradeep and Sreerama (2018), and Shahidi and Chandrasekara (2013)
Little millet	163.5	1082	7.8	32.6	41.4	14.21	63.06	72.19	5.65	8.57	Shahidi and Chandrasekara (2013), Goudar et al. (2023), Pradeep and Sreerama (2017), Pradeep and Sreerama (2018), and Kaur et al. (2019)

their slow release during digestion, which improves their bioavailability and functional benefits in the human body (Chethan et al., 2008b). Ferulic acid (trans-4-hydroxy-3-methoxycinnamic acid) is among the most prevalent phenolic acids found in finger millet. It is mostly found in the aleurone layer, exterior (pericarp), and embryonic cellular walls in different grains, where it plays an important role in cell wall structure and antioxidant activity. However, its presence in trace amounts in the starchy endosperm emphasizes its relevance in protecting the grain's outer layers from environmental stress and oxidative damage, as well as adding to the grain's overall nutritional and functional characteristics (Hoda et al., 2019).

9.2.3 Flavonoids

Flavonoids are the abundant and diversified class of phenolic compounds found in plants. These are characterized by an oxygen-containing heterocyclic structure composed of two benzene rings joined through a 3-carbon chain (Papoutsis et al., 2021). Flavonoids have an array of bioactive properties, including tumor suppressive, antidiabetic, anti-inflammatory, cardioprotective, immunomodulatory, neuroprotective, antiviral, and antiaging properties (Babypriyanka et al., 2024). These compounds play a significant role in promoting health and preventing various aliments due to their strong biological activities. Millets are rich in an array of bioactive flavonoids, which consist of three flavonols (quercetin, myricetin, and kaempferol), two flavones (apigenin and luteolin), along with a representative each of flavan-3-ol (catechin), flavanone (naringenin), and isoflavone (daidzein). Among these, the most important flavonoids present in millets are catechin, gallocatechin, quercetin, and apigenin. These substances are well-known for their anti-inflammatory and antioxidant properties, highlighting millets' nutritional value and functional role in promoting health.

Pradeep and Sreerama (2017) studied the flavonoid profile of foxtail millet (*Setaria italica*) and small millet (*Panicum sumatrense*) across different milling fractions (dehulled, bran, hull, and pearled grains). Quercetin, kaempferol, myricetin, apigenin, luteolin, catechin, naringenin, and daidzein were among the major flavonoids identified. Quercetin was the most abundant compound, while apigenin and kaempferol were prominent in the hull and bran fractions. Notably, daidzein, which is typically associated with legumes, was found in the bran and pearled fractions. The study, emphasizing the hull and bran as the richest sources of bioactive flavonoids, highlighted their potential health benefits (Pradeep & Sreerama, 2017).

9.2.4 Dietary Fibers

Dietary fiber in millets is one of the essential bioactive compounds which contributes to its nutritional and therapeutic benefits. Millets such as pearl millet, finger millet, foxtail millet, tiny millet, barnyard millet, kodo millet, proso millet, and brown top millet are high in soluble and insoluble dietary fiber (Jacob et al., 2024). Among these, finger millet contains the highest fiber content (15%), followed by kodo millet (9%), brown top millet (9%), foxtail millet (8%), pearl millet (8%), barnyard millet (6%), little millet (7%), and proso millet (3%) (Devi et al., 2014; Goudar et al., 2023; Joshi & Srivastava, 2020; Zhu et al., 2018). Soluble fiber such as β-glucans, pectin, and arabinoxylans promote the growth of beneficial gut microbiota like Bifidobacterium and Lactobacillus, contributing to improved gut health and reduced postprandial glucose response (Devi et al., 2014). Whereas insoluble fiber, including cellulose and hemicellulose, increases stool bulk, improving normal bowel movements and boosting digestive health (Jacob et al., 2024). The high fiber content in millets also contributes to their low glycemic index, making them beneficial for managing diabetes and other metabolic disorders (Geetha et al., 2020; Jacob et al., 2024). Additionally, the fermentation of millet fibers in the colon produces short-chain fatty acids (SCFAs) like acetic, propionic, and butyric acids, which play a significant role in reducing inflammation, enhancing insulin sensitivity, and protecting against metabolic disorders (Den Besten et al., 2013). These properties make millets an important dietary component for improving gut health and managing lifestyle-related diseases.

Millets are also gaining attention due to their high amount of resistant starch (RS), a form of dietary fiber with numerous health benefits. Resistant starch in millets is not completely digested in the small intestine instead reaches the large intestine, where it ferments, producing SCFAs that possess health benefits such as reducing diabetes and obesity risks (Meenu & Xu, 2019). The resistant starch (RS) content varies across different types of millets. For instance, foxtail millet has an RS content that varies between 23.6% and 34.33%, while barnyard millet has around 29.90% RS after cooking. Finger millet has a lower RS content in its raw condition, ranging from 0.9% to 1.3%; however, this can be enhanced by processing methods such as autoclaving (Bora et al., 2019; Thippeswamy et al., 2016). RS in millets helps regulate blood sugar levels and improves insulin sensitivity, making them beneficial for diabetes management. It also acts as a prebiotic, supporting gut health by promoting healthy bacteria. Millets, including foxtail and barnyard, have a lower glycemic index (GI) than wheat and rice, making them ideal for low-GI diets (Bora et al., 2019). Traditional millet-based foods, such as flatbreads, have higher RS content compared to wheat-based foods.

A study by Sharma and Gujral (2019) found kodo millet flatbreads contain 1.47 g/100 g of RS, significantly more than wheat flatbreads. With increasing demand for low-GI and high-fiber foods, millet-based products with enhanced RS are becoming more prevalent compared to corn-based products.

9.3 Bioaccessibility of Bioactive Compounds in Millet

9.3.1 Bioaccessibility

The bioaccessibility of bioactive compounds in millets is crucial for determining their health benefits, which include antioxidant, anti-inflammatory, and anticancer properties. Bioaccessibility refers to the fraction of an ingested bioactive compound that is made available for intestinal absorption after being released from its food matrix or delivery system and subsequently incorporated into mixed micelles. This fundamental process relies on the compound's effective passage through the gastrointestinal tract (GIT), where it initially is released from the food matrix via mechanical and enzymatic digestion prior to solubilization in the micellar phase and is calculated as the ratio of the biocomponent solubilized in the micellar phase to the biocomponent present in the raw digested material

(Alminger et al., 2014). Bioaccessibility is defined (Food and Drug Administration [FDA]) as the extent and rate at which the active compound or moiety is absorbed from its product matrix (drug or food) and becomes accessible for intestinal absorption (Fang, 2014).

9.3.2 Bioavailability

The bioavailability of bioactive compounds in millets represents the amount and rate at which these beneficial compounds like polyphenols, flavonoids, tannins, and phytosterols are liberated from the millet matrix, absorbed in the gastrointestinal tract, and available for physiological functions inside the body. Based on a pharmacological viewpoint, bioavailability refers to the rate and extent to which a bioactive molecule or drug is absorbed and becomes available at the site of activity. Whereas in nutritional viewpoint, bioavailability refers to the proportion of a specific food that the body can absorb and utilize, thus indicating an aspect of nutritional efficacy (Rein et al., 2013). Bioavailability refers to whatever is absorbed from ingested food and how much it goes into the bloodstream (Rodríguez et al., 2020). Many factors influence the absorption of nutrients into the bloodstream, for example, processing of food, digestion efficiency of nutrients, and the nature of food.

9.3.3 Mechanism of Bioaccessibility and Bioavailability of Bioactive Compounds

The sequential process of bioaccessibility and bioavailability of bioactive compounds consists of gastrointestinal digestion (liberation), absorption, tissue distribution, metabolism, and excretion (LADME) (Figure 9.2).

Bioaccessibility is the first phase of bioavailability, which is defined as the portion of a molecule that is liberated from the food matrix in the gastrointestinal lumen, making it available for intestinal absorption. The process begins with the mastication that takes place in the mouth. The phytochemicals are released from food through the process called liberation which primarily takes place in the mouth and stomach. The liberation of polyphenols is often incomplete, as bound forms require gut microbial activity for complete liberation. These phytochemicals then move from the digestive tract into the bloodstream via small intestine. About 5–10% of dietary polyphenols are absorbed in the small intestine as aglycones, whereas 90–95% enter the colon intact. However, not all the phytochemicals are absorbed, some remain unabsorbed and move through the digestive system. The unabsorbed polyphenols serve as substrates for colonic bacteria, producing bioactive metabolites such as phenyl-γ-valerolactones. These phytochemicals potentially undergo metabolism in the liver, with their hepatic metabolites released into the circulatory system (D'Archivio et al., 2010; Shahidi & Pan, 2022). The majority of polyphenol metabolites circulate as glucuronide/sulfate conjugates. It takes approximately 1–2 hours after consumption for polyphenol metabolites to reach their peak plasma concentrations. Meanwhile, the transport/diffusion from the intravascular to the extravascular of phytochemicals is referred to as distribution, while biochemical conversion or biotransformation of bioactive compounds carried out in the liver and colon is called metabolism. Phase II enzymes in the liver transform polyphenols into more hydrophilic compounds to facilitate excretion, although some retain bioactivity. Excretion then eliminates these compounds or their metabolites from the

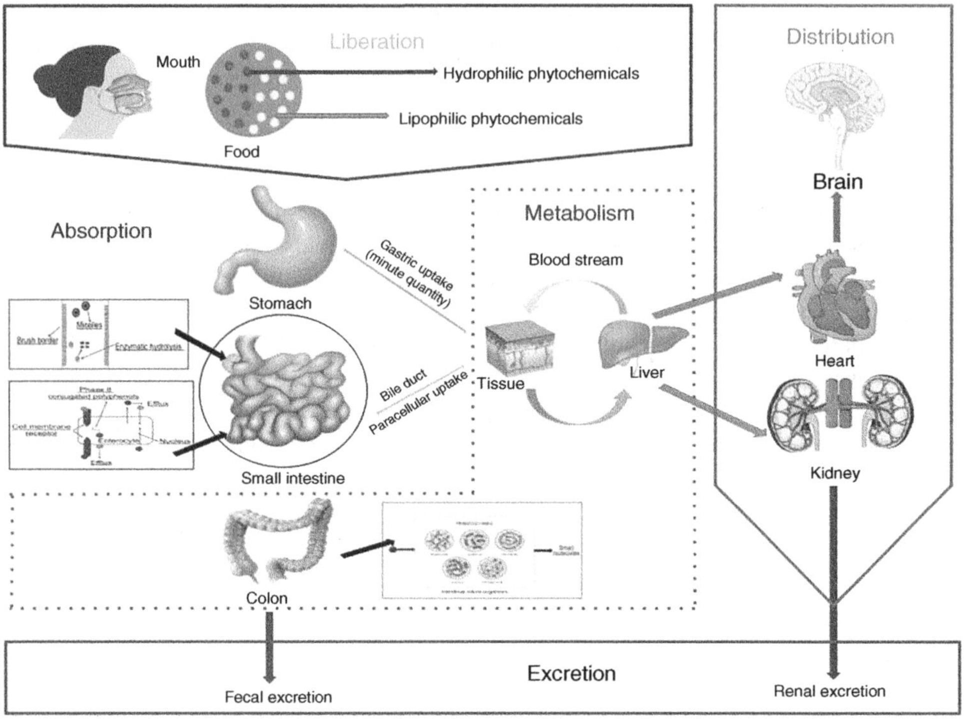

Figure 9.2 Steps of bioaccessibility and bioavailability of bioactives and their three stages. *Source:* Shahidi & Pan (2022).

body through renal or fecal processes. Urinary excretion constitutes 20–50% of ingested polyphenols, whereas biliary excretion facilitates enterohepatic recycling. Hydrophilic and lipophilic phytochemicals show slight variations in their digestion and absorption mechanisms. Polyphenols are hydrophilic compounds, and glycoside hydrolysis serves as the rate-limiting step for their absorption. Once phytochemicals bypass the obstacle of release from food, they become accessible to absorption inside the gastrointestinal tract, as shown in Figure 9.2 (Rein et al., 2013).

The accessibility of polyphenols is dependent upon their molecular weight; monomers exhibit optimal absorption, whereas oligomers and polymers necessitate microbial cleavage for effective absorption. The absorption of these compounds depends upon their solubility, interactions with food components, molecular transformations, cellular transport mechanisms, metabolism, gut microbes, and their bioaccessibility (Neilson & Ferruzzi, 2011). Lipophilic and hydrophilic phytochemicals have distinct absorption pathways attributable to their varying solubility (Rein et al., 2013). Food contains a wide range of lipophilic phytochemicals, including carotenoids, fatty acids, phytosterols, and vitamins. Lipophilic compounds form micelles with bile salts act as an emulsifier. Enterocytes absorb them through passive diffusion or assisted diffusion via transporters. In contrary to hydrophilic compounds, lipophilic phytochemicals are not easily excreted from the body, rather they are either stored in the liver or rereleased into circulation as lipoproteins and then stored in adipose tissue (Golding & Wooster, 2010). Hydrophilic phytochemicals, including polyphenols present in food, typically occur as esters, glycosides, or polymers and require enzymatic hydrolysis at the brush border of the small intestine to release absorbable aglycones. These aglycones pass through enterocytes, where phase II enzymes conjugate them (Del Rio et al., 2010). Numerous compounds are translocated from enterocytes into the intestinal lumen through adenosine triphosphate (ATP)-binding cassette (ABC) transporters. These compounds may be released from the liver either into the bloodstream or the bile (Hostetler et al., 2017). Also, metabolites released into the bile can be reabsorbed in the intestine. Eventually, polyphenol metabolites in the bloodstream are eliminated through urine (Shahidi & Pan, 2022).

9.3.4 Challenges in Bioaccessibility and Bioavailability

Millets possess distinct characteristics due to the enriched presence of dietary fibers, polyphenols, and protein in comparison to the commonly consumed cereals. Current studies indicate that these bioactive compounds including vanillic acid, sinapic acid, quercetin, and catechin show antioxidant and anti-inflammatory properties, which make millets beneficial for modern health-conscious diets. They are increasingly being consumed by the urban and elite population groups of the world due to their unique nutritional profiles. However, there are certain challenges faced by these grains concerning the bioaccessibility and bioavailability of the abovementioned nutrients in the millets. These grains have antinutritional components like tannins and phytates which deter the release of these important constituents from the food matrix. Antinutrients can lead to enzyme inhibition and metal chelating effects which will form poorly soluble salts affecting the bioavailability (Fernando et al., 2011). These antinutritional components not only influence mineral availability but also the absorption and stability of other bioactive compounds throughout digestion. Processing millets by various methods significantly reduces antinutritional factors which also leads to alterations in bioaccessibility and bioavailability. Various factors targeting bioaccessibility and bioavailability are physicochemical properties of food matrices, for example, load degree, size, morphology, electric charge, etc. It is important to know the absorption of these nutrients and to enhance their bioavailability. The efficacy of nutrient liberation and absorption is influenced by the characteristics of the food matrix, particularly in the presence of dietary fibers and polyphenols, which can function as either facilitators or barriers. Adoption of novel and value addition technologies will lead to the improvement in diet and ultimately the bioavailability of the nutrients which will help in safeguarding human health from diseases.

The bioavailability and assimilation of nutrients are affected by gastrointestinal digestion, encompassing catabolic processes and enzymatic hydrolysis, that determine the liberation of nutrients in the intestinal tract. Bioavailability and assimilation of nutrients could be enhanced by various processing techniques such as soaking, germination, and thermal processing. It is reliant on the catabolic processes of foods as well as enzymatic hydrolysis responsible for the release of nutrients into the gastrointestinal tract (Tharifkhan et al., 2021). Because of intricate interactions between bioactive compounds and dietary components, nutrient release remains variable despite such treatments. Heterogeneity in the release of nutrients exists due to the various factors which contribute to the bioavailability of an individual, such as differences in the matrix type in which nutrients are embedded; their chemistry with other foods and constituents of other foods might also affect the absorption, metabolism, and the living organism's health conditions, genetics, age, and lifestyle changes. The intricate bioavailability challenges need an effective nutrition approach, whereby methods of processing can be tailored to the millet type and physiological needs of the target population to maximize their nutritional benefits.

9.4 Processing Techniques for Enhancement of Bioaccessibility and Bioavailability of Millet Bioactives

Millet processing is essential for improving the bioaccessibility and bioavailability of nutrients by modifications in the grain's structural matrix and the reduction of antinutritional factors. The release and subsequent action of bioactive compounds from the foods are dependent on their accessibility to an enzymatic attack from either the food or the gastrointestinal tract. The gastrointestinal release of bioactive compounds in millets is limited due to interactions with cell wall matrix components, including dietary fiber, polyphenols, and antinutritional factors like tannins and phytates (Udeh et al., 2017). Traditional methods like decoration, germination, malting, milling, roasting, and fermentation enhance organoleptic properties as well as nutrient bioavailability. For example, milling alters dietary structure, boosting glycemic index, whereas semi-refined flour decreases antinutrients and improves mineral bioaccessibility (Pradeep & Sreerama, 2015). Hydrothermal treatments (e.g., soaking, cooking) and germination are helpful in lowering antinutrients such as phytates, enhancing iron and zinc bioavailability, as well as in vitro protein digestibility. Similarly, fermentation and malting improve nutrient availability by reducing inhibitory compounds (Chethan et al., 2008b). Overall, the choice and severity of processing methods, whether mechanical (milling, dehulling), biochemical (fermentation, germination), or thermal (cooking, puffing), have a significant impact on the bioavailability of millet bioactives, balancing retention of nutrients with removal of antinutrients. Table 9.2 presents the impact of various processing treatments on the bioaccessibility and bioavailability of bioactive compounds in different millets.

9.4.1 Milling, Pounding, and Grinding

Mechanical processing influences the liberation of phytochemicals from the plant matrix (Tagliazucchi et al., 2012). Processing methods such as milling, pounding, and grinding facilitate particle size reduction and disrupt the interactions between phytochemicals and the other food molecules, hence enhancing the release of bioactive molecules from the food matrix (Hotz & Gibson, 2007). Furthermore, decreasing particle size to the nanoscale can disrupt the covalent linkage within arabinoxylan molecules and phenolic acid moieties in rye and wheat bran, potentially increasing the bioaccessibility of phenolic compounds that are covalently linked to arabinoxylan in the fiber matrix (Van Craeyveld et al., 2009). Liang et al. (2024) studied the effects of superfine grinding on defatted foxtail millet bran (DMB), emphasizing its physicochemical characteristics, phenolic bioaccessibility, and antioxidant properties. The investigators

Table 9.2 Impact of processing treatments on millet bioactive compounds: bioaccessibility and bioavailability

Millet	Treatment	Effect on bioactive compounds	Effect on bioaccessibility/bioavailability	References
Finger millet	Sprouting	Decrease in total polyphenols and flavonoids; increase in some phenolic acids	Significant increase in bioaccessibility of polyphenols and flavonoids	Hithamani and Srinivasan (2014)
	Roasting	Increase in total polyphenols and tannins; appearance of some new phenolic compounds	Slight decrease in polyphenol bioaccessibility	
	Pressure cooking	Significant reduction in polyphenols, flavonoids, and tannins	Decrease in bio accessible polyphenols and flavonoids	
	Microwave heating	Minimal change in polyphenol content; some phenolic acids decreased	Notable decrease in bioaccessibility	
Pearl millet	Sprouting	No major change in total polyphenols; increase in specific phenolic acids	Increase in bioaccessibility of polyphenols and flavonoids	
	Roasting	Increase in total polyphenols, flavonoids, and specific phenolic acids	Increase in bio accessible polyphenols and flavonoids	
	Pressure cooking	Decrease in polyphenols, flavonoids, and tannins	Slight reduction in bioaccessibility	
	Open pan boiling	Gradual decrease in polyphenols and flavonoids with longer boiling time	Mild reduction in bioaccessibility depending on duration	
	Microwave heating	Decrease in total polyphenols and flavonoids	Moderate decrease in bioaccessibility	
Foxtail millet	Hydrothermal treatment soak-boil (SB40)	Increased total phenolic content and moderate increase in antioxidant properties	Improved bioaccessibility of phenolic compounds	Kaur et al. (2025)
	Hydrothermal treatment soak-steam (SS40)	Highest increase in total phenolics and antioxidant activity; better retention at lower drying temperatures	Significantly increase in bioaccessibility of most phenolic compounds	
Foxtail millet	Germination	Increased total phenolics, flavonoids, and antioxidant activity	Enhanced mineral bioavailability (Ca, Mg, Fe, Na); improved nutrient digestibility due to reduced antinutrients	Sharma et al. (2015)

(Continued)

Table 9.2 (Continued)

Millet	Treatment	Effect on bioactive compounds	Effect on bioaccessibility/bioavailability	References
Foxtail millet	Milling	Increased total phenolic content (TPC), antioxidant activity, higher levels of specific phenolic acids (e.g., ferulic, vanillic, syringic acids)	Improved release and bioaccessibility of phenolic compounds during in vitro digestion; enhanced absorption of bound phenolics due to cell wall disruption	Liang et al. (2024)
Little millet	Fermentation	Improves protein content; reduces phytic acid; elevates gallic acid and ascorbic acid	Enhance nutrient absorption and bioactive compound bioavailability	Taylor and Awika (2017)
Kodo millet	Soaking and fermentation	Increases phenolics, flavonoids, and tannin content; decreases antinutrients	Enhances bioavailability by reducing antinutritional factors	Perin and Murano (2017)
Foxtail millet	Fermentation	Increases gallic acid, catechol, p-coumaric acid, and enhances antioxidant activity	Improves bioaccessibility of antioxidants and bioactive compounds	Kumari et al. (2025)
Finger millet	Germination/soaking/ sprouting	Increases phenolics, flavonoids, and tannins; enhances specific phenolic acids and flavonoids	Enhances bioavailability by decreasing antinutrients; alters functional groups of compounds	Srivastava and Sharma (2012)

evaluated coarse (425 μm), fine (125 μm), and superfine (14.85 μm) grinding methods, suggesting that superfine grinding significantly disrupted the bran's cell wall matrix, resulting in a 96.52% breakage ratio. This process enhanced the liberation of bound phenolic compounds, increasing total phenolic content (TPC) by 80%, and enhancing antioxidant activity (2,2-diphenyl-1-picrylhydrazyl [DPPH] and 2,2′-azinobis (3-ethylbe nzothiazoline-6-sulfonic acid) [ABTS] assays). Whereas in vitro digestion demonstrated increased bioaccessibility of important phenolics, including ferulic and syringic acids, in superfine powder, emphasizing its potential as functional food applications. The study highlights the significance of particle size reduction in enhancing the nutritional advantages of millet bran.

9.4.2 Cooking

Traditional domestic cooking practices include diverse preparation methods that vary across different countries. Phytochemicals, including polyphenols and carotenoids, are liberated by the disruption of cell walls, subsequently forming complexes or dissociating molecular connections between the food matrix (Hidalgo & Zamora, 2017). Domestic cooking can lead to insignificant losses in the food matrix and bioaccessibility of certain phytochemicals, including carotenoids. Although cooking may diminish the carotenoid bioaccessibility while, oils can help mitigate this phenomenon (Hornero-Méndez & Mínguez-Mosquera, 2007). When exposed to high temperature, polyphenols exhibit a certain degree of stability. Moreover, only 2% of phenolics in raw cardoon stalks remained during in vitro gastrointestinal digestion, but their retention was significantly higher after roasting (60%) and frying (67%) (Mcclements, 2015). However, some studies indicate that the overall phenolic content decreases after cooking.

Hithamani and Srinivasan (2014) investigated the impact of sprouting, roasting, pressure cooking, open-pan boiling, and microwave heating on the phenolic profile and bioaccessibility of finger millet. Results revealed that except for roasting, other processing methods resulted in reduced total phenolic acid and flavonoid content. The observed decrease in sprouted grains was due to the loss of phenolic compounds during soaking. During germination, endogenous phenolic enzymes are activated, leading to their hydrolysis and subsequent loss during soaking. On the other hand, sprouting and roasting positively influenced the bioaccessibility of phenolic compounds in the grain, resulting in a 67% increase after sprouting. The influence of processing methods on the food matrix and the conversion of compounds into more active forms could contribute to the observed increase in phenolic compounds.

9.4.3 Enzymatic and Chemical Treatments

Enzyme treatments can reduce matrix interactions and enhance the bioaccessibility of phytochemicals. Research indicates that polysaccharide-degrading enzymes disrupt the internal structure of cell walls, hence enhancing the extractability of phytochemicals (Ribas-Agustí et al., 2018). In addition, fermentation utilizing hemicellulase (xylanase), β-glucanase, and α-amylase in wheat bran can enhance the bioaccessibility of phenolic acids (Anson et al., 2011). Chandrasekara and Shahidi (2012a) investigated the bioaccessibility of phenolic compounds in five millet species (foxtail, proso, finger, kodo, and pearl) through simulated digestion. Their findings revealed that millet phenolics remain bioaccessibility throughout the gastrointestinal tract, with notable increases during colonic fermentation. Pearl millet demonstrated the highest phenolic bioaccessibility (34.4 μmol FAE/g), whereas proso millet exhibited the strong antioxidant capacity (81.3 μmol TE/g). Results indicated that fermentation is important for releasing bound phenolics from the insoluble fiber portion of millet. The research highlighted significant varietal differences in phenolic profiles and bioaccessibility, indicating that different millet types may provide unique health benefits. These findings highlight millets as significant sources of bioaccessible antioxidants that could help improve gut health and prevent diseases when incorporated into diets.

9.4.4 Soaking, Milling, and Germination

Soaking, milling, and germination are basic household methods that enhance the bioaccessibility and nutritional quality of millets. These processes improve nutrient availability by decreasing antinutritional factors and simultaneously suppressing unfavorable mousy aroma. Soaking and germination significantly enhance mineral bioavailability, and these benefits are linked to phytates that either leach into the soaking water or are partially degraded by the grain's endogenous enzymes during germination. These approaches enhance the release along with absorption of essential nutrients by breaking down phytate-mineral complexes, making millets more nutritionally available while maintaining their beneficial characteristics (Afify et al., 2011).

Sharma et al. (2015) showed that soaking foxtail millet for 46.5 hours significantly increased protein content, due to the synthesis of new amino acids and a reduction in carbohydrate content upon germination. The total dietary fiber content in foxtail millet increased from 22.20% to 27.42% with extended germination time, due to alterations in seeds cell wall polysaccharides which disrupted protein-carbohydrate interactions and affected the cellular structure of intact tissue. These alterations lead to the formation of new cell walls and variations in dietary fiber. Barnyard millet flour prepared from grains soaked for 60 minutes exhibited higher levels of crude fat and protein. Prolonged soaking time (24–72 hours) resulted in a decrease in protein content by 5–20%. On the other hand, roasting at 70 °C resulted in an increase in iron, calcium, sodium, zinc, and phosphorus content. Millets soaked for 60 minutes had high potassium and manganese, whereas those soaked for 120 minutes had lower the amount of total phenolics (5.12 mg/100 g) and tannins (2.21 mg/100 g) than other cultivars (Anbalagan & Nazni, 2020). Germination of pearl, sorghum, and finger millets for 72 hours has been shown to enhance mineral bioavailability and decrease tannin levels (untreated > soaked > germinated) of millets (Tharifkhan et al., 2021). The activation of endogenous enzymes in millets during soaking and germination enhances their nutritive value and bioavailability. Phytates chelate phosphorus and trace elements, including zinc, manganese, iron, and copper, thereby limiting their absorption. Soaking millets activates the native enzyme phytase, which has resulted in a decrease in the amount of phytate content (Shigihalli et al., 2018). Overall, the millet flour which was produced from grains that had been soaked and germinated had an overall rise in protein availability, swelling, and hygroscopicity, while simultaneously showing a decrease in phytic acid and the capability for the least amount of gelation (Kumar et al., 2010).

9.4.5 Fermentation

Fermentation improves the bioavailability of essential nutrients in millets, such as polyphenols, minerals, and dietary fiber, due to the reduction of antinutritional factors including tannins, phytates, and trypsin inhibitors. This process takes place through the natural microflora in the grains or by the incorporation of starter cultures, with lactic acid bacteria (LAB) being especially significant. LAB fermentation acidifies the substrate, preventing harmful microorganisms while simultaneously enhancing the flavor and nutritional profile (Jaybhaye et al., 2014; Thapliyal & Singh, 2015). The combined action of LAB with yeast, such as *Saccharomyces boulardii*, enhances mineral bioavailability and diminishes phytic acid levels, a process attributed to microbial β-glucosidase activity that degrades sugar conjugates (Di Stefano et al., 2017). Fermentation enhances the enzymatic activity of polyphenol oxidase (PPO) and phytases, leading to the reduction in concentration of polyphenols and activity of phytic acid in fermented pearl millet grains (Elyas et al., 2002). At the industry scale, fermentation is a critical process essential for enhancing the nutritional value and market acceptance of millet-based products. According to reports, enabling the endogenous grain microflora to ferment for 24 hours can greatly reduce phytates and polyphenols (Rani et al., 2018). Fermentation significantly enhanced the nutritional value of foxtail millet flour. Fermentation reduced the antinutritional factors like tannin (30%), phytate (30%), while increasing the phenolics (10%). The process also enhanced protein digestibility (from 69% to 72%) and antioxidant activity (from 18% to 26%). The acidic conditions (pH 3.0) further improved the mineral bioavailability, making millet more digestible (Sharma & Sharma, 2022).

9.4.6 Hydrothermal Treatment

Hydrothermal treatment or parboiling involves three steps, soaking, steaming or boiling, and drying the grains, to safe moisture levels (Bora et al., 2019). Hydrothermal treatment or parboiling is partial cooking of millet grains followed by soaking, steaming, and drying. This treatment enables the exodus of nutrients from the bran to the endosperm (Rocha-Villarreal et al., 2018). This treatment immensely affects various parameters, for example, water absorption, swelling behavior, starch crystallization, and viscosity of flour with the improvement in digestibility and availability of nutrients (Prashanth et al., 2024). Kaur et al. (2025) reported hydrothermal treatment of foxtail millet; they found an increase in the total phenolic content, DPPH, and ferric reducing antioxidant power (FRAP). Furthermore, they also estimated the bioaccessibility index of phenolics, which was found to be higher in hydrothermally treated samples (soak-boil [53.06%] and soak-steam [59.56%]) in comparison to the untreated samples (49.35%).

9.5 Conclusion

Millets are a powerhouse of nutrition abundant in bioactive components such as phenolic acids, flavonoids, and essential minerals which offer considerable health advantages. Although their potential has been limited due to the complex interactions between food matrix components and antinutritional factors, which result in poor bioavailability. Processing techniques that involve traditional methods such as fermentation and germination, along with modern approaches like enzymatic treatments, have shown effectiveness in improving bioavailability and nutrient absorption by degrading inhibitors and protecting sensitive compounds during digestion. Understanding these effects helps bridge traditional knowledge with modern nutritional science. Future research should investigate processing methods which enhance the release as well as absorption of bioactive compounds in millet, maximizing their nutritional impact for improved health and food security.

References

Abhilasha, A., Kaur, L., Monro, J., Hardacre, A., & Singh, J. (2022). Effects of hydrothermal treatment and low-temperature storage of whole wheat grains on *in vitro* starch hydrolysis and flour properties. *Food Chemistry*, *395*, 133516.

Afify, A. E. M. M., El-Beltagi, H. S., Abd El-Salam, S. M., & Omran, A. A. (2011). Bioavailability of iron, zinc, phytate and phytase activity during soaking and germination of white sorghum varieties. *PLoS One*, *6*(10), e25512.

Akanbi, T. O., & Barrow, C. J. (2018). Lipase-produced hydroxytyrosyl eicosapentaenoate is an excellent antioxidant for the stabilization of omega-3 bulk oils, emulsions and microcapsules. *Molecules*, *23*(2), 275.

Alminger, M., Aura, A. M., Bohn, T., Dufour, C., El, S. N., Gomes, A., Karakaya, M. C., Cesta, M., McDougall, G. J., Requena, T., & Santos, C. N. (2014). In vitro models for studying secondary plant metabolite digestion and bio accessibility. *Comprehensive Reviews in Food Science and Food Safety*, *13*(4), 413–436.

Anbalagan, S., & Nazni, P. (2020). Effect of processing techniques on nutritional, viscosity and osmolarity of barnyard millet based diarrheal replacement fluids. *Current Research in Nutrition and Food Science*, *8*(1), 164–173.

Anson, N. M., Aura, A. M., Selinheimo, E., Mattila, I., Poutanen, K., Van Den Berg, R., Havennar, R., Bast, A., & Haenen, G. R. (2011). Bioprocessing of wheat bran in whole wheat bread increases the bioavailability of phenolic acids in men and exerts anti-inflammatory effects ex vivo. *The Journal of Nutrition*, *141*(1), 137–143.

Arts, I. C., & Hollman, P. C. (2005). Polyphenol and disease risk in epidemiologic studies. *The American Journal of Clinical Nutrition*, *81*(1), 317S–325S.

Babypriyanka, S., Hepziba, J. S., Pushpam, K. A., Pillai, A. M., Vijayalakshmi, R., & Theradimani, M. (2024). Exploring the bioactive components of millets for their nutraceutical potential. *Food Science and Biotechnology*, *34*(3), 563–575.

Bora, P., Ragaee, S., & Marcone, M. (2019). Characterisation of several types of millets as functional food ingredients. *International Journal of Food Sciences and Nutrition*, *70*(6), 714–724.

Călinoiu, L. F., & Vodnar, D. C. (2018). Whole grains and phenolic acids: A review on bioactivity, functionality, health benefits and bioavailability. *Nutrients*, *10*(11), 1615.

Chandrasekara, A., & Shahidi, F. (2010). Content of insoluble bound phenolics in millets and their contribution to antioxidant capacity. *Journal of Agricultural and Food Chemistry*, *58*(11), 6706–6714.

Chandrasekara, A., & Shahidi, F. (2012a). Bio accessibility and antioxidant potential of millet grain phenolics as affected by simulated in vitro digestion and microbial fermentation. *Journal of Functional Foods*, *4*(1), 226–237.

Chandrasekara, A., & Shahidi, F. (2012b). Antioxidant phenolics of millet control lipid peroxidation in human LDL cholesterol and food systems. *Journal of the American Oil Chemists' Society*, *89*, 275–285.

Chauhan, M., Sonawane, S. K., & Arya, S. S. (2018). Nutritional and nutraceutical properties of millets: A review. *Clinical Journal of Nutrition and Dietetics*, *1*(1), 1–10.

Chethan, S., Dharmesh, S. M., & Malleshi, N. G. (2008a). Inhibition of aldose reductase from cataracted eye lenses by finger millet (*Eleusine coracana*) polyphenols. *Bioorganic & Medicinal Chemistry*, *16*(23), 10085–10090.

Chethan, S., & Malleshi, N. G. (2007). Finger millet polyphenols: Optimization of extraction and the effect of pH on their stability. *Food Chemistry*, *105*(2), 862–870.

Chethan, S., Sreerama, Y. N., & Malleshi, N. G. (2008b). Mode of inhibition of finger millet malt amylases by the millet phenolics. *Food Chemistry*, *111*(1), 187–191.

D'Archivio, M., Filesi, C., Varì, R., Scazzocchio, B., & Masella, R. (2010). Bioavailability of polyphenols: Status and controversies. *International Journal of Molecular Sciences*, *11*(4), 1321–1342.

Dayakar Rao, B., Ananthan, R., Hariprasanna, K., Bhatt, K. V., Rajeswari, S. S., & Tonapi, V. A. (2018). *Nutritional and health benefits of nutri cereals* (pp. 96). Nutri hub TBI, ICAR_Indian Institute of Millets Research (IIMR) Rajendranagar, Hyderabad.

Del Rio, D., Borges, G., & Crozier, A. (2010). Berry flavonoids and phenolics: bioavailability and evidence of protective effects. *British journal of nutrition*, *104*(S3), S67–S90.

Den Besten, G., Van Eunen, K., Groen, A. K., Venema, K., Reijngoud, D. J., & Bakker, B. M. (2013). The role of short-chain fatty acids in the interplay between diet, gut microbiota, and host energy metabolism. *Journal of Lipid Research*, *54*(9), 2325–2340.

Devi, P. B., Vijayabharathi, R., Sathyabama, S., Malleshi, N. G., & Priyadarisini, V. B. (2014). Health benefits of finger millet (*Eleusine coracana* L.) polyphenols and dietary fiber: A review. *Journal of Food Science and Technology*, *51*, 1021–1040.

Di Stefano, E., White, J., Seney, S., Hekmat, S., McDowell, T., Sumarah, M., & Reid, G. (2017). A novel millet-based probiotic fermented food for the developing world. *Nutrients*, *9*(5), 529.

Elyas, S. H., El Tinay, A. H., Yousif, N. E., & Elsheikh, E. A. (2002). Effect of natural fermentation on nutritive value and in vitro protein digestibility of pearl millet. *Food Chemistry*, *78*(1), 75–79.

Fang, J. (2014). Bioavailability of anthocyanins. *Drug Metabolism Reviews*, *46*(4), 508–520.

Fernando, W. M. A. D. B., Flint, S., Zou, M., Brennan, C. S., Ranaweera, K., & Bamunuarachchi, A. (2011). The effect of rice fibre fractions on the growth of co-cultures of probiotics. *Journal of Food Science and Technology*, *48*, 14–25.

Geetha, K., Yankanchi, G. M., Hulamani, S., & Hiremath, N. (2020). Glycemic index of millet-based food mix and its effect on pre diabetic subjects. *Journal of Food Science and Technology*, *57*, 2732–2738.

Golding, M., & Wooster, T. J. (2010). The influence of emulsion structure and stability on lipid digestion. *Current Opinion in Colloid & Interface Science*, *15*(1–2), 90–101.

Gong, L., Cao, W., Chi, H., Wang, J., Zhang, H., Liu, J., & Sun, B. (2018). Whole cereal grains and potential health effects: Involvement of the gut microbiota. *Food Research International*, *103*, 84–102.

Goudar, G., Manne, M., Sathisha, G. J., Sharma, P., Mokalla, T. R., Kumar, S. B., & Ziouzenkova, O. (2023). Phenolic, nutritional and molecular interaction study among different millet varieties. *Food Chemistry Advances*, *2*, 100150.

Gowda, N. N., Siliveru, K., Prasad, P. V., Bhatt, Y., Netravati, B. P., & Gurikar, C. (2022). Modern processing of Indian millets: A perspective on changes in nutritional properties. *Food*, *11*(4), 499.

Harish, M. S., Bhuker, A., & Chauhan, B. S. (2024). Millet production, challenges, and opportunities in the Asia-Pacific region: A comprehensive review. *Frontiers in Sustainable Food Systems*, *8*, 1386469.

Hassan, Z. M., Sebola, N. A., & Mabelebele, M. (2020). Assessment of the phenolic compounds of pearl and finger millets obtained from South Africa and Zimbabwe. *Food Science & Nutrition*, *8*(9), 4888–4896.

Hegde, P. S., Chandrakasan, G., & Chandra, T. S. (2002). Inhibition of collagen glycation and crosslinking in vitro by methanolic extracts of Finger millet (*Eleusine coracana*) and Kodo millet (*Paspalum scrobiculatum*). *The Journal of Nutritional Biochemistry*, *13*(9), 517–521.

Hidalgo, F. J., & Zamora, R. (2017). Food processing antioxidants. *Advances in Food and Nutrition Research*, *81*, 31–64.

Hithamani, G., & Srinivasan, K. (2014). Effect of domestic processing on the polyphenol content and bio accessibility in finger millet (*Eleusine coracana*) and pearl millet (*Pennisetum glaucum*). *Food Chemistry*, *164*, 55–62.

Hoda, M., Hemaiswarya, S., Doble, M., Hoda, M., Hemaiswarya, S., & Doble, M. (2019). Trends in research and development of phenolic phytochemicals as potential antidiabetic therapeutics. In M. Hoda, S. Hemaiswarya, & M. Doble (Eds.), *Role of phenolic phytochemicals in diabetes management: Phenolic phytochemicals and diabetes* (pp. 175–184). Springer.

Holland, B., Agyei, D., Akanbi, T. O., Wang, B., & Barrow, C. J. (2017). Bioprocessing of plant-derived bioactive phenolic compounds. In A. M. Grumezescu & A. M. Holban (Eds.), *Food biosynthesis* (pp. 135–181). Academic Press.

Hornero-Méndez, D., & Mínguez-Mosquera, M. I. (2007). Bio accessibility of carotenes from carrots: Effect of cooking and addition of oil. *Innovative Food Science & Emerging Technologies*, *8*(3), 407–412.

Hostetler, G. L., Ralston, R. A., & Schwartz, S. J. (2017). Flavones: Food sources, bioavailability, metabolism, and bioactivity. *Advances in Nutrition*, *8*(3), 423–435.

Hotz, C., & Gibson, R. S. (2007). Traditional food-processing and preparation practices to enhance the bioavailability of micronutrients in plant-based diets. *The Journal of Nutrition*, *137*(4), 1097–1100.

Jacob, J., Krishnan, V., Antony, C., Bhavyasri, M., Aruna, C., Mishra, K., Nepolean, T., Satyavathi, C. T., & Visarada, K. B. (2024). The nutrition and therapeutic potential of millets: An updated narrative review. *Frontiers in Nutrition*, *11*, 1346869.

Jaybhaye, R. V., Pardeshi, I. L., Vengaiah, P. C., & Srivastav, P. P. (2014). Processing and technology for millet-based food products: A review. *Journal of Ready to Eat Food*, *1*(2), 32–48.

Joshi, S., & Srivastava, S. (2020). Comparative analysis of nutritional value, glycemic index and sensory quality of barnyard millet and rice varieties and their products. *International Journal of Current Microbiology and Applied Sciences*, *9*, 2044–2056.

Kaur, P., Purewal, S. S., Sandhu, K. S., Kaur, M., & Salar, R. K. (2019). Millets: A cereal grain with potent antioxidants and health benefits. *Journal of Food Measurement and Characterization*, *13*, 793–806.

Kaur, T., Panesar, P. S., & Riar, C. S. (2025). Effect of controlled hydrothermal treatment, drying temperature, de-husking and decortication on phenolics characteristics and functional properties of foxtail millet flour. *Journal of Cereal Science*, *123*, 104170.

Khan, T., Azad, A. A., & Ul Islam, R. (2025). Millets: A comprehensive review of nutritional, antinutritional, health, and processing aspects. *Journal of Food Composition and Analysis*, *141*, 107364.

Khare, P., Maurya, R., Bhatia, R., Mangal, P., Singh, J., Podili, K., Bishnoi, M., & Kondepudi, K. K. (2020). Polyphenol rich extracts of finger millet and kodo millet ameliorate high fat diet-induced metabolic alterations. *Food & Function*, *11*(11), 9833–9847.

Kumar, V., Sinha, A. K., Makkar, H. P., & Becker, K. (2010). Dietary roles of phytate and phytase in human nutrition: A review. *Food Chemistry*, *120*(4), 945–959.

Kumari, A., Sadh, P. K., Kamboj, A., Yadav, B., Kumar, A., Sivakumar, S., Surekha, Saharan, B. S., Brar, B., Goyal, C., Dhull, S. B., & Duhan, J. S. (2025). Exploring the benefits of nutrition of little millet: Unveiling the effect of processing methods on bioactive properties. *Journal of Food Biochemistry*, *2025*(1), 2488816.

Kumari, D., Madhujith, T., & Chandrasekara, A. (2017). Comparison of phenolic content and antioxidant activities of millet varieties grown in different locations in Sri Lanka. *Food Science & Nutrition*, *5*(3), 474–485.

Liang, K., Liang, S., & Zhu, H. (2024). The physicochemical characteristics and phenolic bio accessibility of defatted millet bran powder prepared using superfine grinding. *LWT*, *201*, 116173.

Mawouma, S., Condurache, N. N., Turturică, M., Constantin, O. E., Croitoru, C., & Rapeanu, G. (2022). Chemical composition and antioxidant profile of sorghum (*Sorghum bicolor* (L.) moench) and pearl millet (*Pennisetum glaucum* (L.) R. Br.) grains cultivated in the far-North region of Cameroon. *Foods*, *11*(14), 2026.

McClements, D. J. (2015). Encapsulation, protection, and release of hydrophilic active components: Potential and limitations of colloidal delivery systems. *Advances in Colloid and Interface Science*, *219*, 27–53.

Meenu, M., & Xu, B. (2019). A critical review on anti-diabetic and anti-obesity effects of dietary resistant starch. *Critical Reviews in Food Science and Nutrition*, *59*(18), 3019–3031.

Neilson, A. P., & Ferruzzi, M. G. (2011). Influence of formulation and processing on absorption and metabolism of flavan-3-ols from tea and cocoa. *Annual Review of Food Science and Technology*, *2*(1), 125–151.

Ofosu, F. K., Elahi, F., Daliri, E. B.-M., Chelliah, R., Ham, H. J., Kim, J.-H., Han, S.-I., Hur, J. H., & Oh, D.-H. (2020). Phenolic profile, antioxidant, and antidiabetic potential exerted by millet grain varieties. *Antioxidants*, *9*(3), 254.

Pradeep, P. M., & Sreerama, Y. N. (2017). Soluble and bound phenolics of two different millet genera and their milled fractions: Comparative evaluation of antioxidant properties and inhibitory effects on starch hydrolysing enzyme activities. *Journal of Functional Foods*, *35*, 682–693.

Papoutsis, K., Zhang, J., Bowyer, M. C., Brunton, N., Gibney, E. R., & Lyng, J. (2021). Fruit, vegetables, and mushrooms for the preparation of extracts with α-amylase and α-glucosidase inhibition properties: A review. *Food Chemistry*, *338*, 128119.

Perin, D., & Murano, E. (2017). Starch polysaccharides in the human diet: Effect of the different source and processing on its absorption. *Natural Product Communications*, *12*(6), 1934578X1701200606.

Pradeep, P. M., & Sreerama, Y. N. (2015). Impact of processing on the phenolic profiles of small millets: Evaluation of their antioxidant and enzyme inhibitory properties associated with hyperglycemia. *Food Chemistry*, *169*, 455–463.

Pradeep, P. M., & Sreerama, Y. N. (2018). Phenolic antioxidants of foxtail and little millet cultivars and their inhibitory effects on α-amylase and α-glucosidase activities. *Food Chemistry*, *247*, 46–55.

Prashanth, P., Joshi, T. J., Singh, S. M., & Rao, P. S. (2024). Impact of hydrothermal treatment on functional properties of pearl millet flour: Process modelling and optimisation. *Journal of Food Measurement and Characterization*, *18*(9), 7627–7640.

Priya, K. L., Shobana, S., Sudha, V., Gayathri, R., Beatrice, D. A., Anjana, R. M., Kamala, K., & Mohan, V. (2024). Consumption pattern of millets among south Indian adults. *Journal of Diabetology*, *15*(1), 63–69.

Rani, S., Singh, R., Sehrawat, R., Kaur, B. P., & Upadhyay, A. (2018). Pearl millet processing: A review. *Nutrition & Food Science*, *48*(1), 30–44.

Rein, M. J., Renouf, M., Cruz-Hernandez, C., Actis-Goretta, L., Thakkar, S. K., & da Silva Pinto, M. (2013). Bioavailability of bioactive food compounds: A challenging journey to bioefficacy. *British Journal of Clinical Pharmacology*, *75*(3), 588–602.

Ribas-Agustí, A., Martín-Belloso, O., Soliva-Fortuny, R., & Elez-Martínez, P. (2018). Food processing strategies to enhance phenolic compounds bio accessibility and bioavailability in plant-based foods. *Critical Reviews in Food Science and Nutrition*, *58*(15), 2531–2548.

Rocha-Villarreal, V., Hoffmann, J. F., Vanier, N. L., Serna-Saldivar, S. O., & García-Lara, S. (2018). Hydrothermal treatment of maize: Changes in physical, chemical, and functional properties. *Food Chemistry*, *263*, 225–231.

Rodríguez, J. P., Rahman, H., Thushar, S., & Singh, R. K. (2020). Healthy and resilient cereals and pseudo-cereals for marginal agriculture: Molecular advances for improving nutrient bioavailability. *Frontiers in Genetics*, *11*, 49.

Saleh, A. S., Zhang, Q., Chen, J., & Shen, Q. (2013). Millet grains: Nutritional quality, processing, and potential health benefits. *Comprehensive Reviews in Food Science and Food Safety*, *12*(3), 281–295.

Serna-Saldivar, S. O., & Espinosa-Ramírez, J. (2019). Grain structure and grain chemical composition. In J. R. N. Taylor & K. G. Duodu (Eds.), *Sorghum and millets* (pp. 85–129). AACC International Press.

Shahidi, F., & Chandrasekara, A. (2013). Millet grain phenolics and their role in disease risk reduction and health promotion: A review. *Journal of Functional Foods*, *5*(2), 570–581.

Shahidi, F., & Pan, Y. (2022). Influence of food matrix and food processing on the chemical interaction and bio accessibility of dietary phytochemicals: A review. *Critical Reviews in Food Science and Nutrition*, *62*(23), 6421–6445.

Sharma, B., & Gujral, H. S. (2019). Modulation in quality attributes of dough and starch digestibility of unleavened flat bread on replacing wheat flour with different minor millet flours. *International Journal of Biological Macromolecules, 141*, 117–124.

Sharma, N., Sahu, J. K., Bansal, V., Esua, O. J., Rana, S., Bhardwaj, A., Bangu, S. P., & Adedeji, A. A. (2023). Trends in millet and pseudomillet proteins – Characterization, processing and food applications. *Food Research International, 164*, 112310.

Sharma, R., & Sharma, S. (2022). Anti-nutrient & bioactive profile, in vitro nutrient digestibility, techno-functionality, molecular and structural interactions of foxtail millet (*Setaria italica* L.) as influenced by biological processing techniques. *Food Chemistry, 368*, 130815.

Sharma, S., Saxena, D. C., & Riar, C. S. (2015). Antioxidant activity, total phenolics, flavonoids and antinutritional characteristics of germinated foxtail millet (*Setaria italica*). *Cogent Food & Agriculture, 1*(1), 1081728.

Shigihalli, S., Ravindra, U., & Ravishankar, P. (2018). Effect of processing methods on phytic acid content in selected white finger millet varieties. *International Journal of Current Microbiology and Applied Sciences, 7*(2), 1829–1835.

Singh, R. K., Muthamilarasan, M., & Prasad, M. (2021). Biotechnological approaches to dissect climate-resilient traits in millets and their application in crop improvement. *Journal of Biotechnology, 327*, 64–73.

Srivastava, K., & Sharma, A. K. (2012). Nutraceutical importance of finger millet (*Eleusine coracana*) for improved human health. *European Journal of Plant Science Biotechnology, 6*, 91–95.

Tagliazucchi, D., Verzelloni, E., & Conte, A. (2012). The first tract of alimentary canal as an extractor. Release of phytochemicals from solid food matrices during simulated digestion. *Journal of Food Biochemistry, 36*(5), 555–568.

Taylor, J., & Awika, J. (2017). *Gluten-free ancient grains: Cereals, pseudocereals, and legumes: Sustainable, Nutritious, and health-promoting foods for the 21st century*. Woodhead publishing.

Thapliyal, V., & Singh, K. (2015). Finger millet: Potential millet for food security and powerhouse of nutrients. *International or Research in Agriculture and Forestry, 2*(2), 22–33.

Tharifkhan, S. A., Perumal, A. B., Elumalai, A., Moses, J. A., & Anandharamakrishnan, C. (2021). Improvement of nutrient bioavailability in millets: Emphasis on the application of enzymes. *Journal of the Science of Food and Agriculture, 101*(12), 4869–4878.

Thippeswamy, T. G., Junna, L., & Shinde, M. (2016). Proximate composition, resistant starch and other phytochemical constituents of native finger millet cultivar. *International Journal of Food and Nutrition Science, 5*(3), 67–79.

Tomar, M., Bhardwaj, R., Verma, R., Singh, S. P., Krishnan, V., Kansal, R., Yadav, V. K., Praveen, S., & Sachdev, A. (2022). Interactome of millet-based food matrices: A review. *Food Chemistry, 385*, 132636.

Udeh, H. O., Duodu, K. G., & JIdeAnI, A. I. O. (2017). Finger millet bioactive compounds, bio accessibility, and potential health effects—A review. *Czech Journal of Food Sciences, 35*(1), 7–17.

Van Craeyveld, V., Holopainen, U., Selinheimo, E., Poutanen, K., Delcour, J. A., & Courtin, C. M. (2009). Extensive dry ball milling of wheat and rye bran leads to in situ production of arabinoxylan oligosaccharides through nanoscale fragmentation. *Journal of Agricultural and Food Chemistry*, *57*(18), 8467–8473.

Xiang, J., Zhang, M., Apea-Bah, F. B., & Beta, T. (2019). Hydroxycinnamic acid amide (HCAA) derivatives, flavonoid C-glycosides, phenolic acids and antioxidant properties of foxtail millet. *Food Chemistry*, *295*, 214–223.

Zhu, Y., Chu, J., Lu, Z., Lv, F., Bie, X., Zhang, C., & Zhao, H. (2018). Physicochemical and functional properties of dietary fiber from foxtail millet (*Setaria italic*) bran. *Journal of Cereal Science*, *79*, 456–461.

Section III

Millet Processing and Applications

10

Emerging Technologies in Millet Processing

Shiva Bakshi[1], *Durga S. Bunkar*[1], *Vinod K. Paswan*[1], *Suneel K. Goyal*[2], *and Vishal Kumar*[1]

[1] *Department of Dairy Science and Food Technology, Institute of Agricultural Sciences, Banaras Hindu University, Varanasi, India*
[2] *Department of Agricultural Engineering, Institute of Agricultural Sciences, Banaras Hindu University, Varanasi, India*

10.1 Introduction

Currently, various global challenges are contributing to a decline in farming efficiency. Key factors include environmental shifts, economic inflation, water scarcity, pollution, and social and economic challenges. Specifically, ecological and economic shortcomings impede agricultural progress in many regions, reducing grain production. This decline in production exacerbates food price increases and contributes to extensive food scarcity across the globe (Kumar et al., 2018). Additionally, climatic shifts pose significant challenges for local farmers due to limited materials (Huang et al., 2016). Developing nations rely heavily on cereals such as wheat, rice, corn, barley, oats, and rye to fulfill their dietary needs, more so than developed countries. However, as nutritional demands increase, there is a pressing need for alternative, nutrient-rich grains (Yousaf et al., 2021). Millet exhibits significant potential to serve as a nutritious alternative to meet the dietary needs of a growing global population (Kumar et al., 2018).

Millets, including pearl millet (*Pennisetum glaucum*), proso millet (*Panicum miliaceum*), barnyard millet (*Echinochloa frumentacea*), brown top millet (*Brachiaria ramosa*), finger millet (*Eleusine coracana*), foxtail millet (*Setaria italica*), little millet (*Panicum sumatrense*), sorghum (*Sorghum bicolor*), and khasi millet (*Digitaria cruciata*), are grown in diverse regions around the world. In 2021, global millet production was 30.1 million tons, with India accounting for 41% of the output. Around 97% of the world's millet is produced and consumed by developing African and Asian nations (Joshi, 2025). Millets are a valuable resource of essential nutrients, including proteins, energy, fiber, and micronutrients. They are also rich in phytonutrients and bioactive substances, such as dietary fiber, phenols, flavonoids, and plant sterols (Suri et al., 2022). Millet protein is considered a superior

Nutraceutical Potential of Millet-Based Food Products: Chemistry and Applications, First Edition.
Edited by Parmjit S. Panesar, Charanjit S. Riar, and Sushil Dhital.

Published 2026 by John Wiley & Sons Ltd.

source of essential amino acids, such as methionine, compared to many widely consumed staple cereals. Additionally, consuming millet can reduce the likelihood of several degenerative diseases. In addition to its cholesterol-lowering effects, millet provides several additional health benefits, including antioxidant, antiulcer, hypoglycemic, and anti-inflammatory properties (Schoenlechner et al., 2013).

Millets retain longer shelf life when stored in their whole form rather than ground into flour. Traditional storage methods such as underground pits, mud rhombuses, aerial storage in knotted bundles, and earthen bins effectively preserve whole millets for four to five years by reducing insect infestations and limiting moisture. However, milling disrupts the seed structure, significantly decreasing the shelf life of millet flour and making it more vulnerable to environmental conditions and biological degradation. The primary factors affecting the stability of millet flour are microbial contamination and lipase enzyme activity. Inadequate storage conditions can promote microbial growth, leading to food spoilage (Sruthi & Rao, 2021). Lipase-mediated lipid breakdown in flour results in rancidity and undesirable odors, making it unsuitable for consumption. Research indicates that pearl millet flour, with relatively high-fat content, starts deteriorating within 10–15 days under standard storage conditions (Aher et al., 2022). Various processing techniques, such as mechanical processing, heat treatments, and preservatives, are employed to extend the shelf life of millet flour.

Primary processing means eliminating unwanted seed coats containing antinutritional factors through soaking, wetting, blanching, dehulling, milling, decortication, and roasting. Moreover, malting, soaking, and fermentation enhance nutrient bioavailability while improving storage stability (Gaikwad et al., 2021). In contrast, secondary processing transforms these preprocessed materials into ready-to-eat or ready-to-cook food products using parboiling, extrusion, popping, flaking, and baking (Gowda et al., 2022). Further, various novel thermal and nonthermal secondary processing technologies are used today for millet processing and value addition. These include thermal (infrared heating, microwave, ohmic heating) and nonthermal methods (gamma irradiation, pulsed electric field [PEF], ultrasonication, ozone processing, ultraviolet (UV)-radiations, pulsed light, high-pressure processing [HPP] and cold plasma). They are better than traditional processing methods because of the lower exposure and processing time (Sruthi & Rao, 2021). Recent research suggests that nonthermal processing techniques also exert little influence on the protein content, nutrient profile, flavor, texture, and other functional characteristics of millets (Ramakrishnan et al., 2023). These environmentally friendly techniques contribute to sustainability by decreasing wastewater generation, reducing the need for organic solvents, and shortening processing time (Lohita & Srijaya, 2024).

10.2 Traditional Millet Processing Meets Modern Equipment Innovations

Different techniques and treatments have been used for millet grains at the household level since ancient times to make them suitable for human consumption. These operations include soaking, heating, roasting, fermentation, cooking, etc. Nowadays, various unit operations have been standardized and optimized independently for each type of millet

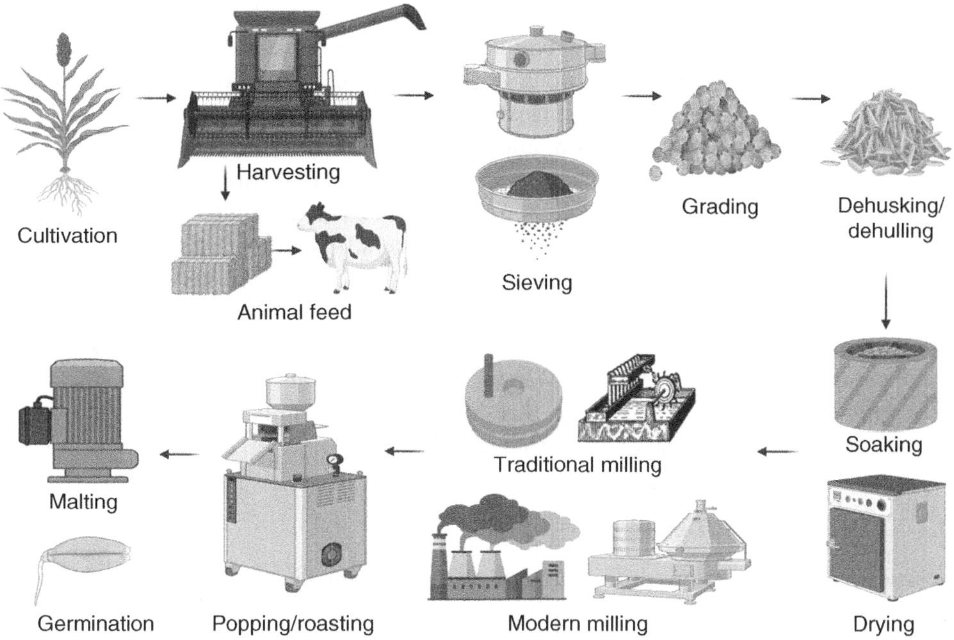

Figure 10.1 Primary processing of millets from cultivation to milling and value addition

(Figure 10.1). All these unit operations, like soaking, dehulling, grinding, roasting, puffing, fermentation, malting, etc., with appropriately designed processing equipment and machines, are well-established in industries for commercial-scale processing of millets. Another innovative agricultural advancement is the automatic sowing machine, which offers an efficient and modern approach to seed planting. The machine has several key components, including a seedbox, frame, furrow openers, guided pipes, ground wheel, and an engine. These components are meticulously designed and assembled to ensure that seeds are sown systematically with uniform spacing, ultimately leading to higher crop yields. Additionally, the machine is cost-effective and serves multiple functions beyond sowing, such as seed removal and land leveling (Yadav & Dhiman, 2023).

10.2.1 Cleaning

Winnowing is a process that separates grains from impurities such as sand, stones, chaff, broken seeds, and other debris through shaking or wind. Manual winnowing by shaking requires significant labor and time, with studies indicating that women can clean up to 100 kg of grains per hour using this method (Song et al., 2013). Mechanization, including using windmills, fans, or aspirators, significantly enhances efficiency, enabling the cleaning of 400–500 kg of grains per hour (Beta & Ndolo, 2019). Kaleemullah et al. (2023) designed a winnower incorporating an air-blast fan and a reciprocating sieve, achieving 91% cleaning efficiency with only 0.75% grain loss at a feed rate of 3.5 kg/s for millet grains. Additionally, a destoner-cum-grader-cum-aspirator was developed, demonstrating high cleaning efficiency, recorded at 93.81% for brown top, 95.66% for foxtail, 95.74% for barnyard, and 96.56% for little millet (Kaleemullah et al., 2023). Rajaiah et al. (2025) recently evaluated a Central Institute of Agricultural Engineering (CIAE)-developed motorized thresher for different millets. Key performance metrics included threshing efficiency, 92.17% for finger, 97.52% for barnyard, and 97.94% for foxtail millet, while cleaning efficiencies were 92.5%, 94.0%, and 99.0%, respectively. An axial-flow threshing and separating device with a front-end rasp bar and rear-end nail tooth structure addressed low threshing efficiency and high losses in millet combine harvesters. A mathematical model showed that the rasp bar's threshing capacity is 1.206 times higher than the nail tooth, while its separation capacity is 0.8532 times lower. Discrete element simulations confirmed that combination rollers enhance both separation and threshing efficiency. The optimized operational parameters—roller speed of 865 r/min, threshing clearance of 7 mm, and strip spacing of 8 mm—achieved a 90.65% clean threshing rate, 8.74% millet petal rate, and 0.61% loss rate, meeting technical standards (Zhang et al., 2022a).

10.2.2 Decortication/Dehulling

Decortication is a meticulous process that involves removing the outer layer of millet seeds while preserving the inner portion. During this process, millet seeds lose about 20–22% of their weight, significantly improving their nutritional quality and market value (Krishnan et al., 2012). Traditional decortication methods, such as wooden mortar and pestle, are labor-intensive and less efficient. They have lower costs and help retain natural fiber, making them valuable in resource-limited areas (Latha Ravi & Rana, 2024). Modern mechanized decortication improves efficiency, consistency, and throughput, making it suitable for large-scale processing despite higher costs (Babiker et al., 2018). It enhances nutrient retention and reduces antinutrients like tannins and phytic acid, improving protein digestibility and mineral bioavailability (Joshi & Rao, 2024). Dehulling is the removal process of the hull (inedible portion) in millets. A double-stage foxtail millet dehuller was designed and tested, featuring a hopper, knurled and rubber rolls, an air blower, and a grader. Performance evaluation considered grain moisture (10–17% wet basis), knurled roll types, and feed rates (100–300 kg/h). Optimal dehulling efficiency (82.24%) and cleaning efficiency (82.47%) were achieved at 12% moisture, a diamond-knurled roll, and a 100 kg/h feed rate. The highest head grain yield (97.82%) with minimal breakage (2.18%) occurred at 12% moisture and 300 kg/h feed rate. Economic analysis confirms its suitability for small-scale millet processing (Chikkaballapur Krishnappa et al., 2022).

A centrifugal dehulling system was developed to overcome the challenges associated with stone grinding, yielding a higher post-dehulling recovery rate (72–76%) than traditional methods (Durairaj et al., 2019). Gowda et al. (2020) designed a tabletop centrifugal dehuller specifically for minor millets, utilizing a combination of centrifugal and impact centrifugal forces for dehulling. This study demonstrated dehulling efficiencies of 80.90% for kodo millet, 92.70% for foxtail millet, 93.50% for proso millet, 91.00% for little millet, and 74.40% for barnyard millet, depending on variations in impeller speed and moisture content. They also noted that higher impeller speeds increased the percentage of broken grains. Similarly, in a recent work, Singh and Rao (2025) examined the dehulling of brown top millet using impact force. Increasing speed from 1,100 to 1,400 rpm improved dehulling efficiency (72.67–94.87%) but reduced head yield (57.24–38.54%). Optimized conditions yielded 76.98% dehulling efficiency and 70.89% milling efficiency. The process also reduced tannins and phytates, confirming its effectiveness. Table 10.1 presents an overview of millet dehulling machines, highlighting their mechanisms, design features, and processing capacities.

Decortication reduces phytic acid and increases the total mineral contents in finger millets (Krishnan et al., 2012). Decortication is generally carried out in three ways: abrasive, roller mill, and CIAE millet mill. An abrasive decorticator is most employed for removing bran and husk from the millet surface in powder or flaky form. However, these are not very successful as they remove different grain layers because of the uniform nature of the millet surface, leading to endosperm loss (Lochte-Watson et al., 2000). In a roller mill, decortication occurs as steel rollers rotate over the grain, rubbing off the pericarp through grain-to-grain and grain-to-roller friction. The grains pass through a series of steel rollers operating at equal or varying speeds in concurrent or counter-current directions (Balasubramanian and Viswanathan, 2010). ICAR-CIAE Bhopal has developed an energy-efficient and ecofriendly millet processing machine, the CIAE-Millet Mill (Models I and II), for sustainable postharvest processing. It operates with a one-horsepower motor and processes 100–110 kg of grain per hour at 10–12% moisture content. The mill efficiently de-husks various minor millets with 95% efficiency while simultaneously collecting husks, minimizing environmental impact. Designed for domestic and industrial use, it requires no specialized labor and produces minimal noise or air pollution. Though decortication affects the nutritional composition, it enhances edibility, sensory quality, and appearance, highlighting the need for innovative high-speed grain decortication methods (Bisht, 2022). Dias-Martins et al. (2024) employed a mechanical abrasion rice de-husking machine (model MT-97, no. 3788-5, Suzuki, Brazil) for the decortication of whole pearl millet grains and achieved a reasonable extraction rate of 87.8%.

10.2.3 Milling

Millets need to undergo dehulling or milling to produce both fine (low-fiber) and coarse (high-fiber) flours, making them suitable for various food applications (Birania et al., 2020). This process involves mechanical grinding, which disrupts the seed structure by breaking the cell wall, making it more susceptible to enzymatic activity, microbial degradation, and environmental factors such as moisture and gases. Ball milling is generally classified into wet ball milling (WBM) and dry ball milling (DBM), depending on the type of grinding media utilized (Bangar et al., 2023). DBM is recognized as a quick and energy-efficient technique. However, it causes intense mechanical stress, leading to significant structural damage and an increase in damaged starch content. Consequently, flour processed through this method demonstrates lower gelatinization temperature and enhanced hydration properties (de Oliveira Barros et al., 2023). On the other hand, WBM involves the addition of water, which effectively reduces particle friction, thus producing finer and more uniform particles with minimized damaged starch (Ge et al., 2023). Wang et al. (2025) investigated the effects of WBM and DBM on the physicochemical properties of foxtail millet flour. WBM (6 : 5 water-to-flour ratio, 300 r/min, 1–5 hours, 40 °C drying) reduced particle size, increased specific surface area, higher whiteness, and improved thermal stability with

Table 10.1 Millet dehulling machines with mechanism, design, and processing capacity

Machine name	Millets processed	Dehulling mechanism	Power (hp)	Capacity (kg/h)	Overall dimensions	Cleaning mechanism	References
Abrasive polisher-mill ($486.47)	Coarse cereals + multi-millets	Force: Abrasive (A-24-grade-vitrified bond carborundum-grinding wheels), number: 10, spacing: 18 mm, diameter: 300 mm, and thickness: 25 mm, with the horizontal axis of rotation	5	96–340	–	The centrifugal blower has forward curved impeller blades, an airflow rate of 15 m^3/min, and a horizontal axis of rotation	Sahay (1990)
CIAE-Millet Mill	Kodo, foxtail, little, barnyard, and proso	Force: Abrasive (A-24-super life V272 features), number: 2 (one stationary and other rotating), 300 × 40 × 38.1 mm, speed: 960 rpm with the vertical axis of rotation	1 single phase, 1440 rpm	100	860 × 842 × 1,460	The blower extended to the cyclone separator, speed: 1920 rpm with the vertical axis of rotation	Balasubramanian et al. (2020)
Barnyard millet-dehuller	Barnyard	Force: Combined shear and impact generating friction between the canvas strip and millet. Dehulling impeller length: 396 mm and diameter: 330 mm	5	40–50	–	Blower (surface area of fins: 56,000 mm^2, length and width of fins: 467 and 120 mm)	Singh et al. (2011)
Electric proso millet-dehusking machine (chino kutak) (Model 1)	Proso	Dehusking drum of cast iron with stoppers (flat belt material) on the periphery, speed: 460 rpm	1420 rpm, 1.5 single phase	33.5	Height: 1,160	2 Round hole screens, oscillating speed: 720 rpm, slope: 30°	Bhandari et al. (2020)

Electric proso millet-dehusking machine (Model 2)	Proso	Impact force (double-shell drum having inner rubber-cylinder), diameter: 300 mm, speed: 2300 rpm	2 single phase, 1440 rpm	52.5	650 × 300 × 650	Metallic screen between the two shells.	Bhandari et al. (2021)
Finger millet-thresher-cum-pearler	Finger	Force: Abrasive and Impact (drum with canvas strips and 8 MS strips), diameter: 200 mm, length: 180 mm, sieve size for pearling: 7.25 m/s	2 single phases	32	1,159 × 710 × 1,250	Sieve walker (250 × 375 mm) and number of blower blades: 4, diameter: 180 mm, length: 180 mm	Powar et al. (2019)
Batch-Compact foxtail millet-deshelling machine($226.49)	Foxtail	Force: Abrasive (disc diameter: 230 mm, thickness: 6 mm, distance between the disks: 18 mm)	1500 rpm, 0.5 single phase	64	700 × 320 × 450	Blower	Ullegaddi et al. (2020)
Minor millet-dehusker	Little and Kodo	Abrasive force (rotating cone with steel wire mesh).	1725 rpm, 1 AC motor	9–15	Height: 1,280	–	Patil et al. (2018)

minimal starch damage. In contrast, DBM (300 r/min, 1–5 hours, 3 : 1 ball-to-powder ratio) degraded the starch crystal structure, enhancing hydration, lowering pasting temperature, and improving viscoelasticity. These findings highlight that the choice of milling method depends on specific production needs to enhance millet flour quality and functionality. Besides, Sunagar and Sreerama (2024) explored that pin milling produced the finest flour (82 μm) but caused the most starch damage (4.64%) compared to roller milling (2.46%) and hammer milling (2.51%) of proso millet. Pasting properties varied, with pin-milled flour showing the highest peak viscosity (2,295 cP). Despite differences in grinding methods, bread quality remained consistent, with specific volumes ranging from 2.40 to 2.52 cm^3/g, supporting proso millet's potential in value-added food products. One of the advanced technologies is air jet milling, which utilizes high-velocity fluid energy to achieve precise flour particle size reduction. This method is widely used to produce uniform flour particles, enhancing their functionality in food applications. Research indicates that air jet milling improves water absorption capacity (WAC) and forms a more compact structure in bakery products (Yadav & Dhiman, 2023).

10.2.4 Polishing

Polishing involves the removal of the outer bran layer or pericarp, which primarily consists of indigestible fiber, through friction after softening with water (Kumari et al., 2025). The bran layer contains antinutrients that hinder nutrient digestibility, absorption, and utilization (Samtiya et al., 2021). Research by Kumar et al. (2016a) indicates that polishing finger millet reduces 2,2-diphenyl-1-picrylhydrazyl (DPPH) activity, total flavonoid content (TFC), and total phenolic content (TPC) by 28%, 32%, and 30%, respectively. In pearl millet and sorghum, polishing reduces lipid and zinc content due to germ removal, with 10% grain abrasion resulting in a 50% decrease in iron content (Hama et al., 2011). Paddy-based millet polishers, such as centrifugal and abrasive, are commonly employed for millet processing (Ravindra et al., 2008). Pearl millet endosperm yield ranges from 83.47% to 90.29%, with a polishing degree of 10.72–19.56%, where increased polishing time enhances the degree of polishing and breakage but reduces endosperm yield (Joshi & Rao, 2024). Lohani et al. (2012) polished barnyard millet using a rice polisher at 8–14% moisture for three to six minutes. At 8% moisture, bran yield was 18.86%, increasing to 19.21% at 14% moisture. The highest head yield (52.97%) occurred at 10% moisture while polishing reduced fat, protein, fiber, and ash with maximum loss at 14% moisture.

10.2.5 Roasting

Roasting uses dry heat from an open flame, oven, or other sources to brown food, enhance flavor, and develop unique aroma compounds in millets (Bi et al., 2020). The optimal roasting temperature ranges from 120 to 140 °C for 5–15 minutes (Obadina et al., 2016). Vitamin E, C, oxalates, and alkaloids content were reported to diminish after roasting Himalayan barnyard millet (Rana et al., 2023). Roasting can cause little millet to develop a more porous structure due to heat-induced expansion (Kumari et al., 2025). Singh et al. (2024) examined the effects of microwave roasting at 360, 720, and 900 W for three and five minutes and pan roasting at 195 °C for five minutes on kodo millet flour. Roasting reduced bulk density while increasing compressibility index and tap density, with higher power (900 W) leading to darker flour. Microwave roasting enhanced WAC but reduced oil absorption capacity (OAC), and high-power roasting (900 W) degraded starch, flattening the pasting curve. Antioxidant activity (AA) increased in all roasted samples except 900 W, while TPC was higher in microwave-roasted flour compared to pan-roasted and control samples. In another work, microwave roasting of sorghum (300–600 W, 5–15 minutes) reduced the L^* value and increased TPC and WAC, while peak viscosity declined with higher time and power (Sharanagat et al., 2019). While infrared roasting has been extensively studied for various grains (van Rooyen et al., 2023), limited research specifically focuses on its application to millets.

10.2.6 Puffing/Popping

Puffing involves simultaneous starch gelatinization and grain expansion, with vapor escaping through pores due to high temperature and pressure. This process enhances taste and aroma, producing ready-to-eat snacks like puffed rice. The impact of puffing and popping depends on grain structure and composition, improving starch accessibility and digestibility (Huang et al., 2018). Sand roasting (150–250 °C) rapidly removes moisture and expands pregelatinized grains, enhancing crispness and aroma, and reducing antinutrients (Jaybhaye et al., 2014). Gun puffing involves steam-cooking grains under high pressure, followed by a sudden release, resulting in light, crispy puffed grains (Mishra et al., 2014). Microwave puffing heats grains volumetrically, causing rapid moisture loss and expansion, improving flake size and expansion with higher moisture levels (Sharma et al., 2014). High-temperature short-time (HTST) fluidized bed puffing (240–270 °C for 7–10 seconds) ensures even heating and mass transfer, minimizing thermal damage while optimizing texture (Tavanandi et al., 2021). A hot air puffing system for barnyard millet was optimized at 310 °C and a 7500 g/h feed rate, yielding 72.61% puffed grains with 4.81% moisture. A second pass increased puffing efficiency to 99.19%, reducing unpuffed grains to 0.81% (Sharad et al., 2024). Mirza et al. (2015) assessed finger millet genotypes for popping quality, with moisture content and grain hardness playing key roles. The brown genotype PRM-6107 showed the best results, with 10.65% moisture, 3.6% fiber, and 9.92% protein, making it ideal for ready-to-eat products.

10.2.7 Germination/Malting

Natural biofortification in millet processing can enhance its nutritional value, addressing micronutrient deficiencies while reducing the cost of external fortification. Sharma et al. (2015) reported that germinated foxtail millet contains higher dietary fiber (27.42 g/100 g), protein (14.32 g/100 g), calcium (25.62 mg/kg), magnesium (107.16 mg/kg), and iron (54.23 mg/kg) compared to nongerminated flour. The increase in mineral content is attributed to enzyme activity that hydrolyzes bound minerals. Controlled sprouting improves the bioavailability of minerals and vitamins in millets (Lemmens et al., 2019). Optimized sprouting conditions for phytase activity enhanced the bioaccessibility of iron and zinc in sorghum and millets by up to 30% (Benincasa et al., 2019). Additionally, germination increases the levels of vitamins B, C, and E, contributing to their daily dietary requirements

(Idowu et al., 2020). White finger millet flour germination at 30 °C for 48 hours exhibited higher lightness (L^*) and the highest hue angle (86.81°), indicating a yellowish color. Bulk density decreased from 773 to 530.06 kg/m^3, favoring its use in geriatric and weaning foods. Germinated flour showed shear-thinning behavior ($n < 1$), with increased emulsifying (21.97%) and foaming (11.40%) capacity (Navyashree et al., 2022). Dispersibility reached 89.20%, similar to creamy finger millet (Ramashia et al., 2018). Proline, tyrosine, and lysine levels increased (Navyashree et al., 2022), aligning with findings in malted pearl millet (Adebiyi et al., 2017). Reduced fat and carbohydrate content may result from enzymatic breakdown during germination (Najdi Hejazi and Orsat, 2017). Malting enhances the quality of cereal grains through enzymatic activity that releases fermentable sugars. During malting, native enzymes such as α- and β-amylase, limit α-glucosidase, and dextrinase are produced and activated during sprouting. Fermentation is more effective than unmalted grains (Panda & Shetty, 2018).

10.2.8 Nixtamalization

Nixtamalization is a process that involves soaking dried grains in an alkaline solution, such as lye ($NaOH$), lime ($Ca[OH]_2$), or wood ashes (KOH), followed by rinsing and hull removal. The alkaline environment helps dissolve components of the grain cell walls, including cellulose and pectin, which loosens the hull and facilitates dehulling (Pandey & Waghmare, 2019). According to Owusu-Kwarteng and Akabanda (2013), lime treatment improves the WAC, hygroscopicity, and swelling power of sorghum flour. Additionally, soaking or heating the grains with lime significantly reduces antinutritional factors such as phytate, tannins, trypsin inhibitors, and cyanide. The reduction in phytate levels occurs due to the breakdown of phytic acid, catalyzed by the activation of the phytase enzyme during lime treatment. Gaytán-Martínez et al. (2017) reported an inverse relationship between total phenols, flavonoids, AA, and lime concentration and cooking duration. The decline in polyphenols is attributed to pericarp removal during dehulling, leaching during soaking, or degradation during boiling, whereas flavonoid reduction results from their breakdown during the treatment process (Rajeswari et al., 2015). Nixtamalization (2% $Ca\ [OH]_2$ at 75 °C for 30 minutes) increased fiber, mineral, and ash mineral content, with solubility rising at higher temperatures (71.86% for dry milled, 50% for wet milled, and 66.25% for untreated) pearl millet flour. Free fatty acids increased significantly, reaching 6.48% in dry milled and 14.89% in wet milled. X-ray analysis showed reduced crystallinity due to structural disruptions in nixtamalized samples (Gopika & Joshi, 2024).

10.3 Novel Thermal Processing Technologies of Millets

Heat processing has been shown to decrease the activity of enzymes such as lipoxygenase and lipase, which helps prolong the shelf life of grains (Bucsella et al., 2016). According to Sirohi et al. (2021), the thermal treatment of grains is associated with several changes, including protein denaturation, alterations in flavor and aroma, modification of starch granule structure, and reduction in microbial load. Table 10.2 represents impact of thermal

Table 10.2 Effects of thermal and nonthermal processing on the quality of millet

Processing	Type of millet	Operational conditions	Inference	Sources
Extrusion cooking	*Eleusine coracana*	Intermeshing and co-rotating twin-screw extruder, length-to-diameter ratio: 20 : 1, and die: 3 mm. Temperature: 103.18–136.82 °C, screw speed: 357.96–442.04 rpm	Extrusion processing significantly improves finger millet flour's expansion ratio, water absorption, and pasting properties	Kumar et al. (2016b)
Microwave	*Pennisetum glaucum*	100, 200, and 300 W for 30, 60, and 90 s	Reduced lipase activity (219.76 to 31.98 μg oleic acid/mL) and antinutrients (phytates by 16.30%, tannins by 87.32%)	Sundaresan et al. (2025)
Infrared heating	*Eleusine coracana*	120–180 °C	Modified finger millet's physical, functional, and microstructural properties, reducing particle size, peak viscosity, and phenolic content, with more significant losses in hydrated samples	Kumari and Jayadeep (2024)
Induction cooking	Millet	820–1300 W with a 1 : 14 millet-to-water ratio, boiled for 40 min	Different electric cookers affected millet porridge composition, enhancing amylose and reducing sugars and protein, while improving sensory quality or retaining more minerals and fatty acids	Zhang et al. (2022c)
Spray drying	*Eleusine coracana*	Millet milk fermented for 5 h with *L. plantarum*, followed by spray drying at 151.68 °C and 70 °C (inlet and outlet temperature), 29.32% maltodextrin, and a 100 mL/h feed rate	The Peleg model best described the type III sorption isotherm of probiotic finger millet milk powder, which is optimally stored at 4 °C (56-day shelf life) and 25 °C (35-day shelf life)	Yadav and Mishra (2023)
Spray and Freeze drying	*Eleusine coracana*	Spray-dried at 200 °C with a 0.48 L/h feed rate using a pressure nozzle atomizer, while freeze-drying conducted at −18 °C for 2days	Freeze-dried ragi retained higher micronutrient levels, with better calcium and iron (349.6 and 0.550 mg/100 g) than those dried by spray or tray	Kabeer et al. (2023)

(*Continued*)

Table 10.2 (Continued)

Processing	Type of millet	Operational conditions	Inference	Sources
Ultrasonication	*Setaria italica*	Amplitude settings of 5% and 10%, with 5, 10, and 20 min intervals	Improved heat denaturation, emulsification, and foaming help prevent secondary structure formation	Sharma et al. (2023)
	Panicum miliaceum bran	Power: 400 watts, with durations of 5, 12.5, or 20 min. Amplitude settings are 60%, 80%, and 100%	The digestibility of bran protein did not change before or after processing. Cooking proso millet proteins causes them to form hydrophobic clumps, reducing their digestibility	Mustač et al. (2019)
	Eleusine coracana	0.29 W/mm^2 intensity for 2.63 min after inoculation of *L. rhamnosus* on probiotic beverage	Lower fermentation time by 2.6-fold compared to untreated sample. During fermentation, lactic acid levels undergo three phases: a rapid increase, a stable phase, and a gradual change	Meena et al. (2024)
Cold plasma	*Eleusine coracana*	20 kV, 20-min, 9 cm	Reduction in tannin and phytic acid by 27% and 43% in flour	Arsha et al. (2025)
	Pennisetum glaucum	20–30 kV for 10–20 min	Enhanced solubility, emulsification and foaming capacity, with reduction in crystallinity of flour	Sarkar et al. (2023)
	Paspalum scrobiculatum	10, 20, and 30 kV for 10, 20, and 30 min	Increased water and oil absorption capacity and whiteness index while reduced amylose (31.06–25.94 g/hg) and crystallinity of starch	Sonkar et al. (2023)
	Echinochloa frumentacea	10–30 kV for 10–30 min	30 kV, 30 min showed enhanced viscosity (180.14 cP), solubility (3.92 g/g), and TPC (280.54 mg GAE/100 g) of flour	Sharma et al. (2024b)

High-pressure processing	*Setaria italica*	Pressure: 300 and 600 MPa, temperature: room temperature, duration: 10 min	As storage progresses, the nitrogen content in the amino acids decreases, likely due to oxidation causing the breakdown of amino acids over time	Zhou et al. (2022)
	Eleusine coracana	Pressure: 350 MPa, temperature: 20 °C, duration: 10 min	Both the Maillard reaction and nonthermal cross-linking reduce the digestibility of proteins	Angioloni and Collar (2012)
	Millet	Pressure: 100, 300, and 500 MPa, duration: 10 min, temperature: 25 °C	HPP increases the average polarity of the polar groups in gliadins. This means that gliadins become less hydrophobic, which shows their solubility in ethanol has reduced	Zhang et al. (2022d)
Ozonation	*Setaria italica*	Ozone treatment 0.06 L/min for 15, 30, 45, and 60 min	Significantly reduced tannin (112.43–64.83 mg/100 g) and TFC (331.46–273.59 mg/100 g), enhanced TPC and phytic acid, water and oil absorption, solubility, swelling power, and paste clarity	Sharma et al. (2022)

and nonthermal processing on millet quality attributes. This section elaborates on different heat methods for processing millet, including microwave, infrared, radio frequency, extrusion, and ohmic heating.

10.3.1 Microwave Processing

Microwave processing works by causing molecules to rotate and move due to electromagnetic radiation, which falls within the 0.3–300 GHz wavelength range. In this method, the material absorbs some energy while the remainder is transmitted through or reflected off the surface. The dielectric constant and dielectric loss factor affect microwave interactions with food systems (Gomez Chandrasekaran et al., 2013). Literature indicates that microwave processing reduces antinutritional factors in millet and products derived from millet (Singh et al., 2017). Antinutrients are deactivated by breaking peptide bonds, disrupting covalent bonds, and altering disulfide bonds (Rahate et al., 2021). Free radicals generated during microwave processing decrease inositol and inositol phosphate levels, contributing to a reduction in phytic acid concentration (Kala & Mohan, 2012). Microwave treatment at 900 W for 40–100 seconds reduced tannin levels in various millets, with reductions of 45.53% in pearl millet, 45.18% in finger millet, and 77.54% in sorghum (Singh et al., 2017).

Microwave technology is used in grain processing for several tasks, such as cooking, drying, sterilizing, roasting, baking, deactivating enzymes, and disinfecting (Sharanagat et al., 2019). Gopal and Bhuvana (2021) investigated the impact of microwave treatment on the chemical and functional properties of kodo millet. They discovered that higher microwave power and longer treatment times increased TPC from 2.64 to 4.67 mg Gallic Acid Equivalent (GAE)/100 g. However, the ability to scavenge free radicals decreased from 91.12% to 87.94%. Microwave treatment also reduced the flour's ability to absorb oil and increased its ability to absorb water (Rao et al., 2021). Additionally, the swelling power of the millet flour decreases with higher microwave power levels and longer treatment times (Adebowale et al., 2020). Microwave treatment process significantly lowered the acidity and anisidine values of the flour during storage. The thermal effects of microwave heating led to more damaged starch in the flour. Microwave treatment can disrupt the intermolecular interactions between starch and proteins, leading to a rise in starch damage (Semwal and Meera, 2021). Further sensory qualities of porridge made from the treated flour are not impacted. Almaiman et al. (2021) demonstrated that microwave heating at 500 and 350 W reduced the total viable fungus count in sorghum by 33.4% and 26.2%, respectively.

Microwave food processing significantly shortens drying time compared to conventional methods due to its efficient energy use and rapid volumetric heating. Sunil and Chacko (2018) observed that a 120 seconds duration at 720 W resulted in the highest dehulling yield of 85.64%. Microwave drying of germinated little millet at 1050 W enhanced AA by 34.54% and increased total phenolic and flavonoid content—the Page kinetic model best described drying behavior, with the rate constant rising to 6.33 times. Higher microwave power also reduced crystallinity and enthalpy, suggesting structural modifications for better bioavailability (Jena et al., 2025). Yadav et al. (2021) assessed how much people liked MW-treated flour samples and found that flour could be stored at room temperature for up to 30 days. In contrast, the control flour began to develop a foul smell and bitter taste by the 10th day. However, the drawback of microwave pretreatment is the formation of hot spots due to uneven temperature distribution, which can negatively impact grain germination rates (Das et al., 2013). The energy-efficient impact of the process needs to be analyzed to commercialize the technology.

10.3.2 Radiofrequency Processing

Radiofrequency (RF) waves are a type of electromagnetic wave that do not ionize atoms. They have frequencies ranging from 1 to 300 MHz. They can penetrate dielectric materials and generate heat throughout the material through ionic polarization or dipole rotation. Due to their longer wavelength, RF waves can penetrate more deeply than microwave radiation. RF treatment is known for various uses, such as controlling pests, killing microbes and enzymes, altering proteins, and drying or roasting grains like rice, wheat, and corn (Hassan et al., 2019). However, research on millet with RF processing is limited. A significant challenge with RF treatment is the uneven heating caused by the inconsistent radiation field distribution between the product and its environment (Fadiji et al., 2021). It has been shown that using RF in combination with hot air enhances both the effectiveness of the treatment and the consistency of heating (Wang et al., 2014). A pilot-scale free-running oscillator-type hot air-assisted RF (10 kW, 40.68 MHz) effectively reduced lipase activity in pearl millet, achieving 97.3% inactivation at 15% moisture for 15 minutes (Yarrakula et al., 2022).

Wang et al. (2021) introduced a novel approach to enhance RF heating uniformity by modifying the sample's bottom and top shapes. Increasing convexity and reducing concavity improved uniformity, while adding polypropylene blocks at the container bottom further optimized heating. Using a 6 kW, 27.12 MHz RF system showed better distribution and adaptability. This simple geometric adjustment offers a practical solution for improving RF processing in the food industry. Hot air-RF (6 kW, 27.12 MHz) heating inactivated peroxidase and lipase in brown millet (BM) and its bran (BB), reducing rancidity and enhancing storage stability. Lower temperatures (95 °C for BM, 105 °C for BB) preserved structure and nutrients, while higher temperatures caused more degradation. Enzyme activity dropped by <13% in BM and <10% in BB, slowing fatty acid and Malondialdehyde formation. Ohmic heating is another method that rapidly heats millet using electrical currents, improving microbial safety and nutrient retention (Chen et al., 2024).

10.3.3 Infrared Processing

Infrared (IR) radiation lies between visible light and microwave wavelengths. It is divided into three regions based on wavelength: near IR (0.78–1.4 μm), mid-IR (1.4–3.0 μm), and far-IR (3.0–1000 μm) (Rastogi, 2012). When IR penetrates, it causes water molecules to vibrate, generating heat. IR technology offers several benefits, including shorter heating times, more uniform heat distribution, lower energy use, faster heat transfer, and improved product quality (Aboud et al., 2019). IR radiation generates thermal energy that can damage microbial cells by causing membrane contraction, disrupting mesosomes, and leaking intracellular contents. Additionally, IR heating effectively inactivates lipase (Li et al., 2016). Swaminathan et al. (2015) evaluated the impact of infrared heating on sorghum flour and revealed reduced free fatty acid content with higher temperatures and longer treatment durations. Specifically, lipase activity in the sorghum flour decreased by 88.8% when subjected to IR heating at 120 °C for 10 minutes. In another study on sorghum grains, Rousta et al. (2014) utilized infrared to treat sorghum samples, resulting in a notable moisture loss of 3–6%. This reduction in moisture led to increased concentrations of protein, fat, fiber, and ash. However, protein digestibility decreased by 6.8–23.2%, likely due to protein denaturation. On the other hand, starch digestibility improved significantly due to gelatinization. Additionally, the treatment significantly lowered phenolic compounds, phytic acid, and tannin levels in the sorghum grains. IR heating at 1760 W (9 minutes) exhibited 100% mortality of *Tribolium castaneum* life stages. Further, Scanning Electron Microscope (SEM) imaging confirmed the surface irregularities on finger millet flour (Manickam et al., 2024). This highlights the potential of IR heating for effective pest control while preserving millet flour quality.

10.3.4 Extrusion

Extrusion technology originated in the 1800s in the polymer industry and was introduced to the food sector in the 1930s for pasta and breakfast cereal production. This HTST process applies thermomechanical energy to create various food products, including ready-to-eat cereals, puffed snacks, pasta, noodles, textured vegetable proteins, and pet food. Ingredients are heated, cooked, and shaped through a die, ensuring precise form and texture. Known for its versatility, automation, cost-effectiveness, and efficiency, extrusion remains a widely used food processing method (Yadav et al., 2022). Cold extrusion is a heating process of food using a single screw to shape food below 100 °C. The raw material undergoes hydration, blending, and molding while maintaining a steady temperature (Shelar & Gaikwad, 2019). Cold-extruded products include pasta, vermicelli, noodles, and flakes. According to Cordelino et al. (2019), proso millet pasta contains more than 20% amylose, making it suitable for gluten-free pasta due to its excellent gel-forming ability and high viscosity in cold paste form. Proso millet pasta exhibits lower total soluble solids loss, higher carotenoid content, and slower starch digestibility than commercial samples. Goel et al. (2021) reported that enriched barnyard millet vermicelli contained significant levels of beta-carotene (1039 μg/100 g) and iron (3.81 mg/100 g). Optimized conditions for formulating millet analogue developed from parboiled rice, sorghum, and pearl millet (28% moisture, 12 rpm screw speed, 80 °C die temperature, 7 rpm feeder speed) improved cooking properties, lower glycemic index, serving as healthier rice alternative suitable for commercialization (Kishore et al., 2024). Recently, Jin et al. (2025) examined the effect of extrusion temperature (70, 90, 110 °C) on proso millet starch. Higher temperatures caused gelatinization, increased particle size, reduced crystallinity, and altered viscosity. WAC, solubility, and swelling power decreased, while freeze–thaw stability and transparency declined over time. Overall, millet-based cold-extruded products offer a superior nutritional profile compared to commercially available alternatives.

Hot extrusion is an HTST process that involves mixing, shearing, shaping, plasticizing, and cooking hydrated materials within a twin-screw chamber. Kharat et al. (2019) found that foxtail millet (4.41) exhibited better extrusion characteristics compared to pearl millet (3.78) and finger millet (3.24). The water absorption index values for foxtail millet, finger millet, and pearl millet were recorded as 4.18, 3.54, and 2.69 g/g, respectively. Overall, extrusion had minimal impact on these millet extrudates. A study explored the use of pretreatments such as sonication, germination, and fermentation to preserve nutrients in foxtail millet during extrusion. Germinated-sonicated and germinated millet extrudates contained 6.15 and 7.24 mg/100 g of gamma-aminobutyric acid, respectively. Additionally, germinated millet extrudates exhibited high DPPH scavenging activity (15.51%) and TPC (30.54 mg GAE/g). Incorporating these pretreatments before extrusion can enhance the nutritional value of foxtail millet extrudates (Mushtaq et al., 2021). Under optimal conditions (140 °C, 150 rpm, 30% moisture), extrusion significantly reduced phytic acid (89%), trypsin inhibitors (~100%), lipase (79.2%), lipoxygenase (77.9%), and tannins (85%) (Pathare et al., 2024). Gaurav et al. (2022) developed a carotenoid-rich breakfast cereal by blending biofortified maize and kodo millet. Optimized extrusion conditions included feed moisture (12.5–17.5% wb), die temperature (150–180 °C), and screw speed (420–470 rpm). The total carotenoid content ranged from 273.17 to 388.65 µg/100 g, with desirable physical properties like hardness (13.21–67.59 N) and expansion index (2.06–4.46). Thus, hot extrusion remains a viable method for producing nutritious and convenient millet-based snacks.

10.3.5 Ohmic Heating

Ohmic heating (OH) operates on Ohm's law, which generates heat in food when an electric current passes through it. The food's electrical conductivity (EC) plays a crucial role in regulating the OH process. Foods with an EC ranging from 0.01 to 10 S/m suit this technique. Incorporating salts enhances the ionic concentration in food, thereby accelerating the heating rate (D'cruz et al., 2023). However, OH use in millet processing remains unexplored mainly (Joshi et al., 2023). Dias-Martins et al. (2019) studied ohmic cooking's effects on pearl millet, including cooking yield, texture, color, WAC, Water Solubility Index (WSI), and pasting properties. They found that ohmic-heated samples had lower cooking yield due to the pericarp layer hindering electrical conductivity and limiting heat transfer to the endosperm. Ohmic cooking also resulted in firmer grains since kodo millet dehulling is challenging due to its small size and compact husk. Pretreatment before milling has increased grain hardness and minimized milling loss. Ohmic parboiling (OP) improved kodo millet's milling properties over conventional parboiling, increasing head grain yield (28.03%) and dehulling efficiency while reducing broken grains. OP also retained more protein (7.23–7.64 g), enhancing nutritional value. The optimized conditions were soaking at 50–70 °C to 30% moisture and parboiling at 14 V/cm for 20 minutes (Kabui & Athmaselvi, 2024). More work is needed to explore its potential across different millet types to enhance their milling, nutritional, and functional properties.

10.3.6 Hydrothermal Processing

Hydrothermal treatment involves exposing grains to high-pressure steam or hot water for a set duration. The high enthalpy of moist heat modifies the absorbed fat globular membranes, preventing lipase release and effectively inhibiting its activity. The final grain quality depends on the treatment method, primarily influenced by starch gelatinization and retrogradation. This process alters water absorption, swelling, starch crystallization, enzyme activity, and flour viscosity (Sruthi & Rao, 2021; Chávez-Murillo et al., 2021). Acquisgrana et al. (2019) found that hydrothermal treatment significantly reduces tannin content in wholegrain sorghum and its flour, making it more suitable for food processing while allowing potential tannin recovery for other applications. The hydrothermal treatment of finger millet involves steeping millet for 10 hours at 30 °C, steaming at 98 ± 1 °C for 30 minutes, and drying at 38 °C resulting in color enhancement, increased sphericity, and sixfold higher hardness. Swelling and solubility power improved while peak viscosity declined, showing zero breakdown viscosity. Microscopic analysis revealed structural changes, with the endosperm becoming a coherent mass and seed coat layers fusing (Dharmaraj et al., 2015). Prashanth et al. (2024) examined hydrothermal treatment on pearl millet, optimizing soaking (35–55 °C, 2–4 hours) and steaming (5–15 minutes). The process reduced phytates (0.36–6.00%), tannins (0.99–5.94%), and lipase activity (up to 10%), enhancing nutrient retention and shelf life, making it suitable for food processing. Future research should optimize hydrothermal processing conditions to enhance millets' nutrient retention and dehulling efficiency. Developing energy-efficient equipment and integrating novel technologies like OH, MW, IR, RF, etc., can improve processing sustainability. Additionally, exploring hydrothermally processed millet in diverse food products can expand its commercial potential.

10.4 Emerging Nonthermal Processing of Millets

Nonthermal methods are innovative techniques that have recently gained significant attention. These methods are considered effective alternatives to traditional thermal processes, offering the advantage of preserving food quality while minimizing impact on its attributes. Nonthermal methods in food processing are used to achieve several benefits, such as cleaning food and changing the properties of ingredients like starches and proteins (Zhang et al., 2019). Nonthermal processing methods that significantly affect the quality and performance of millet protein include ultrasound, irradiation, cold plasma, high-pressure processing, and PEFs (Ramakrishnan et al., 2023). Figure 10.2 illustrates various emerging thermal and nonthermal processing techniques for millets, emphasizing their advantages and limitations in enhancing nutritional quality and shelf life.

10.4.1 Ultrasonication

Ultrasonication is a processing method that utilizes high-frequency sound waves, generally between 1 and 10 MHz. These sound waves create mechanical vibrations that can dismantle plant cell walls, allowing for the release of their components. Ultrasound can assist in extracting phytonutrients, proteins, and bioactive compounds from plant materials, including phenols, flavonoids, and antioxidants (Kumar et al., 2023). High-intensity ultrasound treatment on proso millet bran at 80% amplitude for 12.5 minutes increased AA (15%) and TPC (16%), while 100% amplitude for 20 minutes enhanced dietary fiber solubility (38%). All these parameters are crucial for millet utilization in the pasta and

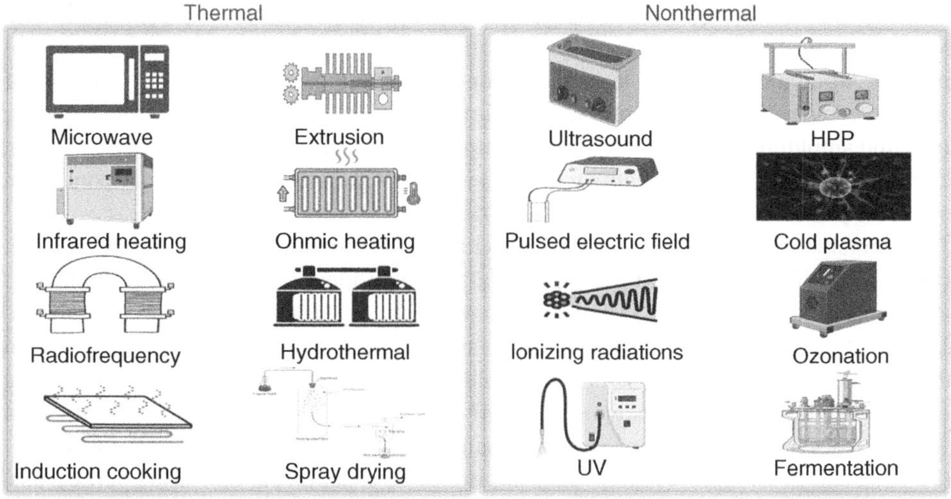

PROS	CONS	PROS	CONS
• Enhanced nutritional retention • Improved digestibility • Better sensory properties • Energy efficiency and sustainability	• High initial cost • Uneven heating • Limited industrial adoption • Excessive exposure cause nutrient degradation	• Nutrient preservation • Improved safety and shelf life • Enhanced functional properties • Energy-efficient and eco-friendly	• High equipment cost • Longer processing time • Variable effectiveness • Limited reduction of antinutritional factors

Figure 10.2 Thermal and nonthermal processing techniques for millets, highlighting their pros and cons

bakery industry (Mustač et al., 2019). Ultrasound (250 W, 40 kHz) at 40 °C followed by fermentation on foxtail and proso millet significantly enhanced protein solubility, carotenoids, phenolics, flavonoids, and gamma-aminobutyric acid (GABA) (Alkaltham et al., 2024). Likewise, in the case of millet milk, Saxena et al. (2023) reported increased TPC (92 mg GAE/mL) and Ferric Reducing Antioxidant Power (33.42%) after sonication (180 W for 20 minutes) and germination. A combination of fermentation (12 hours) and ultrasonication (30 W/cm^2 for 3 minutes) applied to sorghum flour also improves the nutritional and textural properties of extrudates with elevated antioxidant and TPC values (Lohani & Muthukumarappan, 2020).

Ultrasound-assisted hydration (USH) also reduces antinutrients in millet samples, ultimately lowering processing time (Joshi et al., 2023). As the ultrasound amplitude and treatment duration increase, the moisture content of the samples rises. Prolonged exposure to higher amplitudes disrupts cell structures, creating microcavities that facilitate mass transfer through diffusion, thereby improving the hydration properties of the grains (Yildirim et al., 2013). Lately, USH was optimized to reduce antinutrients in finger millet, achieving a 71% and 73% reduction in tannins and phytates, respectively. Artificial neural network-particle swarm optimization identified optimal conditions: 76% amplitude, water : grain ratio of 3.5 : 1, 40 kHz for 17.5 minutes (Dubey & Tripathy, 2024). At higher ultrasound amplitudes, the sample generates heat, leading to a temperature rise. The reduction of phytates occurs through heat-induced chemical degradation, breaking them down into inositol phosphate. Ultrasound also facilitates the breakdown of hydrolysable tannic acid into gallic acid, reducing overall tannin levels, while promoting the leaching of condensed tannins (Bhangu et al., 2018; Diouf et al., 2019). Yadav et al. (2021) observed a decline in the TPC of finger millet with increasing ultrasound amplitude and duration. Greater amplitude and exposure time enhance cavitation within the cellular structure, leading to increased solvent penetration, which may contribute to reducing phenolic compounds (Zhu & Li, 2019).

Food proteins are essential macronutrients with unique functional properties, such as film-forming, emulsifying, and gelling abilities, making them ideal carriers for hydrophilic and lipophilic bioactive compounds (Gharibzahedi & Smith, 2021). Millets serve as an economical, gluten-free protein source. Utilizing nano-sized protein particles for encapsulation can enhance bioactive compounds' stability, solubility, and cellular absorption while protecting them from degradation by digestive enzymes (Neilson et al., 2007). Looking into this, Jhan et al. (2022) developed crocin-loaded protein nanoparticles from sorghum (SRN), foxtail millet (FMN), and pearl millet (PMN) using ultrasonication, achieving nanoscale particle size with high encapsulation efficiency (FMN: 83.78%; SRN: 78.74%; PMN: 70.01%). Attenuated total reflectance – Fourier Transform Infrared Spectroscopy (ATR-FTIR) confirmed crocin encapsulation, and Field Emission-SEM revealed topographical characteristics. The nanocomplex exhibited antimicrobial activity against *Staphylococcus aureus*, *Escherichia coli*, and *Fusarium oxysporium*. In vitro release studies showed controlled crocin release in intestinal conditions, with retained antihypertensive and anticancer properties postdigestion.

Further, US modification of protein is widely used because of its low cost, environment-friendly nature, and high efficiency. US of at 450 W significantly improved the structural, functional, and bioactive properties of foxtail millet bran protein hydrolysates by conversion of β-sheet and α-helix structures into β-turns and random coils, along with the increased exposure of sulfhydryl and hydrophobic groups (Peng et al., 2024). US (20 kHz, 10 and 20 minutes) enhanced pearl millet protein properties, improving solubility, foaming, emulsification, and digestibility while reducing particle size (4.13 to 2.32 μm) (Hegde et al., 2025). The US can also modify finger millet starch by increasing amylose content and improving gelatinization temperature, swelling, and WAC (Amarnath et al., 2023). Overall, ultrasonication improved millet protein and starch functionality for food applications.

10.4.2 Pulsed Electric Field (PEF)

Using nonthermal mechanical methods, such as PEF, to modify proteins is an innovative approach that is gaining traction. This approach is being used to improve the performance of proteins in both food and nonfood industries (Ran et al., 2019). The PEF technique involves applying brief, high-intensity electric field pulses to objects between two electrodes. This nonthermal method has recently gained considerable attention. It utilizes high electric fields between 10 and 80 kV/cm, with processing times usually just microseconds to milliseconds. Due to the brief duration required for PEF treatment—usually less than one second—the technique utilizes very narrow pulse widths and low pulse frequencies. As a result, the process generates minimal heat, with pre- and post-treatment temperatures remaining below 30 °C. The application of high-voltage pulses leads to swelling, the formation and expansion of pores, water influx, membrane rupture, and cell inactivation

(Mohamed & Eissa, 2012; Zhang et al., 2021). PEF treatment can effectively preserve the flavor and nutrients in food. This nonthermal technique is also environmentally friendly (Alkhafaji & Farid, 2007).

Lohani and Muthukumarappan (2016) investigated the use of mild-intensity PEF to enhance the release of bound phenolics in sorghum flour. The optimized conditions were 45% (w/v) flour-to-water ratio, 2 kV/cm electric field intensity, and 875 μs time. Under these conditions, TPC and AA increased by 24.8% and 33.9%, respectively, highlighting the potential of PEF-treated flour for developing antioxidant-rich processed foods. PEF also exhibits bactericidal effects on glutinous millets at 54.43 J/cm^2 fluence. Further, rough surfaces of grains can act as a shield against radiation (Hwang et al., 2018). PEF treatment alters the spatial conformation of foxtail millet prolamin, enhancing its solubility and hydrophobicity without affecting its subunit composition distribution. It also facilitates the formation of soluble aggregates, leading to an increase in protein particle size. Additionally, the treatment reduces α-helix content while increasing disulfide bonds and β-sheet content, resulting in a more structured and orderly protein arrangement, improving its applicability in food formulations (Zhang et al., 2023). Ramesh et al. (2020) optimized Electromagnetic field priming (EMF) to enhance foxtail millet seed germination, achieving a 93.0% germination rate under 10 Hz, 30,007 nT, and 30-minute duration. Magneto-priming improved vigor indices (39.14% and 46.28%), germination efficacy (15.66%), and antioxidant enzyme activity (superoxide dismutase: 19.62%; catalase: 114.63%). These findings highlight EMF priming as an effective method to boost seedling vigor and germination. However, comprehensive evaluations are required to study the effect of PEF on different millet types.

10.4.3 High-Pressure Processing (HPP)

HPP is a nonthermal processing technique that applies pressures ranging from 100 to 1000 MPa and temperatures between 20 and 90 °C for a specified duration. In recent years, it has demonstrated significant potential for enhancing safety and quality (Mirzababaee et al., 2022). High-pressure processing provides an alternative to heat methods by killing harmful microorganisms and enzymes without additional preservatives. When applied at room temperature, high pressure can significantly disrupt noncovalent bonds within proteins and microbial cells, such as hydrogen and ionic bonds. This disruption can cause covalent bonds to form, which helps inactivate microorganisms (Pandiselvam et al., 2024). A study by Zhou et al. (2022) compared HPP and thermal processing on Lion's Mane-millet-composite beverages. While both methods showed no significant differences in total solids, pH, or amino acid nitrogen, Thermally processed (100 °C for 3 minutes) beverages had a longer shelf life (63 days at 4 °C) than those treated with 300 MPa (39 days) and 600 MPa (52 days). However, HPP better preserved ester flavor compounds, particularly ethyl acetate, and showed less change in volatile compounds during storage. HPP has also been used to extract bioactive compounds. Zhang et al. (2022b) utilized a combination of HPP (500 MPa, 15 minutes) and deep eutectic solvent (octanoic acid-linalool) for the extraction of millet polyphenols.

HPP enhances grain protein digestibility while lowering allergenicity by eliminating or reducing allergenic compounds (Estrada-Girón et al., 2005). HPP (200 and 600 MPa) for 5 and 20 minutes significantly enhanced the digestibility of uncooked proso millet flour, which could be due to protein denaturation by breaking internal bonds (Gulati et al., 2018). Soaking of pearl millet at 350 MPa for 90 minutes reduced phytate by 80.61% and increased free iron from 127.73 to 579.24 mg/kg, with minimal nutritional changes except for reduced carbohydrates at longer treatments. OAC and WAC capacities increased by 5.9% and 12.1%, respectively, while thermal and pasting properties were altered (Himashree & Mahendran, 2025). Concurrently, treating foxtail millet with 400 MPa at 40 °C leads to a higher phytic acid reduction (67.87%) (Sharma et al., 2018). The elevated pressure alters the structure of the phytate, primarily quaternary and tertiary, causing entrapment or exposure of active sites (Tribst & Cristianini, 2017). HPP (200–600 MPa, 10 minutes) increased amylose content, OAC and WAC, alkaline water retention, and pasting temperature, while paste clarity decreased over time. Structural analysis (X-ray Diffraction, SEM, FTIR) showed partial gelatinization and deformation at 600 MPa, with reduced crystallinity. These findings highlight HPP as an effective physical modification method for altering millet starch properties (Sherin et al., 2024).

10.4.4 Cold Plasma

Plasma, the fourth state of matter, is classified into two types based on the temperature of the generated gas (Lokeswari et al., 2021). High-temperature plasmas operate in thermal equilibrium at very high temperatures and are generally unsuitable for food applications. In contrast, low-temperature plasmas, also known as cold plasmas, function at room or ambient temperatures. Cold plasma is employed in various food-related applications because it is chemical-free, nontoxic, and environmentally friendly. Moreover, the active species produced in cold plasma through electron collisions provide potential advantages for product sterilization, modification of food properties, and reduction of harmful substances (Pankaj et al., 2018). Cold plasma can also produce oxidizing agents such as ozone and molecular oxygen. These potent oxidizers can enhance the properties of little millet without leaving any residual effects (Jaddu et al., 2022). Cold plasma technology has recently been recognized as an advanced method for inactivating enzymes responsible for degradation, such as polyphenol oxidase, peroxidase, lipoxygenase, alkaline phosphatase, lactate dehydrogenase, and lipase (Mollakhalili-Meybodi et al., 2021). Recently, atmospheric nonthermal plasmas have also been used in the improvement of seed germination. Plasma-treated (8–24 kV) pearl millet seeds showed a 20% higher germination rate, with increased surface roughness and new oxygen functional groups, making them more hydrophilic (50% lower water contact angle). Plasma-activated water (20 kV, 50 Hz, 20 minutes) reduced pH from 7.4 to 4.5 and generated 4.2 ppm H_2O_2 and 1.78 ppm NO_3^-, leading to a 30% improvement in seed germination (Mohandoss et al., 2024).

Cold plasma (2 kV, 5 minutes) on barnyard pearl and millet reduced the phytic acid by 57–61% and tannins by 18–24%. Microbial counts, particularly *E. coli* and *Enterococcus* spp., declined and extended the shelf life. The fat content increased by 43% and 87%, respectively. Despite this, TPC decreased by 2% in pearl millet and 19% in barnyard millet, while TFC decreased by 10% and 7%, respectively (Charu et al., 2024). Further, atmospheric-cold plasma treatment (10 and 20 kV) improved the nutritional and functional properties of little millet flour by enhancing foaming (up to 51.47%) and emulsification (up to 47.02%), while increasing swelling and WAC. It reduced antinutritional factors, like saponins, by 58%, and boosted bioactive compounds, raising TPC from 527.54 to 575.82 mg GAE/100 g and AA by 59% (Jaddu et al., 2024a). With over 4% lipids and 85% polyunsaturated fatty acid (PUFA), millet is more prone to rancidity than rice or corn—oxidation and lipid hydrolysis by endogenous lipoxygenase and lipase lead to foxtail millet rancidity. Lipase breaks down triglycerides into free fatty acids, lowering pH, while lipoxygenase oxidizes linolenic and linoleic acids, producing hydroperoxides and reactive oxygen species, leading to discoloration and an aging odor. Dielectric barrier discharge-air cold plasma treatment decreased lipase and lipoxygenase activity in mizhi millet, reducing 56.0 to 15.1 U/(mg pro) and 44.0 to 18.7 U/g/min, respectively. Treatment at 15 kV reduced millet lipid oxidation and preserved its color, making it a promising storage method (Wang et al., 2022). Atmospheric-cold plasma treatment on little millet flour (10–20 kV) enhanced its loss modulus and storage while confirming its pseudoplastic nature ($R^2 > 0.99$). Temperature and frequency sweep analyses showed structural changes, indicating its potential for viscous and pasted food items (Jaddu et al., 2024b). Jaddu et al. (2023) optimized cold plasma treatment for kodo millet flour using Response Surface Methodology (RSM) and Artificial Neural Network-genetic algorithm (ANN-GA), where ANN-GA (14.1 kV, 28 minutes) showed better functional properties than RSM (17.5 kV, 16 minutes). Further, structural analyses confirmed ANN-GA as the superior optimization model.

Cold plasma processing changes protein structure and function by effectively breaking the covalent bonds within protein fragments. According to the study done by Lokeswari et al. (2021), cold plasma processing at atmospheric pressure (40–45 kV, 5–15 minutes) notably decreased the protein content in pearl millet compared to untreated samples. In a recent study, cold plasma treatment (1 and 2 kV, 25 minutes) improved the functionality of foxtail millet protein, enhancing emulsifying, solubility, and foaming capacities. Surface modifications affected carbonyl, sulphydryl content, particle size, and zeta potential, while high-intensity treatment intensified electrophoretic bands. With a 72.88% protein recovery rate, these modifications make foxtail millet protein suitable for plant-based food applications (Monica et al., 2025). Perumal et al. (2024) explored a novel method for selective generation and extracting energetic nitrogen neutrals from glow discharge plasma. Unlike conventional low-temperature plasma techniques, this approach enables controlled active species generation, improving energy efficiency (~46%) and reducing processing costs. Treatment of native finger millet flour with energetic N_2 neutrals (34.20 eV) enhanced shelf life and functional properties. Cold plasma treatment (9 kV, 20 kHz) enhanced the OAC and WAC of the nutri-mix (malted pearl and finger millet), while reducing carbohydrate content (Sharma et al., 2024a). The findings highlight a sustainable, chemical-free plasma processing technique with industrial potential for quality enhancement and food preservation.

10.4.5 Ozone

Ozone (O_3), a triatomic form of oxygen, is produced by adding a free oxygen radical to molecular oxygen. Ozone is blue when derived from dried air, whereas high-purity oxygen produces colorless gas. It can be utilized in either gaseous or aqueous form and, after application, naturally decomposes into molecular oxygen without leaving harmful residues on food products. The application of ozone in food processing has gained significant attention since receiving GRAS (Generally Recognized as Safe) status in 1997. In 2001, the US Food and Drug Administration (FDA) approved ozone as a secondary direct food additive and antimicrobial agent for all food categories. Its inactivation was outlined by Victorin (1992), showing two primary pathways: (i) oxidation of amino acids and sulfhydryl groups in proteins, enzymes, and peptides, leading to their degradation into smaller peptides; and (ii) oxidation of PUFA into acid peroxides. These oxidative processes cause damage to the microbial cell envelope, resulting in cell lysis and inactivation (Ersoy et al., 2019). Because of the high redox potential of ozone, it cannot be stored for a more extended period and is generally generated at the application site (Sharma et al., 2021).

Minor millets are highly nutritious grains vital for food security, but storage pests cause significant losses. Ozone treatment effectively disinfested *T. castaneum* in proso millet, with eggs being the most susceptible (LT99: 160 minutes at 750 ppmv, 8% moisture). No significant changes were observed in fat, protein, and ash content, while FTIR showed functional modifications, confirming ozone as a safe pest control method (Muniswamy et al., 2023). In another work, sorghum flour was ozonated at 0.06 L/min for 15–45 minutes, leading to a pH decrease and increased lightness (L^*), while reducing yellowness (b^*). Peak viscosity and gluten-free cake volume improved significantly with prolonged exposure, enhancing texture and slice brightness while reducing crumb firmness. However, ozonation did not enhance gluten-free bread volume, resulting in an open, ragged structure. These findings suggest that ozone treatment enhances the quality and acceptability of sorghum-based cakes (Marston et al., 2015). Ozonation (0.06–0.1 L/min for 60 minutes) improved the functionality of pearl millet and sorghum starch, enhancing WAC/OAC, solubility, paste clarity, and freeze–thaw stability. Swelling power declined with prolonged treatment, while starch digestibility decreased. Structural analysis confirmed surface modifications, reduced crystallinity, and minor FTIR peak intensity changes (Thakur et al., 2023). Moreover, ozone treatment (0.02–0.06 L/min for 15–30 minutes) improved ethanol production from tannin-rich sorghum, increasing fermentation efficiency to 90% (8–14% higher at 36 hours, 2–5% higher at 72 hours). Tannin content decreased (3.8% to 2.7%), while starch modification enhanced processing. Ozonated sorghum produced distillers' dried grains with lower residual starch (<1%) and higher crude protein (~35%) (Yan et al., 2012).

10.4.6 Ionizing Radiations

High-energy ionizing radiation, including γ-rays and electron beams, has been utilized at varying doses for food preservation. Low doses (~1 kGy) are effective in preventing sprouting in bulbs and tubers or slowing fruit ripening, while medium doses (1–10 kGy) help reduce food pathogens to extend shelf life. Higher doses (>10 kGy) are used for complete microbial inhibition and sterilization of food items (Mshelia et al., 2023). A 2.5 kGy irradiation dose significantly enhanced the phenolic content in pearl millet, proso millet, and finger millet, with DPPH AA ranging from 42.77% to 72.65%, indicating a positive impact on antioxidant properties and shelf life (Wani et al., 2021). A storage study evaluated the stability of dehulled and whole millets (foxtail millet, sorghum, and pearl millet) over 90 days after treatment. At 0.50 kGy, fungal reduction was negligible, whereas doses of 0.75 kGy or higher effectively inhibited fungal growth. Consequently, ionizing radiation was recommended as a safe postharvest method for whole and dehulled millets (Huang et al., 2021). Finger millet slurry was examined at different γ-radiation doses (2.5, 5, 7.5, and 10 kGy) using a cobalt-60 source at a dosage rate of 2.5 kGy/h. The results indicated a decline in TPC, with CO-14 and CO-15 varieties experiencing reductions of 23.71% and 29.39%, respectively, at 10 kGy compared to untreated samples. Conversely, the TFC of CO-14 and CO-15 flour exhibited an increase of 24.82% and 19.25%, respectively, at the highest dose of 10 kGy (Gowthamraj et al., 2021). γ-Irradiation (2–15 kGy) reduced moisture, pasting properties, malondialdehyde content, and lipoxygenase activity while increasing protein content and antioxidant enzyme activity in finger millet flour (Reddy & Viswanath, 2019). Moreover, a 10 kGy dose reduced ochratoxin A contamination by 74% in pearl millet flour. Although irradiation had minimal impact on fatty acid composition and antioxidant properties, it increased peroxide levels and led to 88.6% loss of vitamin A at 1 kGy (Mustapha et al., 2014). γ-Irradiation (1.5 kGy) of little millet flour significantly reduced peroxide values and delayed fat deterioration over 90 days. It also prevented insect infestation and fungal growth, which were observed in nonirradiated samples after 30–45 days. The findings highlight gamma irradiation as an effective method to extend the shelf life of millet flour with minimal biochemical degradation (Shobha et al., 2024).

Besides, electron beam (EB) irradiation is also a nonthermal, physical processing technique in which a stream of high-energy electrons generated by an electron accelerator is employed to irradiate various products (Liang et al., 2022). Recently, Li et al. (2025) investigated the effectiveness of EB irradiation in removing deoxynivalenol (DON) from standard solutions and DON-contaminated foxtail millet. At 20 kGy, DON removal reached 89.13% in aqueous solutions but was lower in millet grains (21.86–38.34%) and flour (26.71–51.17%). EB irradiation had minimal impact on millet's nutritional composition but enhanced the soluble solid content of cooked samples. Another study found notable reductions in polyphenol content in sorghum grains following gamma and electron beam irradiation, which was linked to enhanced protein complex formation (Rousta et al., 2014). EB irradiation reduced hardness and chewiness while increasing soluble solid content, enhancing its suitability for soft and sticky congee. The starch content, pasting properties, pH, and color were slightly affected at 2 kGy, which preserved volatile compounds better than at higher doses (Yuan et al., 2024).

10.4.7 UV Radiations

UV radiation is widely used for surface disinfection and decontaminating transparent foods, owing to its bactericidal properties and capacity to inactivate enzymes (Kumar et al., 2020). Lately, Prashanth et al. (2025) examined how UV-C treatment (power: 86 W; bed density: 5 mg/cm^2; time: 4 minutes) affects the functional properties of pearl millet flour, including WAC, OAC, WSI, EC, and viscosity. Results showed that WSI and WAC increased with higher UV power and bed density, while WSI decreased with prolonged exposure. OAC followed an opposite trend under similar conditions. Process optimization using ANN-GA and RSM revealed that the ANN model provided the best fit. Elevated UV-B radiation (ambient +7.2 kJ/m^2/day) influenced the physiological and biochemical responses of four tropical pearl millet cultivars. While MPMH-17 and MPMH-21 exhibited reduced grain biomass, HHB-67 and HHB-272 showed yield increases of 23.2% and 28%, respectively. UV-B exposure promoted secondary metabolite synthesis, including flavonoids, phenolics, and bioactive compounds like squalene and β-sitosterol, aiding stress adaptation and potentially offering health benefits (Singh & Choudhary, 2025). Moreover, exposure of foxtail millet to UV-B (43.9 μW/cm^2) was found to enhance the expression of genes responsible for carotenoid synthesis while simultaneously suppressing the expression of genes associated with carotenoid degradation (Huo et al., 2024). Harshitha et al. (2024) investigated the effects of ultraviolet-A (906,120 μWs/cm^2) radiation combined with thermal treatments on foxtail millet flour. Different treatment combinations, including milling, hydrothermal, and autoclaving, significantly influenced flour's physicochemical, rheological, and techno-functional properties. This combination enhanced starch structure, color, and WAC, making the flour suitable for diverse culinary applications such as soups, baking, beverages, and snacks.

10.4.8 Pulsed Light

An intense pulsed light (IPL) system is designed to minimize surface microorganisms by delivering high-intensity light quickly across a broad wavelength spectrum. This method provides a quick and efficient alternative to thermal sterilization. In the food industry, IPL treatment helps extend shelf life while preserving product quality. Several studies have explored the application of IPL in food processing (Hwang et al., 2021). Hwang et al. (2018) evaluated the bactericidal effect of IPL at energy fluences of 64.83–226.8 mJ/cm^2 per pulse, voltages of 1,200–2,400 V, pulse durations of 0.5–3.0 ms, a frequency of 2 Hz, and exposure times of 30–120 seconds (60–240 pulses) on glutinous millet. The belt-type IPL device achieved a maximum microbial reduction of 0.55 log/g, while the cyclone-type IPL reduced microbial load by 0.66 log/g. The effectiveness of IPL treatment varied based on grain size, with larger particles showing more significant reductions. However, limited research has been conducted on the impact of pulsed light treatment on millets.

10.4.9 Fermentation

Pearl millet in vitro protein and starch digestibility improves after 72 hours of natural fermentation at 20–30 °C. The most prominent increase, reaching 110.1%, occurs at 30 °C because of enzymatic hydrolysis (Suma & Urooj, 2017). Another study highlighted the enhancement of protein digestibility in pearl millet flour during fermentation over different periods—0, 8, and 16 hours. The highest In Vitro Protein Digestibility value (45.75–64.38%) was recorded at 16 hours of fermentation, likely due to microbial breakdown of antinutritional factors like phytates, production of proteolytic enzymes, and partial hydrolysis of complex storage proteins into more digestible and soluble forms (Nour et al., 2018). Similarly, fermenting finger millet flour with naturally occurring microflora for 24 hours reduced tannin and phytate content by 52% and 20%, respectively. This process significantly increased the bioavailability of key minerals, with iron rising by 27%, calcium by 20%, zinc by 26%, and phosphorus by 26%. The reduction in phytic acid levels was crucial in enhancing mineral absorption. Additionally, fermentation improved the vitamin content and amino acid profile, remarkably increasing cysteine and methionine levels, suggesting that fermented millet products contribute positively to dietary diversity (Konapur et al., 2014).

In finger millet, traditional fermentation led to a reduction in phenolic compounds by 41%, which was attributed to self-polymerization under acidic conditions (Chen et al., 2022). However, other studies indicated that fermentation can also enhance bioactive compounds. When *Rhizopus azygosporus* was used as a starter culture for a 10-day fermentation, the TPC of millet significantly increased from 6.6 to 21.8 mg GAE/g (Purewal et al., 2019). Furthermore, research by Balli et al. (2020) found that fermenting millet for 72 hours with *Lactobacilli* and yeast led to a 36% rise in phenolic content and a nearly 30% increase in flavonoids, such as vitexin and vitexin 2-O-rhamnoside. These flavonoids were found to have partial inhibitory effects on tyrosine phosphatase, an enzyme linked to type 2 diabetes. Vila-Real et al. (2022) demonstrated the advantages of co-fermenting finger millet with *Lactiplantibacillus plantarum* and *Weissella confusa*. This innovative fermentation method produced a beverage enriched with beneficial probiotic strains and higher levels of essential amino acids such as arginine, glutamine, threonine, and GABA. Joseph et al. (2025) recently formulated eight health drinks using barnyard millet and Spirulina, fermented with *Saccharomyces cerevisiae* and *Lactobacillus rhamnosus*. Protein content ranged from 1.39% to 4.61% in bacterial fermentation and 2.52–4.09% in yeast fermentation, while carbohydrate levels (5–14.8%) made them suitable for diabetics. Gas Chromatography–Mass Spectrometry detected oleic acid (anti-inflammatory, cholesterol-lowering) and hexadecenoic acid (Alzheimer's management).

10.5 Conclusion

Millet grains are important for supporting sustainable food systems and enhancing human health. Their exceptional nutritional value, resilience in diverse growing conditions, and versatility in culinary applications make them crucial in addressing global challenges like malnutrition and food insecurity. To maximize the potential of millets, it is important to embrace technological advancements, promote research, and enhance consumer awareness. Milling, dehulling, and decortication reduce micronutrients, protein, and fiber, with excessive processing causing significant nutrient loss. Proper control is essential to preserve millet's nutritional value. Thermal and nonthermal techniques enhance microbial stability and shelf life by inhibiting lipase activity. These methods improve millet properties while minimizing nutrient loss and optimizing energy use. Innovations in infrared dryers, microwave-assisted systems, and high-pressure processing units enable precise control over heating and microbial decontamination. Recent improvements in processing techniques have expanded the cooking possibilities for millets, resulting in a range of healthy products. Despite extensive research, gaps remain in large-scale production, emphasizing the need for further advancements in processing and industrial applications. Hybrid processing systems integrating automation and artificial intelligence-driven monitoring can optimize milling, polishing, and dehulling, promoting large-scale commercialization while ensuring sustainability and regulatory compliance. By promoting millet farming and new ideas, those involved can help create a stronger food system. Targeted investments and supportive policies will help fully realize the benefits of millets for a healthier, more sustainable future.

References

Aboud, S. A., Altemimi, A. B., Al-Hilphy, R. S., & A., Yi-Chen, L., & Cacciola, F. (2019). A comprehensive review on infrared heating applications in food processing. *Molecules, 24*(22), 4125.

Acquisgrana, M. D. R., Gómez Pamies, L. C., & Benítez, E. I. (2019). *Hydrothermal treatment to remove tannins in wholegrains sorghum, milled grains and flour* (Vol. 3, pp. 122).

Adebiyi, J. A., Obadina, A. O., Adebo, O. A., & Kayitesi, E. (2017). Comparison of nutritional quality and sensory acceptability of biscuits obtained from native, fermented, and malted pearl millet (Pennisetum glaucum) flour. *Food Chemistry, 232*, 210–217.

Adebowale, O. J., Taylor, J. R., & de Kock, H. L. (2020). Stabilization of wholegrain sorghum flour and consequent potential improvement of food product sensory quality by microwave treatment of the kernels. *LWT-Food Science and Technology, 132*, 109827.

Aher, R. R., Reddy, P. S., Bhunia, R. K., Flyckt, K. S., Shankhapal, A. R., Ojha, R., Everard, J. D., Wayne, L. L., Ruddy, B. M., Deonovic, B., Gupta, S. K., Sharma, K. K., & Bhatnagar-Mathur, P. (2022). Loss-of-function of triacylglycerol lipases are associated with low flour rancidity in pearl millet [*Pennisetum glaucum* (L.) R. Br.]. *Frontiers in Plant Science, 13*, 962667.

Alkaltham, M. S., Qasem, A. A., Ibraheem, M. A., & Hassan, A. B. (2024). Exploring ultrasonic energy followed by natural fermentation processing to enhance functional properties and bioactive compounds in millet (*Pennisetum glaucum* L.) grains. *Fermentation, 10*(11), 590.

Alkhafaji, S. R., & Farid, M. (2007). An investigation on pulsed electric fields technology using new treatment chamber design. *Innovative Food Science & Emerging Technologies, 8*(2), 205–212.

Almaiman, S. A., Albadr, N. A., Alsulaim, S., Alhuthayli, H. F., Osman, M. A., & Hassan, A. B. (2021). Effects of microwave heat treatment on fungal growth, functional properties, total phenolic content, and antioxidant activity of sorghum (*Sorghum bicolor* L.) grain. *Food Chemistry, 348*, 128979.

Amarnath, M. S., Muhammed, A., Antony, A. K., Malini, B., & Sunil, C. K. (2023). White finger millet starch: Physical modification (annealing and ultrasound), and its impact on physicochemical, functional, thermal and structural properties. *Food and Humanity, 1*, 599–606.

Angioloni, A., & Collar, C. (2012). Effects of pressure treatment of hydrated oat, finger millet and sorghum flours on the quality and nutritional properties of composite wheat breads. *Journal of Cereal Science, 56*(3), 713–719.

Arsha, S., Abdullah, S., Navya, K. P., Sunooj, K. V., Abhishek, S., & Sudheer, K. P. (2025). Cold plasma-induced modification of finger millet (*Eleusine coracana*) flour: unveiling potential for anti-nutrient reduction and functional enrichment. *European Food Research and Technology, 251*, 1075–1095.

Babiker, E., Abdelseed, B., Hassan, H., & Adiamo, O. (2018). Effect of decortication methods on the chemical composition, antinutrients, Ca, P and Fe contents of two pearl millet cultivars during storage. *World Journal of Science, Technology and Sustainable Development, 15*(3), 278–286.

Balasubramanian, S., Deshpande, S. D., & Bothe, I. R. (2020). Design, development and performance evaluation of CIAE-millet mill. *Agricultural Mechanization in Asia, Africa, and Latin America, 51*(1), 42–48.

Balasubramanian, S., & Viswanathan, R. (2010). Influence of moisture content on physical properties of minor millets. *Journal of Food Science and Technology*, *47*(3), 279–284.

Balli, D., Bellumori, M., Pucci, L., Gabriele, M., Longo, V., Paoli, P., Melani, F., Mulinacci, N., & Innocenti, M. (2020). Does fermentation really increase the phenolic content in cereals? A study on millet. *Foods*, *9*(3), 303.

Bangar, S. P., Singh, A., Ashogbon, A. O., & Bobade, H. (2023). Ball-milling: A sustainable and green approach for starch modification. *International Journal of Biological Macromolecules*, *237*, 124069.

Benincasa, P., Falcinelli, B., Lutts, S., Stagnari, F., & Galieni, A. (2019). Sprouted grains: A comprehensive review. *Nutrients*, *11*(2), 1–29.

Beta, T., & Ndolo, V. U. (2019). Postharvest technologies. In J. R. N. Taylor & K. G. Duodu (Eds.), *Sorghum and millets* (pp. 69–84). AACC International Press.

Bhandari, G. R., Gauchan, D., Bhandari, B., Joshi, B. K., & Panta, S. (2020). Advancement, simplification and piloting of electrical proso millet dehusker (chino kutak) for the mountain farmers. Traditional crop biodiversity for mountain food and nutrition security in Nepal. *Journal of Nepal Agricultural Research Council*, 75–81. https://doi.org/10.3126/jnarc.v7i1.36922

Bhandari, G. R., Gauchan, D., Bhandari, B., Joshi, B. K., & Panta, S. (2021). Advancement, simplification and piloting of electrical proso millet dehusker (chino kutak). *Journal of Nepal Agricultural Research Council*, *7*, 67–74.

Bhangu, S. K., Singla, R., Colombo, E., Ashokkumar, M., & Cavalieri, F. (2018). Sono-transformation of tannic acid into biofunctional ellagic acid micro/nanocrystals with distinct morphologies. *Green Chemistry*, *20*(4), 816–821.

Bi, S., Xu, X., Luo, D., Lao, F., Pang, X., Shen, Q., Hu, X., & Wu, J. (2020). Characterization of key aroma compounds in raw and roasted peas (*Pisum sativum* L.) by application of instrumental and sensory techniques. *Journal of Agricultural and Food Chemistry*, *68*(9), 2718–2727.

Birania, S., Rohilla, P., Kumar, R., & Kumar, N. (2020). Post harvest processing of millets: A review on value added products. *International Journal of Chemical Studies*, *8*(1), 1824–1829.

Bisht, A. (2022). Processing of small millets. In S. Srivastava (Ed.), *Small millet grains: The superfoods in human diet* (pp. 49–77). Springer Nature Singapore.

Bucsella, B., Takács, Á., Vizer, V., Schwendener, U., & Tömösközi, S. (2016). Comparison of the effects of different heat treatment processes on rheological properties of cake and bread wheat flours. *Food Chemistry*, *190*, 990–996.

Charu, C., Vignesh, S., Chidanand, D. V., Mahendran, R., & Baskaran, N. (2024). Impact of cold plasma on pearl and barnyard millets' microbial quality, antioxidant status, and nutritional composition. *Food and Humanity*, *2*, 100238.

Chávez-Murillo, C. E., Aceves-Flores, M. S., Verástegui-Quevedo, M., & de la Rosa-Millán, J. (2021). Influence of starch and protein molecular interactions on the in vitro digestion characteristics of biscuits partially substituted with pulse flours. *International Journal of Food Science and Technology*, *56*(7), 3388–3399.

Chen, C., Yin, J., Jing, P., & Jiao, S. (2024). Effects of hot air-assisted radio frequency stabilization of brown millet and millet bran: Enzyme inactivation, microstructure, nutritional quality, and storage stability. *Innovative Food Science & Emerging Technologies*, *91*, 103529.

Chen, Y., Zhu, Y., & Qin, L. (2022). The cause of germination increases the phenolic compound contents of Tartary buckwheat (Fagopyrum tataricum). *Journal of Future Foods, 2*(4), 372–379.

Chikkaballapur Krishnappa, S., Saravanan, S., & Natarajan, V. (2022). Development and performance evaluation of foxtail millet (*Setaria italica* L.) dehuller. *Journal of Food Process Engineering, 45*(6), e13937.

Cordelino, I. G., Tyl, C., Inamdar, L., Vickers, Z., Marti, A., & Ismail, B. P. (2019). Cooking quality, digestibility, and sensory properties of proso millet pasta as impacted by amylose content and prolamin profile. *LWT-Food Science and Technology, 99*, 1–7.

Das, I., Kumar, G., & Shah, N. G. (2013). Microwave heating as an alternative quarantine method for disinfestation of stored food grains. *International Journal of Food Science, 2013*(1), 926468.

D'cruz, V., Chandran, M., Athmaselvi, K. A., Rawson, A., & Natarajan, V. (2023). Ohmic heating using electrolytes for paddy parboiling: A study on thermal profile, electrical conductivity, milling quality, and nutritional attributes. *Journal of Food Process Engineering, 46*(3), e14276.

Dharmaraj, U., Meera, M. S., Reddy, S. Y., & Malleshi, N. G. (2015). Influence of hydrothermal processing on functional properties and grain morphology of finger millet. *Journal of Food Science and Technology, 52*, 1361–1371.

Dias-Martins, A. M., Cappato, L. P., da Costa, M. M., Rodrigues, F. N., Pacheco, S., & Carvalho, C. W. P. (2019). Impacts of ohmic heating on decorticated and whole pearl millet grains compared to openpan cooking. *Journal of Cereal Science, 85*, 120–129.

Dias-Martins, A. M., Trombete, F. M., Cappato, L. P., Chávez, D. W., Santos, M. B., & Carvalho, C. W. (2024). Processing, composition, and technological properties of decorticated, sprouted, and extruded pearl millet (*Pennisetum glaucum* (L.) R. Br.) flours. *Journal of Food Process Engineering, 47*(2), e14561.

Diouf, A., Sarr, F., Sene, B., Ndiaye, C., Fall, S. M., & Ayessou, N. C. (2019). Pathways for reducing anti-nutritional factors: Prospects for Vigna unguiculata. *Journal of Nutritional Health & Food Science, 7*(2), 1–10.

Dubey, A., & Tripathy, P. P. (2024). Ultrasound-mediated hydration of finger millet: Effects on antinutrients, techno-functional and bioactive properties, with evaluation of ANN-PSO and RSM optimization methods. *Food Chemistry, 435*, 137516.

Durairaj, M., Gurumurthy, G., Nachimuthu, V., Muniappan, K., & Balasubramanian, S. (2019). Dehulled small millets: The promising nutricereals for improving the nutrition of children. *Maternal & Child Nutrition, 15*, e12791.

Ersoy, Z. G., Barisci, S., & Dinc, O. (2019). Mechanisms of the *Escherichia coli* and *Enterococcus faecalis* inactivation by ozone. *LWT-Food Science and Technology, 100*, 306–313.

Estrada-Girón, Y., Swanson, B. G., & Barbosa-Cánovas, G. V. (2005). Advances in the use of high hydrostatic pressure for processing cereal grains and legumes. *Trends in Food Science & Technology, 16*(5), 194–203.

Fadiji, T., Ashtiani, S. H. M., Onwude, D. I., Li, Z., & Opara, U. L. (2021). Finite element method for freezing and thawing industrial food processes. *Foods, 10*(4), 869.

Gaikwad, P. S., Sarma, C., Negi, A., & Pare, A. (2021). Alternate Food Preservation Technology. In *Food chemistry: The role of additives, preservatives and adulteration* (pp. 275–340). Wiley.

Gaurav, A., Panigrahi, S. S., Pradhan, R. C., & Mishra, S. (2022). Co-rotating extrusion cooking impact on product characteristics using hulled kodo millet and hybrid maize flour. *Journal of Food Science and Technology*, *59*(11), 4165–4175.

Gaytán-Martínez, M., Cabrera-Ramírez, Á. H., Morales-Sánchez, E., Ramírez-Jiménez, A. K., Cruz-Ramírez, J., Campos-Vega, R., Velazquez, G., Loarca-Piña, G., & Mendoza, S. (2017). Effect of nixtamalization process on the content and composition of phenolic compounds and antioxidant activity of two sorghums varieties. *Journal of Cereal Science*, *77*, 1–8.

Ge, F., Xia, R., Wu, D., Cheng, W., Meng, L., Wang, Z., & Tang, X. (2023). Toward a comprehensive understanding of various milling methods on the physicochemical properties of highland barley flours and eating quality of corresponding sugar-free cookies. *Food Chemistry*, *413*, 135657.

Gharibzahedi, S. M. T., & Smith, B. (2021). Legume proteins are smart carriers to encapsulate hydrophilic and hydrophobic bioactive compounds and probiotic bacteria: A review. *Comprehensive Reviews in Food Science and Food Safety*, *20*(2), 1250–1279.

Goel, K., Goomer, S., & Aggarwal, D. (2021). Formulation and optimization of value-added barnyard millet vermicelli using response surface methodology. *Asian Journal of Dairy and Food Research*, *40*(1), 55–61.

Gomez Chandrasekaran, S., Ramanathan, S., & Basak, T. (2013). Microwave food processing—A review. *Food Research International*, *52*(1), 243–261.

Gopal, D. S., & Bhuvana, S. (2021). Effect of microwave treatment on phytochemical, functional and rheological property of kodo millet (*Paspalum scrobiculatum*). *Pharma Innovation Journal*, *10*(11), 397–401.

Gopika, M., & Joshi, S. (2024). Elucidating the effect of nixtamalization on the functional, nutritional, and microstructure properties of pearl millet flour. *Journal of Cereal Science*, *117*, 103891.

Gowda, N. N., Siliveru, K., Prasad, P. V., Bhatt, Y., Netravati, B. P., & Gurikar, C. (2022). Modern processing of Indian millets: A perspective on changes in nutritional properties. *Food*, *11*(4), 499.

Gowda, N. N., Taj, F., Subramanya, S., & Ranganna, B. (2020). Development a table top centrifugal dehuller for small millets. *AMA Agricultural Mechanization in Asia, Africa and Latin America*, *51*, 72–78.

Gowthamraj, G., Jubeena, C., & Sangeetha, N. (2021). The effect of γ-irradiation on the physicochemical, functional, proximate, and antinutrient characteristics of finger millet (CO14 & CO15) flours. *Radiation Physics and Chemistry*, *183*, 109403.

Gulati, P., Sabillón, L., & Rose, D. J. (2018). Effects of processing method and solute interactions on pepsin digestibility of cooked proso millet flour. *Food Research International*, *109*, 583–588.

Hama, F., Icard-Vernière, C., Guyot, J. P., Picq, C., Diawara, B., & Mouquet-Rivier, C. (2011). Changes in micro-and macronutrient composition of pearl millet and white sorghum during in field versus laboratory decortication. *Journal of Cereal Science*, *54*(3), 425–433.

Harshitha, T., Ramesh, J., Gawali, P. P., Adusumilli, S., Dasalkar, A. H., & Yannam, S. K. (2024). Impact of ultraviolet radiation coupled with thermal treatment on the foxtail millet flour properties. *Journal of Food Measurement and Characterization*, *18*(8), 7145–7159.

Hassan, A. B., von Hoersten, D., & Ahmed, I. A. M. (2019). Effect of radio frequency heat treatment on protein profile and functional properties of maize grain. *Food Chemistry*, *271*, 142–147.

Hegde, K. R., Buvaneswaran, M., Bhavana, M. R., Sinija, V. R., Rawson, A., & Hema, V. (2025). Effects of ultrasound and high-pressure assisted extraction of pearl millet protein isolate: Functional, digestibility, and structural properties. *International Journal of Biological Macromolecules, 289*, 138877.

Himashree, P., & Mahendran, R. (2025). Effect of high-pressure soaking on the physicochemical, nutritional, and techno-functional properties of pearl millets. *Sustainable Food Technology., 3*, 714–724.

Huang, H. H., Dikkala, P. K., Sridhar, K., Yang, H. T., Lee, J. T., & Tsai, F. J. (2021). Effect of heat and γ-irradiation on fungal load, pasting, and rheological characteristics of three whole and dehulled millets during storage. *Journal of Food Processing and Preservation, 45*(4), e15355.

Huang, J., Yu, H., Guan, X., Wang, G., & Guo, R. (2016). Accelerated dryland expansion under climate change. *Nature Climate Change, 6*(2), 166–171.

Huang, R., Pan, X., Lv, J., Zhong, W., Yan, F., Duan, F., & Jia, L. (2018). Effects of explosion puffing on the nutritional composition and digestibility of grains. *International Journal of Food Properties, 21*(1), 2193–2204.

Huo, Y., Pang, X., Feng, H., Hao, R., Zhang, R., Wang, Y., Wang, H., Zhang, B., & Han, Y. (2024). Effect of ultraviolet B radiation on the biosynthesis of carotenoids in foxtail millet grains. *Agriculture, 14*(2), 289.

Hwang, H. J., Cheigh, C. I., & Chung, M. S. (2018). Comparison of bactericidal effects of two types of pilot-scale intense-pulsed-light devices on cassia seeds and glutinous millet. *Innovative Food Science & Emerging Technologies, 49*, 170–175.

Hwang, H. J., Yee, S. Y., & Chung, M. S. (2021). Decontamination of powdery foods using an intense pulsed light (IPL) device for practical application. *Applied Sciences, 11*(4), 1518.

Idowu, A. T., Olatunde, O. O., Adekoya, A. E., & Idowu, S. (2020). Germination: An alternative source to promote phytonutrients in edible seeds. *Food Quality and Safety, 4*(3), 129–133.

Jaddu, S., Abdullah, S., Dwivedi, M., & Pradhan, R. C. (2023). Optimization of functional properties of plasma treated kodo millet (open air multipin) using response surface methodology (RSM) and artificial neural network with genetic algorithm (ANN-GA). *Journal of Food Process Engineering, 46*(10), e14207.

Jaddu, S., Dwivedi, M., & Pradhan, R. C. (2024b). Exploring the impact of cold plasma treatment on rheological characteristics of little millet flour. *Journal of Food Measurement and Characterization, 18*(7), 5622–5628.

Jaddu, S., Pradhan, R. C., & Dwivedi, M. (2022). Effect of multipin atmospheric cold plasma discharge on functional properties of little millet (*Panicum miliare*) flour. *Innovative Food Science & Emerging Technologies, 77*, 102957.

Jaddu, S., Sahoo, S., Sonkar, S., Alzahrani, K., Dwivedi, M., Misra, N. N., & Pradhan, R. C. (2024a). Cold plasma treatment of little millet flour: Impact on bioactives, antinutritional factors and functional properties. *Plant Foods for Human Nutrition, 79*(2), 503–510.

Jaybhaye, R. V., Pardeshi, I. L., Vengaiah, P. C., & Srivastav, P. P. (2014). Processing and technology for millet based food products: A review. *Journal of Ready to Eat Food, 1*(2), 32–48.

Jena, K., Vairakannu, P., & Singha, S. (2025). Microwave drying of sprouted little millet and assessment of their kinetic and physiochemical properties. *Journal of Cereal Science, 122*, 104122.

Jhan, F., Gani, A., Noor, N., Malla, B. A., & Ashwar, B. A. (2022). Nano reduction coupled with encapsulation as a novel technique for utilising millet proteins as future foods. *Ultrasonics Sonochemistry, 86*, 106006.

Jin, S., Xiao, C., Lu, H., & Deng, X. (2025). Effects of extrusion temperature on structure and physicochemical properties of proso millet starch. *International Journal of Biological Macromolecules, 299*, 140011.

Joseph, B., Bhavadharani, M., Nivetha, S., Lavanya, M., Baskaran, N., & Vignesh, S. (2025). Synbiotic fermented barnyard millet drink: Exploring its nutritional profile, sensory attributes, and bioactive health potentials. *Food Chemistry Advances, 6*, 100872.

Joshi, B. K. (2025). The 7-brother millets: a pathway to global food, nutrition, health, business, and environmental security. *Journal of Crop Improvement, 39*, 167–191.

Joshi, J., & Rao, P. S. (2024). Characterization and multivariate analysis of decortication-induced changes in pearl millet. *Journal of Food Composition and Analysis, 125*, 105788.

Joshi, T. J., Singh, S. M., & Rao, P. S. (2023). Novel thermal and non-thermal millet processing technologies: advances and research trends. *European Food Research and Technology, 249*(5), 1149–1160.

Kabeer, S., Govindarajan, N., Radhakrishnan, P., Essa, M. M., & Qoronfleh, M. W. (2023). Effect of drying technique on physiochemical and nutritional properties of *Eleusine coracana* (finger millet) porridge powder. *Journal of Food Science and Technology, 60*(12), 3024–3034.

Kabui, K. K., & Athmaselvi, K. A. (2024). Ohmic heating parboiling as a novel pretreatment to enhance the milling properties of Kodo millet. *Journal of Cereal Science, 120*, 104049.

Kala, B. K., & Mohan, V. R. (2012). Effect of microwave treatment on the antinutritional factors of two accessions of velvet bean, *Mucuna pruriens* (L.) DC. var. *utilis* (Wall. ex Wight) Bak. ex Burck. *International Food Research Journal, 19*(3), 961–969.

Kaleemullah, S., Reddyand, M. R., & Prabhakar, B. (2023). Performance evaluation of minor millet processing machines suitable for small and medium scale industries. *Pharma Innovation Journal, 12*, 2726–2729.

Kharat, S., Medina-Meza, I. G., Kowalski, R. J., Hosamani, A., Ramachandra, C. T., Hiregoudar, S., & Ganjyal, G. M. (2019). Extrusion processing characteristics of whole grain flours of select major millets (foxtail, finger, and pearl). *Food and Bioproducts Processing, 114*, 60–71.

Kishore, S. G., Dwivedi, M., Dalbhagat, C. G., Thota, N., Singh, C. N., Saravanan, S., Nikitha, G., Kumar, N. V., & Vignesh, V. (2024). Development of extruded millet analogue rice: A comparison of rheological, structural, physiochemical, and cooking properties with conventional rice. *ACS Food Science & Technology, 4*(10), 2504–2516.

Konapur, A., Gavaravarapu, S. R. M., Gupta, S., & Nair, K. M. (2014). Millets in meeting nutrition security: Issues and way forward for India. *Indian Journal of Nutrition and Dietetics, 51*, 306–321.

Krishnan, R., Dharmaraj, U., & Malleshi, N. G. (2012). Influence of decortication, popping and malting on bioaccessibility of calcium, iron and zinc in finger millet. *LWT-Food Science and Technology, 48*(2), 169–174.

Kumar, A., Metwal, M., Kaur, S., Gupta, A. K., Puranik, S., Singh, S., Singh, M., Gupta, S., Babu, B. K., Sood, S., & Yadav, R. (2016a). Nutraceutical value of finger millet [*Eleusine coracana* (L.) Gaertn.], and their improvement using omics approaches. *Frontiers in Plant Science, 7*, 934.

Kumar, A., Rani, P., Purohit, S. R., & Rao, P. S. (2020). Effect of ultraviolet irradiation on wheat (*Triticum aestivum*) flour: Study on protein modification and changes in quality attributes. *Journal of Cereal Science, 96*, 103094.

Kumar, A., Tomer, V., Kaur, A., Kumar, V., & Gupta, K. (2018). Millets: A solution to agrarian and nutritional challenges. *Agriculture & Food Security, 7*(1), 1–15.

Kumar, G., Upadhyay, S., Yadav, D. K., Malakar, S., Dhurve, P., & Suri, S. (2023). Application of ultrasound technology for extraction of color pigments from plant sources and their potential bio-functional properties: A review. *Journal of Food Process Engineering, 46*(6), e14238.

Kumar, P., Rudra, S. G., Varghese, E., & Kaur, C. (2016b). Extrusion conditions effects functional and pasting properties of finger millet. *Vegetos, 29*, 2.

Kumari, A., Sadh, P. K., Kamboj, A., Yadav, B., Kumar, A., Sivakumar, S., Surekha, Saharan, B. S., Brar, B., Goyal, C., & Duhan, J. S. (2025). Exploring the benefits of nutrition of little millet: Unveiling the effect of processing methods on bioactive properties. *Journal of Food Biochemistry, 2025*(1), 2488816.

Kumari, V., & Jayadeep, A. (2024). Effect of infrared heatingon finger millet in unhydrated and hydrated condition on physical, functional, pasting and phenolic profile. *Journal of Cereal Science, 116*, 103834.

Latha Ravi, J., & Rana, S. S. (2024). Maximizing the nutritional benefits and prolonging the shelf life of millets through effective processing techniques: A review. *ACS Omega, 9*(37), 38327–38347.

Lemmens, E., Moroni, A. V., Pagand, J., Heirbaut, P., Ritala, A., Karlen, Y., ... Delcour, J. A. (2019). Impact of cereal seed sprouting on its nutritional and technological properties: A critical review. *Comprehensive Reviews in Food Science and Food Safety, 18*(1), 305–328.

Li, B., Zhao, L., Chen, H., Sun, D., Deng, B., Li, J., ... & Wang, F. (2016). Inactivation of lipase and lipoxygenase of wheat germ with temperature-controlled short wave infrared radiation and its effect on storage stability and quality of wheat germ oil. *PloS One, 11*(12), e0167330.

Li, M., He, S., Yang, S., Peng, N., Shen, H., Liu, X., Zhang, Y., & Zhang, H. (2025). Deoxynivalenol removal and quality evaluation of foxtail millet under electron beam irradiation. *Food Control, 174*, 111245.

Liang, W., Zhao, W., Liu, X., Zheng, J., Sun, Z., Ge, X., ... & Li, W. (2022). Understanding how electron beam irradiation doses and frequencies modify the multiscale structure, physicochemical properties, and in vitro digestibility of potato starch. *Food Research International, 162*, 111947.

Lochte-Watson, K. R., Weller, C. L., & Jackson, D. S. (2000). PH—postharvest technology: fractionation of grain sorghum using abrasive decortication. *Journal of Agricultural Engineering Research, 77*(2), 203–208.

Lohani, U. C., & Muthukumarappan, K. (2016). Application of the pulsed electric field to release bound phenolics in sorghum flour and apple pomace. *Innovative Food Science & Emerging Technologies, 35*, 29–35.

Lohani, U. C., & Muthukumarappan, K. (2020). Influence of fermentation followed by ultrasonication on functional properties of sorghum extrudates. *Journal of Food Process Engineering, 43*(6), e13390.

Lohani, U. C., Pandey, J. P., & Shahi, N. C. (2012). Effect of degree of polishing on milling characteristics and proximate compositions of barnyard millet (*Echinochloa frumentacea*). *Food and Bioprocess Technology, 5*, 1113–1119.

Lohita, B., & Srijaya, M. (2024). Novel technologies for shelf-life extension of food products as a competitive advantage: A review. *Food Production, Diversity, and Safety Under Climate Change*, 285–306.

Lokeswari, R., Sharanyakanth, P. S., Jaspin, S., & Mahendran, R. (2021). Cold plasma effects on changes in physical, nutritional, hydration, and pasting properties of pearl millet (*Pennisetum glaucum*). *IEEE Transactions on Plasma Science*, *49*(5), 1745–1751.

Manickam, S., Manickam, L., Natarajan, V., & Rajamani, M. (2024). Hot air and infrared heating methods as alternative control methods against *Tribolium castaneum* (Herbst.) and their effect on the physiochemical properties of finger millet (*Eleusine corocana*) flour. *Journal of Stored Products Research*, *106*, 102307.

Marston, K., Khouryieh, H., & Aramouni, F. (2015). Evaluation of sorghum flour functionality and quality characteristics of gluten-free bread and cake as influenced by ozone treatment. *Food Science and Technology International*, *21*(8), 631–640.

Meena, L., Malini, B., Byresh, T. S., Sunil, C. K., Rawson, A., & Venkatachalapathy, N. (2024). Ultrasound as a pre-treatment in millet-based probiotic beverage: It's effect on fermentation kinetics and beverage quality. *Food Chemistry Advances*, *4*, 100631.

Mirza, N., Sharma, N., Srivastava, S., & Kumar, A. (2015). Variation in popping quality related to physical, biochemical and nutritional properties of finger millet genotypes. *Proceedings of the National Academy of Sciences, India Section B: Biological Sciences*, *85*, 507–515.

Mirzababaee, S. M., Ozmen, D., Hesarinejad, M. A., Toker, O. S., & Yeganehzad, S. (2022). A study on the structural, physicochemical, rheological and thermal properties of high hydrostatic pressurized pearl millet starch. *International Journal of Biological Macromolecules*, *223*, 511–523.

Mishra, G., Joshi, D. C., & Panda, B. K. (2014). Popping and puffing of cereal grains: A review. *Journal of Grain Processing and Storage*, *1*(2), 34–46.

Mohamed, M. E., & Eissa, A. H. A. (2012). Pulsed electric fields for food processing technology. *Structure and Function of Food Engineering*, *11*, 275–306.

Mohandoss, S., Mohan, H., Balasubramaniyan, N., Assadi, A. A., & Loganathan, S. (2024). Pearl Millet seed surface modification and improved germination by non-thermal plasma discharge: understanding the role of reactive species. *Plasma Chemistry and Plasma Processing*, *44*(2), 1031–1051.

Mollakhalili-Meybodi, N., Yousefi, M., Nematollahi, A., & Khorshidian, N. (2021). Effect of atmospheric cold plasma treatment on technological and nutrition functionality of protein in foods. *European Food Research and Technology*, *247*, 1579–1594.

Monica, V., Anbarasan, R., & Mahendran, R. (2025). Cold plasma-induced changes in the structural and techno-functional properties of sprouted foxtail millet protein concentrate. *Food and Bioprocess Technology*, *18*(1), 850–867.

Mshelia, R. D. Z., Dibal, N. I., & Chiroma, S. M. (2023). Food irradiation: An effective but under-utilized technique for food preservations. *Journal of Food Science and Technology*, *60*(10), 2517–2525.

Muniswamy, K., Sugumar, A., Yarrakula, S., Manickam, L., & Chidanand, D. V. (2023). Effect of ozone fumigation on controlling common storage pest *Tribolium castaneum* (Herbst) in proso millet during storage. *Ozone: Science & Engineering*, *45*(6), 543–559.

Mushtaq, B. S., Waleed, A. A., Dhungle, A., Haq, F., Mahdi, A. A., Walayat, N., Manzoor, M. S., Nawaz, A., Fan, M., Qian, H., Jinxin, L., & Wang, L. (2021). Influence of pretreatments combined with extrusion on γamino butyric acid, nutritional composition and physicochemical properties of foxtail millet (*Setaria italica*). *Journal of Cereal Science, 102*, 103359.

Mustač, N. Č., Voučko, B., Novotni, D., Drakula, S., Gudelj, A., Dujmić, F., & Ćurić, D. (2019). Optimization of high intensity ultrasound treatment of proso millet bran to improve physical and nutritional quality. *Food Technology and Biotechnology, 57*(2), 183–190.

Mustapha, M. B., Bousselmi, M., Jerbi, T., Bettaïeb, N. B., & Fattouch, S. (2014). Gamma radiation effects on microbiological, physico-chemical and antioxidant properties of Tunisian millet (*Pennisetum glaucum* LR Br.). *Food Chemistry, 154*, 230–237.

Najdi Hejazi, S., & Orsat, V. (2017). Optimization of the malting process for nutritional improvement of finger millet and amaranth flours in the infant weaning food industry. *International Journal of Food Sciences and Nutrition, 68*(4), 429–441.

Navyashree, N., Sengar, A. S., Sunil, C. K., & Venkatachalapathy, N. (2022). White Finger Millet (KMR-340): A comparative study to determine the effect of processing and their characterisation. *Food Chemistry, 374*, 131665.

Neilson, A. P., Hopf, A. S., Cooper, B. R., Pereira, M. A., Bomser, J. A., & Ferruzzi, M. G. (2007). Catechin degradation with concurrent formation of homo-and heterocatechin dimers during in vitro digestion. *Journal of Agricultural and Food Chemistry, 55*(22), 8941–8949.

Nour, A. A. M., Mohamed, A. R., Adiamo, O. Q., & Babiker, E. E. (2018). Changes in protein nutritional quality as affected by processing of millet supplemented with Moringa seed flour. *Journal of the Saudi Society of Agricultural Sciences, 17*(3), 275–281.

Obadina, A., Ishola, I. O., Adekoya, I. O., Soares, A. G., de Carvalho, C. W. P., & Barboza, H. T. (2016). Nutritional and physico-chemical properties of flour from native and roasted whole grain pearl millet (*Pennisetum glaucum* [L.] R. Br.). *Journal of Cereal Science, 70*, 247–252.

de Oliveira Barros, M., Mattos, A. L. A., de Almeida, J. S., de Freitas Rosa, M., & de Brito, E. S. (2023). Effect of ball-milling on starch crystalline structure, gelatinization temperature, and rheological properties: Towards enhanced utilization in thermosensitive systems. *Foods, 12*(15), 2924.

Owusu-Kwarteng, J., & Akabanda, F. (2013). Applicability of nixtamalization in the processing of millet-based maasa, a fermented food in Ghana. *Journal of Food Research, 2*(1), 59.

Panda, S. K., & Shetty, P. H. (2018). *Innovations in technologies for fermented food and beverage industries* (pp. 20203174920). Springer.

Pandey, S., & Waghmare, R. (2019). Nixtamalization and its applicability for the processing of food grains particularly maize. *Indian Food Industry Mag, 1*, 55–59.

Pandiselvam, R., Özaslan, Z. T., Sahni, P., Khanashyam, A. C., Kutlu, N., Yilmaz, M. S., Isleroglu, H., Ramniwas, S., & Rustagi, S. (2024). High pressure homogenization for preservation of liquid foods-mechanisms, molecular modifications and recent developments. *Future Foods, 10*, 100488.

Pankaj, S. K., Wan, Z., & Keener, K. M. (2018). Effects of cold plasma on food quality: A review. *Foods, 7*(1), 4.

Pathare, A. M., Singhal, R. S., Rao, B. D., & Gokhale, J. S. (2024). Extrusion inhibits lipolytic enzymes in pearl millet (*Pennisetum glaucum*) flour. *Innovative Food Science & Emerging Technologies, 93,* 103605.

Patil, M., Gupta, A., & Nayak, P. (2018). Design, development and performance evaluation of minor millet de-husker. *International Journal of Farm Sciences, 8*(1), 63.

Peng, Z., Wang, F., Yu, L., Jiang, B., Cao, J., Sun, Z., & Cheng, J. (2024). Effect of ultrasound on the characterization and peptidomics of foxtail millet bran protein hydrolysates. *Ultrasonics Sonochemistry, 110,* 107044.

Perumal, M., Saravanan, A., Kommuguri, S. L., Alex, P., Sunooj, K. V., Mani, M., Bharathi, P., & Sinha, S. K. (2024). Extraction of energetic N2 neutrals for efficient plasma food processing of finger millet flour. *Plasma Chemistry and Plasma Processing, 44*(1), 471–485.

Powar, R. V., Aware, V. V., Patil, S. B., & Shahare, P. U. (2019). Development and evaluation of finger millet thresher-cum-pearler. *Journal of Biosystems Engineering, 44*(4), 194–206.

Prashanth, P., Joshi, T. J., Singh, S. M., & Rao, P. S. (2024). Hydrothermal treatment of pearl millet grains: effects on nutritional composition, antinutrients and flour properties. *Grain & Oil Science and Technology, 7*(2), 87–95.

Prashanth, P., Joshi, T. J., Singh, S. M., & Rao, P. S. (2025). Effect of UV-C treatment on functional properties of pearl millet flour: Process modelling and optimisation. *Journal of Food Measurement and Characterization, 19*(2), 1064–1078.

Purewal, S. S., Sandhu, K. S., Salar, R. K., & Kaur, P. (2019). Fermented pearl millet: A product with enhanced bioactive compounds and DNA damage protection activity. *Journal of Food Measurement and Characterization, 13,* 1479–1488.

Rahate, K. A., Madhumita, M., & Prabhakar, P. K. (2021). Nutritional composition, anti-nutritional factors, pretreatments-cum-processing impact and food formulation potential of faba bean (*Vicia faba* L.): A comprehensive review. *LWT-Food Science and Technology, 138,* 110796.

Rajaiah, P., Kumar, C. S., Laxman, B., Vennela, B., & Reddy, A. P. (2025). Performance evaluation of motorized millet thresher. *Journal of Scientific Research and Reports, 31*(1), 91–96.

Rajeswari, J. R., Guha, M., Jayadeep, A., & Rao, B. S. (2015). Effect of alkaline cooking on proximate, phenolics and antioxidant activity of foxtail millet (*Setaria italica*). *World Applied Sciences Journal, 33*(1), 146–152.

Ramakrishnan, S. R., Antony, U., & Kim, S. J. (2023). Non-thermal process technologies: Influences on nutritional and storage characteristics of millets. *Journal of Food Process Engineering, 46*(10), e14215.

Ramashia, S. E., Gwata, E. T., Meddows-Taylor, S., Anyasi, T. A., & Jideani, A. I. O. (2018). Some physical and functional properties of finger millet (Eleusine coracana) obtained in sub-Saharan Africa. *Food Research International, 104,* 110–118.

Ramesh, B., Kavitha, G., Gokiladevi, S., Balachandar, R. K., Kavitha, K., Gengadharan, A. C., & Puvanakrishnan, R. (2020). Effect of extremely low power time-varying electromagnetic field on germination and other characteristics in foxtail millet (*Setaria italica*) seeds. *Bioelectromagnetics, 41*(7), 526–539.

Ran, X. L., Zhang, M., Wang, Y., & Adhikari, B. (2019). Novel technologies applied for recovery and value addition of high value compounds from plant byproducts: A review. *Critical Reviews in Food Science and Nutrition, 59*(3), 450–461.

Rana, S., Agnihotri, V., Anjum, S., & Bhandari, N. S. (2023). Effect of dehulling, roasting, and cooking on the nutritional composition of Himalayan barnyard millet (*Echinochloa frumentacea*). *JSFA Reports*, *3*(5), 196–206.

Rao, M. V., Akhil, K. G., Sunil, C. K., Venkatachalapathy, N., & Jaganmohan, R. (2021). Effect of microwave treatment on physical and functional properties of foxtail millet flour. *International Journal of Chemical Studies*, *9*(1), 2762–2767.

Rastogi, N. K. (2012). Recent trends and developments in infrared heating in food processing. *Critical Reviews in Food Science and Nutrition*, *52*(9), 737–760.

Ravindra, U., Vijayakumari, J., Sharan, S., Raghuprasad, K. P., & Kavaloor, R. (2008). A comparative study of postharvest processing methods for little Millet (*Panicum miliare* L.). *Tropical Agricultural Research.*, *20*, 115–122.

Reddy, C. K., & Viswanath, K. K. (2019). Impact of γ-irradiation on physicochemical characteristics, lipoxygenase activity and antioxidant properties of finger millet. *Journal of Food Science and Technology*, *56*, 2651–2659.

van Rooyen, J., Marini, F., Orth, S. H., Oyeyinka, S. A., Simsek, S., & Manley, M. (2023). Effect of wheat roasting conditions and wheat type on short-wave infrared (SWIR) spectral data of whole and milled wheat by ANOVA-simultaneous component analysis. *Spectrochimica Acta Part A: Molecular and Biomolecular Spectroscopy*, *303*, 123160.

Rousta, M., Sadeghi, A. A., Shawrang, P., Aimn Afshar, M., & Chamani, M. (2014). Effect of gamma, electron beam and infrared radiation treatment on the nutritional value and anti-nutritional factors of sorghum grain. *Iranian Journal of Applied Animal Science*, *4*(4), 723–731.

Sahay, K. M. (1990). Evaluation of a general purpose abrasive mill for pearling of coarse cereals and dehusking of pulses. *International Journal of Food Science & Technology*, *25*(2), 220–225.

Samtiya, M., Soni, K., Chawla, S., Poonia, A., Sehgal, S., & Dhewa, T. (2021). Key anti-nutrients of millet and their reduction strategies: An overview. *Acta Scientific Nutritional Health*, *5*(12), 68–80.

Sarkar, A., Niranjan, T., Patel, G., Kheto, A., Tiwari, B. K., & Dwivedi, M. (2023). Impact of cold plasma treatment on nutritional, antinutritional, functional, thermal, rheological, and structural properties of pearl millet flour. *Journal of Food Process Engineering*, *46*(5), e14317.

Saxena, S., Vasudevan, H., Saini, S., & Sasmal, S. (2023). Comparative nutritional assessment of millet-based milk produced by ultrasound, germination, and a combined approach. *ACS Food Science & Technology*, *3*(4), 600–607.

Schoenlechner, R., Szatmari, M., Bagdi, A., & Tömösközi, S. (2013). Optimisation of bread quality produced from wheat and proso millet (*Panicum miliaceum* L.) by adding emulsifiers, transglutaminase and xylanase. *LWT-Food Science and Technology*, *51*(1), 361–366.

Semwal, J., & Meera, M. S. (2021). Infrared radiation: Impact on physicochemical and functional characteristics of grain starch. *Starch-Stärke*, *73*(3-4), 2000112.

Sharad, J. S., Pardeshi, I. L., Swami, S. B., Ranveer, R. C., & Kadam, J. H. (2024). Process optimization for hot air puffing of barnyard millet (*Echinochloa frumentacea*). *International Journal of Advanced Biochemistry Research*, *SP-8*, 272–279.

Sharanagat, V. S., Suhag, R., Anand, P., Deswal, G., Kumar, R., Chaudhary, A., Singh, L., Kushwah, O. S., Mani, S., Kumar, Y., & Nema, P. K. (2019). Physico-functional, thermo-pasting and antioxidant properties of microwave roasted sorghum [*Sorghum bicolor* (L.) Moench]. *Journal of Cereal Science*, *85*, 111–119.

Sharma, N., Goyal, S. K., Alam, T., Fatma, S., Chaoruangrit, A., & Niranjan, K. (2018). Effect of high pressure soaking on water absorption, gelatinization, and biochemical properties of germinated and non-germinated foxtail millet grains. *Journal of Cereal Science*, *83*, 162–170.

Sharma, N., Sahu, J. K., Choudhary, A., Meenu, M., & Bansal, V. (2023). High intensity ultrasound (HIU)-induced functionalization of foxtail millet protein and its fractions. *Food Hydrocolloids*, *134*, 108083.

Sharma, R., Bhandari, M., Kaur, K., Singh, A., Sharma, S., & Kaur, P. (2022). Molecular interactome and starch–protein matrix, functional properties, phytochemical constituents, and antioxidant activity of foxtail millet (*Setaria italica*) flour as influenced during gaseous ozonation. *Cereal Chemistry*, *99*(5), 1101–1111.

Sharma, R., Singh, A., & Sharma, S. (2021). Influence of ozonation on cereal flour functionality and dough characteristics: a review. *Ozone: Science & Engineering*, *43*(6), 613–636.

Sharma, S., Pandey, S., Gusain, T., Gautam, P., Mishra, R., Hussain, A., & Kumar, S. (2024a). Investigating the multifaceted changes in physicochemical and nutritional attributes of chicken-based nutri cereal mix under cold plasma treatment. *Journal of Stored Products Research*, *109*, 102416.

Sharma, S., Saxena, D. C., & Riar, C. S. (2015). Antioxidant activity, total phenolics, flavonoids and antinutritional characteristics of germinated foxtail millet (Setaria italica). *Cogent Food & Agriculture*, *1*(1), 1081728.

Sharma, V., Champawat, P. S., & Mudgal, V. D. (2014). Process development for puffing of sorghum. *International Journal of CurrentResearch and Academic Review*, *2*(1), 164–170.

Sharma, V., Tomar, M. S., Sahoo, S., & Pradhan, R. C. (2024b). Development of barnyard millet flour rich in bioactive compounds with enhanced functional properties by application of multipin atmospheric cold plasma. *Journal of Food Process Engineering*, *47*(6), e14667.

Shelar, G. A., & Gaikwad, S. T. (2019). Extrusion in food processing: An overview. *The Pharma Innovation Journal*, *8*(2), 562–568.

Sherin, A. J., Sunil, C. K., Chidanand, D. V., & Venkatachalapathy, N. (2024). Structural, physicochemical and functional properties of high-pressure modified white finger millet starch. *International Journal of Biological Macromolecules*, *261*, 129919.

Shobha, D., Vasudevan, S. N., & Badigannavar, A. (2024). Impact of gamma irradiation and packaging on the storage quality of little millet (*Panicum sumatrense*) flour. *Cereal Research Communications*, *52*(2), 777–787.

Singh, A., Gupta, S., Kaur, R., & Gupta, H. R. (2017). Process optimization for anti-nutrient minimization of millets. *Asian Journal of Dairy and Food Research*, *36*(4), 322–326.

Singh, D. K., Kurichh, R., Bist, Y., Kheto, A., Kumar, Y., Sharma, R., Shikha, D., & Saxena, D. C. (2024). Effect of microwave roasting on the physicochemical, functional, rheological, and antioxidant properties of kodo millet flour. *Journal of Food Processing and Preservation*, *2024*(1), 6861190.

Singh, K. P., Mishra, H. N., & Saha, S. (2011). Design, development and evaluation of barnyard millet dehuller. *Journal of Agricultural Engineering*, *48*(3), 17–25.

Singh, P., & Choudhary, K. K. (2025). Mechanistic insights on physiological, biochemical, and metabolite profiling of pearl millet cultivars focusing bioactive compounds under elevated UV-B radiation. *Plant Physiology and Biochemistry*, *222*, 109682.

Singh, S. M., & Rao, P. S. (2025). Impact dehulling of browntop millet: Its physical and nutritional characterization. *Journal of Cereal Science, 121*, 104078.

Sirohi, R., Tarafdar, A., Gaur, V. K., Singh, S., Sindhu, R., Rajasekharan, R., Madhavan, A., Binod, P., Kumar, S., & Pandey, A. (2021). Technologies for disinfection of food grains: Advances and way forward. *Food Research International, 145*, 110396.

Song, J., Zhao, Z., & Fuller, D. Q. (2013). The archaeobotanical significance of immature millet grains: an experimental case study of Chinese millet crop processing. *Vegetation History and Archaeobotany, 22*, 141–152.

Sonkar, S., Jaddu, S., Pradhan, R. C., Dwivedi, M., Seth, D., Goksen, G., Sarangi, P. K., & Lorenzo, J. M. (2023). Effect of atmospheric cold plasma (pin type) on hydration and structure properties of kodo-millet starch. *LWT-Food Science and Technology, 182*, 114889.

Sruthi, N. U., & Rao, P. S. (2021). Effect of processing on storage stability of millet flour: A review. *Trends in Food Science & Technology, 112*, 58–74.

Suma, F. P., & Urooj, A. (2017). Impact of household processing methods on the nutritional characteristics of pearl millet (*Pennisetum typhoideum*): A review. *Journal of Food Processing and Technology, 4*(1), 28–32.

Sunagar, R. R., & Sreerama, Y. N. (2024). Impact of milling on the nutrients and anti-nutrients in browntop millet (Urochloaramosa) and its milled fractions: Evaluation of their flour functionality. *Journal of the Science of Food and Agriculture, 104*(9), 5504–5512.

Sundaresan, T., Joshi, J., & Rao, P. S. (2025). Characterisation and multivariate analysis of changes in quality attributes of microwave-treated pearl millet flour. *Journal of Cereal Science, 121*, 104086.

Sunil, C. K., & Chacko, N. (2018). Microwave assisted de-husking of foxtail millet. *Journal of Agricultural engineering, 55*(4), 21–33.

Suri, S., Dutta, A., Shahi, N. C., Singh, A., Raghuvanshi, R. S., & Chopra, C. S. (2022). Formulation, process optimization, and quality evaluation of barnyard millet (*Echinochloa frumentacea*)-based ready-to-eat (RTE) extruded snack. *Journal of Food Processing and Preservation, 46*(11), e16936.

Swaminathan, I., Guha, M., Hunglur, U. H., & Rao, D. B. (2015). Optimization of infrared heating conditions of sorghum flour using central composite design. *Food Science and Biotechnology, 24*(5), 1667–1671.

Tavanandi, H. A., Das, A. K., Venkateshmurthy, K., & Raghavarao, K. S. M. S. (2021). Design and development of a machine for continuous popping and puffing of grains. *Journal of Food Science and Technology, 58*, 1703–1714.

Thakur, K., Sharma, S., Sharma, R., Bhardwaj, R., Sohu, R. S., & Singh, B. (2023). Characterization of ozonated pearl millet and sorghum starches: Pasting profile, functional properties, *in vitro*-digestibility, molecular interactions, and morphology. *Starch-Stärke, 75*(11–12), 2300020.

Tribst, A. A. L., & Cristianini, M. (2017). Effect of high-pressure technologies on enzymes applied in food processing. In M. Şentürk (Ed.), *Enzyme inhibitors and activators* (pp. 49). InTech.

Ullegaddi, M. M., Mahandra Babu, N. C., Faisal, A. R., Mohammad, M., Shreenidhi, M. S., & Anjum, S. (2020). Design and development of compact foxtail millet deshelling machine. *Materials Today: Proceedings, 42*, 781–785.

Victorin, K. (1992). Review of the genotoxicity of ozone. *Mutation Research/Reviews in Genetic Toxicology*, *277*(3), 221–238.

Vila-Real, C., Pimenta-Martins, A., Mbugua, S., Hagrétou, S. L., Katina, K., Maina, N. H., Pinto, E., & Gomes, A. M. (2022). Novel synbiotic fermented finger millet-based yoghurt-like beverage: Nutritional, physicochemical, and sensory characterization. *Journal of Functional Foods*, *99*, 105324.

Wang, K., Zhang, C., Zhao, W., Zhang, A., Liu, J., & Sheng, Q. (2025). Effects of dry and wet ball milling on physicochemical properties of foxtail millet. *Food Chemistry*, *483*, 143916.

Wang, L. H., Li, Z., Qin, J., Huang, Y., Zeng, X. A., & Aadil, R. M. (2022). Investigation on the impact of quality characteristics and storage stability of foxtail millet induced by air cold plasma. *Frontiers in Nutrition*, *9*, 1064812.

Wang, P., Liu, J., Mao, Y., Guan, X., & Wang, S. (2021). Improvement of radio frequency heating uniformity for millets by changing shape and adding polypropylene blocks. *Innovative Food Science & Emerging Technologies*, *74*, 102856.

Wang, Y., Zhang, L., Johnson, J., Gao, M., Tang, J., Powers, J. R., & Wang, S. (2014). Developing hot air-assisted radio frequency drying for in-shell macadamia nuts. *Food and Bioprocess Technology*, *7*, 278–288.

Wani, H. M., Sharma, P., Wani, I. A., Kothari, S. L., & Wani, A. A. (2021). Influence of γ-irradiation on antioxidant, thermal and rheological properties of native and irradiated whole grain millet flours. *International Journal of Food Science and Technology*, *56*, 3752–3762. https://doi.org/10.1111/ijfs.15062

Yadav, G. P., Dalbhagat, C. G., & Mishra, H. N. (2022). Effects of extrusion process parameters on cooking characteristics and physicochemical, textural, thermal, pasting, microstructure, and nutritional properties of millet-based extruded products: A review. *Journal of Food Process Engineering*, *45*(9), e14106.

Yadav, S., & Dhiman, N. (2023). Innovations and technology in millet farming. *Agricultural Engineering Today*, *47*(1), 5–8.

Yadav, S., & Mishra, S. (2023). Moisture sorption isotherms and storage study of spray-dried probiotic finger millet milk powder. *Journal of Stored Products Research*, *102*, 102128.

Yadav, S., Mishra, S., & Pradhan, R. C. (2021). Ultrasound-assisted hydration of finger millet (*Eleusine coracana*) and its effects on starch isolates and antinutrients. *Ultrasonics Sonochemistry*, *73*, 105542.

Yan, S., Wu, X., Faubion, J., Bean, S. R., Cai, L., Shi, Y. C., Sun, X. S., & Wang, D. (2012). Ethanol-production performance of ozone-treated tannin grain sorghum flour. *Cereal Chemistry*, *89*(1), 30–37.

Yarrakula, S. A. S., Rehaman, A., & Saravanan, S. (2022). Effect of hot air assisted radio frequency technology on physical and functional properties of pearl millet. *Pharma Innovation Journal*, *11*(5), 633–637.

Yildirim, A., Öner, M. D., & Bayram, M. (2013). Effect of soaking and ultrasound treatments on texture of chickpea. *Journal of Food Science and Technology*, *50*, 455–465.

Yousaf, L., Hou, D., Liaqat, H., & Shen, Q. (2021). Millet: A review of its nutritional and functional changes during processing. *Food Research International*, *142*, 110197.

Yuan, Q., Yang, M., Yue, X., & Xu, H. (2024). Effects of electron beam irradiation on cooking quality and volatile components of waxy and non-waxy proso millet (*Panicum miliaceum* L.). *Journal of Cereal Science*, *119*, 103997.

Zhang, D., Yi, S., Zhang, J., & Bao, Y. (2022a). Establishment of millet threshing and separating model and optimization of harvester parameters. *Alexandria Engineering Journal*, *61*(12), 11251–11265.

Zhang, H., Zhao, W., Bai, T., Fu, L., Chen, Z., Jing, X., & Wang, X. (2022b). Sustainable extraction of polyphenols from millet using switchable deep eutectic solvents. *LWT-Food Science and Technology*, *170*, 114082.

Zhang, J., Li, P., Liu, J., Wang, Y., Zhang, A., Zhao, W., Li, S., Liu, Y., & Liu, J. (2022c). Nutritional components of millet porridge cooked by different electric cookers based on principal component and cluster analyses. *Food*, *11*(18), 2823.

Zhang, S., Sun, L., Ju, H., Bao, Z., Zeng, X. A., & Lin, S. (2021). Research advances and application of pulsed electric field on proteins and peptides in food. *Food Research International*, *139*, 109914.

Zhang, Y., Jing, X., Chen, Z., & Wang, X. (2023). Effects of moderate-intensity pulsed electric field on the structure and physicochemical properties of foxtail millet (*Setaria italica*) prolamin. *Cereal Chemistry*, *100*(2), 360–370.

Zhang, Y., Zhang, X., Zhang, Z., Chen, Z., Jing, X., & Wang, X. (2022d). Effect of high hydrostatic pressure treatment on the structure and physicochemical properties of millet gliadin. *LWT-Food Science and Technology*, *154*, 112755.

Zhang, Z. H., Wang, L. H., Zeng, X. A., Han, Z., & Brennan, C. S. (2019). Non-thermal technologies and its current and future application in the food industry: a review. *International Journal of Food Science & Technology*, *54*(1), 1–13.

Zhou, Z., Shang, X., Wang, Z., Sun, C., & Zhang, L. (2022). Effect of high-pressure treatment on the quality of a *Hericium erinaceus*: Millet composite beverage. *Journal of Food Quality*, *2022*(1), 1–12.

Zhu, F., & Li, H. (2019). Modification of quinoa flour functionality using ultrasound. *Ultrasonics Sonochemistry*, *52*, 305–310.

11

Millet-Based Products and Commercialization

Narender K. Chandla[1], *Nitin Wakchaure*[2], *Gajanan P. Deshmukh*[1], *and Sunil K. Khatkar*[3]

[1] *Department of Dairy Engineering, College of Dairy and Food Science Technology, Guru Angad Dev Veterinary and Animal Sciences University, Ludhiana, India*
[2] *Department of Dairy Economics and Business Management, College of Dairy and Food Science Technology, Guru Angad Dev Veterinary and Animal Sciences University, Ludhiana, India*
[3] *Department of Dairy Technology, College of Dairy and Food Science Technology, Guru Angad Dev Veterinary and Animal Sciences University, Ludhiana, India*

11.1 Introduction

Millets, among the earliest domesticated cereal grains, have been integral to human diets for over 7,000 years, particularly in Asia and Africa, due to their remarkable resilience and adaptability. Despite a historical decline in consumption driven by the rise of refined grains and industrialized food processing, millets have recently regained attention due to their exceptional nutritional value and significant nutraceutical potential in the development of functional foods. Archaeological evidence traces millets' long-standing presence in human diets. Noodles made from "proso and foxtail" millets were consumed in Northern China as early as 2000 BC, and finger millet (*Eleusine coracana*), known as "ragi" in India, has deep agricultural and cultural roots. Ancient Sanskrit texts even refer to it as "nrtta-kondaka," highlighting its cultural relevance. While some theories suggest finger millet's introduction to India via trade routes, archaeological findings point to its African origins, with its presence in India for over 2,000 years.

Millets have historically thrived in semi-arid regions with nutrient-poor soils, making them indispensable dietary staples, particularly in India and parts of Africa. In the 2023–2024 agricultural year, India's millet production reached 12.84 million metric tons, covering 9.07 million hectares. Rajasthan was the leading contributor, producing 33.5% of the national output. However, modern agricultural practices and the increased consumption of refined grains, such as polished rice, have led to a decline in millet consumption. This dietary shift of eating polished rice and wheat, combined with chemical fertilizers, is associated with rising lifestyle diseases like diabetes and hypertension. The nutraceutical potential of millet-based foods is now being recognized as an

Nutraceutical Potential of Millet-Based Food Products: Chemistry and Applications, First Edition.
Edited by Parmjit S. Panesar, Charanjit S. Riar, and Sushil Dhital.

Published 2026 by John Wiley & Sons Ltd.

important strategy to address these health concerns. Recently, the emphasis on health-conscious eating has revived interest in millets. These grains are rich in dietary fiber, antioxidants, B vitamins, and essential minerals like iron, magnesium, and calcium, which play crucial roles in digestive health, cholesterol reduction, and blood glucose regulation, making millets particularly beneficial for managing diabetes and cardiovascular diseases. Additionally, bioactive compounds such as phenolic acids, flavonoids, and tannins provide antioxidant, anti-inflammatory, and anticarcinogenic properties, enhancing their nutraceutical value.

Incorporating millets into modern functional foods is becoming increasingly popular. Millet-based products, such as protein bars, probiotic yogurts, sprouted millet powders, and functional beverages, are gaining traction. Finger millet, with its high calcium content, is used in bone health supplements, while iron-rich pearl millet is included in fortified weaning foods to combat iron deficiency anemia. Furthermore, millet-based products like gluten-free cereals, high-fiber cookies, prebiotic-enriched bread, and extruded snacks support gut health and metabolic stability. The growing demand for nutraceutical foods focused on immune support, digestive health, and chronic disease management is driving the success of millet-based functional foods. Slow-digesting carbohydrates and prebiotic fibers in millets also promote gut microbiota modulation and provide sustained energy, positioning millets as a key player in health-oriented food innovation.

11.2 History

The history of millet cultivation dates to ancient civilizations, with archeobotanical studies showing that common millet (*Panicum miliaceum*) was a staple in North China as early as 10,000 years ago (Li et al., 2009). Millets, including pearl millet, finger millet, foxtail millet, barnyard millet, and proso millet, are drought-tolerant crops capable of thriving in poor soil conditions, high temperatures, and arid regions where conventional staples like rice and wheat often fail (Dwivedi et al., 2012; Saleh et al., 2013b). Millet is a generic term used for small-sized grains that form a heterogeneous group and, along with maize and sorghum, is referred to as "coarse cereals" (FAO, 1995). Millets are of minor importance in Western countries but remain a staple food in the diets of people in Africa and Asia. Their agricultural importance arises from their hardiness, tolerance to extreme weather, and ability to grow with low inputs in low rainfall areas (Ushakumari et al., 2004). Bajra (pearl millet), ragi (finger millet), navane (foxtail millet), samai (little millet), haraka (kodo millet), panivaragu (proso millet), and banti (barnyard millet) are the key millet varieties cultivated in Asia and Africa. These crops play a crucial role in ensuring food security in resource-constrained and climate-vulnerable areas (Mal et al., 2010). Despite their historical significance, millets have become underutilized in modern agriculture, largely due to the Green Revolution's focus on high-yielding cereals like rice and wheat, which has led to a decline in millet production and consumption (Saleh et al., 2013b). However, growing awareness of the health benefits and sustainability of millets has revived interest in their cultivation. Countries such as India, China, and the United States of America have launched research projects to improve millet-processing technology and promote their large-scale utilization as a food source (Saleh et al., 2013a).

Millets are nutritionally superior to many conventional cereals. They are rich in protein, fiber, vitamins, and minerals, and are gluten-free, making them a great alternative for individuals with celiac disease or gluten intolerance (Chandrasekara & Shahidi, 2011). Millets are packed with antioxidants and have been linked to improved digestion, reduced risk of heart disease, and better blood sugar control (Shahidi & Ambigaipalan, 2015). These grains are versatile and can be used in a variety of dishes, such as porridge, bread, pilaf, and baked goods. In some regions, millet is used to make alcoholic beverages. Millets are considered a crop of food security due to their resilience in adverse agroclimatic conditions and their contribution to broadening genetic diversity in the food basket (Mal et al., 2010; Ushakumari et al., 2004). Despite their many health benefits and culinary uses, millets remain underutilized in many parts of the world in modern agriculture, largely due to the global preference for high-yielding grains like rice and wheat, which has led to a decline in millet production and consumption over time. However, with increasing awareness of climate change and the growing demand for sustainable agriculture, millets are gaining renewed interest for their role in food security, nutritional health, and environmental sustainability. Their ability to grow with minimal water and inputs makes them an important crop for addressing food insecurity and promoting sustainable agriculture in a rapidly changing climate. However, interest is growing, particularly in areas affected by climate change and other environmental factors.

11.3 Nutritional Values

Millets are rich in protein, dietary fiber, and essential micronutrients such as calcium, iron, and magnesium. For example, finger millet contains 300–350 mg of calcium per 100 g, and pearl millet is an excellent source of iron, providing 5–8 mg per 100 g. Moreover, millets are packed with bioactive compounds like phenolic acids, flavonoids, and tannins, which exhibit strong antioxidant and anti-inflammatory properties. These compounds contribute to reducing the risk of chronic diseases, including diabetes, cardiovascular disorders, and cancer (Chandrasekara & Shahidi, 2011; Mal et al., 2010). Millets also require less water and chemical inputs than rice and wheat, making them a sustainable option for global health and environmental sustainability (Shahidi & Ambigaipalan, 2015). The continued study and development of millet-based products are essential to enhancing their nutritional value and expanding their role in global food systems. Researchers are exploring innovative processing technologies to improve the digestibility, taste, and bioavailability of nutrients in millets, while also promoting their broader adoption in modern diets. The objective of this review is to provide a comprehensive overview of the different types of millets, their nutritional value, and the potential health benefits. The review also examines various processing technologies used to prepare millet grains for consumption and their impact on the nutrient content and bioavailability. Additionally, it explores the prospects of millet as a food source for promoting food security, improving public health, and meeting the dietary needs of vulnerable populations.

Millets are a group of highly nutritious grains that provide a range of essential nutrients, including proteins, carbohydrates, vitamins, and minerals. These grains are nutritionally comparable to, or even superior, to major cereals like rice and wheat in terms of protein, energy, vitamins, and minerals (Sehgal & Kawatra, 2003). Millets are particularly rich in minerals, nutraceuticals, and dietary fibers, containing 9–14% protein and 70–80% carbohydrates (Hadimani & Malleshi, 1993). They also provide significant amounts of phytochemicals and micronutrients (Singh et al., 2012b). The nutritional status of a community is recognized as a critical indicator of national development, making millets an important alternative for addressing food insecurity and malnutrition (Singh & Raghuvanshi, 2012). The quality of protein in millets is mainly determined by its essential amino acids. Finger millet, for instance, contains 44.7% essential amino acids (Mbithi et al., 2000), which is higher than the 33.9% found in FAO reference protein (FAO, 1991). Millet proteins are primarily prolamins, with lysine being the most limiting amino acid, followed by cystine. However, millets are relatively high in methionine (Kumar & Parmeswaran, 1999; Monteiro et al., 1987). The true digestibility and biological value of millets range from 95.0 to 99.3 and 48.3 to 56.5, respectively (Geervani & Eggum, 1989b).

Among millets, pearl millet (bajra) stands out for its high content of macronutrients and micronutrients, such as iron, zinc, magnesium, phosphorus, folic acid, and riboflavin, as well as being significantly rich in resistant starch and dietary fibers (Antony et al., 1996; Ragaee et al., 2006). Finger millet (ragi) is also known for its high calcium content. The seed coat of finger millet is an edible by-product rich in dietary fiber, minerals, and phytochemicals and can be used in the production of composite flours for food products like biscuits (Krishnan et al., 2011). Kodo millet and little millet are particularly high in dietary fiber (37–38%), which is the highest among cereals, and are rich in polyunsaturated fatty acids (PUFAs) despite being low in fat (Antony et al., 1996; Malleshi & Hadimani, 1993b). Kodo millet is also notable for its high free radical quenching activity (DPPH), followed by sorghum and finger millet (Hegde & Chandra, 2005). Sorghum is especially high in antioxidant activities compared to other millets. Millets are also valued for their high content of vitamins, particularly B vitamins, folic acid, and minerals like phosphorus, iron, and potassium. Finger millet contains more calcium than maize, while pearl millet has higher niacin content than other cereals. Additionally, millets are gluten-free, easy to digest, and are known for their antioxidant and potential anticarcinogenic properties (Dykes & Rooney, 2006). The lipid content in pearl millet (3–6%) is higher than that of sorghum and most other cereals, with 75% of the fatty acids being unsaturated, particularly linoleic acid (46.3%). This makes the energy content of millet comparable to brown rice, though higher than sorghum. Finger and teff millets are notable sources of dietary calcium, magnesium, and iron. Millets are distinct from other cereals due to their high content of calcium, dietary fiber, polyphenols, and protein (Sehgal & Kawatra, 2003).

Millets are particularly rich in essential amino acids, including sulfur-containing amino acids like methionine and cysteine, and have higher fat content compared to maize, rice, and sorghum (Dykes & Rooney, 2006). Finger millet, for example, contains higher dietary fiber (22.0%) than many other grains, such as wheat, rice, and maize (4.6–12.8%). Pearl millet has dietary fiber content ranging from 8% to 9% (Singh et al., 2012a). Millets are also high in polyphenols, especially phenolic acids and tannins, which act as antioxidants and support the immune system (Singh et al., 2012b). The dietary fiber content in millet-based foods can vary depending on the degree of milling, with whole grains providing the highest fiber content. These grains are a valuable source of carbohydrates and rich in magnesium and phosphorus, which help in reducing the effects of migraines and heart attacks and contribute to energy production in the body (Sehgal & Kawatra, 2003).

11.4 Types of Millets and Their Nutritional Value

Millets are a diverse group of small-seeded grains, each with its unique nutritional profile (Figure 11.1). Here is an overview of some of the commonly grown millets and their nutritional value.

i) Pearl millet (bajra): Pearl millet is widely cultivated in India and is rich in protein, fiber, and essential minerals like iron, calcium, and magnesium. It is particularly high in resistant starch, soluble and insoluble dietary fibers, and antioxidants (Yadav et al., 2011). Nutritionally, it contains 92.5% dry matter, 2.1% ash, 2.8% crude fiber, 7.8% crude fat, 13.6% crude protein, and 63.2% starch (Singh et al., 2008). Pearl millet is commonly used to prepare flatbreads like roti and bhakri, as well as porridge and khichdi.
ii) Finger millet (ragi): Finger millet is rich in protein (5–8%), carbohydrates (65–75%), dietary fiber (15–20%), and minerals (2.5–3.5%). It is particularly noted for its high calcium content (344 mg/100 g) compared to other cereals (Veena, 2003). Finger millet is also high in phytates, polyphenols, and tannins (0.48 g/100 g and 0.61%, respectively) (Krishnan et al., 2015). It is commonly used in Indian dishes such as porridge, dosa, and idli.
iii) Foxtail millet (kangni/kakum): This small, yellowish grain is rich in protein, fiber, and minerals such as iron and zinc (Haug & Lantzsch, 2008). It is used in a variety of dishes, including porridge, pulao, and upma.
iv) Kodo millet (kodo): Kodo millet is a small, red grain high in protein, fiber, and minerals like iron and calcium. It is particularly noted for its high dietary fiber content (37–38%), which is considered a nutraceutical (Kumar et al., 2008). It is often used in porridge, dosa, and idli.

Figure 11.1 Major millets grains of India (self-created)

v) Barnyard millet (jhangora/kuthiravali): Barnyard millet is a small, white grain rich in protein, fiber, and minerals such as iron and phosphorus. It is the richest source of crude fiber and iron among millets and contains functional constituents like γ-amino butyric acid (GABA) and β-glucan, which have antioxidant properties (Bisht et al., 2015). It is commonly used in porridge, upma, and khichdi.

vi) Proso millet (chena/panivaragu): Proso millet is high in protein (12.5%), fiber, and essential minerals such as iron and magnesium (Rao et al., 2012). It is used to prepare porridge, pulao, and upma and has a mild, nutty flavor.

vii) Sorghum (jowar/cholam): Sorghum is rich in protein, fiber, and essential minerals like iron and potassium. It contains 10.4% protein and 1.6% dietary fiber, along with 349 Kcal of energy per 100 g (Patel et al., 2011). It is commonly used in flatbreads like roti and bhakri and in porridge and khichdi. Sorghum also contains 25 mg calcium, 222 mg phosphorus, and 4.1 mg iron per 100 g (Anjum et al., 2017).

viii) Little millet (kutki): Little millet is high in protein, fiber, and essential minerals like iron, calcium, and potassium. It has a mild, slightly nutty flavor and is often used to make porridge, pulao, and upma (Rao et al., 2012).

11.5 Comparison of Millet Grains with Other Staple Grains

Millet grains are nutritionally unique, distinguishing them from other staple grains such as rice, wheat, and corn. These grains are rich in protein, dietary fiber, vitamins, and essential minerals like iron, zinc, and magnesium, making them an excellent source of nutrition. Millet contains higher levels of protein and dietary fiber compared to rice, and it has a lower glycemic index, making it less likely to cause spikes in blood sugar levels. This property makes millet a favorable option for people managing their blood sugar levels, such as individuals with diabetes. Furthermore, millet is a rich source of antioxidants, which help protect against chronic diseases and oxidative stress (Sehgal & Kawatra, 2003). When compared to wheat, millet stands out as a gluten-free alternative, making it suitable for individuals with gluten intolerance or celiac disease. In addition, millet has a lower glycemic index than wheat, offering better blood sugar management, which is important for people with diabetes (Dykes & Rooney, 2006). Millet's gluten-free nature, combined with its lower glycemic index, makes it a healthier choice for those needing a grain that is easy to digest and has minimal impact on blood glucose levels. Millet also surpasses corn in nutritional content, offering higher levels of protein, dietary fiber, and essential amino acids, which are crucial for tissue repair and growth (Sehgal & Kawatra, 2003). It is also abundant in antioxidants and minerals like iron and magnesium. These nutrients contribute to various health benefits, including the prevention of chronic conditions. Studies have shown that regular consumption of whole-grain cereals, including millets, is associated with a reduced risk of diabetes, gastrointestinal diseases, and cardiovascular issues (Singh et al., 2012a). Furthermore, the outer layer of the millet grain, or the seed coat, contains concentrated amounts of dietary fiber, phenolics, minerals, and vitamins, which provide substantial nutritional and health benefits when included in the diet as whole grains (Krishnan et al., 2011). In conclusion, millet grains offer a nutrient-dense alternative to more common staple grains, making them an essential component of a healthy, balanced diet. Their superior nutritional profile, particularly in terms of protein, fiber, and micronutrients, provides a valuable option for improving overall health and combating malnutrition.

11.6 Bioactive Components in Millet and Bioavailability of These Compounds

Millet grains are rich in bioactive compounds that contribute to their nutritional and health benefits. These include polyphenols, dietary fibers, antioxidants, essential amino acids, and resistant starch. Each of these components plays a significant role in promoting health through various mechanisms of action, as well as offering bioavailability in the human body.

11.6.1 Polyphenols

Polyphenols are one of the most studied bioactive compounds in millet grains. These compounds act as potent antioxidants, protecting cells from oxidative stress, and reducing the risk of chronic diseases like cardiovascular diseases, diabetes, and certain cancers. Polyphenols in millet are particularly abundant in the seed coat and are known for their anti-inflammatory, anticancer, and antimicrobial properties (Kumar & Parmeswaran, 1999). The bioavailability of polyphenols depends on their chemical structure, with some polyphenols being more easily absorbed and utilized by the body than others. For example, phenolic acids in millet are readily absorbed in the gut and are known to exert beneficial effects on metabolism (Dykes & Rooney, 2006).

11.6.2 Dietary Fibers

Millet grains are an excellent source of dietary fiber, particularly insoluble fiber, which is essential for maintaining digestive health. The fiber content in millet helps in regulating bowel movements, preventing constipation, and improving gut microbiota composition. Fiber also plays a key role in controlling blood sugar levels and lowering cholesterol, contributing to the prevention of diabetes and heart disease. The bioavailability of dietary fiber depends on its solubility, with soluble fibers fermenting in the colon to produce short-chain fatty acids that provide energy for gut cells (Malleshi & Hadimani, 1993a). Insoluble fibers, on the other hand, pass through the digestive system largely unchanged, promoting healthy digestion.

11.6.3 Antioxidants

Millet grains contain significant amounts of antioxidants, including flavonoids, phenolic acids, and tocopherols, which help neutralize free radicals in the body. The antioxidant properties of millet contribute to reducing oxidative stress, which is implicated in aging, inflammation, and the development of chronic diseases (Ragaee et al., 2006). These antioxidants also exhibit anti-inflammatory and anticancer effects by modulating key cellular pathways involved in disease progression (Kaur et al., 2019). The bioavailability of antioxidants is influenced by the grain's processing and preparation methods, with whole millet grains retaining more antioxidants compared to processed forms.

11.6.4 Essential Amino Acids

Millet is a rich source of essential amino acids, including lysine, threonine, and methionine. These amino acids are crucial for protein synthesis and various metabolic functions in the body. Among millets, finger millet (ragi) has a higher proportion of essential amino acids compared to other grains, which makes it a valuable protein source for individuals who follow plant-based diets (Mbithi-Mwikya et al., 2000). The bioavailability of amino acids in millet is high, particularly when the grains are consumed as part of a whole-grain diet, which ensures better protein quality and absorption.

11.6.5 Resistant Starch

Resistant starch is a type of starch that resists digestion in the small intestine and passes into the colon, where it is fermented by gut bacteria. This fermentation process produces short-chain fatty acids, which have beneficial effects on gut health, including improved insulin sensitivity, reduced appetite, and protection against colorectal cancer (Krishnan et al., 2011). The bioavailability of resistant starch is influenced by the grain's cooking and processing methods. Cooking millet and allowing it to cool before consumption can increase its resistant starch content, providing additional health benefits.

The bioactive compounds in millet—polyphenols, dietary fibers, antioxidants, essential amino acids, and resistant starch—offer significant health benefits. These compounds work synergistically to promote metabolic health, protect against oxidative stress, improve digestive function, and contribute to overall well-being. Millet-based foods are increasingly recognized for their health benefits, particularly in the prevention and management of various diseases. Rich in fiber, essential minerals, and antioxidants, millets offer a holistic approach to improving overall health. Millets are nutrient-dense grains rich in complex carbohydrates, fiber, and essential micronutrients such as iron, magnesium, phosphorus, and potassium (Vijayakumar et al., 2017). Their high fiber content supports digestive health and glycemic regulation, aiding in diabetes and obesity management. With a low glycemic index, millets reduce insulin resistance and enhance glycemic control (Sasikala et al., 2019). Finger millet (ragi) contains bioactive antioxidants that mitigate oxidative stress, a key contributor to diabetes. Magnesium in millets supports cardiovascular health by regulating blood pressure and reducing hypertension risk (Bamigboye et al., 2020). Additionally, their unsaturated fatty acids improve lipid profiles and lower low-density lipoprotein (LDL) cholesterol, mitigating cardiovascular disease risk.

Millets' dietary fiber promotes bowel regularity, lowers colorectal cancer risk, and supports gut microbiota (Rao et al., 2021). Their high satiety index aids in weight management by reducing calorie intake (Ghosh et al., 2021). Slow-digesting carbohydrates prevent blood sugar fluctuations, reducing fat accumulation. Millets, particularly finger millet, exhibit anticancer properties through phenolic compounds and flavonoids, which neutralize free radicals and reduce oxidative damage (Bamigboye et al., 2020). Additionally, their phytochemicals demonstrate potential anticancer effects. Finger millet is an excellent source of calcium, essential for bone density and osteoporosis prevention (Sasikala et al., 2019). Combined with magnesium and phosphorus, it enhances bone strength. Furthermore, millets contain bioactive polyphenols with anti-inflammatory properties, lowering chronic disease risk (Vijayakumar et al., 2017). Regular consumption of millet-based foods offers significant health benefits, including improved metabolic and cardiovascular function, weight regulation, and chronic disease prevention.

Incorporating millet-based foods into the diet provides numerous health benefits, from managing blood sugar levels, preventing cardiovascular diseases, to improving digestive health and promoting bone strength. Their high nutrient density, coupled with their antioxidant and anti-inflammatory properties, makes millets a powerful tool in disease prevention and management.

Millets, such as pearl millet, finger millet, and sorghum, are rich in macronutrients, with protein ranging from 7.3% in finger millet to 12.3% in pearl millet, surpassing rice and wheat (Table 11.1). They also contain moderate fat levels (1.3–4.5%) and are high in fiber (6.1–8.6%), which supports digestion and glycemic control (Bhatt et al., 2023; Ragaee & Abdel-Aal, 2015b). Despite their high carbohydrate content (61.5–74.1%), millets have a low glycemic index (52–72), making them beneficial for diabetes management by promoting better blood sugar control (Maurya et al., 2023). Millets are a great source of minerals. Calcium is notably high in finger millet (344 mg/100 g), aiding bone health, while iron (3.5–9.1 mg) helps combat anemia, and magnesium contributes to heart and muscle health (Kaur et al., 2019) (Table 11.1). Antioxidant capacity of millets is enhanced by phenolic compounds and flavonoids, reducing chronic disease risks. Phytic acid, saponins, and tannins—antinutritional factors—are present in lower amounts, which can be minimized through processing (Ragaee & Abdel-Aal, 2015b). Millets provide significant health benefits, such as improved heart health, better blood sugar regulation, and weight management due to their fiber content (Subramanian, 2020).

11.7 Processing and Formulation of Millet-Based Food

Millets are versatile grains that can be processed using various methods to enhance their nutritional value, texture, and shelf life. Both traditional and modern food processing techniques are employed to create millet-based foods that cater to different consumer needs. These processed millet products can range from bakery items to ready-to-eat snacks, offering a wide variety of value-added products.

11.7.1 Traditional Processing Methods

Traditional methods of processing millets include fermentation, germination, and malting. Fermentation is often used to enhance the digestibility of millet and increase its nutritional content, particularly by enriching the levels of certain vitamins (Ochanda et al., 2010). Germination and malting are also common methods, as they help break down antinutritional factors like phytic acid and increase the availability of bioactive compounds, including

Table 11.1 Chemical and nutraceutical component composition of different types of millets

Component	Pearl millet	Finger millet	Sorghum	Foxtail millet	Proso millet	Barnyard millet	Little millet
Crude protein (%)	11.6	7.3	8.2	12.3	10.0	11.0	9.5
Crude fat (%)	4.5	1.3	3.1	4.0	3.4	3.5	3.0
Dietary fiber (%)	8.5	8.6	6.1	8.0	7.0	8.0	7.5
Carbohydrates (%)	64.0	70.6	72.0	61.5	72.6	73.0	74.1
Ash (%)	2.5	2.1	2.6	2.3	2.5	2.8	2.4
Minerals (Ca, Fe, Mg, etc.)	Ca (42 mg), Fe (9.1 mg), Mg (154 mg)	Ca (344 mg), Fe (3.9 mg), Mg (137 mg)	Ca (24 mg), Fe (4.1 mg), Mg (136 mg)	Ca (31 mg), Fe (5.6 mg), Mg (165 mg)	Ca (15 mg), Fe (4.3 mg), Mg (101 mg)	Ca (35 mg), Fe (4.9 mg), Mg (137 mg)	Ca (15 mg), Fe (3.5 mg), Mg (142 mg)
Phenolic compounds (mg/100 g)	250	90	50	190	160	210	200
Flavonoids (mg/100 g)	4.5	3.5	3.2	4.0	3.0	3.8	4.2
Antioxidant activity (ORAC values)	3,500	1,800	1,500	3,200	2,700	2,900	3,100
Glycemic index	52	71	72	55	65	70	60
Saponins (%)	0.4	0.2	0.1	0.3	0.2	0.3	0.2
Tannins (%)	0.3	0.4	0.3	0.2	0.2	0.3	0.2
Phytic acid (%)	1.4	0.9	1.3	1.0	1.2	1.1	1.0
Total phenolic content (mg/g)	3.5	1.8	1.3	3.0	2.5	2.8	2.9

References: Bhat et al. (2022), Ragaee and Abdel-Aal (2015b), Maurya et al. (2023), Kaur et al. (2019), Kumar and Yadava (2018), and Subramanian (2020).

antioxidants (Rao et al., 2021). These methods improve the functional properties of millet, making it easier for the body to absorb its nutrients.

11.7.2 Modern Food Processing Techniques

In contrast, modern processing techniques such as extrusion, fortification, and baking are widely used for the development of millet-based products. Extrusion cooking, a popular method for preparing millet snacks, involves high heat and pressure, leading to the formation of light, crisp products. This process also enhances the digestibility of the grain and helps in the incorporation of other ingredients, such as vitamins or minerals, through fortification (Patel et al., 2018). Baking is commonly used to prepare millet-based bakery items like biscuits and breads, while fortification allows the addition of micronutrients like iron, calcium, and zinc, which are often low in traditional millet foods (Vijayakumar et al., 2017).

11.8 Millet-Based Value-Added and Nutraceutical Food Products

The development of value-added millet-based food products has gained momentum in recent years, catering to a wide range of dietary preferences and nutritional needs in millet-based foods. Millet-based nutraceutical products have gained popularity due to their rich nutrient profile, which includes high fiber, proteins, vitamins, and minerals.

11.8.1 Millet-Based Baked Food

Millet flours, including those made from finger millet (ragi), pearl millet, and foxtail millet, are used in the preparation of bakery items like biscuits, breads, and cakes. These products are especially beneficial for individuals with gluten intolerance, as millet is naturally gluten-free (Sasikala et al., 2019). Millets are also used in combination with other grains to improve the nutritional profile and texture of these baked goods.

11.8.2 Millet-Based Porridge and Breakfast Products

Millet-based porridge has gained popularity for its nutritional benefits and versatility. Millets like pearl, finger, and foxtail millet are rich in fiber, protein, and essential minerals such as iron, calcium, and magnesium, promoting overall health (Ragaee & Youssef, 2014). The development of millet porridge involves using these grains as a base, often paired with fruits, nuts, and milk (dairy or nondairy) to enhance flavor and nutrition. It can be made both savory and sweet to suit various tastes. The process typically includes roasting or boiling millet, then cooking it with water or milk to achieve a smooth, creamy texture (Sanjay Kumar et al., 2024).

11.8.3 Ready-to-Eat Millet Incorporated Snacks

Millet-based snacks, such as chips, puffed grains, and energy bars, are gaining popularity due to their convenience and nutritional value. Extrusion cooking is often employed to create these snacks, providing a crunchy texture while retaining the nutritional benefits of millets (Ghosh et al., 2021).

11.8.4 Millet-Based Beverages

Millets are used to create a range of beverages, including smoothies and malted drinks. These beverages are made by blending millet flour or powder with other ingredients like fruits, nuts, and yogurt, creating nutritious and refreshing drinks that are rich in fiber and micronutrients (Vijayakumar et al., 2017).

11.8.5 Gluten-Free Food Formulations

With the growing demand for gluten-free foods, millets are increasingly being used to create gluten-free formulations. Millet flours are naturally gluten-free, making them an ideal choice for individuals with celiac disease or gluten sensitivity. These gluten-free millet products, such as cookies and cakes, are often fortified with additional vitamins and minerals to enhance their nutritional value (Sasikala et al., 2019).

11.8.6 Millet-Based Energy Bars and Snacks

Companies like Bajra Foods and Millets of India have developed energy bars and snacks using millet as a key ingredient. These products cater to health-conscious consumers seeking nutritious alternatives to traditional snack foods. The high fiber content and low glycemic index of millets make them ideal for inclusion in such products (Vijayakumar et al., 2017). These products often include added nutrients such as vitamins, minerals, and probiotics to enhance their health benefits.

11.8.7 Millet-Based Fortified Breakfast Cereals

Several companies, including Kodo Cereal and True Elements, have launched millet-based breakfast cereals fortified with micronutrients like iron, calcium, and vitamin D. These cereals are marketed as health-promoting foods, particularly for children and adults with dietary deficiencies (Sasikala et al., 2019).

11.8.8 Millet Flour for Gluten-Free Baking

Millets, particularly finger millet, have found success in the gluten-free market. Brands like Ragi Roti and EcoMillets produce millet flours that can be used in the preparation of gluten-free bread, cookies, and cakes. These products cater to individuals with celiac disease or gluten sensitivity, providing them with nutritious and tasty alternatives (Patel et al., 2018).

11.8.9 Millet-Based Prebiotics

Millet-based prebiotic foods utilize the high fiber and bioactive compound content of millets to promote gut health. Prebiotics, as nondigestible fibers, stimulate beneficial gut microbiota, enhancing digestion, immunity, and metabolism. Fermented millet beverages like kanji and probiotic-enriched smoothies (Sharma et al., 2020), millet-based yogurt and kefir fortified with prebiotic fiber (Katara et al., 2025), and resistant starch-enriched millet bread and baked goods (Kumar et al., 2019). Extruded millet snacks, including prebiotic energy bars and crackers (Singh et al., 2022), along with sprouted millet products, further support gut microbiota (Raj et al., 2023). Synbiotic millet-based foods, combining prebiotic fibers with probiotics, offer an effective means to improve gut health (Gupta et al., 2020).

11.8.10 Millet-Based Extrudates

Millets such as pearl, finger, and foxtail millet are commonly used in extrudate formulations due to their high fiber, protein, and essential micronutrient content (Ragaee & Youssef, 2014). The development of millet-based extrudates involves optimizing processing parameters like temperature, pressure, and moisture to enhance texture, flavor, and nutritional value (Henry et al., 2015). Extrusion has been shown to improve nutrient bioavailability, sensory properties, and reduce antinutritional factors, making millets a more viable ingredient for snack products (Bhat & Tharakan, 2020).

11.8.11 Milk-Based Millet Nutraceutical Foods

Milk enriched millet foods have gained attention for their potential health benefits, combining the nutritional strengths of both milk and millet.

11.8.11.1 Milk-Based Millet Nutraceutical Food Preparation

i) Preparation of millets: Various types of millets such as pearl millet (*Pennisetum glaucum*), finger millet (*E. coracana*), and foxtail millet (*Setaria italica*) are used in the preparation of these functional foods. Millets are first cleaned, dehulled (if required), and milled into flour. This flour can be processed further using techniques like fermentation, extrusion, or cooking.

ii) Fermentation: Fermentation is commonly used to enhance the bioavailability of nutrients in millets. The process involves the use of beneficial microorganisms like lactic acid bacteria or yeasts, which not only improve the digestibility of millets but also enhance their nutrient profile. This process can increase the availability of vitamins, minerals, and amino acids (Kadam & Salunkhe, 1991). In the context of milk-based millet foods, fermented millet dough or batter can be incorporated into milk-based preparations such as puddings or soups.

iii) Extrusion cooking: Extrusion is a high-temperature, short-time cooking process that can improve the functional properties of both millets and milk. During extrusion, starch in the millet is gelatinized, and proteins in the milk undergo partial denaturation, which may improve their digestibility and bioavailability (Bhandari & Haug, 2008). The resulting product can be in the form of extruded snacks, breakfast cereals, or even pasta-like foods when combined with milk.

iv) Milk integration: The addition of milk enhances the nutritional profile of the millet product, as milk is rich in proteins, calcium, and vitamins, especially B vitamins and vitamin D. Milk also serves as a medium to dissolve fat-soluble vitamins from millets, enhancing their bioavailability (Zhou et al., 2020). The typical milk used is either cow's milk, goat's milk, or plant-based milk (such as almond or soy milk), depending on the target consumer group. Whole milk is often preferred for its fat content, which aids in the absorption of fat-soluble nutrients.

v) Formulation and fortification: Nutraceutical formulations can be further enhanced by incorporating additional bioactive compounds such as probiotics, prebiotics, antioxidants, and functional ingredients like omega-3 fatty acids. The fortification of milk with nutrients such as vitamins (A, D, B12), calcium, or iron can be particularly beneficial in addressing deficiencies in certain populations (Prasad et al., 2016).
vi) Packaging and storage: To retain the nutritional value of the milk-based millet nutraceutical foods, it is crucial to use proper packaging techniques. Vacuum sealing, controlled-atmosphere packaging, or freeze-drying are often employed to minimize nutrient degradation and extend shelf life (Liu et al., 2014).

11.8.11.2 Health Benefits of Milk-Based Millet Nutraceutical Foods

i) Improved digestion: The high fiber content in millets aids digestion and promotes gut health by acting as a prebiotic. The combination with milk helps to enhance gut flora, improving overall digestive health (Madhusudhan et al., 2015).
ii) Enhanced immunity: Both milk and millets contain bioactive compounds that support immune function. The high calcium content in milk is vital for maintaining healthy immune cells, while the antioxidants in millets (such as phenolic acids) provide additional immune-boosting effects (Bhandari & Haug, 2008).
iii) Better gut health: Fermented milk-based millet foods are particularly beneficial for gut health, as the fermentation process increases the levels of beneficial bacteria (lactic acid bacteria) and improves the digestibility of both the millet and milk proteins (Yang et al., 2015).

11.9 Market Trends and Consumer Perspectives

As global awareness of the health benefits of various foods continues to rise, consumer demand for functional and health-focused foods has significantly increased. This shift is largely driven by a growing focus on nutrition, sustainability, and the prevention of lifestyle-related diseases. Millets, due to their rich nutritional profile, are increasingly being recognized as a key ingredient in health-focused products, particularly in the form of millet-based nutraceuticals.

Table 11.2 presents a comprehensive overview of various millet-based and millet-milk-based nutraceutical products, highlighting their key ingredients, associated health benefits, and relevant academic references. These products encompass a wide range of formulations, including protein bars, probiotic yogurts, functional beverages, gluten-free cereals, and fortified porridges, all of which leverage the nutritional richness of millet grains. The inclusion of bioactive compounds such as antioxidants, dietary fiber, and essential minerals (iron, calcium, magnesium) in these products provides significant health benefits, particularly in managing conditions such as diabetes, cardiovascular diseases, iron deficiency anemia, and digestive health issues. Furthermore, the integration of millet with dairy or plant-based milks in several products offers additional functional advantages, including improved calcium bioavailability for bone health and enhanced protein content for muscle recovery and growth. These

Table 11.2 Millet-based functional foods and their nutraceutical benefits

Nutraceutical millet food products	Key ingredients	Nutraceutical benefits	References
Millet-based protein bars	Millet flour, nuts (almonds, walnuts), seeds (chia, flax), honey, and protein isolates (pea, soy)	High protein content, supports muscle recovery, boosts energy levels, and rich in omega-3 fatty acids and antioxidants	Samuel and Peerkhan (2020); Singh et al., (2021)
Probiotic millet yogurt	Millet flour, probiotics (Lactobacillus, Bifidobacterium), fruit puree (mango, strawberry), sugar, or stevia	Supports gut health, enhances digestion, boosts immunity, and rich in calcium for bone health	Chandra and Desai (2022); Rai (2016)
Millet-based functional beverages	Millet flour, fruit extracts (lemon, pomegranate), natural sweeteners (stevia, honey), vitamin C, and magnesium	Hydrates, replenishes electrolytes, promotes detoxification, and enhances energy levels	Ravi and Rani (2021)
Millet-based energy drink	Millet flour, electrolytes (sodium, potassium), natural flavorings (lime, mint), and B vitamins	Provides sustained energy, supports hydration, and enhances stamina during exercise	Kumar and Yadav (2020)
Millet-based gluten-free cereal	Millet flakes, dried fruits (raisins, cranberries), nuts (cashews, almonds), seeds (pumpkin, sunflower), and honey	Provides fiber, antioxidants, supports heart health, and maintains stable blood sugar levels	Verma and Srivastava (2022)
Millet-based high-fiber cookies	Millet flour, oats, dark chocolate, chia seeds, sunflower seeds, and coconut oil	Rich in dietary fiber, aids in digestion, lowers cholesterol, and supports weight management	Zaveri and Patel (2025))
Prebiotic-enriched millet bread	Millet flour, whole grains (barley, oats), prebiotics (inulin), yeast, and olive oil	Supports gut microbiota, improves digestive health, and stabilizes blood sugar levels	Singh and Yadav (2023)
Millet-based extruded snacks	Millet flour, vegetable powders (spinach, beetroot), spices (paprika, cumin), salt, and olive oil	Offers healthy, low-calorie snack option, promotes satiety, and enhances digestive function	Meena et al. (2025)
Iron-fortified millet porridge	Millet flour, iron fortificants, milk or plant-based milk, and natural sweeteners	Combats iron deficiency anemia, improves cognitive function, and supports energy levels	Patil and Deshmukh (2021)
Sprouted millet powder	Sprouted millet (finger, pearl, foxtail), turmeric, black pepper, and spices (cinnamon, cardamom)	Rich in bioactive compounds, promotes digestion, boosts metabolism, and reduces inflammation	Ghosh et al. (2019)
Millet-milk-based energy drink	Millet flour, milk, honey, natural flavors (mint, lime), B vitamins, and electrolytes	Provides hydration, supports muscle recovery, and boosts energy levels	Yadav and Sharma (2022)
Millet-milk smoothie	Millet flour, milk, berries (strawberries, blueberries), yogurt, and honey	Promotes gut health, supports bone health (calcium), and antioxidant-rich	Rani et al. (2016)
Millet and milk porridge	Millet flour, milk, cinnamon, cardamom, and sweeteners (jaggery, stevia)	Aids digestion, promotes weight loss, and high in calcium and fiber	Sanjana Reddy (2017)

innovations align with the growing demand for health-oriented, sustainable food solutions, positioning millet as a versatile and nutritionally superior ingredient in functional food development. References from recent studies substantiate the effectiveness of these products in promoting overall wellness and addressing contemporary health concerns, such as gut microbiota modulation, immune support, and chronic disease prevention.

11.10 Current Market Trends for Millet-Based Nutraceuticals

The market for millet-based nutraceutical products has been expanding in response to the growing demand for functional foods. Key market trends include the following.

11.10.1 Gluten-Free and Allergy-Friendly Products

With the rise of gluten intolerance, celiac disease, and other food sensitivities, millet-based products are gaining popularity in the gluten-free food market. Millets, being naturally gluten-free, are used in various nutraceutical products, such as gluten-free cereals, biscuits, and snacks, catering to a significant segment of the population with dietary restrictions (Patel et al., 2018).

11.10.2 Fortification of Millet-Based Products

As consumers seek fortified foods with added vitamins and minerals, millet-based products are increasingly being enriched with essential nutrients. Fortification of millet products with vitamins like vitamin D, calcium, and iron is a growing trend, especially in breakfast cereals, energy bars, and nutritional supplements (Ghosh et al., 2021). This trend is particularly prevalent in regions where micronutrient deficiencies are common.

11.10.3 Functional Snacks

The growing demand for convenient, on-the-go snacks with added health benefits has led to the development of millet-based functional snacks. These include products like millet-based energy bars, trail mixes, and ready-to-eat millet-based snacks. These products appeal to busy consumers looking for a nutritious snack option with functional benefits, such as weight management and energy boosting (Sasikala et al., 2019).

11.10.4 Plant-Based and Sustainable Foods

With the increasing popularity of plant-based diets and sustainable food choices, millet-based products align well with these trends. Millets, being hardy, drought-resistant crops, are considered more environmentally sustainable than many other grains. Their ability to grow in arid regions with minimal water resources makes them an attractive choice for environmentally conscious consumers (Vijayakumar et al., 2017). This aligns with the broader trend of plant-based and sustainable food consumption, which has seen increased interest in recent years.

11.10.5 Millet Beverages

Another emerging trend is the development of millet-based beverages. Companies are incorporating millets into smoothies, malted drinks, and health drinks. These beverages are marketed as functional drinks that provide hydration while offering nutritional benefits like improved gut health and enhanced energy levels (Patel et al., 2018). As consumers increasingly seek plant-based alternatives to dairy and sugary drinks, millet-based beverages are gaining popularity.

11.10.6 Expansion of Millet-Based Dietary Supplements

The market for dietary supplements is also seeing the emergence of millet-based products, such as millet protein powders and millet-based probiotics. These products are gaining traction among fitness enthusiasts, athletes, and individuals looking for plant-based sources of protein and other nutrients. Millet protein, in particular, is a high-quality, digestible protein that appeals to those seeking sustainable and plant-based protein sources (Sasikala et al., 2019).

11.11 Market of Millet-Based Nutraceutical Foods

The market for millet-based nutraceuticals is rapidly growing, driven by increased consumer demand for functional, health-focused foods. Trends such as the rise of gluten-free products, fortified foods, functional snacks, and plant-based offerings are shaping the future of the millet market. With their exceptional nutritional value and environmental sustainability, millets are poised to become a key player in the functional food industry, catering to the evolving needs of health-conscious and environmentally aware consumers. While millets are known for their rich nutritional profile, the bioavailability of certain nutrients in millets, such as iron, zinc, and calcium, can be limited due to the presence of antinutritional factors like phytic acid, which can inhibit the absorption of these minerals (Rao et al., 2021). Future research should focus on developing strategies to enhance the bioavailability of these essential nutrients. Some promising strategies for enhancing bioavailability of nutrients approaches include as the following.

11.11.1 Fermentation and Germination

These traditional processing techniques have been shown to reduce antinutritional factors and improve the bioavailability of minerals in millets (Patel et al., 2018). Further studies on optimizing these processes could enhance their efficacy in millet-based food production.

11.11.2 Fortification

Another strategy is fortifying millet-based products with bioavailable forms of essential nutrients. For example, fortifying millets with iron in a chelated form or adding vitamin D and calcium in a more bioavailable format could help improve their nutritional profile and health benefits (Vijayakumar et al., 2017).

11.11.3 Nanoencapsulation

This emerging technology can be used to enhance the bioavailability of micronutrients in millet-based foods. By encapsulating nutrients in nanoparticles, the absorption of nutrients can be increased, thus providing enhanced health benefits (Ghosh et al., 2021).

11.12 Innovations in Product Development and Processing Technologies

The millet market has seen steady growth, but there is considerable potential for innovation in product development and processing technologies. Several research directions could significantly improve the value of millet-based foods.

11.12.1 Extrusion and 3D Printing

Modern food processing technologies like extrusion and 3D printing offer the potential to create novel millet-based products with improved texture, shape, and nutritional value. For instance, extrusion cooking can be used to produce millet-based snacks, while 3D printing can be employed to create customized millet-based food items, such as fortified snacks or dietary supplements that cater to specific health needs (Sasikala et al., 2019).

11.12.2 Functional Ingredients

Integrating other functional ingredients such as probiotics, prebiotics, and omega-3 fatty acids into millet-based products could enhance their health benefits. The development of millet-based functional foods that combine multiple health-promoting ingredients may play a significant role in the growing market for nutraceuticals (Patel et al., 2018).

11.12.3 Enhanced Processing Methods

Research into newer processing methods, such as high-pressure processing (HPP) and ultrasound-assisted extraction, could help preserve the nutritional integrity of millets while also improving their taste and texture (Vijayakumar et al., 2017). These innovations would help overcome some of the limitations posed by traditional processing methods and open up new possibilities for millet-based food products.

11.13 Scope for Millet-Based Nutraceuticals in Addressing Diseases

Millets offer significant promise in addressing global challenges like malnutrition and lifestyle diseases. However, there is a need for more targeted research to fully realize their potential in these areas:

11.13.1 Addressing Micronutrient Deficiencies

Millets are an excellent source of essential micronutrients such as iron, magnesium, and zinc. Further research into how millet-based foods can help address global micronutrient deficiencies, particularly in developing countries, could prove invaluable. Developing millet-based fortified foods targeted at vulnerable populations, such as children and pregnant women, could help reduce the burden of malnutrition (Rao et al., 2021).

11.13.2 Combating Lifestyle Diseases

With the rising prevalence of lifestyle diseases such as obesity, diabetes, and cardiovascular diseases, millet-based nutraceuticals have the potential to play a key role in prevention and management. The high fiber content of millets has been linked to better blood sugar control and heart health, making them ideal candidates for inclusion in functional foods aimed at preventing or managing chronic conditions (Ghosh et al., 2021). Future studies could focus on developing millet-based foods with proven health benefits for these specific diseases.

11.13.3 Personalized Nutrition

There is also an emerging opportunity in the field of personalized nutrition, where millet-based products can be tailored to individual health needs. Research into how different varieties of millets affect various health conditions could lead to the development of targeted millet-based nutraceuticals that address specific health challenges, such as weight management, digestive health, or metabolic disorders (Sasikala et al., 2019).

Research in the field of millet-based nutraceuticals is poised to drive significant advancements in nutrition and health. By addressing research gaps in enhancing bioavailability, innovating product development, and exploring the potential for millets in tackling malnutrition and lifestyle diseases, the future prospects for millet-based foods are promising. Ongoing studies and innovations are crucial in ensuring that millets can fulfill their role as a key component in sustainable, health-promoting diets globally.

11.14 Research Gaps and Future Prospects

As the potential of millets as functional foods and nutraceuticals continues to be explored, there are several key areas where research is needed to further enhance their health benefits, marketability, and global adoption. These areas include improving the bioavailability of nutrients, innovating product development and processing technologies, and exploring the scope for millet-based nutraceuticals in addressing malnutrition and lifestyle diseases.

11.15 Challenges in Millet Food Product Development

Despite the many benefits of millet-based products, several challenges need to be addressed in the development process.

11.15.1 Antinutritional Factors

Millets, while rich in essential nutrients, also contain certain antinutritional factors that can impact nutrient absorption. These compounds are naturally occurring in millets and can interfere with the bioavailability of minerals and other nutrients. The following are the main antinutritional factors found in millets.

i) Phytic acid: Phytic acid is known to bind with essential minerals such as iron, zinc, and calcium, forming insoluble complexes that reduce their absorption in the digestive tract. This can potentially lead to mineral deficiencies, especially in populations with limited access to diverse food sources. Phytic acid content in millets ranges from 0.9% in finger millet to 1.4% in pearl millet (Ragaee & Abdel-Aal, 2015a).
ii) Tannins: Tannins are polyphenolic compounds that can inhibit the absorption of proteins and iron by binding to these nutrients and forming complexes. They also have antidigestive enzyme properties, potentially affecting the overall nutritional value of millets. Tannin levels in millets are relatively low, but they still contribute to reduced bioavailability of nutrients. The levels range between 0.2% and 0.4% (Ragaee & Abdel-Aal, 2015a).
iii) Saponins: Saponins can act as antinutritional factors by forming complexes with proteins and lipids, reducing their digestibility and absorption. Saponins also have detergent-like properties that can interfere with the gut, causing irritation in sensitive individuals. Saponin content in millets is low, ranging from 0.2% to 0.4% (Ragaee & Abdel-Aal, 2015a).
iv) Oxalates: Oxalates can form insoluble salts with calcium, reducing the bioavailability of calcium, and possibly contributing to kidney stone formation in individuals prone to oxalate stones. Oxalate levels in millets are moderate, though they are not as high as in certain other grains and vegetables.
v) Gluten: Gluten itself is not an antinutrient; the absence of gluten in millets is beneficial for individuals with celiac disease or gluten intolerance. However, some people may be sensitive to the specific proteins present in millets.
vi) Sensory attributes: Sensory attributes like taste, texture, and appearance are crucial factors in consumer acceptance. Millet-based products often have a distinctive flavor that may not be preferred by all consumers. In addition, the texture of some millet products can be coarse or gritty, which may affect their palatability (Patel et al., 2018). Therefore, improving the sensory qualities of millet-based foods through better processing methods and ingredient combinations is a priority for researchers. Consumer acceptance is another challenge, as many people are not familiar with millet or may not consider it a mainstream food. This can limit the market for millet-based products. To increase acceptance, there is a need for education on the health benefits of millets and their versatility in cooking, as well as innovations in product formulations that appeal to a wider audience (Ghosh et al., 2021).

vii) Shelf-life issues: One of the main challenges is the relatively short shelf life of millet-based products due to their high fat content, which can lead to rancidity. Proper packaging and preservation techniques, such as the use of antioxidants and refrigeration, are essential to increase the shelf life of these products (Rao et al., 2021).

Millets, when processed using both traditional and modern techniques, can be transformed into a wide range of value-added food products that cater to various consumer preferences. The development of millet-based foods offers numerous health benefits, but challenges related to shelf life, sensory attributes, and consumer acceptance must be overcome to maximize their potential in the global food market. Millets, due to their nutritional benefits and versatility, have gained traction globally as a healthy alternative to other grains. Several success stories and initiatives highlight the growing awareness and adoption of millet-based products in diverse markets. These case studies demonstrate the potential of millets in addressing nutritional deficiencies, promoting health, and contributing to sustainable food systems.

11.16 Global Initiatives Promoting Millet Consumption

Various regional and global initiatives are promoting millet consumption as a means of addressing health and nutrition challenges, especially in developing countries.

11.16.1 The International Year of Millets 2023

The United Nations declared 2023 as the International Year of Millets, aiming to raise global awareness about the nutritional benefits of millets and to promote their cultivation and consumption. The initiative has garnered support from governments, international organizations, and the private sector, and it focuses on improving millet-based product innovation, as well as creating awareness in schools, markets, and communities (Ghosh et al., 2021).

11.16.2 India's National Millet Mission

India, one of the world's largest producers of millets, has implemented the National Millet Mission, which aims to increase the production and consumption of millets through research, market linkages, and promotion. The government is actively encouraging millet-based food products in schools, healthcare institutions, and public distribution systems (Sasikala et al., 2019). These efforts aim to address nutritional deficiencies while supporting farmers who grow millets in rainfed areas.

11.16.3 Africa's Millet Initiatives

In countries like Kenya and Ethiopia, millet is being promoted as a resilient crop in response to climate change. Non-governmental organizations (NGOs) and local governments are working together to create value chains around millet and to encourage its use in both rural and urban settings. Initiatives focus on both traditional food uses and new millet-based innovations, such as fortified porridge and snack foods, to improve the nutrition of vulnerable populations (Vijayakumar et al., 2017).

11.16.4 Government Initiatives

Governmental bodies at national and regional levels are promoting millet cultivation and consumption through subsidies, training programs, and policy advocacy. For example, India's government supports millet farmers with subsidies for growing millets and offers a minimum support price for the crops. Additionally, millet-based food products are being introduced into government-sponsored food programs, such as the mid-day meal scheme in schools (Rao et al., 2021).

11.16.5 Non-governmental Organizations

NGOs like The Millet Network of India (MINI) and International Crops Research Institute for the Semi-Arid Tropics (ICRISAT) play a significant role in promoting millet-based nutrition through campaigns, workshops, and research. These NGOs work closely with farmers to improve production practices, market access, and the development of new millet-based products. They also collaborate with local governments to improve the availability of millet-based food products in schools, hospitals, and rural areas (Patel et al., 2018).

11.16.6 Public–Private Partnerships Promoting Millets

Many successful initiatives are the result of collaborations between the public and private sectors. For example, the partnership between government agencies and private food companies has led to the development of millet-based nutraceutical products and the establishment of millet processing plants. These efforts have increased the availability and affordability of millet-based products, contributing to wider acceptance in urban and rural communities alike (Ghosh et al., 2021).

The promotion of millet-based foods has gained significant momentum through various successful case studies and global initiatives. Nutraceutical products, such as energy bars and fortified cereals, have contributed to the growing market for millet-based foods. Regional and global efforts, like the International Year of Millets and India's National Millet Mission, are driving awareness and consumption. Governmental and non-governmental organizations play a crucial role in supporting millet cultivation, developing value-added products, and promoting millet's role in addressing global health challenges.

11.17 Conclusions

The nutraceutical potential of millet-based food products is increasingly recognized for their role in addressing various health concerns while providing substantial nutritional benefits. Millets, including finger millet, pearl millet, foxtail millet, and sorghum, are rich in proteins, dietary fiber, essential minerals, and bioactive compounds, exhibiting antioxidant, anti-inflammatory, and disease-preventive properties. Millets' contributions to digestive health, glycemic regulation, cardiovascular protection, and bone health underscore their significance in contemporary dietary frameworks. The versatility of millets in gluten-free snacks, fortified cereals, protein bars, probiotic yogurts, and functional beverages highlights their growing importance in nutraceutical food development. Finger millet, with its high calcium content, supports bone health, while iron-rich pearl millet helps mitigate anemia, making these grains vital for targeted nutrition. Furthermore, slow-digesting carbohydrates and prebiotic fibers enhance gut microbiota balance and metabolic stability, reinforcing their functional food applications. Despite challenges such as nutrient bioavailability and sensory appeal, advancements in fermentation, extrusion, and biofortification have improved the nutritional efficacy of millet-based nutraceutical food products. As global demand for functional foods rises, millets offer a sustainable, health-promoting solution for combating malnutrition and lifestyle-related diseases, positioning them as essential components of future nutrition strategies.

References

Anjum, F. M., Haseeb, M., & Rehman, R. (2017). Nutritional and functional properties of sorghum. *Food Chemistry*, *228*, 47–52.

Antony, A., Haug, W., & Lantzsch, H. J. (1996). Nutritional composition and processing of pearl millet. *Food Research International*, *29*(2), 91–96.

Bamigboye, A. S., Olorunsogo, O. O., & Akinmoladun, F. O. (2020). Millet and its health benefits: A review. *Journal of Food Science and Technology*, *57*(6), 2173–2183.

Bhandari, B. M., & Haug, W. (2008). Functional foods and nutraceuticals: An overview. *Food Research International*, *41*(6), 644–653.

Bhat, R., & Tharakan, P. J. (2020). Extrusion processing of millets for improved nutrient bioavailability and sensory properties: A review. *Food Research International*, *128*, 108744. https://doi.org/10.1016/j.foodres.2019.108744

Bhatt, D., Fairos, M., & Mazumdar, A. (2022). Millets: Nutritional composition, production and significance: A review. *Journal of Pharmaceutical Innovation*, *11*, 1577–1582.

Bhatt, D., Rasane, P., Singh, J., Kaur, S., Fairos, M., Kaur, J., ... Sharma, N. (2023). Nutritional advantages of barnyard millet and opportunities for its processing as value-added foods. *Journal of Food Science and Technology*, *60*(11), 2748–2760.

Bisht, N. S., Kumar, S., & Sharma, R. (2015). Health benefits of barnyard millet and its potential for health promotion. *Food Science and Technology International*, *21*(5), 413–421.

Chandra, R., & Desai, M. (2022). Probiotic millet yoghurt: Potential health benefits. *Food Research International*, *141*, 111597.

Chandrasekara, A., & Shahidi, F. (2011). Antioxidant properties of millet grain phenolics and their effects on lipid peroxidation and human LDL cholesterol oxidation. *Journal of Agricultural and Food Chemistry*, *59*(3), 1124–1130.

Dwivedi, S. L., Upadhyaya, H. D., Senthilvel, S., Hash, C. T., Fukunaga, K., Diao, X., & Santra, D. K. (2012). Millets: Genetic and genomic resources. *Plant Breeding Reviews*, *35*, 247–375.

Dykes, L., & Rooney, W. (2006). Millets: Nutritional value and health benefits. *Critical Reviews in Food Science and Nutrition*, *46*(7), 486–504.

FAO. (1991). *Protein quality evaluation of millet grains*. Food and Agriculture Organization of the United Nations.

FAO. (1995). *Sorghum and millets in human nutrition*. Food and Agriculture Organization of the United Nations.

Geervani, P., & Eggum, B. O. (1989b). Protein quality and digestibility of millet. *Food Chemistry*, *32*(1), 83–89.

Ghosh, K., Vinodkumar, T., Singh, R., Singh, A., & Ilavarasan, R. (2024). Promising potential of millets as the staple of our plate: An Indian perspective. *Discover Food*, *4*(1), 103.

Ghosh, S., Biswas, S., & Ghosh, S. (2021). Millets in the management of obesity: A comprehensive review. *Clinical Nutrition*, *40*(2), 462–470.

Gupta, S., Kapoor, S., & Mishra, S. (2020). Synbiotic millet-based foods: Combining prebiotic fibres with probiotics for effective gut health improvement. *Food Research International*, *132*, 109036. https://doi.org/10.1016/j.foodres.2020.109036

Hadimani, D., & Malleshi, N. G. (1993). Nutritional composition of millets. *Journal of the Science of Food and Agriculture*, *62*(4), 289–294.

Haug, W., & Lantzsch, H. J. (2008). Sensitive method for rapid determination of phytic acid in cereals and legumes. *Journal of the Science of Food and Agriculture*, *59*(2), 1–8.

Hegde, M., & Chandra, S. (2005). Antioxidant activity of millets. *Food Chemistry*, *91*(4), 715–718.

Henry, J., Man, C., & Moughan, P. J. (2015). The development of millet-based extrudates: Optimization of processing parameters for enhanced texture, flavor, and nutritional value. *Journal of Food Science and Technology*, *52*(4), 2411–2420.

Kadam, S. S., & Salunkhe, D. K. (1991). Fermentation of millets and their utilization in food products. *Critical Reviews in Food Science and Nutrition*, *30*(4), 309–335.

Katara, N., Kumar, A., Meena, K. K., Choudhary, R., Meena, S., & Goel, S. (2025). Fortification of traditional Indian dairy sweets with pearl millet and jaggery: A comprehensive review. *European Journal of Nutrition and Food Safety*, *17*(7), 258–270.

Kaur, P., Purewal, S. S., Sandhu, K. S., Kaur, M., & Salar, R. K. (2019). Millets: A cereal grain with potent antioxidants and health benefits. *Journal of Food Measurement and Characterization*, *13*(1), 793–806.

Krishnan, M., Rao, M. S., & Reddy, V. S. (2015). Nutritional properties of finger millet and its role in health promotion. *Food and Bioprocess Technology*, *8*(6), 1535–1542.

Krishnan, R., Suresh, K., & Bhaskar, P. (2011). Finger millet seed coat: A potential food ingredient. *Food Research International*, *44*(3), 937–940.

Kumar, A., Gupta, V., & Yadav, S. (2019). Resistant starch-enriched millet bread and baked goods for enhanced gut health. *Food Research International*, *115*, 311–318. https://doi.org/10.1016/j.foodres.2018.11.050

Kumar, A., & Yadav, R. (2020). Energy drink development from millet grains: Nutritional and functional insights. *Food Chemistry*, *327*, 127302.

Kumar, P., & Yadava, S. (2018). Millets as functional food. *Food and Bioprocess Technology*, *11*(7), 1184–1195.

Kumar, S., & Parmeswaran, K. (1999). Proteins and amino acids in millets. *Food Chemistry*, *65*(4), 370–375.

Kumar, V., Chauhan, S., & Yadav, S. K. (2008). Nutritional aspects of kodo millet (*Paspalum scrobiculatum*) and its therapeutic applications. *Journal of Food Science, 73*(9), 1261–1269.

Li, Y., Lu, Z., & Zhang, H. (2009). Archeobotanical evidence of millet domestication in China. *Chinese Science Bulletin, 54*(16), 2774–2782.

Liu, Y., Sun, W., & Wang, S. (2014). Packaging and shelf life of functional foods: A review. *Food Control, 46*, 62–71.

Madhusudhan, S. R., Krishnan, K., & Sundaram, V. (2015). Fermented millet-based probiotic food for gut health. *International Journal of Food Science and Technology, 50*(4), 748–757.

Mal, B., Padulosi, S., & Ravi, S. B. (2010). *Minor millets in South Asia: Learnings from IFAD-NUS project in India and Nepal*. Bioversity International.

Malleshi, N. G., & Hadimani, A. S. (1993a). Dietary fibre in millets: A comparative study. *Food Chemistry, 45*(2), 235–239.

Malleshi, N. G., & Hadimani, D. (1993b). The potential health benefits of millets. *Journal of Food Science, 58*(2), 276–278.

Maurya, R., Boini, T., Misro, L., Radhakrishnan, T., Sreedharan, A. P., & Gaidhani, D. (2023). Comprehensive review on millets: Nutritional values, effect of food processing and dietary aspects. *Journal of Drug Research in Ayurvedic Sciences, 8*(Suppl 1), S82–S98.

Mbithi, P. M., Dendy, D. A. V., & Haug, W. (2000). Amino acid composition and protein quality of finger millet. *Journal of the Science of Food and Agriculture, 80*(3), 331–335.

Mbithi-Mwikya, S., Van Camp, J., & Huyghebaert, A. (2000). Nutrient and antinutrient changes in finger millet (Eleusine coracana) during sprouting. *LWT — Food Science and Technology, 33*, 9–14. https://doi.org/10.1006/fstl.1999.0605

Meena, N., Sruthi, N. U., Dhungana, P., Al-Ali, H., Srinivasa Rao, P., & Bhattarai, R. R. (2025). A concise review of millet milk development. *Journal of Food Science, 90*(3), e70126.

Monteiro, A., Sudharshana, G., & Kumar, S. (1987). Protein quality of millet grains. *Indian Journal of Agricultural Sciences, 57*(7), 505–510.

Ochanda, S. O., Mwasaru, A. M., et al. (2010). Effects of malting and fermentation treatments on group B-vitamins of red sorghum, white sorghum and pearl millets in Kenya. *Journal of Applied Biosciences, 34*, 2128–2134.

Patel, K., Nair, R. R., & Shah, N. (2011). Nutritional and functional evaluation of sorghum varieties. *Journal of Food Science, 76*(7), S99–S106.

Patel, P., Ghosh, R., & Kaur, S. (2018). Extrusion cooking of millet-based snacks: A review. *Journal of Food Science and Technology, 55*(12), 4814–4825.

Patil, M., & Deshmukh, P. (2021). Iron-fortified millet porridge: A strategy for addressing iron deficiency anemia. *Nutrition, 81*, 111299.

Prasad, S. K., Sundararajan, S., & Venkataraman, A. (2016). Nutritional enhancement and health benefits of fermented milk-based foods. *Food Research International, 89*, 150–157.

Ragaee, S., & Abdel-Aal, E.-S. M. (2015a). Antioxidant and antinutritional factors in millets. *Food Chemistry, 172*, 437–445.

Ragaee, S., & Abdel-Aal, E.-S. M. (2015b). Nutritional quality of millets and their role in human health: A review. *Food Research International, 69*, 1–9.

Ragaee, S., Mohamed, A., & El-Faize, M. (2006). Nutritional and antioxidant properties of pearl millet. *Food Research International, 39*(4), 445–450.

Ragaee, S., & Youssef, M. (2014). Potential of whole grain wheat and whole grain millet to improve gut health. *Cereal Chemistry, 91*(3), 244–249.

Rai, D., Pandey, R. K., Maurya, A. K., Rai, D. C., Kumar, D., & Tiwari, M. (2016). Characterization and antioxidant property of cereal enriched bio-yoghurt. *Journal of Pure and Applied Microbiology, 10*(4), 3071–3078.

Raj, R., Kaur, G., & Sidhu, J. (2023). Sprouted millet products: Beneficial effects on gut health. *Journal of Food Science and Agriculture, 102*(2), 5000–5010. https://doi.org/10.1002/jsfa.11015

Rani, R., Kumar, M. H. S., & Sabikhi, L. (2016). Process optimisation for a ready-to-serve breakfast smoothie from a composite milk–sorghum base. *International Journal of Dairy Technology, 69*(3), 372–379.

Rao, P. S., Reddy, P. S., & Rajanna, L. (2021). Millets and their role in digestive health. *International Journal of Food Science, 56*(3), 207–214.

Rao, S. S., Reddy, R. S., & Naidu, M. (2012). Nutritional profile of little millet (*Panicum miliare*) and its application in food products. *Indian Journal of Food Science and Technology, 49*(5), 387–394.

Ravi, K., & Rani, P. (2021). Functional beverages from millet grains. *Journal of Functional Foods, 81*, 104470.

Saleh, A. S. M., Zhang, Q., Chen, J., & Shen, Q. (2013a). Millet grains: Nutritional quality, processing, and potential health benefits. *Comprehensive Reviews in Food Science and Food Safety, 12*(3), 281–295.

Saleh, M., Rahman, M. A., & Mahboob, S. (2013b). Millets as potential crops for food security in arid regions. *Agricultural Systems, 118*, 85–92.

Samuel, K. S., & Peerkhan, N. (2020). Pearl millet protein bar: nutritional, organoleptic, textural characterization, and in-vitro protein and starch digestibility. *Journal of Food Science and Technology, 57*(9), 3467–3473.

Sanjana Reddy, P. (2017). Pearl millet, Pennisetum glaucum (L.) R. Br. In *Millets and sorghum: Biology and genetic improvement* (pp. 49–86). Wiley.

Sanjay Kumar, T., Nageswari, R., Somasundaram, S., Anantharaju, P., Baskar, M., Ramesh, T., Rathika, S., Vanniarajan, C., & Subrahmaniyan, K. (2024). Significance of millets for food and nutritional security—An overview. *Discover Food, 4*(1), 73.

Sasikala, P., Prathima, S., & Sriram, S. (2019). Millets in diabetes management: A review of their glycemic index and effects on blood sugar regulation. *Diabetes and Metabolic Syndrome: Clinical Research & Reviews, 13*(3), 2219–2226.

Sehgal, S., & Kawatra, A. (2003). Nutritional composition of millets: Implications for health and nutrition. *Nutrition and Food Science, 33*(5), 212–215.

Shahidi, F., & Ambigaipalan, P. (2015). Phenolics and polyphenolics in foods, beverages, and spices: Antioxidant activity and health effects – A review. *Journal of Functional Foods, 18*, 820–897.

Sharma, P., Kumar, V., & Singh, R. (2020). Fermented millet beverages: Kanji and probiotic-enriched smoothies for gut health. *Journal of Food Science and Technology, 57*(4), 1260–1267. https://doi.org/10.1007/s11483-020-01731-9

Singh, P., Sharma, S., & Dey, A. (2022). Extruded millet snacks: Prebiotic energy bars and crackers for gut microbiota modulation. *Journal of Cereal Science, 101*, 103250. https://doi.org/10.1016/j.jcs.2022.103250

Singh, P., & Yadav, R. (2023). Prebiotic-enriched millet bread: A new approach to gut health. *Journal of Nutrition, 153*(1), 150–157.

Singh, R., Patel, V., & Jha, A. (2012a). Dietary fibre content in millets. *International Journal of Food Science and Technology, 47*(1), 75–82.

Singh, R., & Raghuvanshi, R. (2012). Nutritional value of millet grains. *Food Research International, 45*(7), 323–328.

Singh, R., Singh, K., & Nain, M. S. (2021). Nutritional evaluation and storage stability of popped pearl millet bar. *Current Science, 120*(8), 1374–1381.

Singh, S., Bansal, V., & Pandey, M. (2008). Pearl millet (*Pennisetum glaucum*): Nutritional and health benefits. *Food Research International, 41*(6), 653–657.

Singh, S., Patel, V., & Jha, A. (2012b). Polyphenol and antioxidant properties of millet grains. *Food Chemistry, 132*(2), 1099–1107.

Subramanian, S. (2020). Nutritional and antioxidant properties of millets. *Food Reviews International, 36*(6), 744–758.

Ushakumari, S. R., Lathika Devi, G., & Malleshi, N. G. (2004). Milling of millets: Present status and future scope. *Current Science, 87*(4), 475–481.

Veena, K. (2003). The role of finger millet in health and nutrition. *Indian Food Research Journal, 6*(1), 55–60.

Verma, R., & Srivastava, A. (2022). Millet-based gluten-free cereals: Health benefits and applications. *Nutrition and Food Science, 52*(4), 453–462.

Vijayakumar, S., Rani, M., & Krishna, M. (2017). Health benefits of millets: A comprehensive review. *International Journal of Food Science, 42*(2), 130–137.

Yadav, K., Saini, S., & Prakash, V. (2011). Millets as a sustainable and nutritious alternative. *Critical Reviews in Food Science and Nutrition, 51*(7), 643–654.

Yadav, R., & Sharma, S. (2022). Millet-milk-based energy drink: Nutritional benefits and formulation. *Nutritional Advances, 24*(2), 150–160.

Yang, Y., Wu, W., & Zhou, S. (2015). The role of fermented foods in gut microbiota modulation. *Food Research International, 69*, 5–14.

Zaveri, D., & Patel, P. (2025). Millets: Importance and Role for Nutritive Improvement of Bakery Products. *Biological Outlook to Improve the Nutritive Quality of Bakery Products*, 125–139.

Zhou, L., Li, F., & Zhang, C. (2020). The bioavailability of nutrients in dairy-based functional foods. *Journal of Dairy Science, 103*(2), 1389–1397.

12

Millet-Based Fermented Food Products: Nutritional and Health Benefits

Priyanka Thakur[1], *Sachin Sharma*[2], *Vikas Kumar*[3], *Ritesh Verma*[1], *Tanveen Kaur*[1], *Abhishek Rana*[2], *and Satish Kumar*[2]

[1] *Department of Agriculture and Environmental Science, National Institute of Food Technology Entrepreneurship and Management (NIFTEM), Kundli, Sonipat, Haryana, India*
[2] *Department of Food Science & Technology, Dr. Y.S. Parmar University of Horticulture and Forestry, Nauni, Solan, Himachal Pradesh, India*
[3] *Indian Council of Agricultural Research, Indian Institute of Millets Research (IIMR), Rajendranagar, Hyderabad, Telangana State, India*

12.1 Introduction

Millets are small-seeded grains cultivated mainly by the rural population across the globe, primarily for harvesting the seeds as a food source. Since they can be climate-resilient and can flourish in various weathers, they have been a staple meal for many societies for several thousand years, especially in Asia and Africa. The importance of millets can be assessed from the reference of these coarse grains in ancient literature such as the Old Testament and works from Greece and Rome as staple food items. They are vital for food security in areas confronting the challenges of climate change. Millets are nutrient-dense foods with vital nutrients like vitamins, minerals like magnesium, iron, dietary fiber, protein, and carbohydrates. Additionally, they are gluten-free, so people with celiac disease or gluten intolerance can also enjoy them as healthy alternates over other coarse grains. They have been reported effective in decreasing blood cholesterol, promoting digestive health, and helping to regulate blood sugar (Adebiyi et al., 2018). Processing millets has been under practice for ages also, which enhances their quality and bioavailability beyond ensuring their unrestricted supply throughout the year. Some primary preprocessing methods include dehulling, germination, roasting, milling, extrusion, and soaking. Fermentation is one of the oldest processing methods for enhancing their nutritional and market value (Saeed et al., 2022). Traditional fermentation has been recognized for improving the nutritional value and digestibility of dishes made from millet. Microorganisms like bacteria and yeast convert complex carbohydrates into simpler ones during the fermentation, resulting in the production of

Nutraceutical Potential of Millet-Based Food Products: Chemistry and Applications, First Edition.
Edited by Parmjit S. Panesar, Charanjit S. Riar, and Sushil Dhital.

Published 2026 by John Wiley & Sons Ltd.

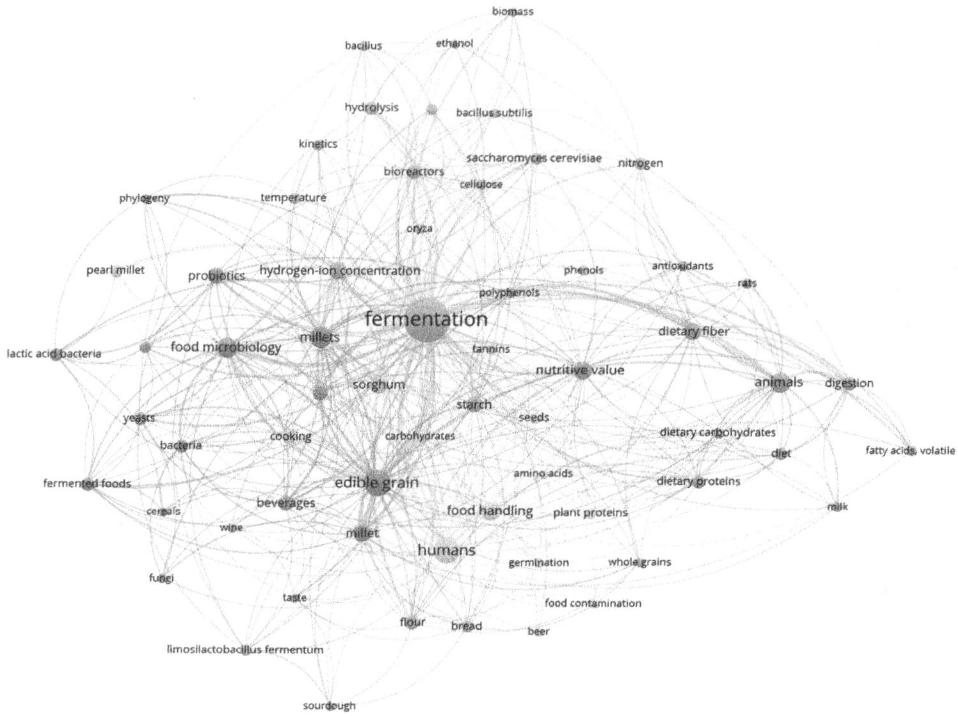

Figure 12.1 Status of fermentation in the production of millet-based products with enhanced nutritional and functional profile by mapping of search key terms cooccurrence system using VOSviewer

organic acids, gases, and alcohol, which provide specific flavors and taste to traditional dishes beyond extended shelf life due to wide-spectrum antimicrobial properties of the secondary metabolites produced during the process. Further, the fermentation of millets also increases the food product's protein content, digestibility, relative nutritional value, mineral availability, and functional qualities. The cooccurrence of words related to recent research on fermented millet-based product data from 2015 to 2024 using VOSviewer has been depicted in Figure 12.1.

They are appropriate for people with diabetes because of their fiber content, which slows down the digestion of carbohydrates and prevents blood sugar increase. Millets are a great complement to contemporary diets as their regular consumption can promote digestive health, control blood sugar levels, and supply vital nutrients. The value added products made from millet can satisfy diverse requirements related to the food forms and at the same time consumer preferences while supporting the sustainable food systems worldwide as the demand for gluten-free or probiotic-rich food keeps growing (Taylor & Duodu, 2018).

The relevance of millets in tackling issues related to global food security is further highlighted by their commemoration during events such as the International Year of Millets. Nutraceutical foods and health-based foods are important consumer considerations because the modern concept of food consumption goes beyond taste, flexibility, energy, or palatability to include maintaining one's health and overall well-being. This chapter discusses in detail the innovations in fermented millet-based value-added food products, their production technology, nutritional profile, and therapeutic properties with insights on the future researchable areas for enhancing the functionality of the regular diet and achieving sustainability goals.

12.2 Primary Processing of Millets

Millets have drawn significant attention of the food processing industry in the past few years owing to their exceptional functional profile and unparalleled nutritional value. Though, they are highly nutritious yet need processing before being consumed. Processing millets results in enhanced quality, improves bioavailability of nutrients, organoleptic properties, and lowers antinutritional compounds. The processing of millets includes different techniques, starting from decortication, milling, soaking, germination, and fermentation to cooking and roasting. Before processing, basic operations like cleaning, destoning, and grading are performed to make millets ready for primary processing and to reduce the antinutritional compounds to some extent. The techniques involved in the primary processing of millets are discussed hereunder.

12.2.1 Decortication

The outermost layer of grain is removed which majorly contains hull and pericarp. This process is known as decortication or dehulling. The frictional force is generated between the grain's surface and the roller removes the hull from the grain. Traditionally, dehulling was performed using stone rollers; however, these have been replaced by modern energy-efficient and timesaving dehullers. Dehulling has been reported to enhance the bioavailability of minerals and decrease the amount of antinutritional compounds in grain. Dehulling can remove 12–30% of the kernel, while dehulling beyond this limit might result in loss of nutritional value.

12.2.2 Malting and Soaking

Malting is the process of steeping grains in water followed by draining and spreading to let the grains germinate. Soaking of grains is done for a specific period which varies from grain to grain (malting). Germination and sprouting are alternative terms which are covered in the process of malting, but both have precise functions to perform and can have a significant impact on the finished quality of any food material. It also serves a similar purpose to that of soaking, i.e., to improve nutritional and functional properties and to decrease antinutritional compounds in millets. Germination reduces the calorific value of millet as carbohydrates are consumed as an energy source. Soaking is the process of introducing grains into water which is responsible for the leaching of bioactive compounds in water. Leaching action due to soaking brings down antinutritional compounds and improves the digestibility of food (Adebiyi et al., 2017).

12.2.3 Milling

A millet grain is made up of bran, endosperm, and germ. Milling of millets is carried out with the purpose of diminution of the grains in order to produce flour. During milling, bran is removed which decreases the nutritional value of flour compared to whole millet. But, milling also improves protein and starch digestibility as antinutritional compounds present in bran are also removed.

12.2.4 Fermentation

The most common and majorly practiced method for processing millets is fermentation. It is the process of breakdown of sugars present in millets in order to preserve them and improve their nutritional value. Fermentation can proceed naturally by the bacteria already present in food or can be forced by introducing previously fermented food in fresh food, referred to as back sloping. Fermentation improves flavor and adds diversity to the food, while decreasing the antinutritional compounds in food. The food products processed by fermenting millets majorly include bakery products, alcoholic drinks, and nonalcoholic beverages.

12.2.5 Roasting and Puffing

Roasting involves the application of heat in order to reduce moisture content, to enhance flavor, and to impart a specific odor to a food product. Roasting is mostly performed using dry heat which alters the nutritional value of food. Puffing also performs the same purpose as roasting, but it involves the application of high-temperature short time in order to produce puffed foods. The majority of food products prepared using puffing includes ready-to-eat food products and breakfast foods which are mostly consumed by mixing milk or water in them (Yousaf et al., 2021).

12.3 Millets as Nutrient-Dense Fermented Food Ingredient

Many developing countries still struggle with malnutrition, particularly in sub-Saharan Africa's areas. Compared to the developed nations, the developing countries experience 15 times higher deaths of children below the age of five years, counting for around 98 deaths for every 1,000 live births (Di Stefano et al., 2017). According to the United Nations Standing Committee on Nutrition, supplementary meals are a crucial component of the nutritional health status of young children.

To increase dietary variety and fight malnutrition, it is possible to improve a diet high in carbohydrates with micronutrients and crops high in vitamins. Millets have great potential to be used as alternative grains to provide food and nutritional security in most regions of the globe since they are rich sources of micronutrients, dietary fiber, vitamins, and phytochemicals with functional health benefits (Kumar et al., 2018). Millets are small-seeded grains that are grown and consumed as food across the world for their nutritional benefits and resistance to harsh environmental conditions. Millets are packed with physiologically active chemicals, low glycemic index, and proteins and are high in fiber and antioxidants. These small-seed crops are a great source of protein, energy, and minerals. Vitamins including niacin, B vitamins, thiamin, riboflavin, and folic acid are all found in millet grains. The average carbohydrate content of millet has been reported to be between 56 and 72 g/100 g, while the protein content has been reported to be 9.8% (Kumar et al., 2018). Nearly 60% of millet is constituted of carbohydrates, which are majorly present in soluble form and the rest of part is dietary fiber (Kumar et al., 2021). They are also rich in various phytochemicals, i.e., polyphenols (0.2–0.3%).

Mostly planted in hot, arid regions of the world where the yields of most common cereals are lower, millets are significant crops in addition to rice, wheat, and maize. Millets have been recognized for their micro- and macronutrients and also contain some antinutritional factors such as phytates, polyphenols, and tannins (Tharifkhan et al., 2021). These factors exhibit adverse effects on the nutritional value of the millet as they significantly reduce mineral absorption and bioavailability. Food processing has a great impact on the nutrient availability of food. Multiple unit operations are involved in the processing and value addition of millets. Traditional food processing techniques including decortication, milling, and fermentation can improve the nutritional and organoleptic properties. Dehulling of millets can result in 80% reduction in the phenolic content of the foods. Additionally, soaking and cooking of millets can help reduce their antinutritional content by decreasing the phytates and polyphenols, which improves their nutrient bioavailability levels (Pawar & Machewad, 2006). Fermentation is reported to enhance the nutrient bioavailability in millets and millet-based products, as depicted in Table 12.1. Fermentation takes place by the actions of microorganisms that are already present in food. These microorganisms enhance the bioavailability of nutrients by altering the acidity levels of the substrate causing the pH to reduce, thereby preventing the growth of harmful organisms. The primary mechanism involved in this functionality is the generation of lactic acid bacteria, which causes a lowering of pH levels below 4.0.

A certain proportion of the normal gut flora is made up of probiotics, which can inhibit the growth of hazardous bacteria, control the human immune system, and are intimately associated with human well-being. Another cofactor that can affect the gut flora and probiotic strains is prebiotic. Probiotics need a carbon source for their growth and survival. Human digestive juices cannot break down prebiotics, and these indigestible prebiotics could be utilized as a carbon source by the gut bacteria, which promotes the growth of beneficial bacteria and controls the makeup of probiotics. These prebiotics often are inulin, fructooligosaccharides (FOS), galactooligosaccharides (GOS), and xylooligosaccharides (XOS) and have been reported to have multiple health benefits as enhanced bioavailability and absorption, enhanced immunological functions and colonic health (Mondal et al., 2022). The kodo and kutki millet positively has been reported to modulate the gut microbial diversity, which is directly linked to metabolite production. The pure cultures of *Lactobacillus* and *Bifidobacterium* sp. have been observed to break down the 60–80% fiber in pearl millet, resulting in the production of high amount of acetate followed by propionate and butyrate.

Budhwar et al. (2020) studied the effect of probiotic fermentation on several components of nutrient-dense millets and reported that fermentation with different probiotic strains has various functional and nutritional effects on the millets. In particular, *Lactobacillus plantarum* increased the antioxidant activity, and *Lactobacillus rhamnosus* is reported to elevate the folic acid content. Sindhu and Khetarpaul (2001) also reported that fermentation with various probiotics resulted in a reduction of antinutritional factors such as phytic acid and tannins in different millets. One of the most researched probiotic genera is *Bifidobacterium*, which is found in significant amounts in the gut as a component of the

Table 12.1 Nutritional composition of millet-based fermented products

S. No.	Millet-based fermented product	Nutritional profile			References
		Macronutrients	Mineral composition	Other major phytochemicals/bioactives	
1.	Nilamadana	• Moisture 32.6% • Protein 6.4% • Fat 2.1% • Ash 2.6% • Carbohydrates 88.9%	• Calcium 82.93 mg/100 g • Potassium 324.9 mg/100 g • Sodium 27.10 mg/100 g • Magnesium 79.30 mg/100 g • Iron 3.86 mg/100 g • Manganese 0.95 mg/100 g • Copper 0.58 mg/100 g • Phosphorus 266.38 mg/100 g • Zinc 2.11 mg/100 g	• Flavonoids (rutin, catechin, myricetin, naringenin, genistein, quercetin) • Phenolic acids (3,4 hydroxybenzoic acid, chlorogenic, ferulic acid, sinapic acid, and caffeic acid)	(Dwivedi et al., 2015)
2.	Koozh	• Moisture 78–80% • Fat 1.6–1.8% • Protein 10–10.5% • Carbohydrate 35–40% • Vitamin C 53.73 mg/100 g • Riboflavin 0.272 mg/100 g • Niacin 1.45 mg/100 g • Thiamine 0.625 mg/100 g	• Iron 2.81 mg/100 g	• Arginine 4.3 g/100 • Threonine 5.8 g/100 • Histidine 5.5 g/100 • Isoleucine 1.5 g/100 • Tyrosine 0.1 g/100 • Lysine 1.9 g/100	(Saeed et al., 2022)
3.	Kodo ko jaanr	• Moisture (69%) • Ash (5%) • Protein (9.3%) • Carbohydrate (83.7%) • Fiber (9%)	• Calcium 281 mg/100 g • Magnesium 118 mg/100 g • Copper 2.2 mg/100 g • Iron 24 mg/100 g • Zinc 1.2 mg/100 g • Potassium 398 mg/100 g • Phosphorus 326 mg/100 g • Manganese 9 mg/100 g • Sodium 39 mg/100 g	• Niacin 2 mg • Thiamine 0.18 mg	(Thapa & Tamang, 2004)

4.	Dolo beer	• Alcohol 2.3% • Proteins 0.65 g/100 g • Carbohydrates 0.77 g/100 g	• Calcium 139.3 mg • Iron 31.3 mg • Magnesium 684.1 mg • Phosphorous 1,264.2 mg • Sodium 22.4 mg • Zinc 21.7 mg • Copper 1.6 mg • Maganese 7.1 mg	• Phenolic compounds 506 µgGAE/mL	(Abdoul-Latif & Bassolé, 2013; Daji et al., 2022)
5.	Sorghum-based Mahewu	• Calories 60–70 kcal • Protein 6.1–77 g • Fat 0.5–1.0 g • Carbohydrates 12–15 g • Dietary fiber 5.7–7.9 g	• Calcium 40–50 mg • Iron 1.5–30 mg • Magnesium 30–50 mg • Phosphorous 100–200 mg • Zinc 0.5–1.0 mg	• Total phenolic content 27.37–65.89 GAE/g • Total flavonoid content 0.18–0.30 GAE/g • Antioxidant capacity 44.62–49%	(Daji et al., 2023; Mashau et al., 2024)
6.	Ting	• Alcohol 2% • Vitamins 1%	• Calcium 31–398 mg • Iron 3.3–14.89 mg • Magnesium 78–567 mg • Phosphorous 130–296 mg • Potassium 307–490 mg • Zinc 2.3–3.1 mg • Copper 0.47–1.06 • Magnesium 1.15–48.43 mg	• Phenols 55% • Phytosterols 2% • Amine-related compounds 2% • Amino acids 1% • Terpenoids 2%	(Kewuyemi et al., 2020)
7.	Burukutu	• Moisture 55% • Carbohydrates 23% • Ash 3%	• Ascorbic acid 0.14 µg/100 g • Calcium 3.96 mg/100 g • Magnesium 16.22 mg/100 g • Manganese 4.71 mg/100 g • Iron 3.52 mg/100 g • Potassium 56.16 mg/100 g • Sodium 1.48 mg/100 g • Retinol 0.54 µg/100 g	NR	(Ogbonna et al., 2016)

(Continued)

Table 12.1 (Continued)

S. No.	Millet-based fermented product	Nutritional profile			References
		Macronutrients	Mineral composition	Other major phytochemicals/bioactives	
8.	Finger millet-based yogurt	• Calories 100–130 kcal • Protein 6–8 g • Total fat 2–4 g • Carbohydrate 15–20 g • Dietary fiber 3–5 g	• Iron 0.46 mg/100 g • Magnesium 12.6 mg/100 g • Manganese 1.71 mg/100 g • Potassium 31.2 mg/100 g • Sodium 3.2 mg/100 g • Phosphorus 23.8 mg/100 g • Calcium mg/100 g • Zinc 0.095%	• Aspartic acid • Glutamic acid • Asparagine • Serine • Glutamine • Threonine • Arginine • Alanine • Tyrosine • Valine • Methionine • Phenylalanine • Isoleucine	(Ponne & Selva Brindha, 2020; Vila-Real et al., 2022)
9.	Finger millet-based probiotic beverage	• Moisture 89% • Protein 1.8% • Carbohydrate 8.27% • Fat 0.08% • Fiber 0.91% • Ash 0.55%	• Magnesium 30–50 mg • Phosphorous 100–150 mg • Zinc 0.5–1 mg • Potassium 200–300 mg • Iron 1.5–3 mg • Calcium 150–200 mg	NR	(Fasreen et al., 2017; Mali et al., 2020)
10.	Millet-based probiotic beverage	• Total solids 8%–8.5% • Protein 2.5%–2.7% • Ash 0.5%–6.6% • Fiber 1.5%–2.3% • Fat 0.07%–0.08%	• Calcium 30–400 mg • Iron 1.5–3.0 mg • Magnesium 30–100 mg • Phosphorous 100–200 mg • Potassium 200–300 mg • Zinc 0.5–1.5 mg • Manganese 1.0–2.0 mg	• Flavonoids • Phenolic compounds • Tannins	(Panigrahi et al., 2024)

NR: not reported.

normal human microbiota. They have a positive effect on the host and a variety of prebiotic substrates successfully fosters their growth. In a study, *Bifidobacterium breve* was found to have anti-inflammatory properties, the capacity to colonize the human gut and retain the integrity of the intestinal epithelium, the ability to boost the immune system, fight off infections, and antiobesity effect (Balli et al., 2023). Fermentation has been employed for creating unique food items across the globe since prehistoric times, and a variety of food products are produced from millet fermentation which have exceptional functional values. The production technology of some of the most common and popular fermented millet-based products is depicted in Table 12.2.

12.4 Diversity of Millet-Based Fermented Food Products

12.4.1 Traditional Millet-Based Fermented Products

12.4.1.1 Hussuwa

Hussuwa is a traditional Sudanese millet-based product prepared by fermenting sorghum grain flour or millet malt. It is mainly utilized as a porridge called *acedia*. At the end of three days of amylolytic fermentation and incubation for 15 days, the final product appears as a brown honey-like syrup that tastes sweet and sour. The presence of microorganisms like *Acetobacter xylinum*, *Gluconobacter oxydans*, and *Saccharomyces cerevisiae* have also been reported in the product (Ilango & Antony, 2021). Hussuwa is a rich source of carbohydrates, dietary fiber, and various vitamins and minerals like vitamin B and minerals like iron and zinc. A nominal amount of proteins and fats are also found in this Sudanese product. Higher antioxidant contents and the presence of phytochemicals give this product characteristics of acting against oxidative stresses and lower free radical production (Mashau et al., 2024).

12.4.1.2 Kisra

Kisra is also a Sudanese sorghum-based fermented product often utilized as a pancake, flatbread, or sourdough. The maximum microbial population of *Pediococcus pentosaceus* has been reported, followed by other bacterial populations like *Lactobacillus confuses*, *Lactobacillus* brevis, *Acetobacter* sp., and yeasts like *Candida intermedia*, *Debaryomces hansenii* (Adebo, 2020). Major mineral constituents of kisra include calcium (0.8–3%), phosphorus (82–83%), iron (25–51%), and manganese (3.1–3.7%), and the major macronutrients found are carbohydrates (72%), fat (2%), protein (14%), and ash (3%) (Taylor & Duodu, 2018).

12.4.1.3 Jiu

It is a fermented alcoholic beverage prepared by mixed fermentation of whole sorghum grains for at least a month. The major microorganisms utilized for the production of jiu are *Rhizopus*, *Aspergillus*, *Saccharomyces*, *Candida*, and acetic acid bacteria. Jiu serves as an excellent appetizer, helps in assisting digestion, and is an energy-rich alcoholic beverage (Adebo, 2020). The production of jiu involves cooking of coarsely ground sorghum whole grains followed by cooling. A starter culture known as daque is added to this cooked

Table 12.2 Production technology of millet-based fermented products

S. No.	Millet	Product	Country of origin	Production technology	Reference
1.	Sorghum grains	Dolo beer	Burkina Faso	Steeping sorghum grains for 24 h and germination for 48 h. Germinated grains are then dried and milled by addition of water (1 : 10 w/v), followed by decanting to collect the supernatant. Add water and boil at 100 °C, allow to cool, filter, collect the filtrate and combine the collected supernatant and the filtrate (mashing) to obtain wort @65–70 °C. Cooking the wort @100 °C then cooling and fermentation by adding dried yeasts to the wort (12–14 h). The final product is then filtered.	(Madilo et al., 2024)
2.	Sorghum	Motoko	Lesotho	Add sorghum meal to warm water (1 : 3) and stir thoroughly. Afterward, add tomato as a starting culture in a 1 : 20 ratio to the sorghum meal mixture. Under strict supervision let the mixture ferment for 24 to 27 h. Once fermentation is complete, grind the mixture until it has a uniform consistency. Boil for 20–30 minutes, and let the boiled mixture cool to room temperature.	(Gadaga & Lehohla, 2013)
3.	Sorghum	Mahewu	South Africa	Sorghum grains are decorticated, cooked, and turned into a paste. After adding water to the paste (1 : 10) and boiling it for 10 minutes, mix the broth with cold water (1 : 10) to let it cool. Stir the sorghum malt (0.5%) into the broth and let it ferment for 36 h at 30 °C.	(Taylor & Duodu, 2018)
4.	Finger millet	Koozh	South India	Millet is cleaned, ground, and then mixed with water to form slurry. Allow slurry to ferment for 12–15 h at room temperature. Crushed rice is cooked, and added on the next day in slurry. For the mixture to become porridge, let it ferment for 24 h to make koozh, combine the porridge with salt and water (1:6 W/V).	(Saeed et al., 2022)
5.	Pearl millet	Fura	Nigeria	Dehulling of pearl millet followed by cleaning and washing of millet. Addition of spices and water followed by mixing and grinding to paste. Let the mixture ferment for 48 h at room temperature. From the paste into balls, steam them for 1 h at 100 °C, let them dry for 1 h at 60 °C, and then grind the dried balls into powder to make fura.	(Amadou et al., 2011)
6.	Pearl millet	Oti-Oka	Nigeria	Clean pear millet grains, soaked for 18 h (1 : 3 w/v), followed by draining and rinsing with water. Germinate the grains for 36 h at 25 °C for 48 h, followed by milled, sieving (diluted 1 : 3), and boiling in water (3 h). Let it cool and ferment at 30 °C for 72 h.	(Temilope & Taiyese, 2012)
7.	Finger millet	Kodo ko jaanr	India	Finger millet seeds dried, cleaned, and washed. After 30 min of cooking, drain extra water. Spread cooked seed on a mandro (a bamboo mat) to cool. Cover with sach cloths, mix, and ack in a bamboo basket and sprinkle dry starter culture (marcha, 2%). Saccharification takes two to four days. Transfer to closed earthen pots ferment for three to five days at 25–30 °C (kodo ko jaanr). Afterward, keep it in a tightly closed bamboo vessel for 10–20 minutes. Using a bamboo straw, sip the white extract.	(Thapa & Tamang, 2004)

mixture and then mixed properly. This mixture is then transferred to containers and allowed to ferment for at least one month (Ilango & Antony, 2021). The finished product beyond containing alcohol is also believed to have potent functional properties with presence of various phytonutrients and high mineral content.

12.4.1.4 Ting

Ting finds its origin from Botswana in South Africa and is majorly utilized as porridge. Due to its unique flavor, high nutritional properties, and sour taste, it is likely consumed by people of South Africa and adjoining countries. It is also known as "letting" in certain regions of South Africa. The preparation of ting involves lactic acid fermentation by microorganisms like *Lactobacillus casei*, *Lactobacillus fermentum*, *L. rhamnosus*, and many more for up to 2–3 days (Kewuyemi et al., 2020). Spontaneous fermentation of sorghum flour in warm water is done to produce ting. The slurry obtained is fermented by keeping at a warm place for at least two to three days and the pH of the final product ranges from 3.5 to 4 for obtaining the highest flavor and taste. The presence of lactic acid bacteria makes it a nutritionally rich product that enhances gut health and helps in maintain a healthy digestive system (Ilango & Antony, 2021).

12.4.1.5 Tchoukoutou

Tchoukoutou is a fermented alcoholic beverage prepared by malting of red sorghum. This alcoholic beverage is highly consumed in Benin region. It is characterized by its strong acidic flavor and is a good source of energy, protein, iron, and fat. Major microorganisms identified are *Saccharomyces cerevisiae*, *Lactobacillus fermentum*, *Lactobacillus fructivorans*, etc. The alcohol percentage of the beverage increases rapidly which causes deleterious effects on the quality of the produce hence it is consumed within two to three days of preparation (Ilango & Antony, 2021). A few steps involved in manufacturing of this beverage are soaking, germinating, drying, and mashing, followed by boiling and then allowing it to ferment. It becomes suitable for consumption of both adults and teenagers as it contains very low amount of alcohol and good amount of dry matter (Kayodé et al., 2007).

12.4.1.6 Bensaalga

Bensaalga is a pearl millet-based nonalcoholic beverage consumed as a morning meal in parts of Africa. This nonalcoholic beverage contains a high amount of protein and is highly acidic with properties like good digestibility, good taste, and pleasant flavors. Good amounts of lactic acid bacteria in the final product have also been reported (Ilango & Antony, 2021). The production technology incudes washing, soaking of pearl millet grains, and then grinding them followed by kneading, sieving, and cooking. Many spices like black pepper, mint, and ginger are added to enhance the flavor and taste of bensaalga before grinding it. Cooking of the supernatant for 40 minutes after the settling process is done for final preparation of the product (Amadou, 2019).

12.4.1.7 Koko

Koko is processed porridge prepared by adding spices to pearl millet grains and fermenting the leftover liquid until high viscosity is attained. It is a traditional food of Ghana in Africa. The final product contains *Weisella confuse*, *Lactobacillus fermentum*, *Lactobacillus salivarius*, and *Pediococcus* and the diversity of the other microbes which further adds to the flavor and nutritional profile of the finished products. Koko water which is the supernatant obtained while processing koko has beneficial medicinal properties and is utilized in treating and curing diarrhea. The final product is often consumed with sugar or milk (Atter et al., 2021). It is a nutrient-rich product with good amounts of fiber, proteins, and carbohydrates. Significant amount of minerals like calcium, iron, and vitamin B are also major constituents of this fermented porridge (Ajagekar et al., 2023).

12.4.1.8 Burukuttu

It is a traditional fermented beverage of the sub-Saharan African continent brewed with red or white sorghum varieties. The action of *Lactobacillus*, the lactic acid bacteria, results in the sour taste of this traditional beverage, and its characteristic opaque appearance is due to the presence of suspended solid particles and yeast materials. The people of northern Guinea, Nigeria, Ghana, and Benin most commonly consume this conventional beverage. Consuming burukuttu is very beneficial for better gut health as it improves and promotes digestive health. It is also rich in antioxidant properties which help in preventing oxidative stresses (Amadou, 2019).

12.4.1.9 Koozh

A fermented conventional product prepared by lactic acid fermentation of finger millet (*Eleucine corocana* Gaertn.). Mostly consumed as a beverage or porridge in southern parts of India or Sri Lanka. Koozh has been historically associated as a traditional fermented food with the local tribes. It is well known for its unique and identical flavor and nutritive value and was a regular breakfast meal incorporation until was slowly replaced by more staple food items like rice-based foods. Major microorganism species involved in the production are *Lactobacillus*, *Lactococcus*, and *Leuconostoc*. The drink has many health-promoting characteristics including antidiabetic, anticarcinogenic, antiallergic, etc., properties with a pleasant flavor and taste (Saeed et al., 2022). Due to high nutritional profile and sustainability, the product has recently become more popular providing an alternative carbohydrate source as such or in blended form. Attempts are underway to standardize the primary fermentation which takes 12–15 hours to provide the product a characteristic quality and unique sensory properties.

12.4.1.10 Kodo Ko Jar

Kodo ko jar is a traditional mild-alcoholic sweet-flavored beverage most commonly consumed in parts of the Eastern Himalayas. It is prepared by fermenting dried finger millet seeds. A sweet taste with a tinge of alcoholic flavor is characteristic of this high-quality fermented product. Major species of microorganisms involved in the preparation of kodo ko jar are *Pichia*, *Candida*, and *Saccharomycopsis* (Ilango & Antony, 2021). Kodo ko jar has high calorific value with the potential to boost immediate energy, thus ailing persons and postnatal women have been traditionally recommended to consume this beverage to regain strength. The traditional production methodology makes use of septum, toongbaa, or fresh fern which significantly increases the mineral content of the beverage, while kodo ko jar is well recognized for its richness in iron, calcium, magnesium, and phosphorus. The leftover grits of kodo ko jaanr after completion of the fermentation process are a good source of proteins and residual minerals which are commonly used as fodder to the animals representing the upcycling and complete utilization of the process residue.

12.4.2 Improved Innovative Millet-Based Fermented Product

12.4.2.1 Millet-Based Yogurt

Millets can be an excellent alternative for producing nondairy options due to the high incidence and prevalence of lactose intolerance and dairy-based allergies among the masses. Millets offer an exceptional nutritional profile that serves as a perfect alternative to traditional yogurt. Fermentation of millet grains by the addition of probiotic strains like *Lactobacillus*, *Streptococcus*, or *Bifidobacterium* strains can provide diversified product range with enhanced bioactive profile and functionality of the products. Di Stefano et al. (2017) developed a millet-based probiotic yogurt using *Lactobacillus rhamnosus* and *Streptococcus thermophilus* strains and concluded that these probiotic strains can be useful in the preparation of tasty and nutritional fermented products with better health management properties. It is a nutritional alternative to traditional yogurt as it contains protein (6–8 g), dietary fiber (3–5 g), and minerals including calcium (150–200 mg), iron (1.5–3 mg), magnesium (30–50 mg), phosphorus (100–150 mg), and zinc (0.5–1 mg). Millets also contribute to enhance the antioxidant activity thus reducing oxidative stress (Ponne & Selva Brindha, 2020).

12.4.2.2 Probiotic Millet-Based Fruit Bar

The higher nutritional composition of millet and the effectiveness of probiotics have led to the production of probiotic-based millet bars which offer a convenient snack option for health-conscious consumers. This innovative product combines the nutritional benefits of millets and the addition of probiotic strains to help increase the bioavailability of these nutrients. Combining probiotics with millets offers consumers good gut health with improved digestibility. Vila-Real et al. (2022) developed millet fruit bars by incorporation of probiotic cultures *Lactobacillus delbrukeii*, *Lactobacillus acidophilus*, and *Streptococcus thermophilus* and concluded that a combination of millets like pearl millet and finger millet with fruit pulps and probiotics enhanced the bioavailability of nutrients like carbohydrates, proteins, fats, vitamin C, and minerals. Kavitha et al. (2018) prepared a millet-based fruit bar and reported 69.80 g/100 g and 352.60 Kcal/Kg of carbohydrate and energy value in the bar. Minerals including calcium and iron were found 80.12 and 4.10 mg/100 g, respectively. Additionally, the viability of probiotic strain was 3.5×10^{8} CFU/g. Moreover, this formulation with probiotic could be beneficial for gut microbiota and human health.

12.5 Health Benefit/Functional Properties

The health benefits of millets have been a point of focus for researchers, industrialists, and healthcare professionals alike for quite a while now, owing to millets being a superfood. For a long time, the vitamins, minerals, essential fatty acids, and fiber found in whole grains were thought to be the major health contributing factors, but recent studies indicate that a combination of different bioactive compounds also contributes to these benefits and more precisely are primary disease preventing a managing diet components. They consist of lipids, oligosaccharides, resistant starch, antioxidants like flavonoids, phenolic acids, and avenanthramides, hormonally active substances like lignans and phytosterols (Amadou, 2019). As previously discussed, fermented millets are a rich source of nondigestible fibers like inulin, FOS, GOS, and XOS which act as prebiotics, and from fermentation, the fermented product becomes a rich source of probiotics (like *Bifidobacterium*, *Lactobacillus* sp., etc.). This combination of prebiotics and probiotics results in a synergistic pair known as synbiotics. Moreover, these have the potential to benefit human wellness in a variety of ailments. They are especially promising in tackling two of the targets in the UN Sustainable Development Goals (SDGs) Goal 3: preventing premature deaths from non-transmissible illnesses, promoting well-being and mental wellness, and preventing neonatal mortality. The health benefits of these probiotic strains are believed to be due to the generation of lactic acid in the fermentation process, which has been found to impact the host physiology in a variety of ways, including an inhibition in the secretion of proinflammatory cytokines from toll like receptors (TLR)-activated dendritic cells and macrophages, as well as a decline in the free radical burden on intestinal enterocytes (Gurry, 2017). Thus, the prebiotic acts as a substrate for the gut microbiome to form physiologically active compounds.

The gut microbiome has also been associated with the emergence of metabolic syndromes. Metabolic syndrome (MetS) is a complex illness defined by interconnected metabolic abnormalities such as obesity, insulin resistance, dyslipidemia, hypertension, and mild inflammation. MetS occurs when three or more of the listed requirements are met: (i) higher circumference of the abdomen; (ii) fasting triglyceride levels greater than 150 mg/dL; (iii) high-density lipoprotein cholesterol less than 50 mg/dL for women and 40 mg/dL for men; (iv) fasting glucose levels greater than 100 mg/dL; and (v) blood pressure greater than 130/85 mmHg (Núñez-Sánchez et al., 2021). A key part of a healthy gut, with connection to both the categories (i.e., probiotic and prebiotic), is the production of the short-chain fatty acids (SCFAs) like acetate, propionate, and butyrate from the fermentation of the dietary fibers in the colon. In particular, the depletion of butyrate production has been previously linked to inflammatory bowel disease (IBD). Thus, synbiotic treatments that combine butyrate-producing probiotics with their preferred fermentation substrate are promising prospects for reducing the worldwide prevalence of gut inflammation-associated diseases. Furthermore, propionate and butyrate stimulate intestinal gluconeogenesis, which is thought to be responsible for the beneficial effects of a fiber-rich diet on glucose and energy homeostasis. Meanwhile, it has been demonstrated that elevated microbial acetate synthesis and turnover trigger glucose-stimulated insulin secretion, which can result in resistance to insulin and consequent obesity. Therefore, altering a patient's current microbiota SCFA production profile (for example, by increasing the microbial net conversion of acetate to butyrate) may be crucial in tackling the rising prevalence of metabolic disorders like type 2 diabetes and obesity worldwide (Gurry, 2017). Additionally, *Bifidobacterium adolescentis*, a butyric acid-producing probiotic, has been positively linked to a healthy metabolic status and adversely linked to metabolic disorders and may be involved in potential microbial treatments. Furthermore, fermented foods made from millets produce antioxidants, organic acids (like lactic acid, etc.), vitamins (like folate, etc.), and other essential elements that strengthen the body's defenses against infections. Antimicrobial peptides synthesized in fermented foxtail millet demonstrate resistance to digestive enzymes, which increases the impact of their activity. Bacteriocins generated in millet-based fermented drinks provide benefits while being produced from harmless ingredients (Amadou, 2019). Recently, it has been reported that when processed pearl millet grains are fermented, the grains' antinutritional characteristics are significantly reduced, and their protein digestibility is much improved. Additionally, prior research suggests that consuming grains and their byproducts reduces the incidence of noncommunicable diseases such as diabetes, heart disease, stroke, and several types of cancer associated with oxidative stress (Budhwar et al., 2020).

Fermented millet products have been providing nutrition to the world since ancient times, but in more recent years, many other innovative products have been developed with a focus on more potent health benefits, as depicted in Figure 12.2. So, keeping in mind these health benefits, the fermented millet product can play a crucial role in promoting a sustainable and health-focused diet globally.

12.6 Antinutritional Factors (ANFs) and Related Safety Aspects

12.6.1 Phytic Acid

It is one of the most important antinutrients in millets. It binds to vital minerals like calcium, magnesium, iron, and zinc, decreasing their bioavailability and possibly causing shortages, particularly in populations where millet is a major staple diet. It is mainly present in finger millet (248–291 mg/100 g) and pearl millet (0.713–1.02 mg/100 g). Process like fermentation has showed positive effects on the reduction of phytic acid content up to 88.3%. Microbes like *Lactobacillus* species help in the breakdown of phytic acid into inositol phosphate which ultimately enhances the bioavailability of minerals. To optimize the reduction of phytic acid, fermentation is used in conjugation with other methods including soaking and germination. Before fermentation initiate, soaking helps in activation of phytases naturally present in grains which results in the breakdown of phytic acid (Budhwar et al., 2020; Saeed et al., 2022). According to the recent research findings, phytic acid concentration can be significantly decreased by using various processing methods. Soaking millets for 12–48 hours, for example, can reduce their phytic acid content by 16.9%. Reduction from 11.7% to 50.6% can result with malting. Additionally, popping techniques have been shown to lower phytic acid level to 17.1% (Bheemaiah Balyatanda et al., 2024).

12.6.2 Trypsin Inhibitors

Trypsin inhibitors, which are found in a variety of millets, prevent the digestive system from breaking down protein by preventing trypsin from accomplishing its task. Mainly bulrush and foxtail millet contain trypsin inhibitor. Level of trypsin inhibitor can also be reduced by soaking millets before cooking. By dissolving the proteins that cause inhibition, cooking intensifies this reduction. Another method is fermentation that can also aid in lowering trypsin inhibitor levels, just like phytic acid elimination does. Microbial action throughout the procedure may break down these inhibitors, improving the millet's nutritional value (Samtiya et al., 2021). Various methods, including thermal processing, roasting, fermentation, and sprouting, have been shown to reduce the antinutritional effect of trypsin inhibitors.

Mainly bulrush and foxtail millet contain trypsin inhibitors. Level of trypsin inhibitor can also be reduced by soaking millets before cooking. By dissolving the proteins that cause inhibition, cooking intensifies this reduction. Another method is fermentation that can also aid in lowering trypsin inhibitor levels, just like phytic acid elimination does. Microbial action throughout the procedure may break down these inhibitors, improving the millet's nutritional value (Bheemaiah Balyatanda et al., 2024).

12.6.3 Tannins

These polyphenolic components may obstruct the absorption of minerals and the digestion of proteins. It is well known that tannin can combine with minerals and protein to produce complexes that make it more difficult to absorb nutrients. It is mainly found in finger millet and barnyard millet. The absorption of vital nutrients, especially protein, and minerals like iron and zinc, might be impeded by high tannin level. This is particularly pertinent for people who primarily rely on millet as a main meal. Potential adverse effects on nutrient absorption can be lessened by hybrid consumption pattern of millet with other foods or by using processing techniques that lower their incorporation levels. Reduction of tannins can be successfully achieved by fermentation, soaking and cooking, and decortication (removal of bran and testa) (Saeed et al., 2022). Soaking of finger millet (24 hours) has been found to reduce from 2.07 to 0.55 mg/g of tannins, while malting reduces its level by up to 54% (Sahoo et al., 2024).

12.6.4 Oxalates

Many millets include oxalates, which are antinutritional elements, especially in proso millet (21–23 mg/100 g) and finger millet (45.7 mg/100 g). These substances can bind with calcium and other minerals, decreasing their absorption and possibly eventually causing a shortage. Because excessive consumption of high-oxalate-containing food, such as millet, might lead to kidney stone formation. However, certain methods including cooking, soaking, and fermentation were effective in the reduction of oxalate levels in millets. Soluble oxalates can be extracted from millet grains by soaking them in water before cooking (Sheethal et al., 2022). Additionally, fermentation also introduces active microorganism that help in reduction of oxalates and also increases bioavailability of minerals.

Despite being high in nutrients, millets have antinutritional characteristics that make it difficult for the body to absorb nutrients as efficiently as possible. Because fermentation with microbial or enzyme activity leads to the breakdown/degradation of phytic acid, trypsin inhibitors, and tannins while improving nutritional quality, it becomes a potent tool to address these problems. Incorporating fermentation with millet processing could be crucial to enhancing the nutritional outcomes for the population that depends on these ancient grains as knowledge of the health advantages of millet and fermented foods expands.

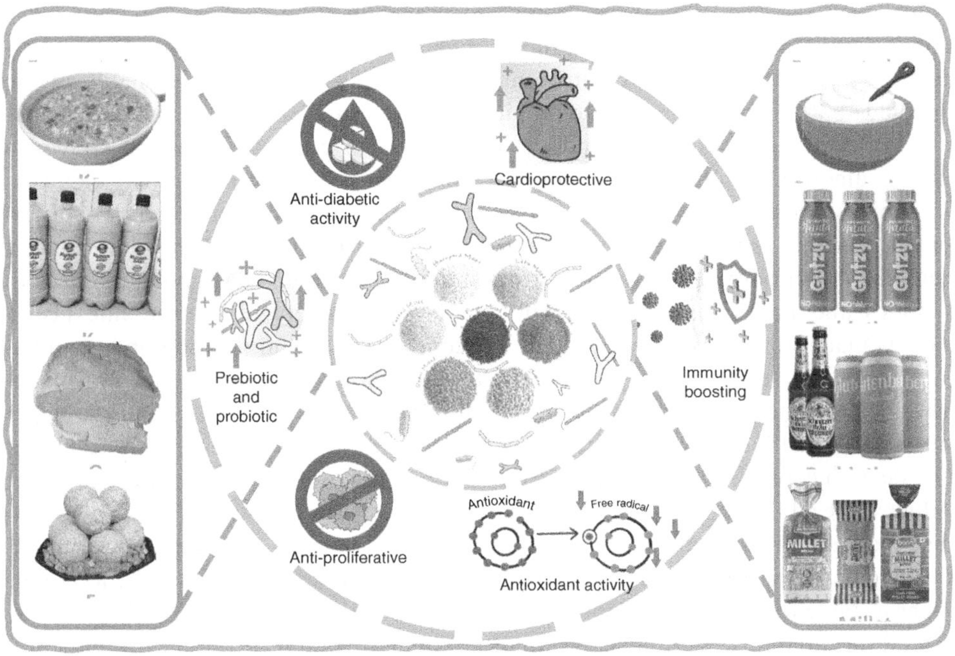

Figure 12.2 An exposition on various fermented millet products, along with their potential health benefits

12.7 Conclusion and Future Perspectives

Millets are renowned for their ability to withstand severe weather conditions and for using less water and resources than other cereal crops making them more climate resilient and sustainable. In arid areas, this provides them with a sustainable option for food production. Several SDGs that address nutrition and food security are in line with the use of millet in agricultural practices. Consumers are becoming more conscious of the advantages of gluten-free, low glycemic index, high protein, etc., diets to promote their overall well-being and health. Millet-based products are becoming increasingly popular in the market as more consumers look for alternatives to wheat-based food products. Millets are naturally gluten-free grains with a nutritious composition that includes high amounts of vitamins, dietary fiber, protein, and vital minerals like magnesium, calcium, and iron. For people who have been suffering from celiac disease or gluten intolerance, this makes them very advantageous. Millet's nutritional value is further increased by the fermentation process, which also introduces healthy bacteria that support gut health and increase the bioavailability of minerals and vitamins. Awareness in fermented millet-based products is being driven by the move toward functional meals, or foods that offer health advantages beyond simple nourishment. In addition to being gluten-free, these foods are promoted as being high in probiotics and other substances that are beneficial for health. Numerous advances in millet fermentation have the potential to revolutionize the way that these grains are seen and used in contemporary diets. Furthermore, current fermentation process research attempts to maximize flavor characteristics while maintaining a nutritional profile and enhanced functionality of the value-added products. In addition to making millets more digestible, fermentation process also enriches the finished products with potential probiotics that are known to promote digestive health. By reducing phytates and other antinutritional elements found in raw millets, this procedure can enhance mineral absorption. A variety of fermented products such as beverages, baked goods, and traditional dishes like koko, a fermented millet porridge is prepared from millet. These items have potential uses in prebiotics and probiotics in addition to satisfying a variety of gastronomic tastes. Products made from millet are expected to become major prominent contenders in the health food industry by utilizing innovative fermentation techniques, concentrating on functional advantages, and stressing sustainability. Millets will probably play a bigger part in the global food system as more benefits are discovered by research, which will coincide with consumer tendencies toward better eating. Additionally, industrial millet product value addition gives impoverished people access to reasonably priced, nutrient-dense food and gives farmers a chance to boost their revenue. To increase nutritional security going forward, attention must continue to be paid to millet consumption's efficacy as a staple food. Millet's adaptability makes it possible to incorporate them in varied food formulations, with good acceptance. Standardization of the local production practices for reduced fermentation time while keeping the nutritional profile intact and the highest functionality of the bioactive compounds, their characterization and clinical documentation for the proposed health benefits shall keep motivating the innovative food processing industry to come up with unique and novel millet-based fermented food combinations.

References

Abdoul-Latif, F. M., & Bassolé, I. H. N. (2013). Proximate composition of traditional local sorghum beer "dolo" manufactured in Ouagadougou. *African Journal of Biotechnology, 12*(13), 1517–1522. https://doi.org/10.5897/AJB12.1694

Adebiyi, J. A., Obadina, A. O., Adebo, O. A., & Kayitesi, E. (2016). Fermented and malted millet products in Africa: Expedition from traditional/ethnic foods to industrial value-added products. *Critical Reviews in Food Science and Nutrition, 58*(3), 463–474. https://doi.org/10.1080/10408398.2016.1188056

Adebiyi, J. A., Obadina, A. O., Adebo, O. A., & Kayitesi, E. (2017). Comparison of nutritional quality and sensory acceptability of biscuits obtained from native, fermented, and malted pearl millet (*Pennisetum glaucum*) flour. *Food Chemistry, 232*, 210–217. https://doi.org/10.1016/j.foodchem.2017.04.020

Adebo, O. A. (2020). African sorghum-based fermented foods: Past, current and future prospects. *Nutrients, 12*(4), 1111. https://doi.org/10.3390/nu12041111

Ajagekar, A., Sali, S., Borse, O., Patil, A., Suri, S., & Patil, A. (2023). Millets based fermented products: A review. *Acta Scientifci Nutritional Health, 7*(6), 69–81. https://doi.org/10.31080/asnh.2023.07.1251

Amadou, I. (2019). Millet based fermented beverages processing. In A. Grumezescu & A. M. Holban (Eds.), *Fermented beverages: Volume 5. The science of beverages* (pp. 433–472). Elsevier Inc. https://doi.org/10.1016/B978-0-12-815271-3.00011-7

Amadou, I., Gbadamosi, O. S., & Le, G. W. (2011). Millet-based traditional processed foods and beverages—A review. *Cereal Foods World*, *56*(3), 115–121. https://doi.org/10.1094/CFW-56-3-0115

Atter, A., Diaz, M., Tano-Debrah, K., Kunadu, A. P. H., Mayer, M. J., Colquhoun, I. J., Nielsen, D. S., Baker, D., Narbad, A., & Amoa-Awua, W. (2021). Microbial diversity and metabolite profile of fermenting millet in the production of Hausa koko, a Ghanaian fermented cereal porridge. *Frontiers in Microbiology*, *12*, 1–15. https://doi.org/10.3389/fmicb.2021.681983

Balli, D., Cecchi, L., Pieraccini, G., Venturi, M., Galli, V., Reggio, M., Di Gioia, D., Furlanetto, S., Orlandini, S., Innocenti, M., & Mulinacci, N. (2023). Millet fermented by different combinations of yeasts and lactobacilli: Effects on phenolic composition, starch, mineral content and prebiotic activity. *Foods*, *12*(4), 748. https://doi.org/10.3390/foods12040748

Bheemaiah Balyatanda, S., Gowda, N. A. N., Subbiah, J., Chakraborty, S., Prasad, P. V. V., & Siliveru, K. (2024). Physiochemical, bio, thermal, and non-thermal processing of major and minor millets: A comprehensive review on antinutritional and antioxidant properties. *Food*, *13*(22), 3684. https://doi.org/10.3390/foods13223684

Budhwar, S., Sethi, K., & Chakraborty, M. (2020). Efficacy of germination and probiotic fermentation on underutilized cereal and millet grains. *Food Production, Processing and Nutrition*, *2*(1), 12. https://doi.org/10.1186/s43014-020-00026-w

Daji, G. A., Green, E., Abrahams, A., Oyedeji, A. B., Masenya, K., Kondiah, K., & Adebo, O. A. (2022). Physicochemical properties and bacterial community profiling of optimal Mahewu (a fermented food product) prepared using white and yellow maize with different inocula. *Food*, *11*(20), 1–14. https://doi.org/10.3390/foods11203171

Daji, G. A., Green, E., & Adebo, O. A. (2023). Nutritional and phytochemical composition of Mahewu (a southern African fermented food product) derived from white and yellow maize (*Zea mays*) with different inocula. *Fermentation*, *9*(1), 1–18. https://doi.org/10.3390/fermentation9010058

Di Stefano, E., White, J., Seney, S., Hekmat, S., McDowell, T., Sumarah, M., & Reid, G. (2017). A novel millet-based probiotic fermented food for the developing world. *Nutrients*, *9*(5), 529. https://doi.org/10.3390/nu9050529

Dwivedi, M., Vasantha, K. Y., Sreerama, Y. N., Haware, D. J., Singh, R. P., & Sattur, A. P. (2015). Nilamadana, new fungal fermented cereal based food. *Journal of Functional Foods*, *15*, 217–224. https://doi.org/10.1016/j.jff.2015.03.023

Fasreen, M. M. F., Perera, O. D. A. N., & Weerahewa, H. L. D. (2017). Development of finger millet based probiotic beverage using *Lactobacillus casei* 431®. *OUSL Journal*, *12*(1), 128. https://doi.org/10.4038/ouslj.v12i1.7384

Gadaga, T. H., & Lehohla, M. (2013). Traditional fermented foods of Lesotho. *Journal of Microbiology, Biotechnology and Food Sciences*, *2*(6), 2387–2391.

Gurry, T. (2017). Synbiotic approaches to human health and well-being. *Microbial Biotechnology*, *10*(5), 1070–1073. https://doi.org/10.1111/1751-7915.12789

Ilango, S., & Antony, U. (2021). Probiotic microorganisms from non-dairy traditional fermented foods. *Trends in Food Science and Technology*, *118*, 617–638. https://doi.org/10.1016/j.tifs.2021.05.034

Kavitha, B., Vijayalakshmi, R., Poorna, C. R., Yalagala, I. M., & Sugasini, D. (2018). Nutritional evaluation and cell viability of formulated probiotic millet fruit bar. *Journal of Food and Nutritional Disorders*, *7*(2), 2–10. https://doi.org/10.4172/2324-9323.1000246

Kayodé, A. P. P., Hounhouigan, D. J., Nout, M. J. R., & Niehof, A. (2007). Household production of sorghum beer in Benin: Technological and socio-economic aspects. *International Journal of Consumer Studies*, *31*(3), 258–264. https://doi.org/10.1111/j.1470-6431.2006.00546.x

Kewuyemi, Y. O., Njobeh, P. B., Kayitesi, E., Adebiyi, J. A., Oyedeji, A. B., Adefisoye, M. A., & Adebo, O. A. (2020). Metabolite profile of whole grain ting (a Southern African fermented product) obtained using two strains of *Lactobacillus fermentum*. *Journal of Cereal Science*, *95*, 103042. https://doi.org/10.1016/j.jcs.2020.103042

Kumar, A., Tomer, V., Kaur, A., Kumar, V., & Gupta, K. (2018). Millets: A solution to agrarian and nutritional challenges. *Agriculture and Food Security*, *7*(1), 1–15. https://doi.org/10.1186/s40066-018-0183-3

Kumar, A., Tripathi, M. K., Joshi, D., & Kumar, V. (2021). *Millets and millet technology*. Springer. https://doi.org/10.1007/978-981-16-0676-2

Madilo, F. K., Parry-Hanson Kunadu, A., Tano-Debrah, K., Saalia, F. K., & Kolanisi, U. (2024). Diversity of production techniques and microbiology of African cereal-based traditional fermented beverages. *Journal of Food Quality*, *2024*(1), 1241614. https://doi.org/10.1155/2024/1241614

Mali, S. D., Deshpande, H. W., & Katke, S. D. (2020). Standardization and quality evaluation of non-dairy probiotic beverage prepared from finger millet and foxtail millet. *International Journal of Current Microbiology and Applied Sciences*, *9*(8), 2118–2123. https://doi.org/10.20546/ijcmas.2020.908.241

Mashau, M. E., Muluvhu, D., & Ramashia, S. E. (2024). Comparative evaluation of health-promoting compounds, physicochemical and microbiological properties of sorghum [*Sorghum bicolor* (L.) Moench] based Mahewu produced by different traditional brewers in Thohoyandou, South Africa. *Fermentation*, *10*(5), 236. https://doi.org/10.3390/fermentation10050236

Mondal, S., Balasubramanian, A., Biswas, P., Agrawal, S., Ghosh, S., & Dey, S. (2022). Characterization of pearl millet oligosaccharides and evaluation of their prebiotic potential. *Bioactive Carbohydrates and Dietary Fibre*, *28*, 100324. https://doi.org/10.1016/j.bcdf.2022.100324

Núñez-Sánchez, M. A., Herisson, F. M., Cluzel, G. L., & Caplice, N. M. (2021). Metabolic syndrome and synbiotic targeting of the gut microbiome. *Current Opinion in Food Science*, *41*, 60–69. https://doi.org/10.1016/j.cofs.2021.02.014

Ogbonna, A., Abuajah, C., & Umanah, I. (2016). Burukutu: Healthy and superior indigenous African traditional opaque beverage. *American Journal of Advanced Food Science and Technology*, *4*(1), 29–37. https://doi.org/10.7726/ajafst.2016.1004

Panigrahi, S., Nanda, A., & Sagar, P. (2024). Characterization and standardization of a millet-based probiotic beverage via physicochemical and microbial analysis. *3*(2), 371–380.

Pawar, V. D., & Machewad, G. M. (2006). Processing of foxtail millet for improved nutrient availability. *Journal of Food Processing and Preservation*, *30*(3), 269–279. https://doi.org/10.1111/j.1745-4549.2006.00064.x

Ponne, S., & Selva Brindha, S. (2020). Formulation and evaluation of millet extract enriched yoghurt. *FoodSci: Indian Journal of Research in Food Science and Nutrition*, *7*(2), 1. https://doi.org/10.15613/fijrfn/2020/v7i2/207175

Saeed, F., Afzaal, M., Hanif, H., Islam, F., Hussain, M., Shah, Y. A., & Ikram, A. (2022). Nutritional composition and functional properties of a fermented product (KOOZH): A review. *Journal of Food Processing and Preservation*, *46*(11), 1–9. https://doi.org/10.1111/jfpp.16969

Sahoo, S. S., Agarkar, B. S., & R.K., A. (2024). Effect of malting on composition of nutrients and anti-nutrients in finger millet flour. *Plant Archives*, *24*(SI), 169–174. https://doi.org/10.51470/PLANTARCHIVES.2024.v24.SP-GABELS.026

Samtiya, M., Soni, K., Chawla, S., Poonia, A., Sehgal, S., & Dhewa, T. (2021). Key anti-nutrients of millet and their reduction strategies: An overview. *Acta Scientifci Nutritional Health*, *5*(12), 68–80. https://doi.org/10.31080/asnh.2020.05.0963

Sheethal, H. V., Baruah, C., Subhash, K., Ananthan, R., & Longvah, T. (2022). Insights of nutritional and anti-nutritional retention in traditionally processed millets. *Frontiers in Sustainable Food Systems*, *5*(February), 1–14. https://doi.org/10.3389/fsufs.2021.735356

Sindhu, S. C., & Khetarpaul, N. (2001). Probiotic fermentation of indigenous food mixture: Effect on antinutrients and digestibility of starch and protein. *Journal of Food Composition and Analysis*, *14*(6), 601–609. https://doi.org/10.1006/jfca.2001.1022

Taylor, J. R. N., & Duodu, K. G. (2018). Traditional sorghum and millet food and beverage products and their technologies. In J. Taylor, S. R. Bean, & K. G. Duodu (Eds.), *Sorghum and millets: Chemistry, technology, and nutritional attributes* (pp. 391–424). AACCI. https://doi.org/10.1016/B978-0-12-811527-5.00009-5

Temilope, O. S., & Taiyese, O. B. (2012). Quality assessment of "Oti-Oka" like beverage produced from pearl millet. *Journal of Applied Biosciences*, *51*, 3608–3618.

Thapa, S., & Tamang, J. P. (2004). Product characterization of kodo ko jaanr: Fermented finger millet beverage of the Himalayas. *Food Microbiology*, *21*(5), 617–622. https://doi.org/10.1016/j.fm.2004.01.004

Tharifkhan, S. A., Perumal, A. B., Elumalai, A., Moses, J. A., & Anandharamakrishnan, C. (2021). Improvement of nutrient bioavailability in millets: Emphasis on the application of enzymes. *Journal of the Science of Food and Agriculture*, *101*(12), 4869–4878. https://doi.org/10.1002/jsfa.11228

Vila-Real, C., Pimenta-Martins, A., Mbugua, S., Hagrétou, S. L., Katina, K., Maina, N. H., Pinto, E., & Gomes, A. M. P. (2022). Novel synbiotic fermented finger millet-based yoghurt-like beverage: Nutritional, physicochemical, and sensory characterization. *Journal of Functional Foods*, *99*(July), 105324. https://doi.org/10.1016/j.jff.2022.105324

Yousaf, L., Hou, D., Liaqat, H., & Shen, Q. (2021). Millet: A review of its nutritional and functional changes during processing. *Food Research International*, *142*(17), 110197. https://doi.org/10.1016/j.foodres.2021.110197

13

Millet By-products: Bioactive Composition and Value Addition

Neha Goyal[1], Parmjit S. Panesar[1], Charanjit S. Riar[1], and Durga S. Bunkar[2]

[1] *Department of Food Engineering and Technology, Sant Longowal Institute of Engineering and Technology, Longowal, Punjab, India*
[2] *Department of Dairy Science and Food Technology, Institute of Agricultural Sciences, Banaras Hindu University, Varanasi, India*

13.1 Introduction

Millets are small-seeded grains that require less amount of water for growth as compared to other cereal grains. This category of grains is also referred to as "nutri-cereals," due to the presence of high fiber, low glycemic index, nonglutinous, and the presence of diverse type of bioactive compounds that possess the ability to mitigate various chronic diseases (Saini et al., 2021). The remarkable nutritional value and favorable agronomic conditions associated with millets have attracted attention, particularly in the context of increasing global populations and diminishing water resources.

In order to obtain the millet grains during the harvesting process, a notable production of by-products takes place, specifically including cobs, stalks, and straw (Tagade & Sawarkar, 2023). Typically, millets are consumed either after cooking or processed into different products like flatbreads, idli, dosa, alcoholic beverages (tella, opaque beers, katikalla, busa or bouza, chhang, etc.), and nonalcoholic beverages (Majumder et al., 2024; Paschapur et al., 2021). For the preparation of these products, millets are dehulled and ground mechanically before processing. Different steps of primary processing of millets include cleaning, destoning, dehusking, debranning, etc., which are responsible for the generation of husk and bran as by-products (Singh et al., 2024a; Srinivas, 2022). The separation of these by-products is known to negatively affect the nutritional composition of the whole millet grain due to reduction in fiber, vitamins, and minerals content of grain (Singh et al., 2024a). Primarily, millet by-products are utilized as animal feedstock or are disposed/ burnt as such by the farmers. Such practices cause environmental pollution and many health complications such as increased risk of acute respiratory, eyes related diseases, and many other health problems (Kaur et al. 2025; Tagade & Sawarkar, 2023). However, these by-products are underutilized, which offers many nutritional and functional benefits,

Nutraceutical Potential of Millet-Based Food Products: Chemistry and Applications, First Edition.
Edited by Parmjit S. Panesar, Charanjit S. Riar, and Sushil Dhital.

Published 2026 by John Wiley & Sons Ltd.

rendering them a crucial component of the millet processing chain. If compared with by-products of other cereals, their utilization remains limited regardless of their significant production. Unfortunately, so far, there has been limited research conducted on millet by-products. The increased demand for highly nutritious food and eco-friendly raw materials for domestic and industrial use makes it an important source for food and nonfood applications, especially in developing nations. This chapter discusses the bioactive composition of different millet by-products with a special focus on their value-addition for the development of different food and nonfood products.

13.2 Millet By-products

The primary processing of millets to obtain endosperm part includes different steps which generate numerous by-products (Figure 13.1). First of all, harvesting and cleaning of millets results in the generation of by-products such as cob, straw, stalks, and foreign particles. Following that, millet grain is covered by an outer hard layer called husk, which is removed

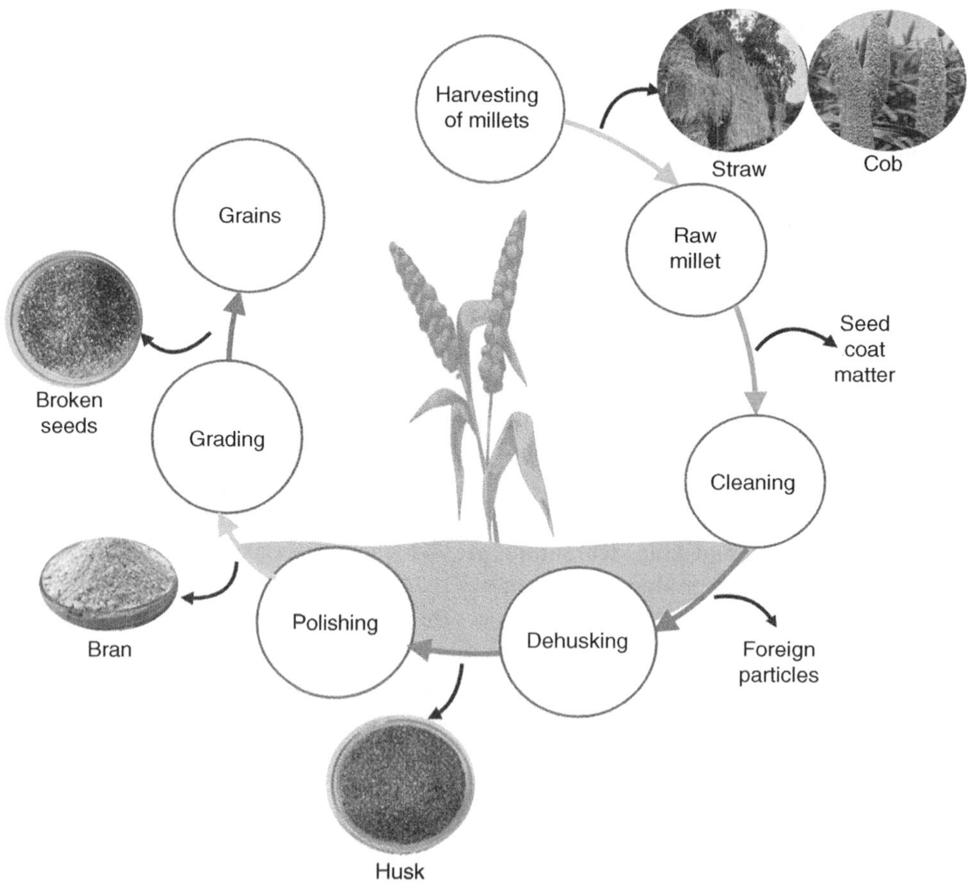

Figure 13.1 Generation of different millet by-products during primary processing

through the process of dehulling or dehusking (Rani et al., 2018). Further, bran layer or pericarp of grain is removed by polishing, which significantly contains indigestible fiber. To efficiently separate the husk and bran, different pretreatment methods such as traditional (soaking, tempering, drying, etc.), thermal (microwave drying, hydrothermal, infrared drying, ohmic heating, etc.), and nonthermal techniques (pulse electric field, high-pressure processing, cold plasma, etc.) are employed, which reduces the antinutritional content (Singh et al., 2024a). In the last step, grading is done to obtain the uniform size of millet grains, which separates broken seeds.

Millet by-products contain substantial amounts of carbohydrates, fats, proteins, vitamins, minerals, and various bioactive compounds such as phenolics, dietary fiber, carotenoids, flavonoids, tannins, etc., which provide anti-inflammatory, antioxidant, anticarcinogenic, and antidiabetic properties (Cao et al., 2021; Ji et al., 2019; Mustač et al., 2019). The proximate and mineral composition of different millet by-products are mentioned in Tables 13.1 and 13.2, respectively.

13.2.1 Millet Husk

Every year, tons of millets are harvested all over the world, resulting in the generation of husk in huge amounts. Around 40% of the millet weight is removed as outer fibrous part, which is also known as hull (Kuhe et al., 2021). It is a lightweight by-product, which is tightly attached to the endosperm and present in multiple layers in some minor millet. For example, the husk of kodo millet exists in many layers, which is difficult to dehull and thus requires a specific dehuller with high efficiency and less grain loss (Nayak et al., 2019). Husk is known to contain about 69–72% carbohydrates, 7–10% protein, 41–53% crude fiber, 6–10% crude fat, and 6–13% ash (Srinivasan et al., 2019; Suri et al., 2024). Although, the nutritional and bioactive composition of millet husk varies according to the type of millet, millet cultivar, area of cultivation, soil type, climatic conditions, etc. For example, husk of red proso millet contains a high amount of crude protein (10.4%), followed by black (10.1%), white (8.45%), and yellow (7.45%) proso millet. On the other hand, white proso millet comprises more starch (5.94%) and ash (13.2%) than other cultivars. Similarly, functional properties of proso millet husk also differ with varying cultivars (Li et al., 2022a).

13.2.2 Millet Bran

It is principal by-product that contributes 9–12% (w/w) of millet weight, also known as seed coat (Chen et al., 2024). It comprises small proportion of embryo, endosperm, testa, and aleurone layer (Suri et al., 2024). The distribution of bran content varies depending on the seed size and pericarp type. For example, foxtail millet contains less amount of bran as compared to kodo and barnyard millet. The nutritional composition of bran contains 30–60% complex carbohydrates, 10–18% protein, and approximately 3–9% fat (Suri et al., 2024; Vashishth & Maurya, 2024). Different types of protein including albumin, globulin, and prolamin are present in bran (Sachdev et al., 2023). Even the content of protein in foxtail millet bran is high as compared to wheat bran (15–18%), which makes it an excellent and economical source of protein (Li et al., 2024a). However, the crude fat of foxtail millet bran is almost similar to wheat

Table 13.1 Nutritional composition of different millet by-products

Millet	Millet by-products	Protein	Crude fiber	Carbohydrates	Fat	Moisture	Ash	References
		(%)						
Pearl millet	Bran	11.02	1.20	64.3	9.38	10.5	3.60	Kaur et al. (2023)
	Stalk	0.32–6.62	—	—	—	—	9.46–13.26	Yadav et al. (2019)
	Husk	9.48	1.30	69.57	6.42	11.27	2.05	Srinivasan et al. (2019)
Finger millet		8.42	2.51	73.32	5.58	7.94	2.51	
	Bran	11.64–13.6	—	6.42	3.2–3.4	11–11.3	5.1–5.6	Onipe and Ramashia (2022)
Proso millet	Bran	1.78–26.33	—	—	1.92–8.96	1.88–7.32	5.77–8.44	Dega and Barbhai (2023)
	Husk	10.40	48	4.76	10.83	8.24	9.72	Li et al. (2022a)
	Bran	14.75	35.80	6.36	20.43	7.62	8.54	
	Bran (>1,000 μm)	4.86	—	4.52	1.92	—	7.84	Mustač et al. (2020)
	Bran (500–1,000 μm)	4.97	—	1.78	3.45	—	8.44	
	Bran (<500 μm)	13.90	—	26.33	8.96	—	5.77	
	Bran	4.06	—	7.67	2.76	6.27	6.06	Liu et al. (2016)
Foxtail millet	Bran	10.49–12.48	—	65.10	9.39–9.87	2.39–8.29	7.5–12.15	Dega and Barbhai (2023)
	Bran	10.23	5.73	10.17	6.6	6.1	6.21	Zhu et al. (2018)
White sorghum	Bran	11.58	8.25	64.18	5.342	8.29	2.36	Adebo and Kesa (2023)
Brown sorghum	Bran	13.7	6.06	63.98	5.71	8.31	2.23	
Sorghum	Stalk	1.3–1.9	—	2.4–3.9	4.8–5.1	4.1–4.7	4.4–5.4	Luna et al. (2022)
Sorghum	Bran	6.46	—	32.87	5.34	8.71	3.75	Qiu et al. (2017)
Barnyard millet	Bran	5.15–7.7	1.66	72.03	4.04–7.85	2.41–6.79	9.11–10.02	Dega and Barbhai (2023)
Kodo millet	Bran	4.92	—	79.84	2.83	1.76–7.07	5.33	

Table 13.2 Mineral composition of millet by-products

Millet	Millet by-products	Iron	Magnesium	Calcium	Zinc	Phosphorus	Potassium	References
		(mg/100 g)						
White sorghum	Bran	0.42	226.0	995.17	1.03	0.57	3.65	Adebo and Kesa (2023)
Brown sorghum		0.5	292.25	1,020.91	1.41	0.66	3.48	
Kodo millet		20.45	—	76.09	2.13	—	—	Dega and Barbhai (2023)
Proso millet		60.74	—	37.87	5.59	—	630.83	
Foxtail millet		65.58	—	94.63	4.71	—	—	
Barnyard millet		8.85	—	62.26	3.83	—	—	
Finger millet		5.9–6.5	—	711–830	2.7–2.86	369–526	5.2	Onipe and Ramashia (2022)
	Husk	—	—	49	—	14	—	Gairola et al. (2022)

bran (11–18%) and less than rice bran (17.3–27.4%). In addition, essential oil (15–20%) is also present in millet bran, which contains tocopherol, tocotrienol, oryzanol, stearic acid, palmitic acid, linoleic acid, oleic acid, linolenic acid, arachidic acid, etc. (Kalsi & Bhasin, 2023). Among these, linoleic acid is most abundant fatty acid, which makes 53.6% of total fatty acid (Li et al., 2022a). In addition, the content and proportion of linoleic acid in millet bran is higher than common cereal sources like rice bran, which makes it great source of polyunsaturated fatty acids. The presence of fatty acids within bran induces rancidity, consequently resulting in a diminishing shelf life. Therefore, it is important to extract essential oil from bran which not only prolongs its shelf life but serves as an efficient source of beneficial fatty acids. The yield of oil recovered from millet bran is dependent on the particular extraction technique and type of millet (Shi et al., 2015). For example, essential oil from foxtail millet bran was extracted using conventional solvent extraction (SE), supercritical carbon dioxide extraction (SCE), and subcritical propane extraction (SPE). Among which, SPE (21.79%) provides high extraction yield of oil, followed by SCE (19.65%) and SE (7.14%). The utilization of different extraction technique significantly influences the physicochemical characteristics, including peroxide value, saponification index, and yield, whereas the differences observed in the fatty acid composition was insignificant according to gas chromatography (Shi et al., 2015). Besides, the oil extracted from millet bran provides protection from reactive oxygen species and various chronic diseases (Yin et al., 2022).

The bran with large particle size is not completely digested by human digestive system, whereas bran having small particle size exhibits better interaction with digestive organs, which increases interaction of enzyme and substrate, as well as the release of beneficial compounds in the digestive tract (Li et al., 2022c). For example, coarse powder of foxtail millet bran contained lower amount of phenolics (2,684.53 μg/g), whereas an increase of around 55% and 80% in total phenolic content was observed in case of fine and superfine powder, respectively (Liang et al., 2024). This was correlated with increase in surface area of small-sized particles, thus facilitating the release of phytochemicals (including phenolic compounds) embedded within the fibrous matrix during extraction. Additionally, superfine ground bran exhibited better antioxidant activity and bioaccessibility of phenolic compounds (syringic, vanillic, ferulic, protocatechuic, isoferulic acid, and 2,3-dihydroxybenzoic acid).

Whole millet flour-based products provided more nutritional compounds and health benefits, but bran covering endosperm imparted dark color that has low consumer acceptability with chewy and gritty texture. To enhance the acceptability among consumers, bran is usually removed. Different pretreatment methods are employed that affect the yield and functional properties of bran. For example, hydrothermal treatment of millets involves pearling, which decreases the antinutritional content (phytic acid, trypsin inhibitor, lipases, etc.) present in bran (Onipe & Ramashia, 2022). Likewise, solid-state fermentation of millet bran also reduces the antinutritional factors. After fermentation, the alkaloid content reduced by 31.1%, phytate by 43.28%, tannins by 30.2%, and oxalate by 22.27% in millet bran. In addition, fermentation treatment also increased the amount of phenols by 33.76%, flavonoids by 9.4%, and crude protein by 114.2%. However, a reduction in crude dietary fiber content by 18.75% was observed following fermentation of millet bran (Ali et al., 2024).

13.2.3 Other By-products

Beside husk and bran, other millet by-products involve straw, cob, stalks, and broken seeds. The process to obtain millet grains (endosperm part) necessitates the removal of cob and straw, which are considered as residues. The quantity of straw produced after postharvesting of millet grains is nearly twice that of the grain yield. These by-products exhibit significant potential for application in bioenergy production due to the abundant presence of lignocellulosic biomass (Aggarwal et al., 2024). Total amount of residue generation after crop harvesting is measured by using residue-to-product ratio (RPR). It states the ratio of crop residue weight to total yield of agro-product of a particular crop. RPR of millets varies on the basis of their production and type of millet. For example, RPR of little millet straw and ragi straw is 1.2 and 1.3, respectively. In addition to this, pearl millet stalk, husk, and cob have RPR of 0.2, 0.3, and 0.33, whereas RPR of sorghum straw, cob, and husk is 0.2, 0.5, and 0.85, respectively (Tagade & Sawarkar, 2023).

Millet stalk is another important by-product, which is basically the stem of millet plant and is typically tall and rigid. It is a fibrous part, which majorly comprises sheath and core. In case of pearl millet stalk, 80% sheath and 20% (w/w) of core is present. A high amount of cellulose (52.49%), hemicellulose (25.4%), and protein (2.06%) is present within the core of pearl millet stalk over sheath (Yadav et al., 2019). Both the acid-soluble and insoluble lignin is present more in sheath part (around 29%) compared to the core (around 10%). Lignin present within the sheath provides impermeability, structural support, and protection against microbial attack. The composition of pearl millet stalk is nearly similar to sugarcane and sorghum bagasse (Yadav et al., 2019). The amount of lignocellulosic materials of millet stalks varies with variety, species, soil type, and climatic conditions.

During the milling of millet grains, broken seeds are obtained as a major fraction. Usually, broken seeds are considered inferior due to their fragmented nature, but they comprise a good amount of fiber, protein, and minerals. The polishing of grains also generate significant amount of broken grains; for example, in case of foxtail millet, around 9.26% of broken seeds are generated, even under optimized polishing conditions (Shivashankar et al., 2024). Being a part of grain, the nutritional properties of broken grains are almost similar to the whole grain, and, therefore, it becomes crucial to look at their potential applications with the aim to enhance their utilization and reduce the amount of waste generated (Suri et al., 2024).

13.3 Bioactive Composition of Millet By-products

Millet by-products are excellent sources of various phytochemicals such as phenolic acids, flavonoids, dietary fiber, peptides, etc. These compounds have the ability to prevent various diseases as well as maintain cellular health that promotes longevity (Akanbi et al., 2019). Major bioactive compounds present in millet by-products are illustrated in Figure 13.2.

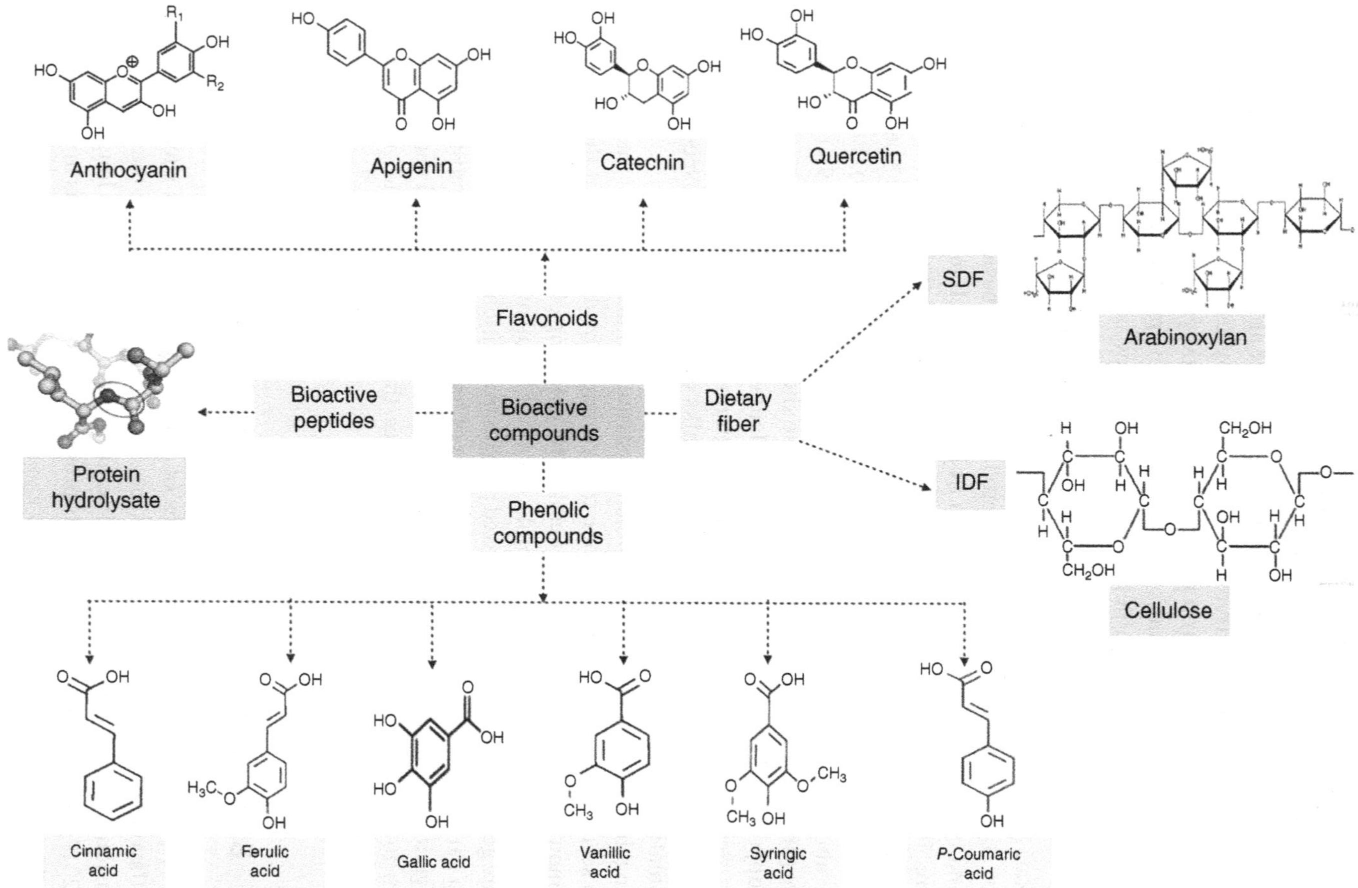

Figure 13.2 Major bioactive compounds present in millet by-products. *Source:* Adapted from Akanbi et al. (2019)

13.3.1 Polyphenols

These compounds contain one or more aromatic rings that are attached to various hydroxyl groups. Most of the polyphenols are found in the husk and bran of millet grain and exhibit high antioxidant activity, whereas endosperm contains a negligible amount of polyphenols (Zhang et al., 2021). For example, extract of finger millet bran contained higher antioxidant activity (86%) as compared to whole finger millet extract (27%). The quantity of polyphenols is also affected by different processing methods like fermentation, soaking, cooking, roasting, etc. For instance, after millet bran fermentation by *Bacillus natto*, the content of polyphenols and flavones was increased by 32.5% and 16.4%, respectively (Zhang et al., 2024). Similarly, fermentation also enhanced the total phenolic content (3.05–7.02 mg GAE/g) of dietary fiber extracted from foxtail millet bran (Chu et al., 2019). It has been found that increase in polyphenols during fermentation might be due to microorganisms, which utilized soluble and fermentable fiber for their growth and released mechanically trapped polyphenols (Wang et al., 2022). Besides this, the concentration of polyphenols also depends on the type of organic solvents (methanol, ethanol, acetone, etc.) and different techniques (conventional and green methods) used for extraction (Priya et al., 2023). There is a need to conduct research on the determination and extraction of polyphenols from other millet by-products rather than husk and bran.

Polyphenols exhibit high antioxidant activity, antiaging, and antihyperglycemic properties, which provide the ability to prevent from several diseases like chronic fatigue syndrome, cardiovascular diseases, diabetes, inflammation, cancer, and sickle cell anemia (Figure 13.3) (Rudrapal et al., 2022). The bran and husk of millets contain substantial amounts of polyphenols, which are mainly composed of phenolic acids and flavonoids. The quantities of polyphenols found in millet bran and husk are presented in Table 13.3.

13.3.1.1 Phenolic Acids

These secondary metabolites within millet bran and husk mainly exist in bound state, which includes hydroxycinnamic and hydroxybenzoic acids (Kaur et al., 2025). The former contains vanillic, syringic acid, protocatechuic, and genistic, whereas the latter mainly includes sinapic, *p*-coumaric, cinnamic, and ferulic acids (Xiang et al., 2019). Particularly, free phenolics are concentrated more in bran, whereas bound phenolics exist more in millet husk. For example, finger millet bran contains high amount of free phenolics (2.98 mg FAE/g dry weight), whereas husk exhibits more amount of bound phenolics (8.6 mg FAE/g dry weight). Among them, *p*-coumaric and ferulic acids are major bound phenolics found in millet husk (Sunil et al., 2024). The bound phenolics of millet bran exhibit higher pharmacological activities such as antioxidant, antitumor, immunomodulation, antifungal, and antihyperglycemia effects.

13.3.1.2 Flavonoids

These compounds contained two or more benzene rings, which are attached to heterocyclic pyran. Different types of flavonoids were found in millets such as anthocyanins, flavanols, flavanones, and flavones. Majorly, flavonoids are present in millet bran and husk that exist in free state, which includes catechin, gallocatechin, quercetin, taxifolin, and

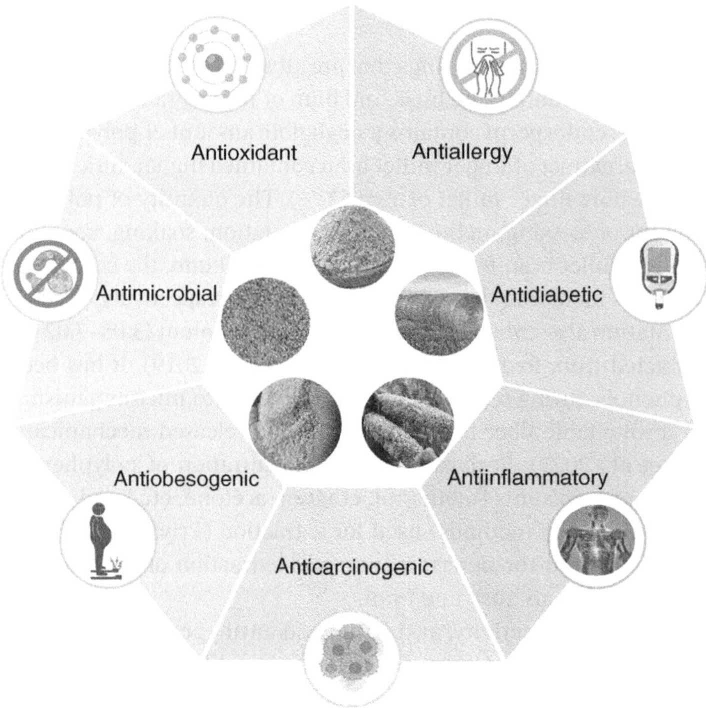

Figure 13.3 Pharmacological properties of millet by-products

apigenin derivatives (Wang et al., 2022). Among these, anthocyanins are specifically present in sorghum that contribute to different colors (white, brown, red, orange, etc.) exhibited by bran. Further, anthocyanins are classified into cyanidin, pelargonidin, delphinidin, malvinidin, petunidin, and peonidin that possess anti-inflammatory, antioxidant, antiallergic, anticancer, and gastroprotective properties (Ahmad et al., 2018). All the millets contain proanthocyanidins, except sorghum, which contains anthocyanin. Further, anthocyanins are subdivided into two classes, namely, 3-oxyanthocyanins and 3-deoxyanthocyanins, which is distinguished by genin. The latter one comprises compounds which impart yellow and red color in acidic and basic medium, respectively. It was found that extract of sorghum husk is one of the rich sources of 3-deoxyanthocyanins. The total content of 3-deoxyanthocyanins in sorghum husk extract comprises 99.33 ± 110.05 mg ELt/g and 4.16 mg EApi/g of luteolinidin and apigeninidin equivalents, respectively (Sawadogo et al., 2024). Particularly, 3-deoxyanthocyanins prevented heart disease and have the potential to act as an anticancer and antidiabetic agent.

Qualitative and quantitative analysis revealed that millet and their by-products contain varying amounts of flavonoids. For example, finger millet bran contains 2,305 mg/100 g of total flavonoid content which comprises naringenin, apigenin, epicatechin, catechin, and kaempferol (Onipe & Ramashia, 2022). On the other hand, proso millet husk and bran contain kaempferol, rutin, quercetin-3-O galactoside, kaempferitrin, and naringenin. It has been determined that proso millet husk comprises a high amount of flavonoids over

Table 13.3 Total polyphenolic content in millet by-products

Millet	By-product	Polyphenols	Content (dry weight)	References
Sorghum	Bran	TPC	12.6–75.5 mg GAE/g	Duke et al. (2024)
		TFC	14.2 mg QE/g	Tyagi et al. (2021)
Foxtail millet	Husk	TPC	7.38–17.99 mg FAE/g	Wang et al. (2023a) and Pradeep and Sreerama (2017)
		TFC	4.30 mg RE/g	
	Bran	TPC	7.80–10.44 mg FAE/g	Zheng et al. (2022) and Pradeep and Sreerama (2017)
		TFC	8.89 mg RE/g	
Sorghum	Husk	TPC	237.73 ± 2.12 mg GAE/g	Sawadogo et al. (2024)
		TPC	606–779 mg GAE/g	Shukla et al. (2022)
		TFC	172–195 mg CE/g	
		Anthocyanin	8.55–10.5 mg LE/g	
	Bran	TPC	12.6–150 mg GAE/g	Tuhanioglu and Ubeyitogullari (2022) and Tyagi et al. (2021)
		TFC	14.2–99.6 CE/g	
Proso millet	Bran	TPC	3.87–9.57 mg GAE/g	Li et al. (2022a)
		TFC	1.61–6.04 mg RE/g	
	Husk	TPC	4.69–16.25 mg GAE/g	
		TFC	5.44–14.87 mg RE/g	
Finger millet	Seed coat	TPC	20.15–21.20 mg GAE/g	Velur Balasubramaniam et al. (2022)
		TFC	483 CE mg/100 g	Balasubramaniam et al. (2019)
Little millet	Husk	TPC	16.82 mg FAE/g	Pradeep and Sreerama (2017)
		TFC	17.14 mg CE/g	
	Bran	TPC	6.23 mg FAE/g	
		TFC	2.82 mg CE/g	

TPC, total phenolic content; GAE, gallic acid equivalent; TFC, total flavonoid content; FAE, ferulic acid equivalent; RE, rutin equivalent; CE, catechin equivalent.

bran. Besides this, different varieties of proso millet contain varying concentrations of flavonoids (Li et al., 2022a). For example, a high quantity of flavonoids was found in black proso millet husk (14.87 mg RE/g) over white proso millet husk (5.44 mg RE/g).

13.3.2 Dietary Fiber

Dietary fiber is a predominant complex nonstarch polysaccharide found within millet by-products, which is neither absorbed nor digested by the small intestine of the humans. However, it can be completely or partially fermented by gut microbiota present in the large intestine (Williams et al., 2019). On the basis of water solubility, it is classified into soluble (SDF) and insoluble dietary fiber (IDF). The former has the ability to control cholesterol level and metabolism in humans, which may help in the treatment of hyperglycemia, atherosclerosis, and other chronic diseases (Siddiqui et al., 2023). The latter one helps in water absorption and healthy intestinal functioning.

Millet by-products contain different forms of dietary fiber such as cellulose, arabinoxylan, oligosaccharides, β-glucan, hemicellulose, and lignin in varying proportions (Figure 13.2). Similar to polyphenols, the amount and composition of dietary fiber depend on extraction method, extraction time, extraction temperature, type of millet, and its by-product. For example, around 38.4–51% and 1.2–6.6% of IDF and SDF were present within finger millet seed coat, whereas the foxtail millet bran contained 54.3% of IDF and 4.24% of SDF (Li et al., 2022a; Onipe & Ramashia, 2022). Furthermore, it has also been observed that foxtail millet husk contains more amount of IDF, whereas bran comprises high amount of SDF (Li et al., 2024a). The high content of SDF in bran indicates the presence of large amount of oligosaccharides. Around 79% and 95% of total dietary fiber (TDF) were obtained from foxtail millet bran and husk, respectively, using enzyme-assisted extraction (Li et al., 2024a; Zheng et al., 2023). The extracted dietary fiber comprises around 4.49% fat, 5.2% ash, 2.81% protein, 19.56% cellulose, 57% hemicellulose, and 10.16% lignin (Zheng et al., 2022). A diverse range of monosaccharides has been identified in dietary fiber fractions (soluble and insoluble dietary fiber) of foxtail millet bran, primarily including xylose, glucose, and arabinose. Among identified monosaccharides, around 45.7% and 48.3% of xylose were present in TDF and IDF, whereas SDF contained high amount of glucose (34.5%) over xylose (30.9%). In addition, a high content of total phenolic content was observed in TDF (1.45 g/kg), while 0.54 and 0.72 g/kg phenolic content were present in SDF and IDF, respectively (Dong et al., 2019). It has been determined that when TDF was obtained from pearl millet bran using hydrothermal and ultrasound method, the resultant yields differ significantly at 74.5% and 78.8%, respectively (Kaur et al., 2023, 2024). The yield and type of dietary fiber obtained from millet by-products are presented in Table 13.4. In addition, utilization of traditional processing methods also affects the extraction yield of dietary fiber present in millet by-products. For instance, the SDF and β-glucan content of millet bran were increased by 92.0% and 164.4%, respectively, when fermented with *Bacillus natto* (Zhang et al., 2024).

The concentration of SDF is generally lower in plant carbohydrates and so in millet by-products as compared to IDF present in higher amounts (>50%). The properties of IDF and a fraction of it can be converted into SDF by following appropriate modification treatments. Different extraction and modification techniques are also known to affect the functional, thermal, and structural properties of dietary fiber obtained from millet by-products. For example, xylanase enzyme modified the cholesterol absorption capacity of dietary fiber extracted from foxtail millet bran due to formation of more soluble polysaccharides. In addition, the composition and microstructure of dietary fiber also changed after enzyme treatment (Zhu et al., 2019). Li et al. (2024b) modified dietary fiber by heating and enzyme hydrolysis (cellulase and xylanase) in combination with carboxymethylation, crosslinking, or acetylation, which enhances soluble dietary fiber content, viscosity, water expansion, and retention properties. The combined treatment of ultrasound and microwave caused the SDF to become porous, loose, and rough. After modification, the crystallinity of the IDF increased, whereas no change was observed in the chemical functional groups, thus preserving the native nature of the product (Cao LongKui et al., 2018). The bran of

Table 13.4 Dietary fiber content in millet by-products

Millet	Millet by-products	Dietary fiber	Yield (w/w)	References
Pearl millet	Bran	Total dietary fiber	74.5–78.8%	Kaur et al. (2023, 2024)
		Arabinoxylan	15.96%	Singh et al. (2024b)
Foxtail millet	Husk	Total dietary fiber	95% (IDF, 93.18%; SDF, 1.82%)	Li et al. (2024a)
	Bran	Total dietary fiber	57.46–79.22%	Zhu et al. (2018) and Chu et al. (2019)
Proso millet	Bran	Soluble dietary fiber	7.6%	Li et al. (2022b)
		Total dietary fiber	36.2–71.75%	Mustač et al. (2020)
Finger millet	Bran	Total dietary fiber	39.6–57.38% (IDF, 38.4–51%; SDF, 1.2–6.38%)	Onipe and Ramashia (2022)
	Seed coat	Xylan	4.8%	Palaniappan et al. (2017)
Sorghum	Bran	Total dietary fiber	35.1%	Qiu et al. (2017)

different millets contained a high amount of TDF, SDF, and IDF as compared to rice and wheat bran (Kumari et al., 2018). Even the contents of TDF, SDF, and IDF are less in polished millet grain as compared to millet bran and husk. For example, polished foxtail millet contains approximately 15%, 13%, and 1% of TDF, IDF, and SDF, respectively, which is substantially lower than the concentrations found in foxtail millet bran (Li et al., 2024a). With the utilization of different techniques (ultrasound, microwave, hydrothermal, alkali, and enzymes), dietary fiber has been successfully extracted from bran of sorghum, barnyard, proso, kodo, pearl, and finger millets (Chu et al., 2019; Li et al., 2022b; Singh et al., 2024b). It was observed that extracted dietary fiber has the ability to bind glucose and reduce the digestion of starch, which provides hypoglycemic effect (Dega & Barbhai, 2023). Additionally, the consumption of dietary fiber present in millet husk and bran-contained food can reduce the risk of colon cancer, type II diabetes, cardiovascular diseases, and intestine problems (14.3) (Cheng et al., 2022; Ren et al., 2023).

13.3.2.1 Arabinoxylan

Arabinoxylan (AX) is the most predominantly found dietary fiber found in millet bran. It constitutes the backbone of xylose with β (1–4) linkage and comprises arabinose substitutions on either C-2 or C-3 of xylose. Around 15% of arabinoxylan was successfully extracted from bran of pearl millet using alkali-assisted extraction, which is mainly composed of arabinose, glucose, galactose, xylose, and mannose (Singh et al., 2024b). It was found that arabinoxylan possesses prebiotic potential and exhibits the capability to lower serum cholesterol and blood glucose level. Additionally, the antioxidant properties of arabinoxylan could be helpful for the treatment of colon cancer (Onipe & Ramashia, 2022).

The extracted arabinoxylan was converted into feruloylated xylooligosaccharides (FXOS) by enzymatic hydrolysis that possess antiglycating, antioxidant, anticancer, and anti-inflammatory properties. The recovered FXOS from pearl millet bran contained 25.19 mg/g of bound ferulic acid, monosaccharides (227.5 mg/g arabinose, 359.9 mg/g xylose, 25.8 mg/g glucose, 27.2 mg/g galactose, and 14 mg/g glucuronic acid), with the total arabinoxylan content of 51.5% (Singh & Eligar, 2021). Arabinose was linked to ferulic acid by an ester bond that formed feruloylated arabinoxylan, which exhibits antioxidant, anti-inflammatory, and anticancer properties. It was found that feruloylated arabinoxylan from cereal bran was incorporated into bread, which does not affect the microstructure or dough rheology, but improved the overall nutritional properties of the bread (Pietiäinen et al., 2024). In order to fulfill the demand of consumers and prevent from various chronic diseases, there is necessity to conduct more research on the extraction of arabinoxylan, FXOS, and feruloylated arabinoxylan from other millet by-products. Besides nutrition, other applications of arabinoxylan extracted from millet by-products such as packaging material, nanomaterial, food emulsifier, etc., need to be explored to widen its scope of applications (Bilal et al., 2024).

13.3.2.2 Cellulose

Cellulose, the most abundant polymer, is made up of D-glucopyranose units joined by 1,4-glycosidic linkage present in plant cell wall (Dominic et al., 2022). All millet by-products exhibit a substantial concentration of cellulose, classified as insoluble dietary fiber, and its amount varies greatly among different millet by-products. An average of around 30% of cellulose is known to be present in millets (Wang et al. 2023b). Among different millet by-products, pearl millet husk has the highest amount of cellulose, followed by foxtail millet husk, finger millet husk, barnyard millet husk, proso millet straw, etc. (Table 13.5).

Table 13.5 Percentage of cellulose, hemicellulose, and lignin in different millet by-products

Millet	By-product	Cellulose (%)	Hemicellulose (%)	Lignin (%)	References
Pearl millet	Husk	50.4–55.6	16.2–23.1	13.9–18.8	Abba et al. (2017) and Dominic et al. (2022)
Finger millet	Husk	38.01	31.48	16.25	Gairola et al. (2022)
	Straw	35.8	33.5	—	Jamaldheen et al. (2019)
Barnyard millet	Husk	34.5	32.16	21.3	Gairola et al. (2022)
Proso millet	Straw	36.68	17.31	19.38	Zhang et al. (2019)
Foxtail millet	Husk	38.44	33.71	20.84	Kiran and Prasanna (2023)
	Bran	17.30	47.83	8.63	Zhu et al. (2018)
Pearl millet	Straw	37.7	29.6	18.8	Vijayanand (2016)
Sorghum	Stalk	28.8–29.4	23.4–24.6	1.7–1.9	Luna et al. (2022)
	Straw	35.4–49.8	19.4–19.18	30.42	Saeed et al. (2017) and Alves et al. (2019)

Additionally, sorghum bran comprises a higher amount of hemicellulose than cellulose and lignin. The cellulose extracted using different approaches finds many applications in different forms like nanofibrils, nanocellulose, nanocrystals, etc., after the removal of noncellulosic parts (hemicellulose, lignin, and waxes) with the help of pretreatment, which can be transformed into a diverse array of products for both commercial and biomedical uses (Srinivasan & Venkatachalam, 2024). For example, cellulose nanofibers (65–70% w/w) were isolated from millet husk using a chlorine-free technique with a crystallinity index of 58.5%, which can be used as a prominent alternative for synthetic fibers for the manufacturing of green composites (Dominic et al., 2022).

13.3.3 Bioactive Peptides

Peptides are basically small strings of fragmented proteins (<10 kDa), which may present naturally or be synthesized from their native protein during hydrolysis (Yaghoubzadeh et al., 2025). Peptides are composed of 2–20 amino acid subunits that are joined through peptide bonds (Gupta et al., 2023). Peptides derived from millet by-products possess various therapeutic properties, which provide many health benefits to humans (Majid & Priyadarshini, 2020). For example, bioactive peptide derived from millet bran exhibits anti-inflammatory properties (He et al., 2022). Xu et al. (2022a) has prepared globulin hydrolysate from millet bran that comprises ferrous chelating activity, which can be utilized as an iron fortification ingredient in the food and pharmaceutical industry. Some researchers have found several bioactive fragments in the protein hydrolysate synthesized from foxtail millet bran, which exhibit anti-inflammatory activity, angiotensin-converting enzyme (ACE) inhibitory activity, and strong metal ion chelation properties. Further, the emulsifying properties of protein hydrolysate can be enhanced by appropriate enzymatic treatment. However, enzymatic method provides low protein hydrolysis efficiency and, as a result, low yield of peptides. To overcome these limitations, green techniques such as ultrasonication have been used for the synthesis of protein hydrolysate from foxtail millet bran, improving their foaming, solubility, emulsifying, and in vitro antioxidant activities by altering peptide bonds. It exhibits the potential for the development of highly active functional food derived from millet bran (Peng et al., 2024).

13.4 Value Addition of Millet By-products

The presence of numerous bioactive compounds in millet by-products makes it a potential source for the development of different value-added products applicable to both food and nonfood sectors.

13.4.1 Application of Millet By-products in Food Sector

13.4.1.1 Bakery and Confectionary

A substantial proportion of bakery products exhibited high calorie content while simultaneously lacking essential nutrients. Therefore, millet bran may serve as a viable component in the formulation of bakery products that are nutritionally dense. For example, Mustač et al. (2020) have developed gluten-free bread by incorporating 10% of proso millet bran, which enhanced the amount of dietary fiber and total phenolic content in bread. Similarly, proso millet bran was also used along with buckwheat hulls for the development of bioactive-rich cookies. Likewise, sorghum bran (14.2%) with gum (1%) and water (145%) were incorporated for the preparation of bread. It was found that the addition of these ingredients resulted in enhancement of dietary fiber and antioxidant activity by 13.4% and 61.6 µmol TE/g, respectively, in bread (Ardoin et al., 2023a). Voučko et al. (2024) utilized starter culture (*Limosilactobacillus fermentum* and *Kluyveromyces marxianus*) for sourdough fermentation by blending proso millet flour (21.7%), proso millet bran (52.5%), and carob flour (25.8%) to prepare gluten-free bread. Many desirable changes in bread, i.e., increase in volume (35%) and reduction in hardness (32%) as well as chewiness (28%), was observed. Bran from different cereals like wheat, millets, and oats were modified by different methods (sizing, autoclaving, and enzymatic treatment) for their application in bakery products. It has been found that 1 mm sized bran exhibits the lowest content of antinutritional factors, whereas cellulase-treated bran possesses high amount of phenolic acids, flavonoids, and antioxidant ability. The results concluded that bread prepared from modified bran provides high nutritional and sensory profile (Usman et al., 2024). The abovementioned studies indicated that millet bran not only enhanced the nutritional content in bakery products but also improved the physical properties. However, further investigation is required to evaluate the effect of millet bran addition in bakery products.

Radoš et al. (2022) found that biscuits containing by-products of proso millet and buckwheat possess a higher amount of dietary fiber, polyphenols, antioxidants, and minerals than whole-wheat flour-based biscuits. Other bakery products (rusk, muffins, and buns) were prepared by adding bran of barnyard, foxtail, and kodo millet. An increase in fiber, calcium, iron, zinc, and antioxidant capacity has been found in bran comprising muffins and buns (Barbhai et al., 2022). On the other hand, a decline in available carbohydrate content was observed in both muffins (46.76–43.87 g/100 g) and buns (51.11–47.85 g/100 g). The presence of antinutritional factors (phytic acid, trypsin inhibitor, lipases, etc.) in barnyard millet bran was decreased by treating with acetic acid, citric acid, and a combination of calcium oxide and water (Nazni & Karuna, 2016).

The proso millet flour and fermented dietary fiber from proso millet bran was incorporated into cake, and an increase in polyphenol as well as antioxidant activity has been observed (Xiao et al., 2023). Bread was formulated with the polyphenolic extract of sorghum bran that showed acrylamide inhibitory effect and increase in antioxidant activity (Chen et al., 2022). Besides millet bran, there is a need to conduct research on the usage of other millet by-products for the formulations of bakery products.

Nutrient-dense healthy bar was formulated by using foxnut and broken seeds of barnyard millet. It was observed that nutritional bar with 15% of each by-product had 29 different bioactive metabolites, major being the glycitein, palmitoyl serinol, persin, apigenin, carvone, and bufagargarizin as detected by high-resolution mass spectrometry. These compounds promote heart health and have potential therapeutic values, namely, antimicrobial, anti-inflammatory, antidiabetic, and anticancer properties, desirable for developing a functional food (Kumar et al., 2024).

13.4.1.2 Prebiotics

Prebiotics are nondigestible substances predominantly consisting of indigestible fibers, with particular emphasis on low molecular weight oligosaccharides (Buitron et al., 2018; Valladares-Diestra et al., 2023). Among oligosaccharides, xylo-oligosaccharides (XOs) are mixtures of sugar polymers that contain xylose units bound by β $(1 \rightarrow 4)$ xylosidic linkages. XOs exhibit a remarkable resistance to gastrointestinal enzymes and gastric acids due to their physicochemical characteristics, thereby enabling their passage through the upper gastrointestinal tract without undergoing digestion, until they enter the lower intestine and are metabolized by probiotic bacteria (*Lactobacillus* and *Bifidobacterium* genera). A diet enriched with XOs increases the metabolic rate of probiotic bacteria, resulting in the production of beneficial components (propionate, primary butyrate, and secondary butyrate). Despite this, the presence of ester-linked phenolics (ferulic, coumaric, and caffeic acids) in XOs provides antioxidant activity (Chen et al., 2021; Palaniappan et al., 2021). Xylan was successfully extracted from finger millet bran to produce XOs, which exhibit higher antioxidant activity than XOs produced from wheat, rice, and corn bran (Samanta et al., 2015). In addition, finger millet bran-based XOs can be used as an efficient source for accelerating the growth rate and cell mass of *Lactobacillus plantarum*, in contrast to marketable XOs and dextrose (Onipe & Ramashia, 2022). XOs exhibit the potential to reduce the chances of colon cancer, blood cholesterol level, and increase the absorption of calcium in the body (Santibáñez et al., 2021). Besides this, XOs prepared from millet bran can be used as a stabilizer in cosmetics, immune-stimulating agent, food, and pharmaceuticals ingredient (Valladares-Diestra et al., 2023).

13.4.1.3 Extruded Product

Extruded products are classified as ready-to-eat snacks, which are becoming popular among consumers due to their affordability, shelf-stability, appealing appearance, and availability in various sizes and shapes. A fiber-enriched extruded product with a good expansion ratio was successfully developed by using enzymatically treated sorghum bran along with sorghum flour. Extruded product having higher expansion ratio resulted in product superior in texture and sensory characteristics (lighter, crispier texture, and appealing appearance). In addition, enzymatic treatment improved the functional properties of the final extrudate, as compared to snacks containing untreated bran (Antwi et al., 2024).

13.4.1.4 Fermented Product

One of the most significant fermented and staple foods of China is Chinese steamed bread. To improve the resistance of consumers against inflammatory and anti-inflammatory diseases, there is a need to enhance functional properties of Chinese steamed bread. To tackle this, sorghum bran has been incorporated into steamed bread, which elevated the total phenolic content and antioxidant activity by 4.5–10 times. Further, the color and overall appearance of bran-fortified bread was maintained, thus scoring high sensory scores. A little effect on some attributes of bread like texture and mouth-feel was observed when sorghum bran was incorporated, which requires further investigation or standardization to achieve optimum concentration of by-product to be added to counter these issues (Zhang & Awika, 2024).

13.4.1.5 Beverages

Millet bran exhibits the ability to be utilized as a functional ingredient for the preparation of different types of beverages. Wine has been prepared by adding millet bran, which contains six times more polyphenols (gallic acid, syringic acid, and vanillic acid) and total amino acids than that found in wine devoid of bran. In addition, millet bran wine is well acceptable as a beverage based on sensory analysis (Guo et al., 2018). Acidified cold-brew was prepared by adding sorghum bran that contains 0.33 mg catechin equivalents per milliliter of tannins, phenols, and flavonoids (Ardoin et al., 2023b).

13.4.1.6 Essential Oils

Oil was extracted from foxtail millet bran using different techniques (Soxhlet method, SPE, and supercritical extraction methods), which contains 80% unsaturated fatty acid (linoleic and oleic acid) and 14% saturated fatty acids. In addition, it constitutes a substantial amount of sterols, tocols, squalene, and γ-oryzanol (Ji et al., 2019). The nutritional profile of foxtail millet bran-based oil makes it a high-grade oil, as it can be used for cooking purposes, medicine, cosmetics, healthcare, etc. The bran of other millets can also be used for the oil extraction, which provides a substitute for vegetable oil, and increases the value of agricultural products.

13.4.1.7 Enzymes

Most of the enzymes used in the food and biotechnology industries are not cost-effective due to their expensive raw material. Therefore, lignocellulosic rich by-products can be utilized as suitable substrates for the enzyme production (Panesar et al. 2016). For example, pearl millet residue and other agriculture by-products (finger millet residues, black gram peels, mustard oil cake, neem oil cake, green gram peels, grass pea peels, groundnut oil cake, wheat bran, broken rice, apple pomace, banana, and orange peels) were fermented by solid-state method for α-amylase production. Among which, pearl millet residue was considered as most suitable substrate for enzyme production (Sethi et al., 2016). Similarly, pesticide degrading laccase enzyme was produced by using pearl millet and finger millet husk (Srinivasan et al., 2019). The maximum activity of laccase (402 U/mL) was attained after four days with substrates of finger millet husk (1.5 g/L) and pearl millet husk (2.0 g/L). Xylanase has been synthesized from pearl millet stover, wheat bran, corn cob, rice husk, sugarcane bagasse, rice and wheat straw, etc., using *Aspergillus flavus* under solid-state fermentation. Among which, pearl millet stover provided maximum yield of enzyme (1345.44 IU/gds) due to better degradability and presence of substantial amount of carbon, nitrogen, as well as other nutrients (Gautam et al., 2015).

13.4.1.8 Food Additives

Mostly, synthetic food additives are used to enhance shelf life of several food products including fat spreads (mayonnaise), in which additives retard lipid oxidation. Therefore, polyphenolic extract (1.0 mg/g) of finger millet bran has been incorporated into mayonnaise that prevents oxidative rancidity more effectively for up to seven days at 4 °C as compared to synthetic antioxidants (Velur Balasubramaniam et al., 2022). Similar effect of extract has been observed when incorporated in peanut oil at two different concentrations (800 and 1000 ppm) that was effective in reducing the primary and secondary oxidation in oil for up to three months (Balasubramaniam et al., 2020).

13.4.1.9 Biodegradable Packaging

Millet bran or husk can be used as polysaccharide substrate for the development of sustainable and biodegradable packaging. Hammajam et al. (2019) have prepared film by mixing the combination of millet husk, polylactic acid, and water that results in improved mechanical strength and biodegradation properties with increase in amount of millet husk. The film developed through the incorporation of finger millet bran extract (1%) demonstrated superior mechanical properties, evidenced by a 65% increase in thickness. An improvement in film thickness enhanced the film opacity along with a decrease in light transmittance rate, water vapor permeability, and absorption rate. Furthermore, films synthesized from finger millet bran exhibited elevated levels of antimicrobial and antioxidant activities in comparison to those lacking bran extract (Xu et al., 2022b). Kodo millet bran proved as an efficient feedstock for the development of lactic acid to be used in green packaging after optimizing the conditions of acid and enzymatic hydrolysis (Balakrishnan et al., 2020).

13.4.2 Application of Millet By-products in Nonfood Industry

13.4.2.1 Antimicrobial Agents

To reduce the chlorophyll level in water caused by algae, silver nanoparticles comprising pearl millet husk were synthesized, which decreased around 41.3% of chlorophyll content (Musere et al., 2021). Similarly, silver nanoparticles containing foxtail millet husk exhibited significant antibacterial efficacy against both gram-negative and gram-positive bacterial strains (Wang et al., 2023a). Millet husk has the ability to develop an eco-friendly and cost-effective method that eliminates algae in wastewater regions experiencing water scarcity.

13.4.2.2 Bioenergy

The anaerobic digestion of agro-residues followed by biological pretreatment may be the most economical, eco-friendly, and energy-efficient method to produce methane as biogas (Kaur et al., 2022). An appreciable amount of biogas was produced using millet husk ($6.53\,dm^3$), when compared with cow dung slurry ($8.55\,dm^3$). In addition, millet husk possesses the ability to be used as an economical source for ethanol production due to the presence of an appropriate amount of reducing sugars. Around 79.80 ± 0.93 g/L volume of ethanol has been produced by millet husk after biomass hydrolysis using *Aspergillus niger* and fermentation by *Saccharomyces cerevisiae* (Gwandu et al., 2021). Likewise, guinea husk and millet husk were treated with acid and fermented (*Aspergillus niger* and *Zymomonas mobilis*) for bioethanol production. Biomass pellets have been synthesized from pearl millet cob, in combination with corn cob and eucalyptus sawdust, which can be used as an alternative for fossil fuel (Kumar et al., 2022).

13.4.2.3 Concrete Material

Across the globe, concrete is the fundamental structural unit prepared by Portland cement, which is used for the development of infrastructure. However, the production process of concrete is expensive as well as affects the ecological condition of the environment due to release of greenhouse gases (Kamau & Ahmed, 2017). In order to protect the environment, agro-residues can be used as alternative cementing material used in concrete. The incineration of millet husk reveals cementitious attributes similar to those found in conventional cement, which makes millet husk ash as a potential binding ingredient for preparation of concrete. It has been found that 10% replacement of cement with millet husk ash provides better compressive and tensile strength while reducing the water absorption and drying shrinkage (Bheel et al., 2023). The concrete was prepared by mixing sorghum husk ash (20%) and palm kernel shell (40%), which enhanced compressive strength, permeability, and porosity. In addition, the prepared mixture decreases the embodied energy, embodied carbon, and cost by 20%, 23%, and 24%, respectively (Tijani et al., 2022).

13.4.2.4 Miscellaneous

Different color cultivars of sorghum husk have the potential to be used as an alternative to natural dye in textile industry, which also provides antimicrobial properties, deodorizing effect, and protection against UV (Wizi et al., 2018). The stalk core of pearl millet majorly contains cellulose that provides a high absorption rate, due to which it can be used as cellulosic fluff for the manufacturing of hygiene products (Yadav et al., 2019). Millet straw can be used as an efficient source for mushroom cultivation as it provides a yield of 20 kg/m^2, which exhibits a comparable outcome to that of wheat straw (Zhang et al., 2019). Besides this, an adsorbent was developed from millet straw, which serves as a proficient method for the remediation of wastewater tainted with methylene blue and ciprofloxacin (Akpotu et al., 2024). Porous nanocomposites were developed by coating of hydrated ferric oxide on the surface of sorghum stalks, which have the ability to reduce copper levels from 5 mg/L to nearly 0 mg/L in simulated wastewater. In addition, sorghum stalk-based adsorbent exhibited great mechanical, chemical, and thermal stability due to the coating of hydrated ferric oxide (Song et al., 2024). Sorghum husk was also used to develop the magnetic-tuned biosorbent by coprecipitation of Fe_3O_4, which eliminates crystal violet and methylene blue dye from aqueous solution. After the optimization of parameters, around 95% of the dyes can be effectively eliminated at an adsorbent dosage of 1.0 g at 50 °C and pH 4.05 in the presence of an initial dye concentration of 50 mg/L (Adeogun et al., 2019). The phenolic extract of different varieties of sorghum bran (Sumac and Burgundy) and hard red winter wheat bran was used to synthesize gold-coated nanoparticles. Around 2.5, 1.8, and 0.3 mg GAE/g of total phenolic content of sorghum-Sumac bran, sorghum-Burgundy bran, and wheat bran, respectively, was determined in the gold-coated nanoparticles, which might possess the ability to fight against colon cancer cells (Norton et al., 2024).

13.5 Conclusions and Future Perspectives

Different stages of millet processing generate various millet by-products, which are improperly discarded or remain underutilized. These by-products are excellent sources of bioactive compounds including phenolic acids, flavonoids, arabinoxylan, bioactive peptides, etc., which have the potential to prevent numerous chronic diseases. Mostly, bran has been utilized in different applications including the development of bread, cake, cookies, wine, essential oil, packaging material, and additives. On the other hand, husk has emerged as a suitable material for the development of nonedible products such as enzymes, packaging material, and cementing material. However, most of the research findings and applications are confined to the laboratory or small scale only. Therefore, efforts are needed to scale up the processing and utilization of millet by-products in both food and non-food sectors. Extensive research is required to examine the bioactive composition of millet by-products and their extraction strategies to harness their rich phytochemical profile. In this view, this chapter highlights the importance of millet by-products with the aim to broaden their scope of applications and develop new cost-effective strategies to mitigate adverse environmental impact and fulfill the nutritional requirements of mankind from an industrial viewpoint.

References

Abba, H. A., Zahari, I. N., Sapuan, S. M., & Leman, Z. (2017). Characterization of millet (*Pennisetum glaucum*) husk fiber (MHF) and its use as filler for high density polyethylene (HDPE) composites. *BioResources, 12*, 9287–9301. https://doi.org/10.15376/biores.12.4.9287-9301

Adebo, J. A., & Kesa, H. (2023). Evaluation of nutritional and functional properties of anatomical parts of two sorghum (*Sorghum bicolor*) varieties. *Heliyon, 9*(6), e17296. https://doi.org/10.1016/j.heliyon.2023.e17296

Adeogun, A. I., Akande, J. A., Idowu, M. A., & Kareem, S. O. (2019). Magnetic tuned sorghum husk biosorbent for effective removal of cationic dyes from aqueous solution: Isotherm, kinetics, thermodynamics and optimization studies. *Applied Water Science, 9*, 1–17. https://doi.org/10.1007/s13201-019-1037-2

Aggarwal, P. R., Muthamilarasan, M., & Choudhary, P. (2024). Millet as a promising C4 model crop for sustainable biofuel production. *Journal of Biotechnology, 395*, 110–121. https://doi.org/10.1016/j.jbiotec.2024.09.019

Ahmad, F., Pasha, I., Saeed, M., & Asgher, M. (2018). Biochemical profiling of Pakistani sorghum and millet varieties with special reference to anthocyanins and condensed tannins. *International Journal of Food Properties, 21*(1), 1586–1597. https://doi.org/10.1080/10942912.2018.1502198

Akanbi, T. O., Timilsena, Y., & Dhital, S. (2019). Bioactives from millet: Properties and effects of processing on bioavailability. In S. Baoguo, J. Wang, & R. Tsao (Eds.), *Bioactive factors and processing technology for cereal foods* (pp. 171–183). Springer Singapore. https://doi.org/10.1007/978-981-13-6167-8_10

Akpotu, S. O., Diagboya, P. N., Osabohien, E., Ajibola, B. A., Nelana, S. S., & Mtunzi, F. (2024). Beyond the boundaries of organics biosorption: Reusable and cost-effective millet-straw-based carboxylate-functionalized mesoporous MCM-41. *Desalination and Water Treatment, 320*, 100805. https://doi.org/10.1016/j.dwt.2024.100805

Ali, M. E., Yusuf, Z. A., Inah, N. U., Yusuf, S. K., Idris, H. A., Salisu, A. S., & Ishaku, A. H. (2024). Effect of solid-state fermentation on phytochemical and proximate composition of millet bran meal. *African Journal of Agricultural Science and Food Research, 15*(1), 1–8.

Alves, J. A. A., Dos Santos, M. D. L., Morais, C. C., Ascheri, J. L. R., Signini, R., Dos Santos, D. M., & Ascheri, D. P. R. (2019). Sorghum straw: Pulping and bleaching process optimization and synthesis of cellulose acetate. *International Journal of Biological Macromolecules, 135*, 877–886. https://doi.org/10.1016/j.ij.iomac.2019.05.014

Antwi, C. K., Rosa-Sibakov, N., & Emmambux, M. N. (2024). Impact of added enzyme-treated bran on the techno-functional properties of puffed-extruded sorghum snack. *Journal of Cereal Science, 120*, 104051. https://doi.org/10.1016/j.jcs.2024.104051

Ardoin, R., Smith, B., Bean, S., & Aramouni, F. (2023a). Optimization of tannin-containing sorghum bran addition to gluten-free bread. *Journal of Food Science, 88*(3), 952–961. https://doi.org/10.1111/1750-3841.16477

Ardoin, R., Smith, B., Lea, J., Boue, S., Smolensky, D., Santana, A. L., & Peterson, J. (2023b). Consumer perceptions and antioxidant profiling of acidified cold-brewed sorghum bran beverages. *Journal of Food Science, 88*(6), 2301–2312. https://doi.org/10.1111/1750-3841.16589

Balakrishnan, R., Tadi, S. R. R., Pavan, A. S. S., Sivaprakasam, S., & Rajaram, S. (2020). Effect of nitrogen sources and neutralizing agents on D-lactic acid production from kodo millet bran hydrolysate: Comparative study and kinetic analysis. *Journal of Food Science and Technology, 57*, 915–926. https://doi.org/10.1016/j.indcrop.2017.11.041

Balasubramaniam, V. G., Ayyappan, P., Sathvika, S., & Antony, U. (2019). Effect of enzyme pretreatment in the ultrasound assisted extraction of finger millet polyphenols. *Journal of Food Science and Technology, 56*, 1583–1594. https://doi.org/10.1007/s13197-019-03672-2

Balasubramaniam, V. G., Sathvika, S., Ayyappan, P., & Antony, U. (2020). Improved oxidative stability of peanut oil through addition of finger millet (*Eleusine coracana*) seed coat polyphenols. *Journal of Food Process Engineering, 43*(3), e13194. https://doi.org/10.1111/jfpe.13194

Barbhai, M. D., Hymavathi, T. V., Kuna, A., Mulinti, S., & Voliveru, S. R. (2022). Quality assessment of nutri-cereal bran rich fraction enriched buns and muffins. *Journal of Food Science and Technology, 59*(6), 2231–2242. https://doi.org/10.1007/s13197-021-05236-9

Bheel, N., Ali, M. O. A., Shafiq, N., Almujibah, H. R., Awoyera, P., Benjeddou, O., & Olalusi, O. B. (2023, March). Utilization of millet husk ash as a supplementary cementitious material in eco-friendly concrete: RSM modelling and optimization. *Structures, 49*, 826–841. https://doi.org/10.1016/j.istruc.2023.02.015

Bilal, M., Li, D., Xie, C., Yang, R., Gu, Z., Jiang, D., & Wang, P. (2024). Valorization of wheat bran arabinoxylan: A review on nutritional and materials perspectives. *Grain & Oil Science and Technology, 7*, 196–208. https://doi.org/10.1016/j.gaost.2024.05.006

Buitron, D. I., Sepulveda, L., Aguilar, C. N., Rodriguez-Herrera, R., & Flores-Gallegos, A. C. (2018). Biotechnological approach for the production of prebiotics and search for new probiotics and their application in the food industry. *Applied Food Biotechnology, 5*(4), 185–192. https://doi.org/10.22037/afb.v4i3.15014

Cao LongKui, C. L., Kang Li Jun, K. L., Kou Fang, K. F., Shen Meng, S. M., Ge YunFei, G. Y., & Wang WeiHao, W. W. (2018). Structural analysis and in vitro inhibitory effect on α-glucosidase activity of millet bran dietary fiber before and after modification.

Cao, R. A., Palanisamy, S., Ma, N., Talapphet, N., Zhang, J., Wang, C., & You, S. (2021). Extraction, structural characterization, and immunostimulatory activity of soluble non-starch polysaccharides of finger millet. *Process Biochemistry, 111*, 40–50. https://doi.org/10.1016/j.procbio.2021.08.011

Chen, C., Yin, J., Jing, P., & Jiao, S. (2024). Effects of hot air-assisted radio frequency stabilization of brown millet and millet bran: Enzyme inactivation, microstructure, nutritional quality, and storage stability. *Innovative Food Science & Emerging Technologies, 91*, 103529. https://doi.org/10.1016/j.ifset.2023.103529

Chen, G., Wang, J., & Li, Y. (2022). Extracts of sorghum bran, grape seed, and green tea: Chromatographic comparison of phenolic profiles and mitigation effect on acrylamide in antioxidant-fortified bread. *Food Chemistry Advances, 1*, 100082. https://doi.org/10.1016/j.focha.2022.100082

Chen, Y., Xie, Y., Ajuwon, K. M., Zhong, R., Li, T., Chen, L., & Everaert, N. (2021). Xylo-oligosaccharides, preparation and application to human and animal health: a review. *Frontiers in Nutrition, 8*, 731930. https://doi.org/10.3389/fnut.2021.731930

Cheng, W., Sun, Y., Fan, M., Li, Y., Wang, L., & Qian, H. (2022). Wheat bran, as the resource of dietary fiber: A review. *Critical Reviews in Food Science and Nutrition, 62*(26), 7269–7281. https://doi.org/10.1016/j.fochx.2024.101541

Chu, J., Zhao, H., Lu, Z., Lu, F., Bie, X., & Zhang, C. (2019). Improved physicochemical and functional properties of dietary fiber from millet bran fermented by *Bacillus natto*. *Food Chemistry, 294*, 79–86.

Dega, V., & Barbhai, M. D. (2023). Exploring the underutilized novel foods and starches for formulation of low glycemic therapeutic foods: A review. *Frontiers in Nutrition, 10*, 1162462. https://doi.org/10.3389/fnut.2023.1162462

Dominic, C. M., Raj, V., Neenu, K. V., Begum, P. S., Formela, K., Saeb, M. R., & Parameswaranpillai, J. (2022). Chlorine-free extraction and structural characterization of cellulose nanofibers from waste husk of millet (*Pennisetum glaucum*). *International Journal of Biological Macromolecules, 206*, 92–104. https://doi.org/10.1016/j.ij.iomac.2022.02.078

Dong, J. L., Wang, L., Lü, J., Zhu, Y. Y., & Shen, R. L. (2019). Structural, antioxidant and adsorption properties of dietary fiber from foxtail millet (*Setaria italica*) bran. *Journal of the Science of Food and Agriculture, 99*(8), 3886–3894. https://doi.org/10.1002/jsfa.9611

Duke, K., Syeunda, C., Brantsen, J. F., Nindawat, S., & Awika, J. M. (2024). Polyphenol recovery from sorghum bran waste by microwave assisted extraction: Structural transformations as affected by grain phenolic profile. *Food Chemistry, 444*, 138645. https://doi.org/10.1016/j.foodchem.2024.138645

Gairola, S., Sinha, S., & Singh, I. (2022). Novel millet husk crop-residue based thermoplastic composites: Waste to value creation. *Industrial Crops and Products, 182*, 114891. https://doi.org/10.1016/j.indcrop.2022.114891

Gautam, A., Kumar, A., & Dutt, D. (2015). Production of cellulase-free xylanase by *Aspergillus flavus* ARC-12 using pearl millet stover as the substrate under solid-state fermentation. *Journal of Advanced Enzymes Research, 1*, 1–9.

Guo, X., Sha, X., Rahman, E., Wang, Y., Ji, B., Wu, W., & Zhou, F. (2018). Antioxidant capacity and amino acid profile of millet bran wine and the synergistic interaction between major polyphenols. *Journal of Food Science and Technology, 55*, 1010–1020. https://doi.org/10.1007/s13197-017-3014-9

Gupta, M., Asfaha, D. M., & Ponnaiah, G. (2023). Millets: A nutritional powerhouse with anti-cancer potential. *Cureus, 15*(10), e47769. https://doi.org/10.7759/cureus.47769

Gwandu, A. Z., Farouq, A. A., Baki, A. S., & Peni, D. N. (2021). Production of bioethanol from rice and millet husks. *International Journal of Advanced Academic Research, 7*(1), 1–11. https://doi.org/10.46654/ij.24889849.e717

Hammajam, A. A., El-Jummah, A. M., & Ismarrubie, Z. N. (2019). The green composites: Millet husk fiber (MHF) filled poly lactic acid (PLA) and degradability effects on environment. *Open Journal of Composite Materials*, *9*(3), 300–311. https://doi.org/10.4236/ojcm.2019.93018

He, R., Liu, M., Zou, Z., Wang, M., Wang, Z., Ju, X., & Hao, G. (2022). Anti-inflammatory activity of peptides derived from millet bran in vitro and in vivo. *Food & Function*, *13*(4), 1881–1889. 10.1039/D1FO03711K

Jamaldheen, S. B., Thakur, A., Moholkar, V. S., & Goyal, A. (2019). Enzymatic hydrolysis of hemicellulose from pretreated Finger millet (*Eleusine coracana*) straw by recombinant endo-1, 4-β-xylanase and exo-1, 4-β-xylosidase. *International Journal of Biological Macromolecules*, *135*, 1098–1106. https://doi.org/10.1016/j.ij.iomac.2019.06.010

Ji, J., Liu, Y., Ge, Z., Zhang, Y., & Wang, X. (2019). Oleochemical properties for different fractions of foxtail millet bran. *Journal of Oleo Science*, *68*(8), 709–718. https://doi.org/10.5650/jos.ess19063

Kalsi, R., & Bhasin, J. K. (2023). Nutritional exploration of foxtail millet (*Setaria italica*) in addressing food security and its utilization trends in food system. *eFood*, *4*(5), e111. https://doi.org/10.1002/efd2.111

Kamau, J., & Ahmed, A. (2017). Suitability of maize cob ash as a partial cement replacement. *Juniper Online Journal Material Science*, *2*(5), 001–003.

Kaur, B., Panesar, P. S., Anal, A. K., & Chu-Ky, S. (2022). Recent trends in the management of mango by-products. *Food Reviews International*, *39*, 4159–4179.

Kaur, R., Goyal, N., Panesar, G., & Panesar, P. S. (2025). Unraveling the nutritional potential of millet by-products through extraction of high value compounds for the development of novel food products. *Food Chemistry*, *472*, 142983. https://doi.org/10.1016/j.foodchem.2025.142983

Kaur, R., Panesar, P. S., Kaur, B., & Riar, C. S. (2024). Hydrothermal extraction of dietary fiber from pearl millet bran: Optimization, physico-chemical, structural and functional characterization. *Journal of Food Science and Technology*, *61*, 1536–1546. https://doi.org/10.1007/s13197-023-05921-x

Kaur, R., Panesar, P. S., & Riar, C. S. (2023). Green extraction of dietary fiber concentrate from pearl millet bran and evaluation of its microstructural and functional properties. *Biomass Conversion and Biorefinery*, *14*, 30269–30278. https://doi.org/10.1007/s13399-023-04627-6

Kiran, K. C., & Prasanna, K. T. (2023). Foxtail millet (*Setaria italica*) husk: A potential source for bioethanol production. *The Pharma Innovation Journal*, *12*, 560–567.

Kuhe, A., Terhemba, A. V., & Iortyer, H. (2021). Biomass valorization for energy applications: A preliminary study on millet husk. *Heliyon*, *7*(8), e07802. https://doi.org/10.1016/j.heliyon.2021.e07802

Kumar, P., Subbarao, P. M. V., Kala, L., & Vijay, V. K. (2022). Influence of physical, mechanical, and thermal properties of biomass pellets from agriculture residue: Pearl millet cob and mix. *Bioresource Technology Reports*, *20*, 101278. https://doi.org/10.1016/j.biteb.2022.101278

Kumar, V., Kumar, A., Singh, M. K., Dhyani, P., Mishra, H., & Rai, D. C. (2024). Bioactive metabolites identification of the foxnut and broken millet-based nutritional bar using HR-MS. *Food Chemistry: Molecular Sciences*, *9*, 100214. https://doi.org/10.1016/j.fochms.2024.100214

Kumari, N., Vinita, N. K., & Rani, P. (2018). Nutrient composition of full fat and defatted rice bran. *Asian Journal of Dairy and Food Research*, *37*(1), 77–80. https://doi.org/10.18805/ajdfr.DR-1277

Li, M., Chang, L., Ren, J., Jiang, F., Zhao, N., Liu, Y., & Du, S. K. (2022a). Nutritional, physical, functional properties and antioxidant potential of different colors proso millet husks and brans. *LWT*, *171*, 114092. https://doi.org/10.1016/j.lwt.2022.114092

Li, M., Chang, L., Ren, J., Jiang, F., Zhao, N., Liu, Y., & Du, S. K. (2022b). Biomass valorization for energy applications: A preliminary study on millet husk. *LWT*, *171*, 114092. https://doi.org/10.1016/j.lwt.2022.114092

Li, P., Cai, X., Li, S., Zhao, W., Liu, J., Zhang, X., & Liu, J. (2024a). Nutrient and metabolite characteristics of the husk, bran and millet isolated from the foxtail millet (*Setaria italica* L.) during polishing. *Food Chemistry: X*, *23*, 101541. https://doi.org/10.1016/j.fochx.2024.101541

Li, Y., Feng, C., Wang, X., Zheng, Y., Song, X., Wang, N., & Liu, D. (2024b). Millet bran dietary fibers modified by heating and enzymolysis combined with carboxymethylation, acetylation, or crosslinking: Influences on properties of heat-induced egg white protein gel. *Food*, *13*(17). https://doi.org/10.3390/foods13172827

Li, Y., Li, M., Wang, L., & Li, Z. (2022c). Effect of particle size on the release behavior and functional properties of wheat bran phenolic compounds during in vitro gastrointestinal digestion. *Food Chemistry*, *367*, 130751. https://doi.org/10.1016/j.foodchem.2021.130751

Liang, K., Liang, S., & Zhu, H. (2024). The physicochemical characteristics and phenolic bioaccessibility of defatted millet bran powder prepared using superfine grinding. *LWT*, *201*, 116173. https://doi.org/10.1016/j.lwt.2024.116173

Liu, M., Xu, Y., He, J., Zhang, S., Wang, Y., & Lu, P. (2016). Genetic diversity and population structure of broomcorn millet (*Panicum miliaceum* L.) cultivars and landraces in China based on microsatellite markers. *International Journal of Molecular Sciences*, *17*(3), 370. https://doi.org/10.3390/ijms17030370

Luna, P., Charalampopoulos, D., & Chatzifragkou, A. (2022). Fractionation of carbohydrate polymers from Indonesian sorghum by-products. *Food and Bioproducts Processing*, *135*, 114–122. https://doi.org/10.1016/j.fbp.2022.07.007

Majid, A., & Priyadarshini, C. G. (2020). Millet derived bioactive peptides: A review on their functional properties and health benefits. *Critical Reviews in Food Science and Nutrition*, *60*(19), 3342–3351. https://doi.org/10.1080/10408398.2019.1686342

Majumder, R., Miatur, S., Roy, S., Saha, A., Manik, S., Biswas, N., & Hossain, S. (2024). Millet starch-based film: A Review. *Trends in Food Science & Technology*, *151*, 104618. https://doi.org/10.1016/j.tifs.2024.104618

Musere, P. S., Rahman, A., Uahengo, V., Naimhwaka, J., Daniel, L., Bhaskurani, S. V., & Jonnalagadda, S. B. (2021). Biosynthesis of silver nanoparticles using pearl millet (*Pennisetum glaucum*) husk to remove algae in the water and catalytic oxidation of benzyl alcohol. *Journal of Cleaner Production*, *312*, 127581. https://doi.org/10.1016/j.jclepro.2021.127581

Mustač, N. Č., Novotni, D., Habuš, M., Drakula, S., Nanjara, L., Voučko, B., & Ćurić, D. (2020). Storage stability, micronisation, and application of nutrient-dense fraction of proso millet bran in gluten-free bread. *Journal of Cereal Science*, *91*, 102864. https://doi.org/10.1016/j.jcs.2019.102864

Mustač, N. Č., Voučko, B., Novotni, D., Drakula, S., Gudelj, A., Dujmić, F., & Ćurić, D. (2019). Optimization of high intensity ultrasound treatment of proso millet bran to improve physical and nutritional quality. *Food Technology and Biotechnology*, *57*(2), 183–190. https://doi.org/10.17113/ftb.57.02.19.6100

Nayak, P., Gupta, A. K., Jain, P., & Pandey, S. (2019). Effect of moisture and machine parameters on de-husking efficiency of kodo millet. *International Journal of Current Microbiology and Applied Sciences*, *8*(2), 1792–1804. https://doi.org/10.20546/ijcmas.2019.802.211

Nazni, P., & Karuna, T. D. (2016). Development and quality evaluation of barnyard millet bran incorporated rusk and muffin. *Journal of Food and Industrial Microbiology*, *2*(116), 2. https://doi.org/10.4172/2572-4134.1000116

Norton, A. E., Kim, M. J., Peiris, K. H., Cox, S., Tilley, M., Smolensky, D., & Bean, S. R. (2024). Synthesis and characterization of hybrid gold-coated cereal particles from sorghum bran flour and wheat bran flour. *Cereal Chemistry.*, *102*, 23–33. https://doi.org/10.1002/cche.10840

Onipe, O. O., & Ramashia, S. E. (2022). Finger millet seed coat—A functional nutrient-rich cereal by-product. *Molecules*, *27*(22), 7837. https://doi.org/10.3390/molecules27227837

Palaniappan, A., Antony, U., & Emmambux, M. N. (2021). Current status of xylooligosaccharides: Production, characterization, health benefits and food application. *Trends in Food Science & Technology*, *111*, 506–519. https://doi.org/10.1016/j.tifs.2021.02.047

Palaniappan, A., Yuvaraj, S. S., Sonaimuthu, S., & Antony, U. (2017). Characterization of xylan from rice bran and finger millet seed coat for functional food applications. *Journal of Cereal Science*, *75*, 296–305. https://doi.org/10.1016/j.jcs.2017.03.032

Panesar, P. S., Kaur, R., Singla, G., & Sangwan, R. S. (2016). Bio-processing of agro-industrial wastes for production of food-grade enzymes: Progress and prospects. *Applied Food Biotechnology*, *3*(4), 208–227. https://doi.org/10.22037/afb.v3i4.13458

Paschapur, A. U., Joshi, D., Mishra, K. K., Kant, L., Kumar, V., & Kumar, A. (2021). Millets for life: A brief introduction. In A. Kumar, M. K. Tripathi, D. Joshi, & V. Kumar (Eds.), *Millets and millet technology* (pp. 1–32). Springer Singapore. https://doi.org/10.1007/978-981-16-0676-2_1

Peng, Z., Wang, F., Yu, L., Jiang, B., Cao, J., Sun, Z., & Cheng, J. (2024). Effect of ultrasound on the characterization and peptidomics of foxtail millet bran protein hydrolysates. *Ultrasonics Sonochemistry*, *110*, 107044. https://doi.org/10.1016/j.ultsonch.2024.107044

Pietiäinen, S., Lee, Y., Jimenez-Quero, A., Katina, K., Maina, N. H., Hansson, H., & Langton, M. (2024). Feruloylation and hydrolysis of arabinoxylan extracted from wheat bran: Effect on dough rheology and microstructure. *Food*, *13*(15), 2309. https://doi.org/10.3390/foods13152309

Pradeep, P. M., & Sreerama, Y. N. (2017). Soluble and bound phenolics of two different millet genera and their milled fractions: Comparative evaluation of antioxidant properties and inhibitory effects on starch hydrolysing enzyme activities. *Journal of Functional Foods*, *35*, 682–693. https://doi.org/10.1016/j.jff.2017.06.033

Priya, Verma, R. K., Lakhawat, S., Yadav, V. K., Gacem, A., Abbas, M., & Mishra, S. (2023). Millets: Sustainable treasure house of bioactive components. *International Journal of Food Properties*, *26*(1), 1822–1840. https://doi.org/10.1080/10942912.2023.2236317

Qiu, S., Yadav, M. P., & Yin, L. (2017). Characterization and functionalities study of hemicellulose and cellulose components isolated from sorghum bran, bagasse and biomass. *Food Chemistry*, *230*, 225–233. https://doi.org/10.1016/j.foodchem.2017.03.028

Radoš, K., Čukelj Mustač, N., Benković, M., Kuzmić, I., Novotni, D., Drakula, S., & Ćurić, D. (2022). The quality and shelf life of biscuits with cryo-ground proso millet and buckwheat by-products. *Journal of Food Processing and Preservation*, *46*(10), e15532. https://doi.org/10.1111/jfpp.15532

Rani, S., Singh, R., Sehrawat, R., Kaur, B. P., & Upadhyay, A. (2018). Pearl millet processing: A review. *Nutrition & Food Science*, *48*(1), 30–44. https://doi.org/10.1108/NFS-04-2017-0070

Ren, A., Chen, L., Zhao, W., Shan, S., Li, Z., & Tang, Z. (2023). Extraction optimization and structural characterization of soluble dietary fiber in foxtail millet and its inhibitory activities on colon cancer. *Journal of Functional Foods*, *107*, 105659. https://doi.org/10.1016/j.jff.2023.105659

Rudrapal, M., Khairnar, S. J., Khan, J., Dukhyil, A. B., Ansari, M. A., Alomary, M. N., & Devi, R. (2022). Dietary polyphenols and their role in oxidative stress-induced human diseases: Insights into protective effects, antioxidant potentials and mechanism(s) of action. *Frontiers in Pharmacology*, *13*, 806470. https://doi.org/10.3389/fphar.2022.806470

Sachdev, N., Goomer, S., Singh, L. R., Pathak, V. M., Aggarwal, D., & Chowhan, R. K. (2023). Current status of millet seed proteins and its applications: A comprehensive review. *Applied Food Research*, *3*(1), 100288. https://doi.org/10.1016/j.afres.2023.100288

Saeed, H. A., Liu, Y., Lucia, L. A., & Chen, H. (2017). Evaluation of Sudanese sorghum and bagasse as a pulp and paper feedstock. *Bioresources*, *12*(3), 5212–5222. https://doi.org/10.15376/biores.12.3.5212-5222

Saini, S., Saxena, S., Samtiya, M., Puniya, M., & Dhewa, T. (2021). Potential of underutilized millets as nutri-cereal: An overview. *Journal of Food Science and Technology*, *58*, 4465–4477. https://doi.org/10.1007/s13197-021-04985-x

Samanta, A. K., Jayapal, N., Jayaram, C., Roy, S., Kolte, A. P., Senani, S., & Sridhar, M. (2015). Xylooligosaccharides as prebiotics from agricultural by-products: Production and applications. *Bioactive Carbohydrates and Dietary Fibre*, *5*(1), 62–71. https://doi.org/10.1016/j.bcdf.2014.12.003

Santibáñez, L., Henríquez, C., Corro-Tejeda, R., Bernal, S., Armijo, B., & Salazar, O. (2021). Xylooligosaccharides from lignocellulosic biomass: A comprehensive review. *Carbohydrate Polymers*, *251*, 117118. https://doi.org/10.1016/j.carbpol.2020.117118

Sawadogo, O., Kaboré, D. S., Ouedraogo, M., Noba, A., Bationo, R. K., Diawara, H. Z., & Palé, E. (2024). Antioxidant activity and identification by HPLC-DAD-MS of 3-deoxyanthocyanins in the husks of *Sorghum caudatum* (L.) Moench grains from Burkina Faso. *Pharmacological Research-Natural Products*, *5*, 100096. https://doi.org/10.1016/j.prenap.2024.100096

Sethi, B. K., Jana, A., Nanda, P. K., DasMohapatra, P. K., Sahoo, S. L., & Patra, J. K. (2016). Production of α-amylase by *Aspergillus terreus* NCFT 4269.10 using pearl millet and its structural characterization. *Frontiers in Plant Science*, *7*, 639. https://doi.org/10.3389/fpls.2016.00639

Shi, Y., Ma, Y., Zhang, R., Ma, H., & Liu, B. (2015). Preparation and characterization of foxtail millet bran oil using subcritical propane and supercritical carbon dioxide extraction. *Journal of Food Science and Technology*, *52*, 3099–3104. https://doi.org/10.1007/s13197-014-1311-0

Shivashankar, S., Shahi, N. C., Rao, B. D., Suresh, K., Lohani, U. C., Niranjan, T., & Mahanti, N. K. (2024). Development of polisher and process optimization for polishing of foxtail millet. *Journal of Food Processing and Preservation*, *2024*(1), 5202552. https://doi.org/10.1155/2024/5202552

Shukla, S., Lohani, U. C., Shahi, N. C., & Dubey, A. (2022). Extraction of natural pigments from red sorghum (*Sorghum bicolor*) husk by ultrasound and microwave assisted extraction: A comparative study through response surface analysis. *Journal of Food Process Engineering, 45*(10), e14130. https://doi.org/10.1111/jfpe.14130

Siddiqui, H., Sultan, Z., Yousuf, O., Malik, M., & Younis, K. (2023). A review of the health benefits, functional properties, and ultrasound-assisted dietary fiber extraction. *Bioactive Carbohydrates and Dietary Fibre, 30*, 100356. https://doi.org/10.1016/j.bcdf.2023.100356

Singh, A., & Eligar, S. M. (2021). Bioactive feruloylated xylooligosaccharides derived from pearl millet (*Pennisetum glaucum*) bran with antiglycation and antioxidant properties. *Journal of Food Measurement and Characterization, 15*, 5695–5706. https://doi.org/10.1007/s11694-021-01139-7

Singh, A., Rajoriya, D., Obalesh, I. S., Prashantha, K. H., Chaudhari, S. R., Mutturi, S., & Eligar, S. M. (2024b). Arabinoxylan from pearl millet bran: Optimized extraction, structural characterization, and its bioactivities. *International Journal of Biological Macromolecules, 279*, 135247. https://doi.org/10.1016/j.ij.iomac.2024.135247

Singh, S. M., Joshi, J., & Rao, P. S. (2024a). Technological advancements in millet dehulling and polishing process: An insight into pretreatment methods, machineries and impact on nutritional quality. *Grain & Oil Science and Technology, 7*(3), 186–195. https://doi.org/10.1016/j.gaost.2024.05.007

Song, W., Chen, Y., Zhang, L., Yin, D., Ding, D., & Ruan, H. (2024). Sheet nanocomposites of sorghum stalk loaded with hydrated ferric oxide for Cu(II) deep removal from water. *Journal of Environmental Chemical Engineering, 12*, 114415. https://doi.org/10.1016/j.jece.2024.114415

Srinivas, A. (2022). Millet milling technologies. In C. Anandharamakrishnan, A. Rawson, & C. K. Sunil (Eds.), *Handbook of millets-processing, quality, and nutrition status* (pp. 173–203). Springer Nature Singapore. https://doi.org/10.1007/978-981-16-7224-8_9

Srinivasan, P., Selvankumar, T., Kamala-Kannan, S., Mythili, R., Sengottaiyan, A., Govarthanan, M., & Selvam, K. (2019). Production and purification of laccase by *Bacillus* sp. using millet husks and its pesticide degradation application. *3 Biotech, 9*(11), 396. https://doi.org/10.1007/s13205-019-1900-8

Srinivasan, S., & Venkatachalam, S. (2024). One pot green process for facile fractionation of sorghum biomass to lignin, cellulose and hemicellulose nanoparticles using deep eutectic solvent. *International Journal of Biological Macromolecules, 277*, 134295. https://doi.org/10.1016/j.ij.iomac.2024.134295

Sunil, C. K., Gowda, N. N., Nayak, N., & Rawson, A. (2024). Unveiling the effect of processing on bioactive compounds in millets: Implications for health benefits and risks. *Process Biochemistry, 138*, 79–96. https://doi.org/10.1016/j.procbio.2024.01.010

Suri, S., Grover, A., Nagpal, K., & Patil, A. G. (2024). Millet processing by-products: Sustainable applications in food and non-food sectors. *Nutrire, 49*(2), 32. https://doi.org/10.1186/s41110-024-00274-7

Tagade, A., & Sawarkar, A. N. (2023). Valorization of millet agro-residues for bioenergy production through pyrolysis: Recent inroads, technological bottlenecks, possible remedies, and future directions. *Bioresource Technology, 384*, 129335. https://doi.org/10.1016/j.biortech.2023.129335

Tijani, M. A., Ajagbe, W. O., & Agbede, O. A. (2022). Recycling sorghum husk and palm kernel shell wastes for pervious concrete production. *Journal of Cleaner Production, 380*, 134976. https://doi.org/10.1016/j.jclepro.2022.134976

Tuhanioglu, A., & Ubeyitogullari, A. (2022). Extraction of high-value lipids and phenolic compounds from sorghum bran via a sequential supercritical carbon dioxide approach. *ACS Food Science & Technology*, *2*(12), 1879–1887. https://doi.org/10.1021/acsfoodscitech.2c00266

Tyagi, V., Saravanan, C., Wang, Y., & Bhattacharya, B. (2021). Solvent dependency of sorghum bran phytochemicals acting as potential antioxidants and antibacterial agents. *Food Technology and Biotechnology*, *59*(1), 31–43. https://doi.org/10.17113/ftb.59.01.21.6878

Usman, M., Ahmad, F., Pasha, I., Nasir, M. A., Subtain, M., & Ahmad, S. (2024). Characterization and end-use perspectives of modified cereal brans: A step towards waste valorization. *Biomass Conversion and Biorefinery*, *15*, 17727–17737. https://doi.org/10.1007/s13399-024-06283-w

Valladares-Diestra, K. K., de Souza Vandenberghe, L. P., Vieira, S., Goyzueta-Mamani, L. D., de Mattos, P. B. G., Manzoki, M. C., & Soccol, C. R. (2023). The potential of xylooligosaccharides as prebiotics and their sustainable production from agro-industrial by-products. *Food*, *12*(14), 2681. https://doi.org/10.3390/foods12142681

Vashishth, N., & Maurya, N. K. (2024). Millet bran: Unveiling its potential health benefits. *International Journal of Nutritions*, *1*(2), 11–19. 10.37591/IJN

Velur Balasubramaniam, G., Sukumar, S., Alagarsamy, M., Velpandi, P. S. D., Palaniappan, A., Roy, L. M., & Antony, U. (2022). Inhibition of lipid oxidation in mayonnaise by finger millet seed coat polyphenols. *Journal of the American Oil Chemists' Society*, *99*(1), 79–90. https://doi.org/10.1002/aocs.12557

Vijayanand, C. (2016). Characterization of indigenous biomass. *International Journal of Agriculture Sciences*, *8*(50), 2124–2127.

Voučko, B., Čukelj Mustač, N., Nanjara, L., Drakula, S., Grgić, T., Ćurić, D., & Novotni, D. (2024). Fermentation performance of carob flour, proso millet flour and bran for gluten-free flat-bread. *Food*, *13*(21), 3458. https://doi.org/10.3390/foods13213458

Wang, C., Li, Z., Xiang, J., Johnson, J. B., Zheng, B., Luo, L., & Beta, T. (2023a). From foxtail millet husk (waste) to bioactive phenolic extracts using deep eutectic solvent extraction and evaluation of antioxidant, acetylcholinesterase, and α-glucosidase inhibitory activities. *Foods*, *12*(6), 1144. https://doi.org/10.3390/foods12061144

Wang, H., Fu, Y., Zhao, Q., Hou, D., Yang, X., Bai, S., & Shen, Q. (2022). Effect of different processing methods on the millet polyphenols and their anti-diabetic potential. *Frontiers in Nutrition*, *9*, 780499. https://doi.org/10.3389/fnut.2022.780499

Wang, Z., Tang, H., Liu, G., Gong, H., Li, Y., Chen, Y., & Yang, Y. (2023b). Compound probiotics producing cellulase could replace cellulase preparations during solid-state fermentation of millet bran. *Bioresource Technology*, *385*, 129457. https://doi.org/10.1016/j.biortech.2023.129457

Williams, B. A., Mikkelsen, D., Flanagan, B. M., & Gidley, M. J. (2019). "Dietary fibre": Moving beyond the "soluble/insoluble" classification for monogastric nutrition, with an emphasis on humans and pigs. *Journal of Animal Science and Biotechnology*, *10*, 1–12. https://doi.org/10.1186/s40104-019-0350-9

Wizi, J., Wang, L., Hou, X., Tao, Y., Ma, B., & Yang, Y. (2018). Ultrasound-microwave assisted extraction of natural colorants from sorghum husk with different solvents. *Industrial crops and products*, *120*, 203–213. https://doi.org/10.1016/j.indcrop.2018.04.068

Xiang, J., Apea-Bah, F. B., Ndolo, V. U., Katundu, M. C., & Beta, T. (2019). Profile of phenolic compounds and antioxidant activity of finger millet varieties. *Food Chemistry*, *275*, 361–368. https://doi.org/10.1016/j.foodchem.2018.09.120

Xiao, J., Li, Y., Niu, L., Chen, R., Tang, J., Tong, Z., & Xiao, C. (2023). Effect of adding fermented proso millet bran dietary fiber on micro-structural, physicochemical, and digestive properties of gluten-free proso millet-based dough and cake. *Food*, *12*(15), 2964. https://doi.org/10.3390/foods12152964

Xu, B., Wang, X., Zheng, Y., Shi, P., Zhang, Y., Liu, Y., & Long, N. (2022a). Millet bran globulin hydrolysate derived tetrapeptide-ferrous chelate: Preparation, structural characterization, security prediction in silico, and stability against different food processing conditions. *LWT*, *165*, 113673. https://doi.org/10.1016/j.lwt.2022.113673

Xu, M., Yu, H., Chen, X., & Yuan, G. (2022b). Physico-chemical, biological properties of chitosan/gelatin-based films with Finger Millet bran extract. *Journal of Food Measurement and Characterization*, *16*(4), 2939–2947. https://doi.org/10.1007/s11694-022-01406-1

Yadav, M., Rengasamy, R. S., & Gupta, D. (2019). Characterization of pearl millet (*Pennisetum glaucum*) waste. *Carbohydrate Polymers*, *212*, 160–168. https://doi.org/10.1016/j.carbpol.2019.02.034

Yaghoubzadeh, Z., Safari, R., & Soheili, M. (2025). Cytotoxicity activity of peptides derived from enzymatic hydrolysis of *Chlorella vulgaris* proteins. *Applied Food Biotechnology*, *12*(1), e1. https://doi.org/10.22037/afb.v12i1.46685

Yin, R., Fu, Y., Yousaf, L., Xue, Y., Hu, J., Hu, X., & Shen, Q. (2022). Crude and refined millet bran oil alleviate lipid metabolism disorders, oxidative stress and affect the gut microbiota composition in high-fat diet-induced mice. *International Journal of Food Science & Technology*, *57*(5), 2600–2610. https://doi.org/10.1111/ijfs.15063

Zhang, H. L., Wei, J. K., Wang, Q. H., Yang, R., Gao, X. J., Sang, Y. X., & Chen, Q. J. (2019). Lignocellulose utilization and bacterial communities of millet straw based mushroom (*Agaricus bisporus*) production. *Scientific Reports*, *9*(1), 1151. https://doi.org/10.1038/s41598-018-37681-6

Zhang, M., Xu, Y., Xiang, J., Zheng, B., Yuan, Y., Luo, D., & Fan, J. (2021). Comparative evaluation on phenolic profiles, antioxidant properties and α-glucosidase inhibitory effects of different milling fractions of foxtail millet. *Journal of Cereal Science*, *99*, 103217. https://doi.org/10.1016/j.jcs.2021.103217

Zhang, P., & Awika, J. M. (2024). Effect of sorghum bran addition on antioxidant activities, sensory properties, and in vitro starch digestibility of Chinese southern-style steamed bread. *Journal of the Science of Food and Agriculture*, *104*, 9652–9659. https://doi.org/10.1002/jsfa.13790

Zhang, S., Wang, A., Lu, Z., Lu, F., & Zhao, H. (2024). Fermentation of millet bran with *Bacillus natto*: enhancement of bioactivity levels and the bioactivity of bran extract. *Journal of the Science of Food and Agriculture*, *104*(10), 6196–6207. https://doi.org/10.1002/jsfa.13455

Zheng, B., Yuan, Y., Xiang, J., Jin, W., Johnson, J. B., Li, Z., & Luo, D. (2022). Green extraction of phenolic compounds from foxtail millet bran by ultrasonic-assisted deep eutectic solvent extraction: Optimization, comparison and bioactivities. *LWT*, *154*, 112740. https://doi.org/10.1016/j.lwt.2021.112740

Zheng, Y., Li, J., Wang, X., Guo, M., Cheng, C., & Zhang, Y. (2023). Effects of three biological combined with chemical methods on the microstructure, physicochemical properties and antioxidant activity of millet bran dietary fibre. *Food Chemistry*, *411*, 135503. https://doi.org/10.1016/j.foodchem.2023.135503

Zhu, Y., Chu, J., Lu, Z., Lv, F., Bie, X., Zhang, C., & Zhao, H. (2018). Physicochemical and functional properties of dietary fiber from foxtail millet (*Setaria italic*) bran. *Journal of Cereal Science*, *79*, 456–461. https://doi.org/10.1016/j.jcs.2017.12.011

Zhu, Y., He, C., Fan, H., Lu, Z., Lu, F., & Zhao, H. (2019). Modification of foxtail millet (*Setaria italica*) bran dietary fiber by xylanase-catalyzed hydrolysis improves its cholesterol-binding capacity. *LWT*, *101*, 463–468. https://doi.org/10.1016/j.lwt.2018.11.052

Index

Nutraceutical Potential of Millet-Based Food Products: Chemistry and Applications, First Edition.
Edited by Parmjit S. Panesar, Charanjit S. Riar, and Sushil Dhital.

Published 2026 by John Wiley & Sons Ltd.

c

d

n

o

Printed and bound by CPI Group (UK) Ltd, Croydon, CR0 4YY
16/08/2026
14942675-0001